Springer Japan KK

L.E. Parker, G. Bekey, J. Barhen (Eds.)

# Distributed Autonomous Robotic Systems 4

With 282 Figures

Springer

LYNNE E. PARKER
Senior Research Staff
Oak Ridge National Laboratory
Center for Engineering Science Advanced Research
P. O. Box 2008, Oak Ridge, TN 37831-6355, USA

GEORGE BEKEY
Gordon Marshall Professor of Computer Science
University of Southern California
Department of Computer Science
941 West 37th Place
Los Angeles, CA 90089-0781, USA

JACOB BARHEN
UT-Battelle Corporate Fellow
Oak Ridge National Laboratory
Center for Engineering Science Advanced Research
P. O. Box 2008, Oak Ridge, TN 37831-6355, USA

ISBN 978-4-431-67991-2 ISBN 978-4-431-67919-6 (eBook)
DOI 10.1007/978-4-431-67919-6

Library of Congress Cataloging-in-Publication Data
Distributed autonomous robotic systems 4 / L.E. Parker, G. Bekey, J. Barhen (eds.).
p. cm.
Includes bibliographical references.
ISBN 978-4-431-67991-2
1. Autonomous robots. I. Parker, Lynne E. II. Bekey, George A., 1928- III. Barhen, J. (Jacob), 1947- IV. International Symposium on Distributed Autonomous Robotic Systems (5th : 2000 : Knoxville, Tenn.)

TJ211.495 .D57 2000 00-058355
629.8'92--dc21
Printed on acid-free paper



Typesetting: Camera-ready by the editors and authors
Printing and binding: Best-set Typesetter Ltd., Hong Kong
SPIN: 10776815

# Preface

The Fifth International Symposium on Distributed Autonomous Robotic Systems (DARS 2000) dealt with new strategies to realize complex, modular, robust, and fault-tolerant robotic systems. Technologies, algorithms, and system architectures for distributed autonomous robotic systems were presented and discussed during the meeting. DARS 2000 was truly an international event, with participants representing eleven countries from Europe, Asia, and the Americas. All of the papers in this volume were presented at DARS 2000, and were selected on the basis of peer reviews to ensure quality and relevance. These papers have the common goal of contributing solutions to realize robust and intelligent multirobot systems.

The topics of the symposium address a wide range of issues that are important in the development of decentralized robotic systems. These topics include architectures, communication, biological inspirations, reconfigurable robots, localization, exploration and mapping, distributed sensing, multirobot motion coordination, target assignment and tracking, multirobot learning, and cooperative object transport. DARS clearly requires a broad area of interdisciplinary technologies related not only to robotics and computer engineering, but also to biology and psychology.

The DARS symposium is the leading established conference on distributed autonomous systems. The First, Second, and Third International Symposia on Distributed Autonomous Robotic Systems (DARS '92, DARS '94, and DARS '96) were held at the Institute of Physical and Chemical Research (RIKEN), Saitama, Japan. In 1998, the event was first held in Europe, at the Institute for Process Control and Robotics, University of Karlsruhe, Germany. From October 4–6, 2000, the symposium (DARS 2000) was held for the first time in the United States, in Knoxville, Tennessee, hosted by the Oak Ridge National Laboratory.

We would like to thank the sponsors and cosponsors of the DARS 2000 symposium, which was sponsored by the Oak Ridge National Laboratory and the Engineering Research Program of the Office of Basic Energy Sciences in the U. S. Department of Energy. The cosponsors of the symposium were the IEEE Robotics and Automation Society, the IEEE Industrial Electronics Society, the Robotics Society of Japan (RSJ), the Japan Society of Mechanical Engineers (JSME), and the Society of Instrument and Control Engineers of Japan (SICE).

We are grateful to the organizing committee members for the successful organization of DARS 2000, and to the program committee members for reviewing papers and for program arrangement. Finally, we would like to thank Ms. Thelma Garrett, Ms. Tina Lanning, and Ms. Carole Holbrook for their help with local arrangements and secretariat administration

Lynne E. Parker
George W. Bekey
Jacob Barhen

# Table of Contents

## Chapter 4: Biological Inspirations



## Chapter 5: Reconfigurable Robots



## Chapter 6: Localization



## Chapter 7: Exploration, Mapping, and Model Acquisition

## Chapter 8: Distributed Sensing



## Chapter 9: Multi-Robot Motion Coordination and Tracking

## Chapter 10: Multi-Robot Learning



## Chapter 11: Cooperative Object Transport



## Chapter 12: Related Research Topics

# DARS 2000 Organization

**General Co-Chairs:**

George Bekey (University of Southern California, USA)
Jacob Barhen (Oak Ridge National Laboratory, USA)

**Advisory Committee:**

Hajime Asama (RIKEN, Japan)
Toshio Fukuda (Nagoya University, Japan)
Tim Lueth (Humboldt University, Germany)

**Program Committee**

**Chair:**

Lynne Parker (Oak Ridge National Laboratory, USA)

**Members:**

Arvin Agah (Univ. of Kansas, USA)
Rachid Alami (LAAS, France)
Ron Arkin (Ga. Tech, USA)
Tucker Balch (CMU, USA)
Gerardo Beni (U. California Riverside, USA)
Richard Brooks (Penn State, USA)
Hans-Dieter Burkhard (Humboldt Univ., Germany)
Paolo Dario (Scuola Speriore S. Anna, Italy)
Marco Dorigo (IRIDIA, Belgium)
Alexis Drogoul (Paris VI Univ., France)
Aarne Halme (Helsinki Univ. of Tech., Finland)
Terry Huntsberger (JPL, USA)
Jeffrey Johnson (Open University, UK)
David Jung (ORNL, USA)
Yukinori Kakazu (Hokkaido Univ., Japan)
Kazuhiro Kosuge (Tohoku Univ., Japan)
Christian Laugier (INRIA, France)
Paul Levi (Univ. Stuttgart, Germany)
Akihiro Matsumoto (Toyo Univ., Japan)
Francesco Mondada (K-Team, Switzerland)
Satoshi Murata (MEL, Japan)
Robin R. Murphy (USF, USA)
Jun Ota (Univ. Tokyo, Japan)
Enrico Pagello (U. Padua, Italy)
Chris Paredis (CMU, USA)
Suparerk Premvuti (Thammasat Univ., Thailand)
Daniela Rus (Dartmouth College, USA)
Arthur Sanderson (Rensselear Polytechnic Inst., USA)
Wei-Min Shen (USC, USA)
Takanori Shibata (MEL, Japan)
Peter Stone (AT&T Labs – Research, USA)
Gaurav Sukhatme (USC, USA)
Claude Touzet (ORNL, USA)
Manuela Veloso (CMU, USA)
Ljubo Vlacic (Griffith Univ., Australia)
Jing Wang (TransDimension Int'l Corp., USA)
Zhidong Wang (Tohoku Univ., Japan)
Alexander Zelinsky (Australian National Univ., Australia)

# Chapter 1

# Introduction

# Current State of the Art in Distributed Autonomous Mobile Robotics

Lynne E. Parker

Center for Engineering Science Advanced Research
Computer Science and Mathematics Division
Oak Ridge National Laboratory, Oak Ridge TN 37831-6355, USA
email: ParkerLE@ornl.gov

**Keywords:** Distributed robotics, survey, cooperative robotics, multi-robot systems

**Abstract.** As research progresses in distributed robotic systems, more and more aspects of multi-robot systems are being explored. This article surveys the current state of the art in distributed mobile robot systems. Our focus is principally on research that has been demonstrated in physical robot implementations. We have identified eight primary research topics within multi-robot systems — biological inspirations, communication, architectures, localization/mapping/exploration, object transport and manipulation, motion coordination, reconfigurable robots, and learning - and discuss the current state of research in these areas. As we describe each research area, we identify some key open issues in multi-robot team research. We conclude by identifying several additional open research issues in distributed mobile robotic systems.

## 1 Introduction

The field of distributed robotics has its origins in the late-1980's, when several researchers began investigating issues in multiple mobile robot systems. Prior to this time, research had concentrated on either single robot systems or distributed problem-solving systems that did not involve robotic components. The topics of particular interest in this early distributed robotics work include:

- *Cellular (or reconfigurable) robot systems*, such as the work by Fukuda, *et al.* [20] on the Cellular Robotic System (CEBOT) and the work on cyclic swarms by Beni [8],
- *Multi-robot motion planning*, such as the work by Premvuti and Yuta [38] on traffic control and the work on movement in formations by Arai, *et al.* [2] and Wang [48], and
- *Architectures for multi-robot cooperation*, such as the work on ACTRESS by Asama, *et al.* [4].

Since this early research in distributed mobile robotics, the field has grown dramatically, with a much wider variety of topics being addressed. This paper examines the current state of the art in autonomous multiple mobile robotic systems. The field of cooperative autonomous mobile robotics is still so new that no topic area within this domain can be considered mature. Some areas have been explored more extensively, however, and the community is beginning to understand how to develop and control certain aspects of multi-robot teams. Thus, rather than summarize the research into a taxonomy of cooperative systems (see Dudek [18] and Cao [12] for previous related summaries), we instead organize this research by the principal topic areas that have generated significant levels of study, to the extent possible in a limited space. As we present the review, we identify key open research issues within each topic area. We conclude by suggesting additional research issues that have not yet been extensively studied, but appear to be of growing interest and need in distributed autonomous multi-robot systems.

## 2 Biological Inspirations

Nearly all of the work in cooperative mobile robotics began after the introduction of the new robotics paradigm of behavior-based control [10,3]. This behavior-based paradigm has had a strong influence in much of the cooperative mobile robotics research. Because the behavior-based paradigm for mobile robotics is rooted in biological inspirations, many cooperative robotics researchers have also found it instructive to examine the social characteristics of insects and animals, and to apply these findings to the design of multi-robot systems[1].

The most common application of this knowledge is in the use of the simple local control rules of various biological societies – particularly ants, bees, and birds – to the development of similar behaviors in cooperative robot systems. Work in this vein has demonstrated the ability for multi-robot teams to flock, disperse, aggregate, forage, and follow trails (e.g., [30,15,17]). The application of the dynamics of ecosystems has also been applied to the development of multi-robot teams that demonstrate emergent cooperation as a result of acting on selfish interests [32]. To some extent, cooperation in higher animals, such as wolf packs, has generated advances in cooperative control. Significant study in predator-prey systems has occurred, although primarily in simulation [7,21]. Competition in multi-robot systems, such as found in higher animals including humans, is beginning to be studied in domains such as multi-robot soccer [46,29].

These areas of biological inspiration and their applicability to multi-robot teams seem to be fairly well understood. More recently identified, less well understood, biological topics of relevance include the use of imitation in higher

[1] For a more detailed analysis of various types of biological systems – what Tinbergen called *differentiating* (e.g., ants) and *integrative* (e.g., birds) – and their relationship to cooperative robotics work (i.e., "swarm" vs. "intentional"), see [35].

animals to learn new behaviors, and the physical interconnectivity demonstrated by insects such as ants to enable collective navigation over challenging terrains.

## 3 Communication

The issue of communication in multi-robot teams has been extensively studied since the inception of distributed robotics research. Distinctions between implicit and explicit communication are usually made, in which implicit communication occurs as a side-effect of other actions, or "through the world", whereas explicit communication is an specific act designed solely to convey information to other robots on the team. Several researchers have studied the effect of communication on the performance of multi-robot teams in a variety of tasks, and have concluded that communication provides certain benefit for particular types of tasks [27,6]. Additionally, these researchers have found that, in many cases, communication of even a small amount of information can lead to great benefit [6].

More recent work in multi-robot communication has focused on representations of languages and the grounding of these representations in the physical world [22,23]. Additionally, work has extended to achieving fault tolerance in multi-robot communication, such as setting up and maintaining distributed communications networks [51] and ensuring reliability in multi-robot communications [34]. While progress is being made in these more recent issues of communication, much work remains to enable multi-robot teams to operate reliably amidst faulty communication environments.

## 4 Architectures, Task Planning, and Control

A great deal of research in distributed robotics has focused on the development of architectures, task planning capabilities, and control. This research area addresses the issues of action selection, delegation of authority and control, the communication structure, heterogeneity versus homogeneity of robots, achieving coherence amidst local actions, resolution of conflicts, and other related issues. Each architecture that has been developed for multi-robot teams tends to focus on providing a specific type of capability to the distributed robot team. Capabilities that have been of particular emphasis include task planning [1], fault tolerance [36], swarm control [31], human design of mission plans [26], and so forth.

A general research question in this vein is whether specialized architectures for each type of robot team and/or application domain are needed, or whether a more general architecture can be developed that can be easily tailored to fit a wider range of multi-robot systems. Relatively little of the previous work has been aimed at unifying these architectures. Perhaps an all-encompassing architecture would be too unwieldy to implement in practical applications. It remains to be seen if a single general architecture for multi-robot teams can be developed that is applicable to a much wider variety of domains than is currently possible with existing architectures.

## 5 Localization, Mapping, and Exploration

An extensive amount of research has been carried out in the area of localization, mapping, and exploration for single autonomous robots. Only fairly recently has much of this work been applied to multi-robot teams. Almost all of the work has been aimed at 2D environments. Additionally, nearly all of this research takes an existing algorithm developed for single robot mapping, localization, or exploration, and extends it to multiple robots, as opposed to developing a new algorithm that is fundamentally distributed. One exception is some of the work in multi-robot localization, which takes advantage of multiple robots to improve positioning accuracy beyond what is possible with single robots [42,19].

As is the case with single robot approaches to localization, mapping, and exploration, research into the multi-robot version can be described using the familiar categories based on the use of landmarks [14], scan-matching [11], and/or graphs [40,39], and which use either range sensors (such as sonar or laser) or vision sensors. While the single robot version of this problem is fairly well understood, much remains to be studied in the multi-robot version. For example, one question is the effectiveness of multi-robot teams over single-robot versions, and to what extent adding additional robots brings diminishing returns. This issue has begun to be studied (see [39]), but much much remains to be determined for the variety of approaches available for localization, mapping, and exploration.

## 6 Object Transport and Manipulation

Enabling multiple robots to cooperatively carry, push, or manipulate common objects has been a long-standing, yet difficult, goal of multi-robot systems. Many research projects have dealt with this topic area; fewer of these projects have been demonstrated on physical robot systems. This research area has a number of practical applications that make it of particular interest for study.

Numerous variations on this task area have been studied, including constrained and unconstrained motions, two-robot teams versus "swarm"-type teams, compliant versus non-compliant grasping mechanisms, cluttered versus uncluttered environments, global system models versus distributed models, and so forth. Perhaps the most demonstrated task involving cooperative transport is the pushing of objects by multi-robot teams [43,45]. This task seems inherently easier than the carry task, in which multiple robots must grip common objects and navigate to a destination in a coordinated fashion

[49,24]. A novel form of multi-robot transportation that has been demonstrated is the use of ropes wrapped around objects to move them along desired trajectories [16].

Nearly all of the previous work in this area work involves robots moving across a flat surface. A challenging open issue in this area is cooperative transport over uneven outdoor terrains.

## 7 Motion Coordination

A popular topic of study in multi-robot teams is that of motion coordination. Research themes in this domain that have been particularly well studied include multi-robot path planning [52], traffic control [38], formation generation [2], and formation keeping [5,48]. Most of these issues are now fairly well understood, although demonstration of these techniques in physical multi-robot teams (rather than in simulation) has been limited. More recent issues studied within the motion coordination context are target tracking [37], target search [25], and multi-robot docking [33] behaviors.

Nearly all of the previous work has been aimed at 2D domains, although some work has been aimed at 3D environments [52]. One of the most limiting characteristics of much of the existing path planning work is the computational complexity of the approaches. Perhaps as computing processor speed increases, the computational time will take care of itself. In the meantime, this characteristic is a limiting factor to the applicability of much of the path planning research in dynamic, real-time robot teams.

## 8 Reconfigurable Robotics

Even though some of the earliest research in distributed robotics focused on concepts for reconfigurable distributed systems [20,8], relatively little work has proceeded in this area until the last few years. More recent work has resulted in a number of actual physical robot systems that are able to reconfigure. The motivation of this work is to achieve function from shape, allowing individual modules, or robots, to connect and re-connect in various ways to generate a desired shape to serve a needed function. These systems have the theoretical capability of showing great robustness, versatility, and even self-repair.

Most of the work in this area involves identical modules with interconnection mechanisms that allow either manual or automatic reconfiguration. These systems have been demonstrated to form into various navigation configurations, including a rolling track motion [53], an earthworm or snake motion [53,13], and a spider or hexapod motion [53,13]. Some systems employ a cube-type arrangement, with modules able to connect in various ways to form matrices or lattices for specific functions [9,54,44,47].

Research in this area is still very young, and most of the systems developed are not yet able to perform beyond simple laboratory experiments. While the potential of large numbers of robot modules has, to some extent, been demonstrated in simulation, it is still uncommon to have implementations involving more than a dozen or so physical modules. The practical application of these systems is yet to be demonstrated, although progress is being made in that direction. Clearly, this is a rich area for continuing advances in multi-robot systems.

## 9 Learning

Many multi-robot researchers believe that an approach with more potential for the development of cooperative control mechanisms is autonomous learning. While a considerable amount of work has been done in this area for multi-agent learning [50], somewhat less work has been accomplished to date in multi-robot learning. The types of applications that are typically studied for this area of multi-robot learning vary considerably in their characteristics. Some of the applications include predator/prey [7,21], box pushing [28], foraging [31], multi-robot soccer [46,29,41], and cooperative target observation [37].

Particularly challenging domains for multi-robot learning are those tasks that are *inherently* cooperative – that is, tasks in which the utility of the action of one robot is dependent upon the current actions of the other team members. Inherently cooperative tasks cannot be decomposed into independent subtasks to be solved by a distributed robot team. Instead, the success of the team throughout its execution is measured by the combined actions of the robot team, rather than the individual robot actions. This type of task is a particular challenge in multi-robot learning, due to the difficulty of assigning credit for the individual actions of the robot team members. Multi-robot learning in general, and inherently cooperative task learning in particular, are areas in which significant research for multi-robot systems remains.

## 10 Additional Open Issues in Distributed Autonomous Mobile Robotics

It is clear that since the inception of the field of distributed autonomous mobile robotics less than two decades ago, significant progress has been made on a number of important issues. The field has a good understanding of the biological parallels that can be drawn, the use of communication in multi-robot teams, and the design of architectures for multi-robot control. Considerable progress has been made in multi-robot localization/mapping/exploration, cooperative object transport, and motion coordination. Recent progress is beginning to advance the areas of reconfigurable robotics and multi-robot learning. Of course, all of these areas have not yet been fully studied; we have identified key open research challenges for these areas in the previous sections.

Several other research challenges still remain, including:

- How do we identify and quantify the fundamental advantages and characteristics of multi-robot systems?
- How do we easily enable humans to control multi-robot teams?
- Can we scale up to demonstrations involving more than a dozen or so robots?
- Is passive action recognition in multi-robot teams possible?
- How can we enable physical multi-robot systems to work under hard real-time constraints?
- How does the complexity of the task and of the environment affect the design of multi-robot systems?

These and other issues in multi-robot cooperation should keep the research community busy for many years to come.

## Acknowledgments

This work is sponsored by the Engineering Research Program of the Office of Basic Energy Sciences, U. S. Department of Energy. Accordingly, the U.S. Government retains a nonexclusive, royalty-free license to publish or reproduce the published form of this contribution, or allow others to do so, for U. S. Government purposes. Oak Ridge National Laboratory is managed by UT-Battelle, LLC for the U.S. Dept. of Energy under contract DE-AC05-00OR22725.

## References

1. R. Alami, S. Fleury, M. Herrb, F. Ingrand, and F. Robert. Multi-robot cooperation in the Martha project. *IEEE Robotics and Automation Magazine*, 1997.
2. T. Arai, H. Ogata, and T. Suzuki. Collision avoidance among multiple robots using virtual impedance. In *Proceedings of the Intelligent Robots and Systems (IROS)*, pages 479–485, 1989.
3. Ronald C. Arkin. Integrating behavioral, perceptual, and world knowledge in reactive navigation. *Robotics and Autonomous Systems*, 6:105–122, 1990.
4. Hajime Asama, Akihiro Matsumoto, and Yoshiki Ishida. Design of an autonomous and distributed robot system: ACTRESS. In *Proceedings of IEEERSJ International Workshop on Intelligent Robots and Systems*, pages 283–290, Tsukuba, Japan, 1989.
5. T. Balch and R. Arkin. Behavior-based formation control for multi-robot teams. *IEEE Transactions on Robotics and Automation*, December 1998.
6. Tucker Balch and Ronald C. Arkin. Communication in reactive multiagent robotic systems. *Autonomous Robots*, 1(1):1–25, 1994.

7. M. Benda, V. Jagannathan, and R. Dodhiawalla. On optimal cooperation of knowledge sources. Technical Report BCS-G2010-28, Boeing AI Center, August 1985.
8. Gerardo Beni. The concept of cellular robot. In *Proceedings of Third IEEE Symposium on Intelligent Control*, pages 57–61, Arlington, Virginia, 1988.
9. H. Bojinov, A. Casal, and T. Hogg. Emergent structures in moduluar self-reconfigurable robots. In *Proceedings of the IEEE International Conference on Robotics and Automation*, pages 1734–1741, 2000.
10. Rodney A. Brooks. A robust layered control system for a mobile robot. *IEEE Journal of Robotics and Automation*, RA-2(1):14–23, March 1986.
11. W. Burgard, M. Moors, D. Fox, R. Simmons, and S. Thrun. Collaborative multi-robot exploration. In *Proceedings of the IEEE International Conference on Robotics and Automation*, pages 476–481, 2000.
12. Y. Uny Cao, Alex Fukunaga, Andrew Kahng, and Frank Meng. Cooperative mobile robotics: Antecedents and directions. In *Proceedings of 1995 IEEE/RSJ International Conference on Intelligent Robots and Systems (IROS '95)*, pages 226–234, 1995.
13. A. Castano, R. Chokkalingam, and P. Will. Autonomous and self-sufficient conro modules for reconfigurable robots. In *Proceedings of the Fifth International Symposium on Distributed Autonomous Robotic Systems (DARS 2000 – this volume)*, 2000.
14. G. Dedeoglu and G. Sukhatme. Landmark-based matching algorithm for cooperative mapping by autonomous robots. In *Proceedings of the Fifth International Symposium on Distributed Autonomous Robotic Systems (DARS 2000 – this volume)*, 2000.
15. J. Deneubourg, S. Goss, G. Sandini, F. Ferrari, and P. Dario. Self-organizing collection and transport of objects in unpredictable environments. In *Japan-U.S.A. Symposium on Flexible Automation*, pages 1093–1098, 1990.
16. B. Donald, L. Gariepy, and D. Rus. Distributed manipulation of multiple objects using ropes. In *Proceedings of IEEE International Conference on Robotics and Automation*, pages 450–457, 2000.
17. Alexis Drogoul and Jacques Ferber. From Tom Thumb to the Dockers: Some experiments with foraging robots. In *Proceedings of the Second International Conference on Simulation of Adaptive Behavior*, pages 451–459, 1992.
18. Gregory Dudek et al. A taxonomy for swarm robots. In *Proceedings of 1993 IEEE International Conference on Intelligent Robots and Systems (IROS '93)*, pages 441–447, 1993.
19. D. Fox, W. Burgard, H. Kruppa, and S. Thrun. Collaborative multi-robot exploration. *Autonomous Robots*, 8(3), 2000.
20. T. Fukuda and S. Nakagawa. A dynamically reconfigurable robotic system (concept of a system and optimal configurations). In *Proceedings of IECON*, pages 588–595, 1987.
21. Thomas Haynes and Sandip Sen. Evolving behavioral strategies in predators and prey. In Gerard Weiss and Sandip Sen, editors, *Adaptation and Learning in Multi-Agent Systems*, pages 113–126. Springer, 1986.
22. L. Hugues. Collective grounded representations for robots. In *Proceedings of Fifth International Conference on Distributed Autonomous Robotic Systems (DARS 2000 – this volume)*, 2000.
23. David Jung and Alexander Zelinsky. Grounded symbolic communication between heterogeneous cooperating robots. *Autonomous Robots*, 8(3), July 2000.

24. O. Khatib, K. Yokoi, K. Chang, D. Ruspini, R. Holmberg, and A. Casal. Vehicle/arm coordination and mobile manipulator decentralized cooperation. In *IEEE/RSJ International Conference on Intelligent Robots and Systems*, pages 546–553, 1996.
25. S. M. LaValle, D. Lin, L. J. Guibas, J-C. Latombe, and R. Motwani. Finding an unpredictable target in a workspace with obstacles. In *submitted to 1997 International Conference on Robots and Automation*, 1997.
26. D. MacKenzie, R. Arkin, and J. Cameron. Multiagent mission specification and execution. *Autonomous Robots*, 4(1):29–52, 1997.
27. Bruce MacLennan. Synthetic ethology: An approach to the study of communication. In *Proceedings of the 2nd interdisciplinary workshop on synthesis and simulation of living systems*, pages 631–658, 1991.
28. S. Mahadevan and J. Connell. Automatic programming of behavior-based robots using reinforcement learning. In *Proceedings of AAAI-91*, pages 8–14, 1991.
29. S. Marsella, J. Adibi, Y. Al-Onaizan, G. Kaminka, I. Muslea, and M. Tambe. On being a teammate: Experiences acquired in the design of RoboCup teams. In O. Etzioni, J. Muller, and J. Bradshaw, editors, *Proceedings of the Third Annual Conference on Autonomous Agents*, pages 221–227, 1999.
30. Maja Mataric. Designing emergent behaviors: From local interactions to collective intelligence. In J. Meyer, H. Roitblat, and S. Wilson, editors, *Proc. of the Second Int'l Conf. on Simulation of Adaptive Behavior*, pages 432–441. MIT Press, 1992.
31. Maja Mataric. *Interaction and Intelligent Behavior*. PhD thesis, Massachusetts Institute of Technology, 1994.
32. David McFarland. Towards robot cooperation. In D. Cliff, P. Husbands, J.-A. Meyer, and S. Wilson, editors, *Proceedings of the Third International Conference on Simulation of Adaptive Behavior*, pages 440–444. MIT Press, 1994.
33. B. Minten, R. Murphy, J. Hyams, and M. Micire. A communication-free behavior for docking mobile robots. In *Proceedings of Fifth International Symposium on Distributed Autonomous Robotic Systems (DARS 2000 – this volume)*, 2000.
34. P. Molnar and J. Starke. Communication fault tolerance in distributed robotic systems. In *Proceedings of Fifth International Symposium on Distributed Autonomous Robotic Systems (DARS 2000 – this volume)*, 2000.
35. L. E. Parker. Adaptive action selection for cooperative agent teams. In Jean-Arcady Meyer, Herbert Roitblat, and Stewart Wilson, editors, *Proceedings of the Second International Conference on Simulation of Adaptive Behavior*, pages 442–450. MIT Press, 1992.
36. L. E. Parker. ALLIANCE: An architecture for fault-tolerant multi-robot cooperation. *IEEE Transactions on Robotics and Automation*, 14(2):220–240, 1998.
37. L. E. Parker. Multi-robot learning in a cooperative observation task. In *Proceedings of Fifth International Symposium on Distributed Autonomous Robotic Systems (DARS 2000 – this volume)*, 2000.
38. Suparerk Premvuti and Shin'ichi Yuta. Consideration on the cooperation of multiple autonomous mobile robots. In *Proceedings of the IEEE International Workshop of Intelligent Robots and Systems*, pages 59–63, Tsuchiura, Japan, 1990.

39. N. S. V. Rao. Terrain model acquisition by mobile robot teams and n-connectivity. In *Proceedings of the Fifth International Symposium on Distributed Autonomous Robotic Systems (DARS 2000 – this volume)*, 2000.
40. I. Rekleitis, G. Dudek, and E. Milios. Graph-based exploration using multiple robots. In *Proceedings of the Fifth International Symposium on Distributed Autonomous Robotic Systems (DARS 2000 – this volume)*, 2000.
41. P. RIley and M. Veloso. On behavior classification in adversarial environmentts. In *Proceedings of Fifth International Symposium on Distributed Autonomous Robotic Systems (DARS 2000 – this volume)*, 2000.
42. S. Roumeliotis and G. Bekey. Distributed multi-robot localization. In *Proceedings of the Fifth International Symposium on Distributed Autonomous Robotic Systems (DARS 2000 – this volume)*, 2000.
43. D. Rus, B. Donald, and J. Jennings. Moving furniture with teams of autonomous robots. In *Proceedings of IEEE/RSJ International Conference on Intelligent Robots and Systems*, pages 235–242, 1995.
44. D. Rus and M. Vona. A physical implementation of the self-reconfiguring crystalline robot. In *Proceedings of the IEEE International Conference on Robotics and Automation*, pages 1726–1733, 2000.
45. Daniel Stilwell and John Bay. Toward the development of a material transport system using swarms of ant-like robots. In *Proceedings of IEEE International Conference on Robotics and Automation*, pages 766–771, 1993.
46. P. Stone and M. Veloso. A layered approach to learning client behaviors in the robocup soccer server. *Applied Artificial Intelligence*, 12:165–188, 1998.
47. C. Unsal and P. K. Khosla. Mechatronic design of a modular self-reconfiguring robotic system. In *Proceedings of the IEEE International Conference on Robotics and Automation*, pages 1742–1747, 2000.
48. P. K. C. Wang. Navigation strategies for multiple autonomous mobile robots. In *Proceedings of the IEEE/RSJ International Conference on Intelligent Robots and Systems (IROS)*, pages 486–493, 1989.
49. Z. Wang, Y. Kimura, T. Takahashi, and E. Nakano. A control method of a multiple non-holonomic robot system for cooperative object transportation. In *Proceedings of Fifth International Symposium on Distributed Autonomous Robotic Systems (DARS 2000 – this volume)*, 2000.
50. Gerhard Weiss and Sandip Sen, editors. *Adaption and Learning in Multi-Agent Systems*. Springer, 1996.
51. A. Winfield. Distributed sensing and data collection via broken ad hoc wireless connected networks of mobile robots. In *Proceedings of Fifth International Symposium on Distributed Autonomous Robotic Systems (DARS 2000 – this volume)*, 2000.
52. A. Yamashita, M. Fukuchi, J. Ota, T. Arai, and H. Asama. Motion planning for cooperative transportation of a large object by multiple mobile robots in a 3d environment. In *Proceedings of IEEE International Conference on Robotics and Automation*, pages 3144–3151, 2000.
53. M. Yim, D. G. Duff, and K. D. Roufas. Polybot: a modular reconfigurable robot. In *Proceedings of the IEEE International Conference on Robotics and Automation*, pages 514–520, 2000.
54. E. Yoshida, S. Murata, S. Kokaji, and K. Tomita dn H. Kurokawa. Micro self-reconfigurable robotic system using shape memory alloy. In *Proceedings of the Fifth International Symposium on Distributed Autonomous Robotic Systems (DARS 2000 – this volume)*, 2000.

# Chapter 2

# Architectures and Development Environments

# Control architecture for an underwater robot society

Mika Vainio, Pekka Appelqvist, and Aarne Halme
Automation Technology Laboratory, Helsinki University of Technology

P.O. Box 5400, FIN-02015 HUT, Finland,
Email: mika.vainio@hut.fi

**Abstract.** In this paper a new three layer control architecture for a multi-robot systems is presented. The architecture is tested in an underwater robot society performing distributed operations in closed aquatic environment. In these tests the performance of three and five member societies is compared to the performance of a single robot.

## 1 Introduction

The research results in the following are inspired by the societal structures developed in nature by evolution, see, for example [1]. The term *robot society* is quite rarely used in the literature. Instead, terms like *robot colony*, *distributed autonomous robotic systems*, *cellular robotics* and *collective robotics* are frequently used to describe systems, where multiple agents are working more or less together towards a common goal. See [2] and [3] for pioneering work. The effect of social relationships to the performance of the whole system can be significant as is presented in [4] and [5]. Some valuable benefits are automatically built-in to the concept of robot society. If a large number of homogenous members are used at the same time for the same purposes, the level of redundancy becomes very high. The benefits and drawbacks of homogenous multi-robot systems compared to heterogeneous systems are under extensive study, see [6] and [7].

## 2 Underwater society

The scenario of an underwater robot society, sometimes called the *Bacterium robot society*, was first presented in [8]. Recent papers include [9], [10] and [11]. The main goal of this study is to construct an underwater robot society to work inside 3D processes as an extra feature linked to a traditional automation system, as illustrated in Fig. 1. The practical testing of the society is done in laboratory's demo process environment. It consists of tanks, pipes and a jet-flow pump. The volume of the transparent demo process is 700 liters.

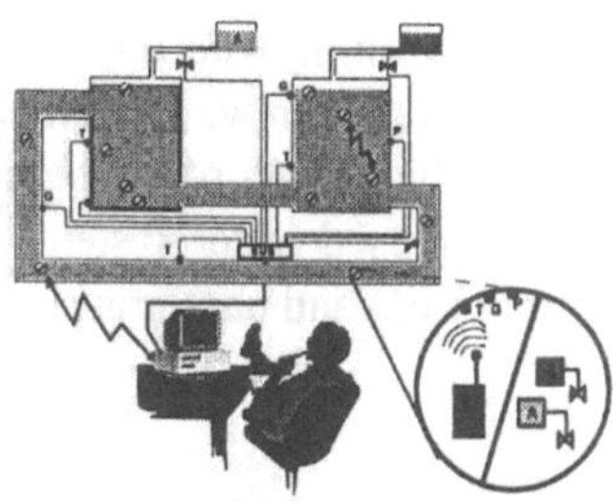

**Fig. 1.** An underwater robot society provides information outside fixed sensory locations and transports chemical agents to wherever they are needed.

### 2.1 Society member

The society consists of small sized (diameter 11 cm) spherical robots called SUBMARs (**S**mart **U**nderwater **B**all for **M**easurement and **A**ctuation **R**outines). These robots move passively along liquid flow or actively by controlling their specific weights. SUBMAR, shown in Figs. 2a-2c, is equipped with a micro-controller, several sensors, tank actuators and a short range radio for communication. See [12] for details. The sensors implemented depend on the application at hand. The tanks can be used for controlling the specific weight, taking samples and carrying a cleaning agent.

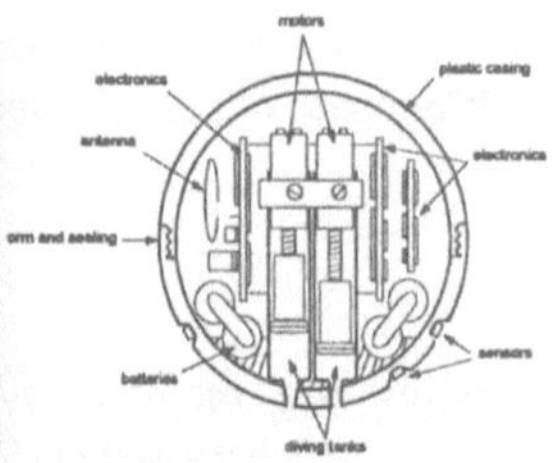


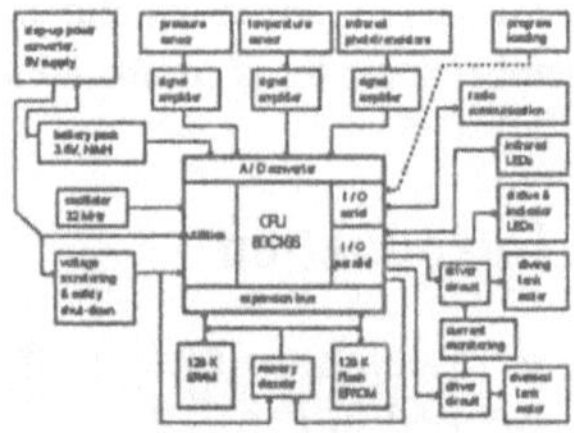

**Figs. 2a.** SUBMAR **2b.** Cross-section **2c.** Electronics[12].

## 3 Task domain

The task studied in the demo process is searching and destroying distributed targets. The targets are supposed to be microbial algae growth spots inside the water system. Algae growth occurs usually in a location where water stands still. A standard solution to the problem is to insert as much cleaning agent as is needed to guarantee a minimum concentration over the whole volume of liquid. The SUBMAR society provides an alternative solution where multiple small robot cooperate and transport the minimum amount of chemical needed straight to the areas where algae growths have been detected.

### 3.1 Emulated Biomass Growths

Because working with real algae in laboratory conditions is impractical, a special LED panel system was developed for the demo process. The system can be used to describe the growth of the biomass, for example bacteria, or algae, which tends to occur in closed water systems. In this system the growth spots are presented by infrared LED and phototransistor matrix panels, illustrated in Fig. 3.

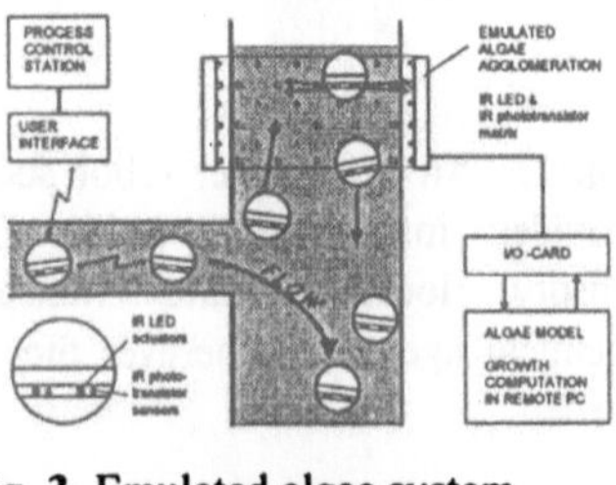


Algae model:

$$\frac{dA}{dt} = (\mu - D) * A$$

where

A status of the algae growth (i.e. the biomass)

$\mu$ growth rate

D death rate

**Fig. 3.** Emulated algae system.

Light emitted from LEDs represent oxygen or some other gas produced by the microbial growth, which in reality could be detected by robots with an appropriate sensor. The emulated organism reacts to the presence of any cleaning agent through its phototransistors. Robots have infrared phototransistors as dissolved gas sensors; while lighting their LEDs, robots emulate the spreading of the cleaning agent for algae removal. There are two independent growth agglomerations installed at different locations in the demo process. Calculation of the current level of biomass and its growth rate at each spot is carried out by a remote PC. The behavior of the biomass is modeled with a generalized growth curve typical to most biological growth processes. If there is a cleaning agent release the value of *D* is related to the concentration of the cleaning agent.

## 4 Mapping and navigation

In order to perform the above task, the members of the society, with a very limited ability to move actively, have to have some kind of spatial representation model of their environment. In a complex underwater environment any absolute localization method is difficult to construct and requires some sort of active beacon system or a highly sophisticated inertia navigation system. The concept of robot society provides a simple way to solve this problem. An individual member's map need not to be highly accurate. It is enough that different process parts can be clearly recognized. The cooperation between society members will then make their maps more detailed. Mapping is based on processing a single variable, pressure. This value is used to detect events (i.e. nodes) where the motion character changes. An event indicates, that motion changes from downwards to certain level, from level downwards, from upwards to level, or from level upwards. Based on an algorithm presented in [9] each robot creates a so called *basic map* of the environment, see Fig. 4. After this these maps are combined into an uniform representation, called as Common Basic Map (CBM) and robust navigation can be demonstrated as is shown in [13].

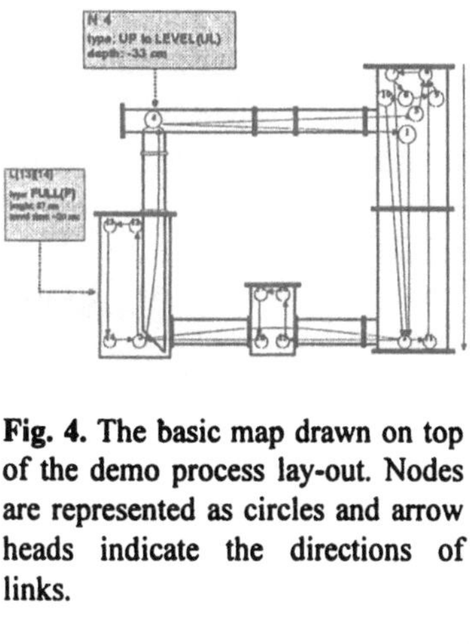


**Fig. 4.** The basic map drawn on top of the demo process lay-out. Nodes are represented as circles and arrow heads indicate the directions of links.

## 5 Control architecture

This paper presents the developed generic control architecture for cooperative robotic systems (i.e. societies). Earlier version of this architecture can be found in [10]. Reactivity is used in the model, as usually, at the lowest level, where a fast response to stimuli is needed. Planning, on the other hand, is related to performing

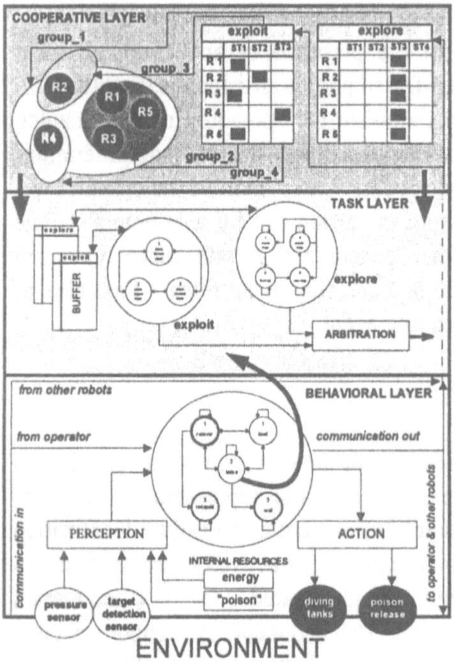


**Fig.5.** The control architecture for the described task.

the task. The robot should be able to plan its actions if it has some sort of symbolic knowledge, i.e. a topological map, of the environment. The architecture, shown in Fig.5, is defined with a three level hierarchy: cooperative, task, and behavioral. Each layer deals with some essential part of the robot's survival and mission completion in a dynamic environment with incomplete sensory information. The layers are described next in detail.

### 5.1 Behavioral Layer

This layer forms an interface between the robot and the environment. The layer receives information through various sensors (internal and external) and receives communication messages from the operator. The information is then fed to the actual control structure represented in the form of a Finite State Automata (FSA). This state machine has two main functions to perform: keeping the robot operational (i.e. self-sufficient) and reaching the given goals (i.e. task performing). Besides these two main groups, there is one further group of states, normally related to the termination of the mission (i.e. miscellaneous). In most scenarios the robot has to perform several tasks simultaneously. This prevents the use of several task states at the behavioral layer's Automata. The output of the FSA is the control of actuators and some communication messages. In the case of the search and destroy mission considered here, the number of states included in the design of the behavioral layer FSA is five, i.e. Q = {*recover, tasks, load, notepad, end*}, as is shown in Fig.6. The states *load* and *recover* represent low level behaviors with highest priority (i.e. self-sufficiency behaviors). The actual task achieving behavior is the *tasks* state. The *recover* state is the initial state, that the robot enters, when the power is turned on. It is also a kind of emergency state, active when the robot detects that something is wrong with its mobility (e.g. its location has not changed even though it should

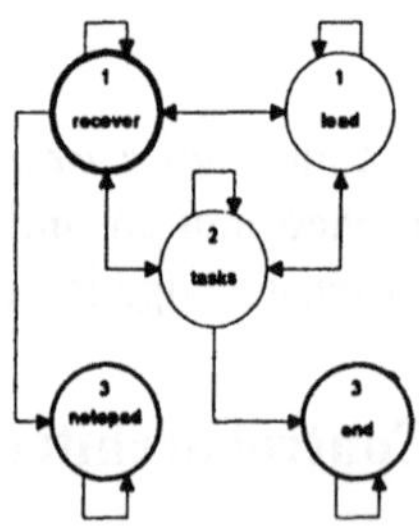


**Fig.6.** The behavioral layer FSA.

have). In this state the robot starts to use its actuators extensively. The robot changes its status to *notepad* (miscellaneous behavior), if *recover* has failed to make it mobile again. In this state the robot no longer moves actively, and its only useful feature is to operate as a kind of beacon or message mediator until its energy resources are used (i.e. abnormal termination). The robot enters *load*, when it detects that either its energy or poison level has reached some threshold value. In the *end* state (miscellaneous behavior), the robot has completed the mission, the mission time is finished or the operator has given the command to abort the mission (i.e. normal termination). Then the robot navigates to a defined location (home) waiting to be removed from the demo process.

### 5.2 Task Layer

The solution for the parallel task performance is to represent all tasks in a single state at the behavioral layer FSA. This state is then represented in detail on the higher layer, called task layer. This layer contains several independent FSAs each representing a single task. Every state machine has access to all the information available on the lower layer. The states in these machines are named as strategies because they contain a behavioral pattern, i.e. how to proceed in reaching an individual task's goal. Strategies and state transitions are based on the performance evaluation. To provide the robot with at least a vague idea of what is going on, the control system should incite the robot to test the initial strategies at the beginning of the mission. Unfortunately, because the environment changes all the time while the mission proceeds, the strategy superior at the beginning can loose its charm later on. Unlike the lower behavioral layer, where only one state has access to the actuators, there has to be an arbitration mechanism on the task layer to solve this problem. There are many possible ways to do this. These include, for example, fixed priority, voting, summing and several others. If a higher priority state at the behavioral layer's FSA becomes active (e.g., the robot is running out of energy), the necessary information about each task is stored in buffers. When the robot program starts to perform the actual task again it retrieves the needed information from these buffers. In this application there are two tasks to be represented on this layer, namely exploit and explore. In explore the robot moves around the process trying to create the basic map. This map is then combined with the other robots' basic maps. The result is the Common Basic Map (CBM), see [13] for details. The robot then uses this map and checks its quality from time to time. In these tests, however, the explore task is fixed to the state *use map*. The fixing was done in order to achieve as identical conditions as possible for the main target of these tests, i.e. exploit. In the exploit task the robot performs the poison release at a specific target location, shown in Fig.7. While operating in the explore task

**Fig.7.** Group of three robots performing algae attack.

(especially when creating a Common Basic Map) a robot can detect an arbitrary number of growth spots. Each spot is registered with an information field containing the location (in the Common Basic Map) and the volume. Based on this information the robot then selects its next strategy. Naturally, there exist several possible operational strategies, but here only three plausible strategies were considered: attacking the closest algal growth(S1), attacking the largest detected algal growth(S2), attacking the smallest detected algal growth(S3). In the exploit task the robot monitors the chosen algae's status by storing the largest sensory value of the target algal growth, i.e. exploit = $algal_{node(i)}$. The location of the growth is always connected (directly to some node or temporally related to two consecutive nodes) to the Common Basic Map. The robot follows how a selected algal growth behaves during the performance evaluation cycle. At the beginning of the mission, the robot tests all available strategies in a row. It receives a Δexploit value for each of the strategies. After this the robot chooses the most profitable one as its next strategy. Each strategy's Δexploit value is updated after testing. Furthermore, if the chosen algae growth is already deceased, the robot heads towards some other known algae location instead.

### 5.3 Cooperative Layer

Even though there is no direct communication between the robots in the society, the members can still receive information about the others through the environment. For example, if a robot detects that the volume of an algae growth is decreasing without its own active operation upon it, it concludes that there must be other robots performing the removal task on the same growth (or alternatively there is some problems with its sensor readings). Nevertheless, when an active communication, no matter how simple, is allowed an individual robot can use task related information received from the others. It is of crucial importance that the amount of inter-robot communication is kept to a minimum, especially when the size of the society grows. This is accomplished by restricting the contents of messages. Only data about one's own personal performance and status is transferred. The individual robot then makes its own decision about the status of the society. The communication used in our system is critical to the distance between sender and receiver, and the locations of the robots in the process. The robot either receives the whole message or otherwise it is omitted due to the protocol. Thus, only part of the transmitted information is available for the other members, while some messages are processed after a long delay. Due to these delays, the value of these messages should be decreased. The cooperative layer contains all the information the robot has gained through inter-robot communication. When a mission starts, the individual robot has no idea of the size of the society. Through communication (messages include senders ID) it gains knowledge of the other robots. A dynamic table is used to store these messages and a time label is attached to the messages upon arrival. This label is decreased as time goes by until it reaches some threshold and is removed from the table. There are several potential types of messages that could be used, but a performance result for the chosen strategy is naturally valuable to the other robots. Other valuable messages include, for example, information about selected topological targets. Common to all is that every task has its own result table, shown in Table 1.

**Table 1.** A dynamic result table for an individual task. It contains information about the success of society members. A time label is set to maximal value, when a message is received and then gradually decreased.

| | $Robot_1$ ($R_1$) | $Robot_2$ ($R_2$) | . . . | $Robot_k$ ($R_k$) |
|---|---|---|---|---|
| $Strategy_1$ ($S_1$) | result[$S_1$][$R_1$] time[$S_1$][$R_1$] | result[$S_1$][$R_2$] time[$S_1$][$R_2$] | | result[$S_1$][$R_k$] time[$S_1$][$R_k$] |
| | | | | |
| $Strategy_n$ ($S_n$) | result[$S_n$][$R_1$] time[$S_n$][$R_1$] | result[$S_n$][$R_2$] time[$S_n$][$R_2$] | | result[$S_n$][$R_k$] time[$S_n$][$R_k$] |

Each robot, represented in the table, belongs to certain groups. A group is formed by robots that have selected the same strategy in a particular task. This means that an individual robot can belong maximally to as many groups as there are different tasks and that the maximal size of a group is the total number of robots in the society. A group can be formed also through some other criteria, such as the spatial distribution (i.e. robots operating at the same location) or the volume of the group (e.g. join the largest or smallest group). These groups are, if not originally fixed, dynamic by nature and change as the mission proceeds. Robot ($R_i$) compares its own performance to the relative performance of the groups. First it calculates its own maximal relative average of success as follows:

$$success_{robot} = \max_{j \in [1,n]} \{Performance_{robot}[S_j]\} \tag{1}$$

where

$$Performance_{robot}[S_j] = \frac{result[S_j][R_i] * time[S_j][R_i]}{\sum_{m=1}^{n} time[S_m][R_i]} \tag{2}$$

It then compares this value to the maximal success values of active groups (i.e. robots operating in the same strategy),

$$success_{group} = \max_{i \in [1,n]} \{Performance_{group}[S_i]\} \tag{3}$$

where

$$Performance_{group}[S_i] = \frac{\sum_{l=1}^{k} ( result[S_i][R_l] * time[S_i][R_l] )}{\sum_{l=1}^{k} time[S_i][R_l]} \tag{4}$$

If $success_{group}$ is greater than $success_{robot}$ it means that the best group is performing relatively better than the individual. Thus the robot should change its operational strategy in that particular task to the strategy indicated by $success_{group}$. On the other hand if $success_{robot}$ is greater than $success_{group}$ then the robot should ignore the group information. Robot performs this procedure each time a certain task's strategy ends. This ensures that the individual robot benefits from the information it has received from the other society members.

## 6 Results

There were actually two main research goals. The first was to study how the size of the society affects the performance, i.e. to find out what kind of balance there would be between the increase in labor vs. increased resource conflicts caused mainly by the lack of space. The second main research goal was to find out how much the cooperation obtained through the described control architecture could help the performance of an individual and thus the whole society. Both of these results were obtained from two types of tests, with or without inter-robot communication. A society of 3 and 5 robots were tested and then compared to the reference case, where only a single robot was operating. In each test the development of both the two algae growth spots and successful cleaning agent (poison) releases were recorded, as is shown in Fig. 8.

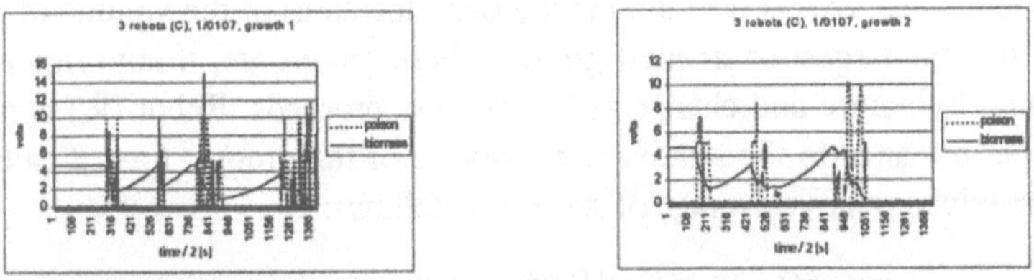


**Fig. 8.** Biomass volume (solid line) in one particular test shown in both growth spots. Poison releases, i.e., attacks, are shown with dashed line.

The test value for the success is the total volume of living biomass, i.e. the sum of biomass volumes in growth spots 1 and 2. The performance for separate cases, i.e. communication and no communication, was calculated by taking an average of the total biomass of the test runs. Due to the fact that there is no actual energy recharging station in the real system, the energy consumption of the robot was based on an energy consumption function. When the energy level drops below an upper limit, the robot starts to navigate towards the energy station immediately after the attack and measurement cycle ends. If the level for some reason drops below a lower limit, the robot stops everything and starts to navigate towards the energy node. The use of a cleaning agent (i.e. poison) is based on an even simpler function. When four full attacks have been completed, the robot heads towards the recharging station. At the beginning of a test both of the algae growths start from the value 4.5 V. After a while, they reach the maximum volume, represented by voltage 4.7 V. Thus, when the robots start their active performance, the volume of the algae in the process is as large as it gets. The maximum mission time was set at 60 minutes after which the robots navigate to the upper part of the largest tank

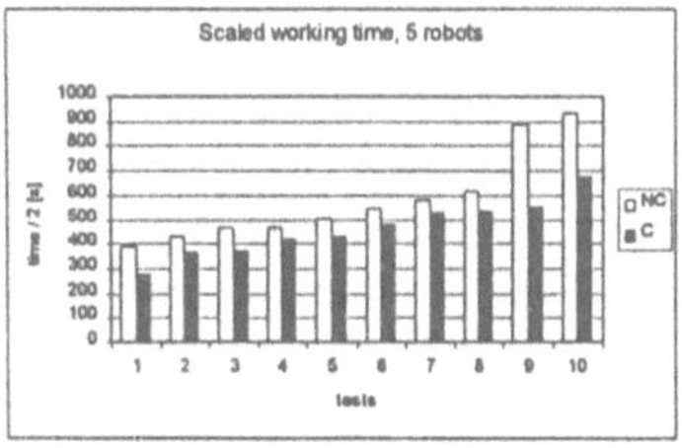


Fig. 9. Combined scaled working times (FI).

from where they could be easily removed from the process. Due to the structure of the demo process (i.e. two algae growths in optimal locations) the recharging station was forced to be set at a non-ideal location at the bottom of the largest tank. This location is problematic due to the strong current from the tank to the horizontal pipe. The robots often fail to refill their resources. This causes long delays during which the robots are out of productive work. In order to avoid this distortion in the results, the time robots used to obtain energy or poison ($T_{refill}$), was removed from the final results. This way the actual working times ($T_{work}=T_{total}-T_{refill}$) could be compared. There are three important factors that should be considered when results are analyzed: mission completion time, number of dead robots and the amount of used poison. In a normal case, i.e. no malfunction in robots, there are no dead robots. The poison on the other hand lasts for only four attacks and must usually be refilled. The actual parameters when judging the performance of the society are the time used and the amount of poison. The goal is naturally to eliminate the algae as fast as possible and with as little poison as possible. To make the individual runs comparable, each run is scaled according to the largest amount of poison used, i.e. $W = poison_{used} / poison_{max}$ and the final value for individual test runs is $FI = W*T_{work}$. These scaled results are combined in Fig.9, where the two cases are shown, sorted according to increasing order. The differences between the cases start to emerge. When non-communication and communication cases are tested with the T-test (equal variances based on the F-test), it can be stated that the mean of mission completion times with communication *is statistically smaller* than the mean of the non-communication case ($\alpha$=0.05). In other words, the used communication considerably improves the performance of the society. The importance of the volume, on the other hand, was clearly demonstrated, as shown in Fig.10. There is no question that if we had added more members to the society, the results have been even better (i.e., the society would have eliminated the algae growths even faster). To what limit, is the key question. At some point the interference between the robots starts to play such a large role that it actually decreases the performance of the whole society. This remains to be tested when we can increase the number of robots in the society.

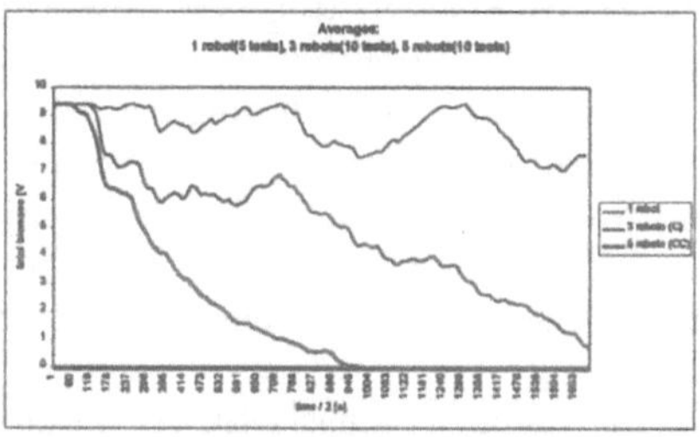

**Fig. 10.** The combined results clearly illustrate that size does matter

## 7. Conclusions

A novel three layer control architecture for an underwater robot society performing distributed operations in closed aquatic environment has been developed. The architecture was verified in tests with societies (max. 5 members). The importance of the size of the society and the usefulness of the communication were demonstrated.

## References

1. B. Hölldobler, E. O. Wilson.(1990) *The Ants*. Belknap Press, Cambridge, Mass.

2. T. Fukuda, S. Nakagawa, K. Kawauchi, M. Buss.(1989) Structure Decision Method for Self Organizing Robots based on Cell Structure -CEBOT, *Proceedings of 1989 IEEE Int. Conference on Robotics and Automation*, 695-700.

3. H. Asama, A. Matsumoto, Y. Ishida.(1989) Design of an Autonomous and Distributed Robot System: "ACTRESS", *Proceedings of IEEE/RSJ Int. Workshop on Intelligent Robots and Systems*, 283-290.

4. K. Dautenhahn (1995) Getting to Know Each Other- Artificial Social Intelligence for Autonomous Robots, *Robotics and Autonomous Systems*, Vol. 16, Nos. 2-4, 333 - 356.

5. M. J. Mataric (1994) Learning to Behave Socially, *Proceedings of From Animals to Animats 3*, (Third Int. Conf. on Simulation of Adaptive Behavior), D. Cliff , P. Husbands, J.-A. Meyer and S. W. Wilson, (Eds.), MIT Press, 453-462.

6. L. E. Parker (1994) Heterogeneous Multi-Robot Cooperation, *Ph.D.Thesis*, Massachusetts Institute of Technology, Artificial Intelligence Laboratory, MA, MIT-AI-TR 1465.

7. T. Balch (1998) Behavioral Diversity in Learning Robot Teams, *Ph.D. Thesis*, College of Computing, Georgia Institute of Technology.

8. A. Halme, P. Jakubik, T. Schönberg, M. Vainio (1993) The concept of robot society and its utilization, Proceedings of the IEEE/Tsukuba International Workshop on Advanced Robotics, 29 - 35.

9. M. Vainio, A. Halme, P. Appelqvist, P. Kähkönen, P. Jakubik, T. Schönberg, Y. Wang (1996) An application concept of an underwater robot society, *Distributed Autonomous Robotic Systems 2*, H. Asama, T. Fukuda, T. Arai, I. Endo, (Eds.), Springer-Verlag, 103 - 114.

10. M. Vainio, P. Appelqvist, A. Halme (1998) Generic Control Architecture for a Cooperative Robot System, 1998 IEEE/RSJ International Conference on Intelligent Robots and Systems, 1119-1125.

11. M. Vainio (1999) Intelligence through Interactions-Underwater robot society for distributed operations in closed aquatic environment, *Ph.D. Thesis*, Department of Automation and Systems Technology, Helsinki University of Technology.

12. P. Appelqvist, M. Vainio, A. Halme (1998) Mechatronical design of underwater sensor/actuator robots for cooperative task execution, *Proceedings of Mechatronics'98*, 249 - 254.

13. M. Vainio, P. Appelqvist, A. Halme (2000) Mobile Robot Society for Distributed Operations in Closed Aquatic Environment, *Robotica,* Vol. 18, Issue 3, 235-250.

# Ayllu: Distributed Port-Arbitrated Behavior-Based Control

Barry Brian Werger

Ullanta Performance Robotics
barry@usc.edu, *http://www-robotics.usc.edu/~barry/ullanta*

**Abstract.** Distributed control of teams of mobile robots presents a number of unique challenges, including unreliable communication, real world task and safety constraints, scalability, dynamic reconfigurability, heterogenous platforms, and lack of standardized tools or techniques. Similar problems plagued development of single robot applications until the "behavior-based" approach introduced new control techniques based on *port-arbitrated* behaviors (PAB). Though there are now a number of systems for behavior-based control of single robots, the potential for distributing such control across robots for multi-agent control has not been fully realized. This paper presents AYLLU[1], a language for distributed multi-robot behavioral control. AYLLU allows standard PAB interaction (message passing, inhibition, and suppression) to take place over IP networks, and extends the PAB paradigm to provide scalability. We give a brief overview of the Broadcast of Local Eligibility, a general technique for constructing such scalable systems in AYLLU, and survey some of the research projects and applications implemented using AYLLU.

**Keywords:** multi-robot, behavior-based, distributed control architectures

## 1 Introduction: Distributed Behavior-Based Systems

There is a strong trend in the robotics community towards distributed robotic systems - the use of groups of robots rather than single robots for greater robustness, scalability, and economy, as well as to achieve ***strongly cooperative*** tasks – those which simply cannot be performed by a solitary robot [14,18,2]. Though it is not uncommon these days to see mobile robots equipped with wireless Ethernet links, there are few that have communications architectures that address the many problems specific to the robotic domain, such as:

- Unreliable communication: Even the most sophisticated radio systems are subject to interference, both loss of bandwidth due to RF noise and competition for bandwidth.
- Real world task and safety constraints: Robotic tasks have time dependencies that are crucial to task performance and system survival. Much

[1] Development of AYLLU was performed under contract to ActivMedia Robotics. Software, manuals, and information available at ***http://robots.activmedia.com***

information has a severely limited useful lifetime (often measured in hundredths of a second) which influences how reliability of communication must be addressed.
- Scalability: Much of the motivation for multi-robot research involves economies of scale; to be able to increase task performance linearly merely by dropping in more robots. A good distributed robotics architecture must allow robots to be efficiently added to the system.
- Dynamic reconfigurability: Robots are prone to failure; communication problems lead to some robots being unreachable by others for varying periods of time; and scalability leads to changes in the makeup of a robot team. Most would benefit from the ability to match available robots to tasks without user programming or configuration.
- Heterogeneity: Robots may be equipped differently, or may differ in reliability of subsystems. Again, applications would benefit from the ability to automatically match available robots to tasks based on individual robot capabilities, which may vary over time.

The rest of this paper describes AYLLU, a language and architecture designed to address the issues of distributed robot control. It allows distribution in many senses: distribution of control of one robot over several machines, coordination of several robots by one machine (in effect using them as a single configurable sensor/actuator), and most interestingly, group interactions without centralized control. First we will discuss the port-arbitrated behavior paradigm on which AYLLU is based; we will outline AYLLU's extensions of this paradigm; we will present an AYLLU-enabled technique for building arbitrarily-scalable systems; and survey some successful AYLLU systems.

## 2 Port-Arbitrated Behavior-Based Control

The behavior-based approach to robot control introduced by Rodney Brooks [4] has been so influential in the field of robotics that the term "revolution" is commonly accepted as a description of its effect. Years after this revolution, the principles of horizontal decomposition, the world as its own model, and "emergent" intelligence (in the eye of the beholder) are Brooks' most salient contributions [1]. A somewhat lesser-known contribution is a set of well-defined abstractions and techniques for behavior interaction, implemented in a number of special-purpose languages which are more flexible successors to the well-known Subsumption Architecture. We refer to these abstractions and techniques as the ***Port-Arbitrated Behavior paradigm***, or PAB.

The Behavior Language [3] has been used in numerous experiments and applications, and has demonstrated scalability of the behavior-based approach to higher levels of competence, particularly in Mataric's demonstration of distributed mapping and "path planning" [9] and social learning [10], and Maes' work in action selection [8]. The MARS/L system [5] maintains the inter-behavior communication methods of the Behavior Language but allows

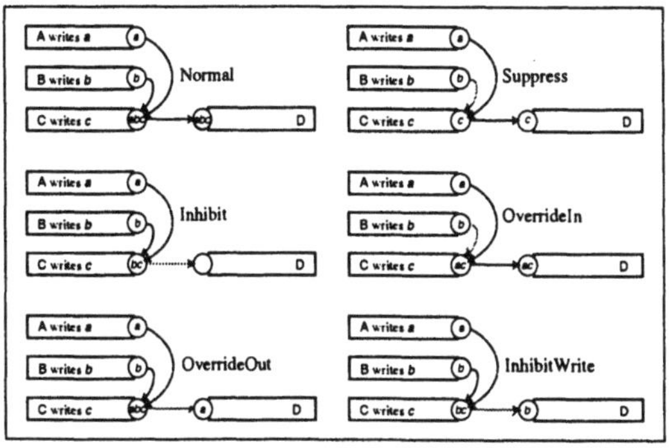


**Fig. 1.** Effects of Connection Type on Propagation. *B*-*C* and *C*-*D* connections are normal; the type of the *A*-*C* connection is specified for each subdiagram. Letters in circles indicate possible values of each port (e.g., for the *Normal* case, *D*'s port might hold any one of *a*, *b*, or *c*, depending on the timing of message arrival; while in the *Inhibit* case, *C*'s port might hold *b* or *c*, but *D* will receive no messages).

behaviors to be coded with all the facilities of Common LISP. This system has been used successfully in such domains as robot soccer [18], multi-robot learning [11], and human-robot interaction [12].

In these PAB systems, controllers are written in terms of behaviors, which are groups of concurrent processes. Each behavior has an interface shared by all of its processes, some parts of which are accessible from outside the behavior. The externally accessible elements are referred to as *ports*; they are registers that hold a single data item (which may be simple or complex, such as an integer or an arbitrary data structure). Internally accessible parts of the interface are *slots*, which are registers like the ports but accessible only to processes within the behavior, and *monostables*, which are boolean variables which remain true for a specified period of time after they are *triggered.*

Ports in different behaviors are linked together by *connections*, unidirectional data paths between a *source port* and a *destination port.* A port can have any number of incoming and outgoing connections. When data arrives at a port, either written directly from a process in the behavior or indirectly through a connection, it is generally propagated along all of that port's outgoing connections. Such data flow can, however, be modified by connections which are specified to be *suppressive, inhibitory,* or *overriding.* Given a connection $C_{s,d}$ from port $s$ to port $d$ with an associated period $p$, the following are the effects on $d$ whenever a message $m$ is propagated from $s$

- If $C_{s,d}$ is *Normal*, then $m$ is written to $d$, and is propagated along all of $d$'s outgoing connections. $p$ is not specified for Normal connections
- If $C_{s,d}$ is *Suppressive*, then $m$ is not written to $d$, and for period $p$, no incoming connections will be able to write to $d$ (that is, for all ports $x$, any connection $C_{x,d}$ will be temporarily disabled)
- If $C_{s,d}$ is *Inhibitory*, then $m$ is not written to $d$, and for period $p$, no messages will be propagated out from $d$ (that is, for all ports $x$, any connection $C_{d,x}$ will be temporarily disabled)

- If $C_{s,d}$ is *Input Overriding*, then $d$ is ***suppressed*** for period $p$, except that *only* messages arriving along $C_{s,d}$ (including $m$) are written to $d$ and propagated as normal
- If $C_{s,d}$ is *Output Overriding*, then $d$ is ***inhibited*** for period $p$, except that *only* messages arriving along $C_{s,d}$ (including $m$) are propagated outward along all $C_{d,x}$ as normal

It is through these mechanisms of suppression and inhibition that subsumption hierarchies, as well as other forms of arbitration, can be efficiently and intuitively implemented. Since connections are external to the behaviors, behavior code is easily re-usable, and interaction between behaviors can be modified dynamically. The port abstraction enforces a data-driven approach to programming that "grounds" computation in sensor readings and effector actions. By placing coordination in the interaction between behaviors (connections) rather than in the behavior code, these systems allow complex controllers to be built "bottom-up" from extremely simple, easily testable behaviors. The PAB approach allows a clean, uniform interface between encapsulated system components (behaviors) at all levels that abstracts away many issues of timing and communication; the "black boxes" of behaviors may contain reactive mappings or deliberative planners. While our research focuses on non-deliberative approaches, we believe that PAB interaction between system components can help reduce the complexity of the components themselves, whatever their type.

Unfortunately, the Behavior Language and MARS/L have been limited to specific target and development platforms, and have not seen widespread use. Other systems built to implement behavior-based control have not captured their elegance and ease of inter-behavior message passing, inhibition, and suppression. We believe that this lack of a good, widely-available implementation of the basic substrate of such behavior-based systems is one of the major reasons that the approach is seen to be problematic for scaling (see, e.g., [1]). Furthermore, neither L nor the Behavior Language has facilities for behaviors distributed across robots. The ALLIANCE architecture [14] and BeRoSH [17], among others, provide PAB-style control within robots and communication between robots, but do not generalize communication so that PAB arbitration techniques can be used between behaviors on different robots. We believe that by extending the notion of Brooksian inter-behavior connections across networks of distributed computers (i.e., over IP), we will be able to open up a wide range of new, simple, robust algorithms for multi-robot coordination. AYLLU has been developed for this purpose.

## 3 Ayllu: Distributed PAB Control

AYLLU is an extension of the C language intended for development of distributed control systems for groups of mobile robots. It facilitates communication between distributed system components and scheduling of tasks. While

AYLLU has many features specialized for behavior-based systems, it is useful for development of a wide range of architectures from reactive to deliberative, and provides the means for coordinating processor-intensive tasks, such as high-level planning and vision processing, with responsive low-level control. A small interface, referred to as AYLLULite[19], can be easily added to non-AYLLU programs, allowing AYLLU to serve as a simple and effective "glue" between heterogeneous components. It is designed to be highly portable, providing inter-operability between operating systems and languages.

AYLLU's principal goal is facilitate implementation of robust multi-robot systems that must cope with the unique challenges of the multi-robot systems described in Sec. 1. Towards this goal, AYLLU extends subsumption-style message passing to the multi-robot domain, provides for a wide variety of behavior-arbitration techniques, and allows a great deal of run-time system flexibility including dynamic reconfiguration of behavior structure and redistribution of tasks across a group of robots as determined by either task constraints or changing availability of resources.

We refer to AYLLU as a language, since it involves syntactic structures foreign to C itself, and some differences in semantics. AYLLU makes extensive use of C's macro facilities, however, and as a result the standard C preprocessor is able to translate all AYLLU code into standard C code. AYLLU therefore consists only of header files and object code (a library) that are included with programs in a standard C compilation process.

### 3.1 Ayllu Extensions to the PAB Paradigm

AYLLU extends the PAB paradigm primarily in two ways: allowing connections over IP networks, and allowing scalable networks of behaviors.

**Connections Over Networks** The extension across IP is straightforward; when making a connection $C_{s,d}$, $s$ and $d$ can each be specified either as a $(behavior, port)$ pair or as a $(host, behavior, port)$ triple. Within the behaviors themselves, no distinction is made. AYLLU also supports IP broadcasting on a local subnet; if a connection is made to $d = (host, behavior, port)$ where *host* is specified as a broadcast address, then any message propagated along the connection is sent to port $(behavior, port)$ in every host on the network except the sender. Such broadcast connections can be special (i.e., suppressive, inhibitory, or overriding).

**Scalability** Given the single-item register nature of ports, in previous PAB systems it is not easy (and often not possible) to arbitrarily add new behaviors to a system; unless new ports are added to behaviors, comparisons cannot be made between data coming from various sources. As new robots are added to a system, either behavior interfaces in all robots must be changed so that data sent by the new robots will be available, or an increasing percentage of all robots' data will be overwritten and lost. This does not allow for flexible group size, or robustness to robot failures. AYLLU addresses this problem through the addition of four specialized port types: when a message $m_{incoming}$ is written to a port $p$ holding a data item $m_{current}$,

- if $p$ is a normal port, $m_{incoming}$ replaces $m_{current}$
- if $p$ is a *MaxPort*, $m_{current}$ becomes $max(m_{incoming}, m_{current})$
- if $p$ is a *MinPort*, $m_{current}$ becomes $min(m_{incoming}, m_{current})$
- if $p$ is a *MinPort*, $m_{current}$ becomes $m_{incoming} + m_{current}$
- if $p$ is a *PriorityPort*, $m_{current}$ becomes $m_{incoming}$ iff $priority(m_{incoming}) > priority(m_{current})$

**Write-Inhibition** AYLLU also adds one new type of connection:

- If $C_{s,d}$ is *Write-Inhibitory*, then $m$ is not written to $d$, and for period $p$, no messages written directly from ***within the behavior*** will be propagated out from $d$, though messages arriving ***through connections*** will be

This provides a "closure" of the control of information flow through ports, and allows construction of multi-dimensional arbitration structures such as the ***cross-subsumption*** we present in Sec. 4.1.

## 4 A Scalable Example: Multi-Target Observation

We now present an example of how effective multi-robot strategies can be implemented with a small number of connections and extremely simple component behaviors using AYLLU. This example is a system that performs the W-CMOMMT [20] cooperative multi-robot observation of multiple moving targets task. The goal is for a group of robots to keep a number of moving targets under observation (that is, in the field of view of the robot and within a meter or so). Each target has a priority, so that when there are more targets than robots, the highest-priority targets should be covered.

In order to construct a system that will maintain observation of the highest priority targets and ensure that robots are properly distributed to separate targets (rather than observing targets redundantly), while adapting to changing numbers of robots and targets and robot failures, we use a technique called the Broadcast of Local Eligibility. Our BLE approach to W-CMOMMT is thoroughly detailed in this volume ([20]).

### 4.1 Broadcast of Local Eligibility (BLE)

The BLE mechanism involves a comparison of locally determined eligibility with the best eligibility calculated by a peer behavior on another robot. When a robot's local eligibility is best for some behavior $B_n$ which performs task $T_n$, it inhibits the peer behaviors (that is, behaviors $B_n$) on all other robots),

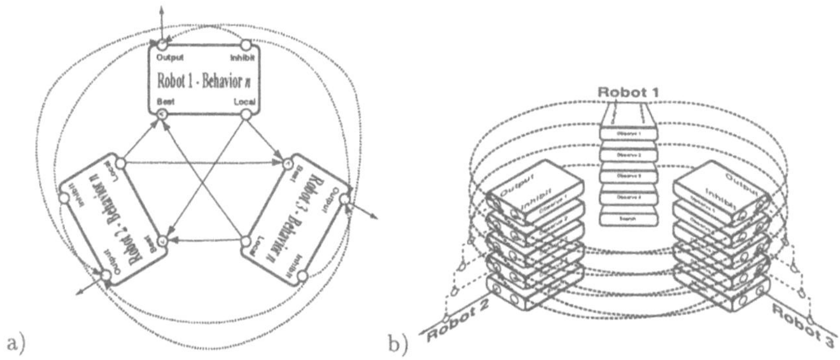


**Fig. 2.** *a) Cross-Inhibition*: A peer group. *b) Cross-Subsumption*: The structure of a system created by the combination of local subsumption with cross-inhibition. Some lines are omitted for clarity; each "layer" is a peer group connected as in a).

thereby "claiming" task $T_n$. Since this inhibition is an active process, failure of a robot which has claimed a task results in the task being immediately "freed" for potential takeover by another robot.

In order for a behavior to participate in BLE arbitration, AYLLU implicitly declares that all behaviors have normal ports named *Local* and *Inhibit* , and a MaxPort named *Best* .

**Cross-Inhibition of Behaviors** Cross-inhibition refers to the process of arbitration between *peer behaviors,* instances of the same behavior on different robots. Given that there is some behavior instance $B_n$ (which performs task $T_n$) on each robot, cross-inhibition results in the selection of at most a single robot to perform $T_n$. As illustrated in Fig. 2a, the *Local* port of each robot's behavior $B_n$ broadcasts a locally-computed eligibility estimate to the *Best* port of each other robot's behavior $B_n$. Each *Best* port maintains the maximum of the eligibility messages it has received in the current cycle. Whichever robot has a local eligibility better than or equal to the *Best* it receives writes to its Inhibit port, causing inhibition of behavior $B_n$ in the other robots.

**Cross-Subsumption** Cross-inhibition arbitrates only between peer behaviors on different robots; some local mechanism must arbitrate between different behaviors on the same robot. Simple subsumption, when combined with cross-inhibition, is sufficient for flexible, scalable, and robust team cooperation in CMOMMT. We call the combination of cross-inhibition and local subsumption *cross-subsumption.*

In cross-subsumption, each robot has a local subsumption hierarchy. Each layer of this hierarchy may be cross-inhibited (as in Fig. 2a), resulting in a system similar to the one diagrammed in Fig. 2b. As a result, each robot is

```
#include <ayllu.h>
#include <ayllustd.h>        /* for AvoidBC and SafeVelocityBC */

ayDefProcess(TurnToTarget) {
  ayLocalPort TargetX, TargetY, Rotate, BLE_Local;
  ayWriteIntPort(Rotate, ayVISCENTERX - ayReadIntPort(TargetX));
  ayWriteIntPort(BLE_Local, ayMAXVISY-ReadPort(TargetY));
}                             /* BLE_Local is local eligibility estimate */

ayDefBehaviorClass(ObserveTarget) {
  ayINTERFACE {
    ayIntPort(TargetX, 0);
    ayIntPort(TargetY, 0);
    ayIntPort(Rotate, 0); }
  ayPROCESSES {
    ayInitProcess(TurnToTarget, ratepersecond(20));
    ayInitProcess(BLE_Select, ratepersecond(1)); }
}

#define MakeObserver(num, next) ayInitBehavior(Observer, OBS##num);  \
  ayConnect(VISION, Target##num##Angle, OBS##num, TargetX);          \
  ayConnect(VISION, Target##num##Bottom, OBS##num, TargetY);         \
  ayBLE_Connect(OBS##num, Peers, Rotate)                              \
  ayOverrideOut(OBS##num, Rotate, next, Rotate, seconds(0.2));

void main () {
  ayBROADCAST Peers = ayMakeBroadcast("10.255.255.255");
  ayInitPioControl("/dev/ttyS0");         /* for Pioneer mobile robot */
  ayInitIPComms();

  ayDefaultInitBehavior(SafeVelocityBC, VEL);
  ayDefaultInitBehavior(AvoidBC, WANDER);
  MakeObserver(5, WANDER); MakeObserver(4, OBS5); MakeObserver(3, OBS4);
  MakeObserver(2, OBS3); MakeObserver(1, OBS2);

  ayRunBehaviors();
}
```

**Fig. 3.** AYLLU code for CMOMMT

controlled by its behavior $B_n$ which has the highest priority of any behavior which is generating output; a behavior that is not generating output fails to do so either because its current input is unsuitable for the task (e.g., some necessary object is not in the field of view), or because its output is cross-inhibited. Thus, each robot claims the highest-priority task that it is most suitable for.

### 4.2 Implementation

Figure 3 shows all of the AYLLU code necessary for the CMOMMT task; the descriptions below refer to behaviors mentioned therein.

A single behavior (VEL) on each robot controls translational motion to maintain a safe velocity based on the distance to sonar-detected obstacles. The task-oriented behaviors only control rotational motion of the robots. Two classes of task-behaviors are implemented: *Observer* behaviors which rotate the robot so that a specific target is centered in its field of view, and a *Search* behavior, which is a random wander (WANDER).

The behavioral coordination consists of a subsumption hierarchy of instantiations of the *Observer* behavior (OBS1...OBS5), each joined (by broadcast connections) into a cross-inhibiting peer group of *Observers* of the same target on each robot (Fig. 2a), producing a cross-subsumption hierarchy (Fig. 2b. The local evaluation function for each *Observer* behavior is a visual approximation of target distance. The highest-priority behavior that is not cross-inhibited controls the robot - that is, each robot approaches and tracks the highest-priority target it sees that is not being observed by another robot.

### 4.3 Discussion

As detailed in [20], a slightly more complicated version of this BLE approach to CMOMMT implemented for three Pioneer robots and four targets has proven to be effective in properly assigning robots to targets. Although robots switch off observation of particular targets, coverage of the highest-priority targets was well-maintained, and the system took advantage of the occasional (due to patterns of motion) ability of robots to observe more than one target time such that the three robots tended to cover all four targets when possible. Further, the task division achieved had interesting beneficial effects that tended to reduce problems of sensor uncertainty and physical interference between the robots.

## 5 Other Ayllu Systems in Practice

In addition to the partial survey of research projects below, Ayllu has been used in real-world applications such as museum installations, film special effects, and roaming sales exhibits at trade shows; in thatrical works and performance art pieces; and in a number of robot competitions including AAAI's and RoboCup Robot Soccer competition.

Ayllu has been used to implement the cooperative behaviors for ground-based robots which cooperate with an autonomous robot helicopter [16] and formation control in outdoor environments. [6] uses Ayllu to build topological mapping systems that run on one robot that communicates with desktop displays, and for work on cooperative mapping with heterogeneous robots. [15] has used Ayllu as a substrate for both fuzzy-logic and multiple-objective behavior control. It is being used in experiments involving artificial emotions in social robotics in continuation of research presented in michaudvuICRA99, and in work on planar object manipulation [7]. [13] uses Ayllu in experiments in which robots learn about cooperation through analysis of the benefits of interaction with humans.

## References

1. R. Arkin. *Behavior-Based Robotics.* MIT Press, 1998.

2. Ronald C. Arkin and Tucker Balch. Cooperative multiagent robotic systems. In David Kortenkamp, R. Peter Bonasso, and Robin Murphy, editors, *Artificial Intelligence and Mobile Robots: Case Studies of Successful Robot Systems*, pages 277–296. AAAI Press/The MIT Press, Cambridge, MA, 1998.
3. R. A. Brooks. The behavior language; user's guide. Memo 1227, MIT AI Lab, April 1990.
4. R. A. Brooks. *Cambrian Intelligence.* MIT Press, 1999.
5. R. A. Brooks and C. Rosenberg. L - a common lisp for embedded systems. In *Proceedings of the Lisp Vendors and Users Conference*, 1995.
6. D. Dedeoglu, M. J. Mataric, and G. S. Sukhatme. Incremental, on-line topological map building with a mobile robot. In *Sensor Fusion and Decentralized Control in Autonomous Robotic Systems, Proceedings of SPIE*, pages 123–139, 1999.
7. Brian Gerkey and Maja Mataric. Murdoch: Publish/subscribe task allocation for heterogeneous agents. In *Proceedings of Autonomous Agents*, 2000.
8. Pattie Maes. Situated agents can have goals. *Robotics and Autonomous Systems*, 6(1–2):49–70, 1990.
9. Maja J Matarić. Navigating with a rat brain: A neurobiologically-inspired model for robot spatial representation. In *Proceedings of the First International Conference on Simulation of Adaptive Behavior: From Animals to Animats.* MIT Press, 1990.
10. Maja J. Matarić. Behavior-based control: Examples from navigation, learning, and group behavior. *Journal of Experimental and Theoretical Artificial Intelligence*, 9(2–3):323–336, 1997.
11. Francois Michaud and Maja J Mataric. Representation of behavioral history for learning in nonstationary conditions. *Robotics and Autonomous Systems*, 29(2):1–14, 1999.
12. Francois Michaud and Minh Tuan Vu. Managing robot autonomy and interactivity using motives and visual communication. In *Proc. Conf. Autonomous Agents*, pages 160–167, 1999.
13. Monica Nicolescu and Maja Mataric. Learning cooperation from human-robot interaction. In *Proceedings of DARS*, 2000.
14. L. E. Parker. Alliance: An architecture for fault tolerant multi-robot cooperation. *IEEE Transactions on Robotics and Automation*, 14, 1998.
15. Paolo Pirjanian and Maja Mataric. Multi-robot target acquisition using multiple objective behavior coordination. In *IEEE International Conference on Robotics and Automation*, San Francisco, April 2000.
16. G. S. Sukhatme, J. F. Montgomery, and M. J. Mataric. Design and implementation of a mechanically heterogeneous robot group. In *Sensor Fusion and Decentralized Control in Autonomous Robotic Systems, Proceedings of SPIE*, pages 111–122, 1999.
17. Z.-D. Wang, E. Nakano, and T. Matsukawa. Designing behavior of a mult-robot system for cooperative object manipulation. In *Proceedings of the International Symposium on Microsystems, Intelligent Materials, and Robots*, 1995.
18. Barry B. Werger. Cooperation without deliberation: A minimal behavior-based approach to multi-robot teams. *Artificial Intelligence*, 110:293–320, 1999.
19. Barry Brian Werger. *Ayllu: Distributed Behavior-Based Control; User's Manual.* ActivMedia Robotics, 1999. available at http://robots.activmedia.com.
20. Barry Brian Werger and Maja J Mataric. Broadcast of local eligibility for multi-target observation. In *Proceedings of DARS*, 2000.

# An Architecture for Reactive Cooperation of Mobile Distributed Robots

O. Simonin, A. Liégeois, and P. Rongier

LIRMM - Université Montpellier II,
161 rue Ada, 34392 Montpellier Cedex 5, France
{simonin,liegeois,rongier}@lirmm.fr

**Abstract.** This paper shows how to include a crude but efficient communication component into reactive robotic multi-agent systems. The basis of the work relies on the concept of self-satisfaction of an agent composed of a selfish tendency and an altruistic one. A signal, representing the variation of an agent's "interactive satisfaction", is emitted in its neighborhood. It measures the expected efficiency of cooperation. The information propagates from agent to agent within a certain range. It is shown how the information is converted into a vector able to alter instantaneously the agent's motion when added to the classical vectors which generate the selfish motion: move-to-goal and avoid-obstacle. Then the paper presents the architecture for implementing such mechanisms to achieve robust adaptive co-operation. The results of simulation experiments are presented which demonstrate the efficiency of the generalized dynamic potential field including communication of satisfaction. In particular, the method allows the agents to escape from a dead-lock situation, and makes the collective behavior converge towards a steady flow of traffic.
**Keywords:** multi-agent systems, cooperating mobile robots, robot communication, reactive architecture, self-organizing robots

## 1 Introduction

In the field of distributed autonomous systems, cooperation appears to be one of the most important way to perform meaningful tasks.

As it is emphasized in [11], many robotic applications are inherently distributed in space, time, or functionality, thus requiring distributed solutions. Adaptive cooperation techniques allow to handle repetitive tasks or tasks which need cooperation of several robots for their completion [3][11].

Many animats which have been implemented with reactive behavioral control techniques have only local interactions with the environment and the other agents. Nevertheless, they can perform complex foraging tasks [4][6][12], multi-robot navigation [1], cooperation tasks [6][9], and distributed problem solving [7].

However, the poor level of communication and cognition of reactive architectures cannot so far handle easily goal-driven behavior [10] and does not allow explicit cooperation [4] to perform complex tasks. The aim of this work is to include a robust but simple intentional cooperation into a reactive architecture without degrading its intrinsic qualities.

In a previous work [13], we have introduced the self satisfaction which expresses the performance of the agent's current task, and the interactive satisfaction which results from agent interactions. This last information is broadcast as a simple signal in order to be directly converted by agents into vectors which may be combined to other vectors (goal attraction, obstacle avoidance).

This paper concentrates on the process of satisfaction propagation and shows that it contributes to adaptive cooperation, while avoiding deadlocks ("intelligent behavior").

The paper is organized as follows: Section 2 presents the model of communication and the related altruistic behavior vector. We then describe the cooperative architecture in section 3. Section 4 shows some typical simulations and discusses experimental results.

## 2 From Satisfaction to Altruism

This section presents a formal description of the components of an agent's satisfaction, and how they are converted into vectors consistent with those resulting from other needs: avoiding collisions and reaching a destination.

### 2.1 Cooperation and Satisfaction

**The satisfaction approach.** From the designer's point of view, the agent behavior must satisfy *goal-oriented*, *conservative* and *cooperative* functions. The principle of our agent model relies on the maximization of agent satisfactions. At any time, the agent tries to maximize

- either its selfish interests: by selecting the optimal task and by emitting repulsive signals to agents which hinder its work or attractive signals to altruistic agents which can help it,
- or the collective interests: by helping other agents, i.e. by reading agent signals and then computing altruistic behaviors.

**Sub-satisfactions.** As introduced in [13], we distinguish the personal satisfaction, the interactive satisfaction and the empathy satisfaction.

The *personal satisfaction* $P$ measures the progress of the agent's task. The *empathy satisfaction* $E$ is the average value of personal satisfaction of its acquaintances, it expresses the *altruism* of the agent. The instantaneous satisfaction of an agent $i$ at time $t$ is $Sat_i(t) = (1-\alpha).P_i(t) + \alpha.E_i(t)$, where $\alpha$ is the *altruistic factor* of the agent ($\alpha \in [0,1]$).

The *interactive satisfaction* $I$ results from agent interactions. An agent computes the interactive satisfaction as a reaction or an intention to other agents.

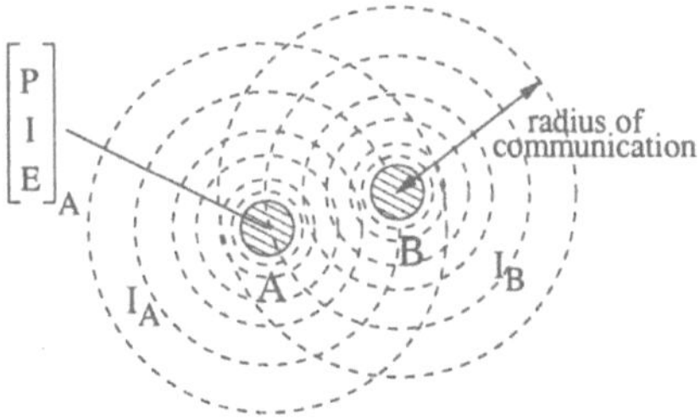


**Fig. 1.** Broadcasting of agent signals

**Signals of Interactive Satisfaction.** Each agent $i$ broadcasts only its interactive satisfaction $I_i(t)$ because it is an intentional signal to alter agent interactions. This signal is broadcast within a bounded distance from the agent (Fig. 1). The *intensity* and the *variation* of the signal define the meaning of the communication.

## 2.2 Altruistic Behavior

Each agent alters its selfish behavior by processing the signals received from its neighbors, that generates a component of its velocity vector.

**Agent-agent altruism:** Agents use interactive satisfaction either to emit a repulsive signal ($\Delta I\searrow$) or to emit an attractive call ($\Delta I\nearrow$). The motion behavior clearly depends on the other agent interactive satisfaction variation $\Delta I_A(t)_T = I_A(t) - I_A(t-T)$.

Thus, in the vector model, this cooperative behavior is defined with the *altruism vector* of an agent (B) relative to another agent (A) $\overrightarrow{\vartheta_{B/A}}(t)$:

$$\overrightarrow{\vartheta_{B/A}}(t) = k.S(\Delta I_A(t)_T).\frac{\mid I_A(t) \mid}{\left\|\overrightarrow{AB}\right\|^2}.\overrightarrow{BA} \tag{1}$$

$$S(x(t)) = \begin{cases} Sign(x(t)) & if\, x(t) \neq 0 \\ Sign(x(t-T)) & if\, x(t) = 0\ (to\, ensure\, continuity) \end{cases}$$

Equation 1, where k is a gain, defines a force field. The interest of equation 1 lies in the fact that it can be used for various situations: i) just one agent moves ii) both agents move iii) none of the agents move but they notice significant interactive satisfactions.

**Multi-agent altruism:** A situated agent close to others can simultaneously perceive several attractions and repulsions. In order to define a new altruism vector we split the multi-agent problem into a set of agent-agent problems. The altruistic reaction of an agent ($B$) close to $N$ acquaintances noted $\Lambda$ is computed as follows

$$\overrightarrow{\vartheta_{B/\Lambda}}(t) = \sum_{j\in\Lambda} \overrightarrow{\vartheta_{B/j}}(t) = \sum_{j\in\Lambda} k.S(\Delta I_j(t)_T).\frac{|I_j(t)|}{\left\|\overrightarrow{BJ}\right\|^2}.\overrightarrow{BJ} \tag{2}$$

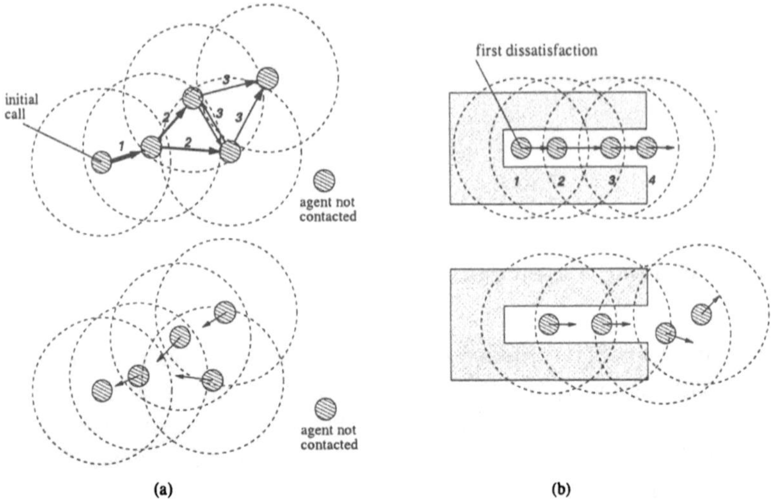


**Fig. 2.** (a) Intentional propagation for attraction (b) Implicit propagation of repulsion (the arrows are numbered following the sequence of emitted signals)

### 2.3 Sequential Signal-Passing

Sequential passing of attractions: When an agent perceives an increasing signal, it may decide to follow the call. In this case, the agent also emits an attractive signal in order to call its own acquaintances (implicit recruitment). The force of the signal variation decreases during the propagation so as to limit the number of attracted agents (see. Fig. 2.(a)).
Sequential passing of repulsions: If an agent is immobilized by another, it can emit a dissatisfaction signal to query a movement of the hindering agent. But, if this other agent is also immobilized, a chain process of dissatisfaction emission is performed until an agent is able to move (Fig. 2.(b)). This implicit passing of dissatisfaction to all involved agents releases the deadlock situation.

## 3 Principle of the Architecture

Our architecture is based on a classical schema-based reactive control [4] using vector combination [1][8][14]. Actions are selected and computed from proprioceptive information (energy, conservative functions) and exteroceptive information (obstacles, goals), with the addition of the new vector called *altruistic vector*. It represents the cooperative part of the agent's behavior.

### 3.1 Primitive Behavior

Each task (primitive behavior) is an *action* which may be triggered by an internal or environmental *stimulus*. The test to release an action is defined by a set of boolean conditions on perceptions, *Cond*(*task*). Moreover, each condition induces a measure of intensity computed from perceptions and called *Int*(*task*) (defined on the interval [0, 1]).

Some tasks are ***functions for survival***, i.e. they are useful for robot working. These tasks have priority on others (with a subsumption control [5]).

## 3.2 Algorithm of Action Selection and Combination

**Data** : $task_c$, $P$, $I$, Sensors

1 **if** *survival function triggered* **then**
  **if** *motion task* **then** compute $\vec{V}_{goal}$
**end**
**else**
2  $task_m \leftarrow \{\forall i\ Int(task_m) \geq Int(task_i)\ and\ Cond(task_m) = true\}$
3  **if** $Int(task_m) > P$ **then** $task_n = task_m$ ; $F$=Int($task_m$)
  **else** $task_n = task_c$ ; $F = P$
  *Read signals:* $I_1$, $I_2$,..
4  **if** $task_c$ *is altruism and sign*($\Delta I_c(t)$) $=$ *sign*($\Delta I_c(t-T)$) ***then***
   *continue* $task_c$
  ***else***
5   $I_{max}$ =*Max*($|I_1|$, $|I_2|$,..)
6   ***if*** $\alpha.|I_{max}| > (1-\alpha).F$ ***then*** $\vec{V}_{goal} = \vec{\vartheta}_{I_{max}}$
   ***else***
    $task_c$=$task_n$
    ***if*** $task_c$ is a motion ***then*** $\vec{V}_{goal} = \vec{V}_{task_c}$
   ***end***
  ***end***
**end**

**if** *moving action* **then**
  Integrate signal repulsions :
7  $R = \{ \bigcup I_i / \Delta I_i < 0\}$ ; $\vec{V}_{alt} = \sum_R \vec{\vartheta}_{R_i}$
  Integrate obstacles sliding :
8  Compute $\vec{V}_{obs}$ for avoiding obstacles
**end**
9 **if** *moving action* **then** Compute $\vec{V} = \gamma_1.\vec{V}_{goal} + \gamma_2.\vec{V}_{obs} + \gamma_3.\vec{V}_{alt}$
10 Perform behavior

**Algorithm 1:** Generic Algorithm of Altruistic Robots

**Task Selection.** If no survival function is released, the algorithm (Alg. 1) selects the next task to perform (a new one or the current one).

First, the task which has the maximum perceived intensity among the triggered tasks is computed (line 2). If this task has an intensity greater than the current task performance index (= personal satisfaction $P$) it becomes the new candidate task, noted $task_n$ (line 3). But this task may be replaced by an altruistic behavior. For this purpose, we compute the intensity $F$ of $task_n$ selection (line 3).

**Altruistic Behavior Selection.** If the agent is already performing an altruistic task and the relative perceived signal ($I_c$) keeps the same sign of variation, it keeps the same behavior (line 4).

Otherwhise, as the agent perceives interactive satisfactions $I$ from others, it can reply to those signals by exhibiting an appropriate behavior. Line 5 computes the absolute value $|I_{max}|$ of the more powerful signal perceived.

An altruistic behavior replaces the potential new task $task_n$ when the signal intensity $|I_{max}|$ is greater than $task_n$ intensity $F$. This comparison is adjusted by using the altruistic factor $\alpha$ (line 6). Then, agents may have different kinds of behavior just according to this factor.

If the altruistic behavior is selected, the new behavior is the altruism vector $\overrightarrow{\vartheta_{I_{max}}}$ computed with equation 1 section 2.2.

**Vector Combination.** If the selected task is a vector motion ($\overrightarrow{V_{goal}}$), it may be added with other compatible agent drives (motion-to-goal, obstacle avoidance and repulsive signals).

From perceived signals which are repulsive (set R computed line 7), the agent computes a repulsive vector using multi-agent altruism equation 2 given above in section 2.2. In practice, this vector $\overrightarrow{V_{alt}}$ is shorter than $\overrightarrow{V_{goal}}$ vector.

To avoid obstacles, a vector for sliding along obstacles is computed line 8, see Fig. 3.(a). The technique is fully presented in [14].

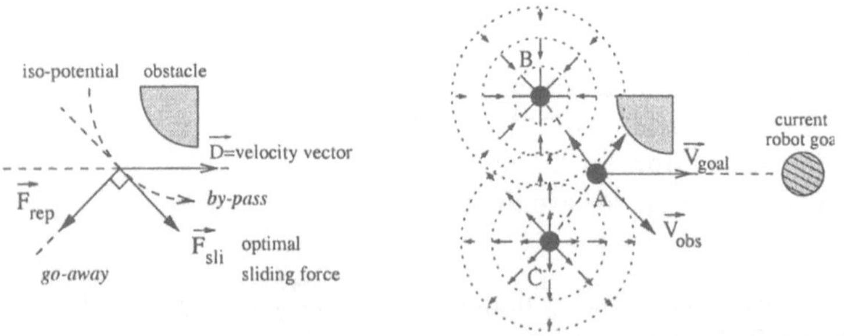


**Fig. 3.** (a) Sliding along an obstacle (b) Multiple influences on agent A

These three vectors are combined line 9 to compute the agent's velocity vector $\overrightarrow{V}$, where $\gamma_1$, $\gamma_2$ and $\gamma_3$ are scalar weights. The principle of combining signal and goal drives is represented in Fig. 3.(b).

Finally, in line 10, the agent performs its current task.

Note that the process of action selection and combination is computed by using only the personal satisfaction $P$, the perceived signals $I$, the altruism factor $\alpha$ and internal or environmental stimuli. Personal satisfaction $P$ and interactive satisfaction $I$ are computed from internal and external perceptions as shown below.

# 4 Experimental Simulations

## 4.1 Extractor Robots

Consider the foraging and consuming application where each robot has to leave a fixed base and explore an area which is *a priori* unknown in order to find a mine, extract raw material and transport this material back to the base. Clearly, the problem is defined by N robots (with a limited rate of extraction $r$), obstacles, attractive zones (mines), and a base which emits a particular signal for its localization. Moreover, each mine has a limited volume of raw material and a maximal rate of extraction. The energy available $E$ on board each robot is bounded by $E_{max}$, and so periodic returns to the base power supply are necessary.

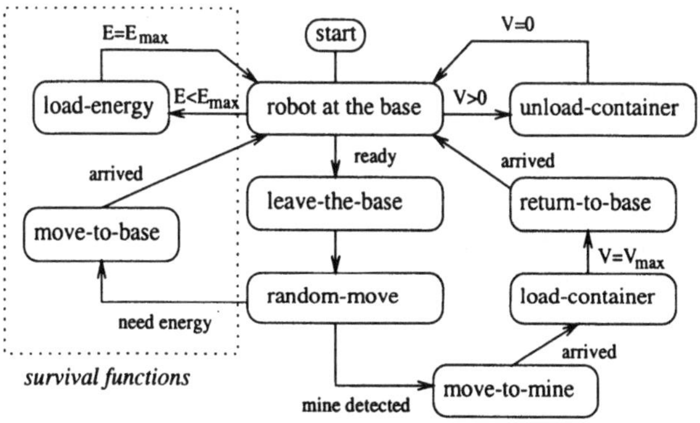


**Fig. 4.** Primitive behaviors of extractor robots

Extractor robots are implemented following our architecture and simple primitive behaviors (tasks). They are able to scan their environment within a short bounded distance. Figure 4 presents the global behavior of an extractor robot by giving its primitive behaviors and their stimulus. The volume of raw material extracted by a robot is noted $V$ (with $V_{max}$ the volume of the container).

**Computation of *personal satisfaction* $P(t)$:** For the moving tasks *move-to-base*, *move-to-mine* and *return-to-base*, the robot computes its progression: if it cannot move, $P(t) = -1$ else $P(t) = cos(\overrightarrow{V}_{motion}(t), \overrightarrow{V}_{goal}(t))$.

For other tasks involving moves: if the robot can move $P(t) = P(t-T) + K.(0.5 - P(t-T))$ (i.e. $P$ increases to 0.5) else $P$ decreases to -1: $P(t) = P(t-T) - K.(1 + P(t-T))$, with $0 < K < 1$ ($K = 0.2$ for the simulations).

For extraction of raw material (*load-container*), $P(t)$ depends on the real rate of extraction $e(t)$: $P = \frac{e(t)}{r}$. With other *static* primitive actions in progress, $P(t) = 1$.

**Computation of *interactive satisfaction* $I(t)$:** Robots compute an increase of interactive satisfaction when they are at an exploitable mine, so that they attract other robots : $I(t) = I(t-T) + K'.(1 - I(t-T))$.

They compute a decrease of their signal when they are hindered by others robots, or when a detected mine is empty or saturated : $I(t) = I(t-T) - K'.(1 + I(t-T))$.

Otherwise, when a robot performs an altruistic behavior upon the request of an increasing signal $I_p(t)$, it also emits an increase of its interactive satisfaction: $I(t) = I_p(t)/2$. Thus, this robot may attract its idle acquaintances.

Note that while a robot has no interaction with others, its interactive satisfaction is not emitted and it tends towards 0 (neùtral value of $I$).

### 4.2 Results

**Cooperation, Altruism and Resolution.** Situations of conflict or of cooperation are visualized as a surface with the X,Y axes reflecting the simulated environment. The Z axis indicates *the negative value* of the sum of interactive satisfaction signals emitted by agents. Peaks of the surface Z(X,Y) indicate blocking zones (because of the emission of negative values) and pits indicate attractive zones between robots. Thus, the altruistic vector is computed as if the robot was a ball rolling on this dynamic surface. Furthermore, as robot locations take a part in surface deformation, their altruistic motions tend to flatten the surface. Thus, robots avoid conflicting trajectories and move towards attractive zones (like attractive fields).

When robots have incompatible trajectories, peaks appear in the dissatisfaction surface (dashed lines 3 and 4 in Fig. 5). The robots at mines (in the left of the environment) broadcast attractive signals (pits in the surface at dashed lines 1 and 2 in Fig. 5).

In every simulation run, robots efficiently explore the environment, slide around obstacles and around other robots by using altruistic vectors. When a robot finds an exploitable mine, its interactive satisfaction increases and, as expected, neighbor robots are attracted, and by propagation they attract remote robots. If a mine is empty or saturated, the robots close to it emit a decreasing interactive satisfaction which repel new arriving robots. Finally, robots adapt their behavior to each environmental evolution: moves of robots, working out of mines and need for energy.

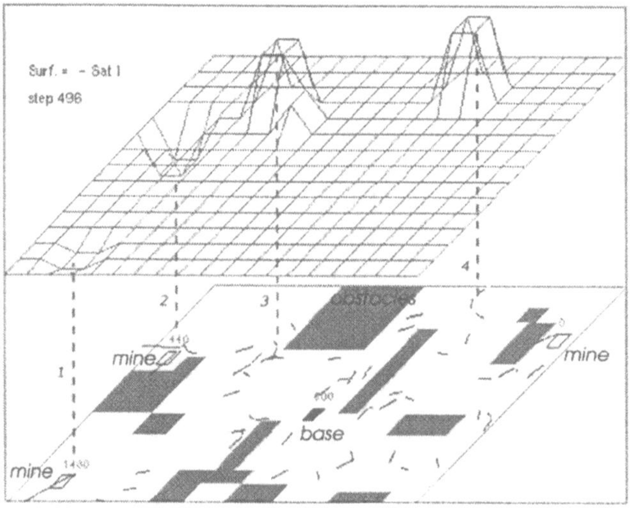


**Fig. 5.** Snapshot of an extractor robots simulation (50 robots, $\alpha = 0.7$).
1 robot=1 pixel, plotted with its velocity vector. Dark areas are the obstacles.

**Study of Repulsion Signal-Passing.** In order to evaluate performance of repulsive signals (situation Fig. 2.(b)), we have designed a long narrow environment (fig. 6). Thus, we measure the average time for at least one robot reach the mine at the right side of the environment. We have compared this time performance for different types of robots (over 500 simulations for each one).

Robots of the first type move randomly without broadcasting signals (time of reference: 226 iterations). The second type of robot avoids obstacles by using sliding vectors: 5.9% of improvement. The third type of robot includes altruism vector as goal behavior: 13.6% of improvement. In the last case, robots use all repulsive signals to compute their trajectory: 17.5% of improvement (see snapshots of these robots in figure 6).

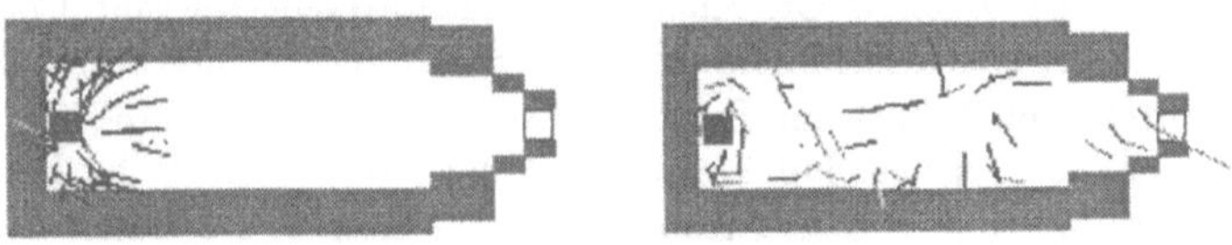

**Fig. 6.** Simulation of 30 altruistic robots (at iterations 12 and 170)

## 5 Conclusion

In this paper, we have described an architecture for multiple mobile robots. It takes into account the concept of satisfaction, and allows an agent to communicate simply its state of satisfaction and its intentions to its neighbors. The received information is directly converted into an altruistic control signal altering the otherwise selfish behavior of each agent, that drastically improves the cooperation. The architecture is simple and robust, it makes the robots immediately react to unexpected events, avoid conflicting situations and exhibit self-organization. We have demonstrated the qualities of the proposed architecture through simulations. These intensive simulations have also shown (not discussed here) that the group performance is not very sensitive to variations of the altruistic factor. Future work will include experiments with real robots and refinement of the processing of the communication signals.

## References


1. R.C. Arkin. (1989) Motor Schema Based Mobile Robot Navigation, International Journal of Robotics Research, 92–112
2. R.C. Arkin. (1995) Reactive Robotic Systems, article in Handbook of Brain Theory and Neural Networks, M. Arbib, MIT Press, 8(4), 793-796
3. R.C. Arkin, T. Balch. (1997) Cooperative Multiagent Robotic Systems, AI-based Mobile Robots: Case studies of Succesful Robot systems, MIT Press
4. T. Balch, R.C. Arkin. (1994) Communication in Reactive Multi-agent Robotic Systems , Autonomous Robots, 1, Kluwer Academic Publishers, Boston, 27–52
5. R.A. Brooks. (1986) A Robust Layered Control System for a Mobile Robot, IEEE Journal of Robotics and Automation, 2(1), 14–23
6. A. Drogoul, J. Ferber. (1992) From Tom Thumb to the Dockers: Some Experiments with Foraging Robots, 2nd International Conference on Simulation of Adaptative behavior, Honolulu, 451–459
7. J. Ferber. (1999) Multi-Agent Systems, an introduction to Distributed Artificial Intelligence, Addison-Wesley
8. O. Khatib. (1985) Real-Time Obstacle Avoidance for Manipulators and Mobile Robots, Proceedings of IEEE International Conference on Robotics and Automation, St. Louis 500–505
9. M.J. Mataric, M. Nilsson, K. Simsarian. (1995) Cooperative Multi-Robot Box-Pushing, Proceedings IROS '95, Pittsburgh, PA, 556-561
10. Jorg. P. Müller. (1996) The design of intelligent Agents, Lecture Notes in Artificial intelligence 1177, Springler-Verlag
11. L. Parker. (1994) ALLIANCE: An Architecture for Fault Tolerant, Cooperative Control of Heterogeneous Mobile Robots, Proceedings of International Conference on Intelligent Robots and Systems (IROS '94), Munich, 776–783
12. P. Rongier, A. Liegeois. (1999) Analysis and Prediction of the Behavior of One Class of Multiple Foraging Robots with the Help of Stochastic Petri Nets, SMC'99, IEEE International Conference on Systems, Man and Cybernetics, Tokyo Japan, 143–148
13. O. Simonin, J. Ferber. (2000) Reactive Real-Time Cooperation as a Combination of Altruism and Self Satisfaction, Proceedings of International Conference on Multi-Agent Systems (ICMAS'2000), accepted (poster), Boston
14. K. Zeghal. (1998) A Comparaison of Different Approaches based on Force Fields for Coordination among Multiple Mobiles, IEEE International Conference on Intelligent Robots and Systems (IROS '98), Victoria, B.C., Canada, 273–278

# A Framework for Distributed Simulation of Multirobot Systems: the VLAB Experience

Stefano Carpin[1], Carlo Ferrari[1], and Enrico Pagello[12]

[1] Intelligent Autonomous Systems Laboratory,
Deparment of Electronics and Informatics,
University of Padua, Italy
{shamano,carlo,epv}@dei.unipd.it
[2] LADSEB-CNR
Padua, Italy

**Abstract.** This paper is devoted to discuss the major features a simulator for multirobot systems must show, in order to be effectively useful for setting up and debug the behavior of a real system. The paper also describe VLAB (Virtual LABoratory), a simulation system that allows the simulation of distributed multirobot systems using a set of computers connected by a LAN. Users can specify the control software and the morphology of more robots that share the same simulation environment. A visual representation of the behavior of all the robots and of the environment is obtained. Users can walk into this virtual laboratory to better observe the system evolution. VLAB allows different users to instantiate their own robotic agents for testing and debugging their coordination and their cooperative strategies.

## 1 Introduction

In recent years multirobot systems are gaining much more attention from the scientific community due to their potentialities of being used in a great variety of contexts. In particular, it looks like civilian applications can benefit from systems that are composed by several (simpler) machines that cooperate to solve a common task. A multirobot system can better respond in less-structured environments, working together with humans that can change the actual world status in sudden and unpredictable ways [1]. The term Service Robotics has been introduced to name all the practical situations where the robotic system is used as assistant of human activities.

Multirobot systems are distributed systems at very different levels. They are physically distributed because their components can be appropriately located in different areas of the working environment. They also shows a temporal distributed behavior because the decomposition of the global task can results in sub-tasks that can be scheduled at different times onto different machines. As the robot fleet can be composed by heterogeneous machines, distribution and parallelism arise also at the functional level. Finally, the software architecture of the system is a distributed architecture, maybe running on different computers, leading the programmer to design distributed multi-threading software for planning and control.

Although multirobot systems show some major advantages compared to single robots (enhanced fault-tolerance, simplicity, high run-time reconfigurability), it should be pointed out how difficult is to correctly program such systems [2]. In fact a major drawback is due to the big complexity in the design and test of the behavior control strategy of the overall robot fleet. The local distribution of intelligent sensing, reasoning and deliberation, asks for robust coordination and communication mechanisms to guarantee better performances. A poor design of such mechanisms results in errors that typically arise at run-time, i.e. while the system is working in its environment. Debugging such errors cannot be easily accomplished looking at the real system, because of the costs, the risk of major damages and the almost impossibility to reproduce the same erroneous situations twice. Moreover, because of the infinitely many world states and transitions we need to leave the idea of a complete test, using instead some statistical method to guarantee the correct behavior of the system in the most prototypical situations, while exceptions can be managed through proper safe and intelligent degradation procedures.

To approach such a problem, it is important to have appropriate simulation tools. Our claim is that multirobot system design strongly rely on the availability of such tools, especially for enhancing the debugging phase at the very early stage of the design. The goal of this paper is to present our experience designing VLAB (Virtual LABoratory), a distributed simulation software for the virtual representation of multirobot system over a LAN. In the next paragraph, we discuss the main requirements and ideas that form a good framework for multirobot systems simulation. In the third paragraph we give more details about the technical aspects of VLAB design, while the fourth paragraph is devoted to the implementation. Examples of experiments using VLAB are given in the fifth paragraph.

We would like to conclude this section pointing out the utility of investigating the problem of simulation to give the robot itself the capability to correctly predict the process evolution over time. As each component of a multirobot systems must show some form of intelligence (spatial and/or logical) in real-time, "short-term simulation" have to be performed continuously during each decision cycle. The robot itself becomes the final user of this simulation activity that have to produce one or more likely evolution of the world according to some proper guess of the persistent behavior of all the other active robots in the system. Obviously the output data format has to be coherent with the internal world representation of the robots, but the main representation mechanisms of the evolution of the world and robots can be the same of standard simulation tools. Then the utility of a distributed simulation tool on board of the component of a real distributed autonomous system.

## 2 The simulation of multirobot systems

Nowadays computer simulations have become tools of common use in many engineering applications. In the field of multiple mobile robots there are numerous examples of simulators and of systems which have been studied using simulators (like those described in [3], [4], [5]), but still there are perplexities and criticisms. Well-known advantages of simulations (no-risk and low-cost set-up and testing, easy tuning of the degree of complexity of the system) are not often considered enough. A real autonomous robotic agent is a system which behavior is deeply influenced from its operating environment. Simulations which ignore the robot-environment strict coupling, will give misleading indications about the performance of the system under exam.

Brooks, in [6], outlined four key-concepts of robotic systems, namely embodiment, situatedness, emergence and intelligence. They have been introduced to stress the importance of robot-environment interaction which comes from sensing and acting capabilities. In fact, the perception of the environment is strongly influenced by the perceptual abilities of the robot, as well as the actions performed in the environment, are strongly influenced by the actuative abilities of the robot. The set of these capabilities form the so-called (robot) morphology. A sound model for a robot (characterized by a particular morphology) must satisfy the embodiment, the situatedness, the emergence and the intelligence requirements.

In a reliable simulator it is then necessary to simulate not only the robots, but even the environment, and their interactions. So the overall simulated system performance strictly depends on a proper representation of the robots capabilities, of the physical laws governing the working environment, that also involve the correct representation of disturbances and noise ([7]). This aspect is particularly important when dealing with sensory devices.

When considering a multirobot system we have to deal with further problems: the internal relations in the fleets, and the interrelations among the fleets. A swarm of robots can be formed of heterogeneous elements which usually have been designed and implemented by different people. They may share the same morphology and differ in the controlling software, or they may also differ in in the morphology (and as consequence in the control) [8]. When these robotic agents act in a shared environment they may not be aware of the presence of other robots belonging to a different groups or fleet. They may cooperate or compete; however they have to deal with the presence of external active entities.

A good simulator for multirobot systems not only has to provide the capability to model embodiment and situatedness of each single robot but also it has to model the emergence and intelligence of each group of robot [9].

Finally, a simulator should not be limited to a tool which outputs a list of data (like position, orientation, speed and so on) for a later plot, but should give also an immediate visual output to let the user to make an immediate qualitative evaluation of the performance.

With all these requirements in mind, we have designed and developed a tool for distributed simulation of multirobot systems, called VLAB (Virtual LABoratory). VLAB provides a virtual laboratory which may be shared by a set of users to test the performance of a group of autonomous robots.

In VLAB the users place their own robots in the (virtual) environment, give them a task and observe how they carry on their task while interacting with the environment and with the other robots. Users can get an immediate visual feedback of what is happening in the laboratory; thanks to the use of virtual reality related tools they may walk into the environment and see what is going on. Users enter the environment via Local Area Network. through a browser, which is used to interact with the environment and to actively visit the virtual laboratory.

Users are free to specify the morphology (i.e. robots' sensing and acting capabilities) and the controlling software (one ore more control software modules) of every robot in the environment. Once the robots have been placed, they start performing in the shared environment; this allows the study of cooperative robotics and of social behaviors. A robotic agent may ignore other robots, but may also percept them (provided it is equipped with suitable sensors), communicate (implicitly or explicitly), cooperate or compete.

The simulator is responsible not only for the physical aspects of the experimentation, but also for the simulation of the sensing (and of the acting) of the robotic agents. Furthermore, the simulator provides a distributed virtual environment (DVE) ([10]) shared not only by the robotic agents but also by the users.

The VLAB software architecture is open; this allows a rapid development of new experimental environments, the re-use of software and a fine tuning of the accuracy of the simulations being performed.

## 3 VLAB software architecture

VLAB has been developed using a client-server paradigm. Simulations are performed by a powerful server which owns the model of the world and continuously updates it; the server executes also the routines controlling the robots inserted in the environment and produces sensorial inputs they require. Users access VLAB server via TCP/IP using an Internet browser; they interact with the server using a Java applet and get back a VRML scene which displays how the simulation is evolving. Users may locally develop controlling routines and are required to send the server only the compiled version. Figure 1 shows this approach.

It is important to point out that the client side may be replied an arbitrary number of times, giving rise to a distributed environment. Users send the server information needed to control robots (morphology and software) and

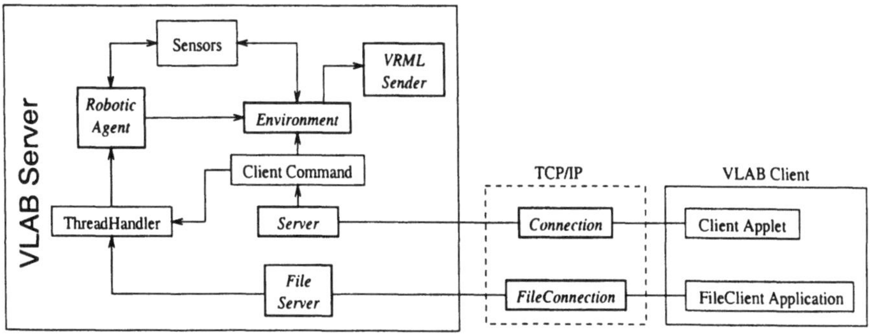


**Fig. 1.** VLAB's architecture

require to insert them; VLAB server creates new virtual robots, put them into the environment and starts executing their controlling software. Whenever a robot requires a sensorial input, VLAB produces what required (an image taken by the camera, for example, or a position estimation based on GPS) and feed the robot with the required information. VLAB server is able to produce such informations since it owns the complete model of the environment and of the robots which live in it.

If the robot performs an action (a move, or a grasp, for example) VLAB server updates the state of the environment keeping it consistently with the law of physics; it is this layer which has the responsibility to produce a reliable simulation. For example it has to test collisions between robots, or between robots and objects; in the simple environments we are currently using, those tests are performed by algorithms developed by us, but, since we are moving towards more complex environments, we are evaluating the possibility to use more sophisticated libraries, such those presented in [11].

A background thread sends users information strings which allows the client applet to update the VRML scene shown. This is the only operation performed by the client host, so it is not necessary for remote users to use powerful computers.

To encourage software reuse and to permit an incremental study of the complexity of the system being studied, VLAB has been developed using a two level architecture, as shown in fig. 2.

The internal level (the kernel) remains fixed and is devoted to the handling of low level tasks as VRML scene management, thread creation, user logging and so on. The outer level is specialized to handle the particular environment being studied; it's this level which establishes the degree of accuracy of the simulation, and which handles its physical aspects.

At the moment, our experimentations have been carried out in a local configuration, i.e. clients and serve were connected with an Ethernet-based LAN (100 Mbit/s); as stated in section 6, we are moving toward the web.

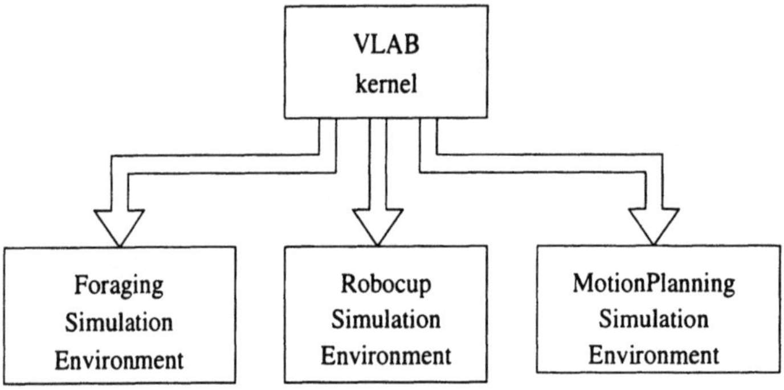


**Fig. 2.** VLAB's modular architecture

## 4 Implementation issues

VLAB has been completely developed using Java. This choice has been made for the following reasons. It's well known that Java is a platform independent language, and this gives some benefits on both the server and the client sides of the application; from the client side this is an important point, because the user is not constrained to a particular platform. All he needs is an Internet browser which supports Java and equipped with a VRML 2.0 compliant plug-in. From the server side the advantages are even more. Again we are not tied to a particular platform (and in fact the software has been run on PC/Linux, Sparc/Solaris and SGI/Irix) but there are a number of further important aspects. Having been developed for Internet applications, Java is naturally integrated in this environment and it is easily interfaceable with other Internet related standards as VRML. One of the interesting aspects of VLAB is that the user is free to ask the server to execute his controlling routines; this is possible thanks to the semi-interpreted nature of Java. In this way the user is enabled to directly send to the server the compiled version of his software; this would not be possible if the routines were compiled into native code of the user's platform. Another important issue is security. Since VLAB server executes pieces of code remotely sent by uncertified users, it is necessary to restrict the execution of some potentially dangerous operations in order to avoid hacking. This task is superbly accomplished by the security manager used in Java 2, which allows a fine tuning of the privileges being granted to the executing code. Finally the Java3D API, released simultaneously to Java2, has been successfully employed as a bridge toward VRML to get a realistic simulation of a camera using the same model used to display visual information to the client.

The major, and perhaps the only, drawback using Java is its performance. It is absolutely true that a program run with actual implementations of JVM is slower than the corresponding written in C and compiled into native code, but this gap is more and more narrowing; the adaptive dynamic compilers recently introduced [12] promise a rapid increase in Java performance.

## 5 Examples

In order to test its capabilities, VLAB is being used in the study of some cooperative tasks performed by a team of robots. In this section we illustrate two examples which highlights some of the more interesting features of our software.

The first system is devoted to foraging: a team robots has to pick a certain number of small items randomly scattered inside a room and has to bring them in a home area. Robots only knows that items are red, walls are green, home-area is yellow and other robots are prevailingly blue; in order to complicate the task of recognizing items, the floor has been covered with an irregular texture. As previously stated, the user may specify the morphology of the robots, the control software end, of course, the number of robots of the team.

```
BODY PNR

CAMERA
        POSITION 0.22 0.27 0
        ORIENTATION 0.5 -0.5 0
        UP 0 1 0
        RESOLUTION 300 300
        FOVX 90
        ASPECT 1

MOTOR
        MAXSPEED 1
        MAXTURN 10

PICK
        POSITION 0

SCANNER
        NOISE 5
```

**Fig. 3.** An example of morphology

Figure 3 shows how we may define the capabilities of a robot; in this case the user specifies a robot with the following characteristics

- the robot will use a mobile platform PNR, which allows inplace-rotations and translations
- the robot is equipped with a fixed camera (and, implicitly, a frame grabber, in order to acquire images to process) with the given optical parameters; the position and the orientation are relative to a fixed point of the platform

- the robot may drive a motor which allows a maximum translation speed of 1 m/s and a maximum rotation speed of 10 degrees/s
- the robot is equipped with a picking device positioned in front of it
- the robot has a range scanner with returns a map of the distances the other robots; each input has a superimposed 5% noise component

Note that the parameters characterizing each device may vary in each simulation; in this way it is possible to test different configurations in order to find the best one.

It's also important to point out that every simulation environment defines the set of devices available for the experimentations and the accuracy of the model. For example it is possible to define a more sophisticated model of the device `MOTOR`, which considers an elaborate speed/acceleration profile; as consequence the model will require more parameters, as is for the `CAMERA`. The control software used is a simple routine written in Java which loops reading sensors, and driving actuators, after having chosen an action in a reactive manner. This approach is oversimplified, since in this stage we are only evaluating the usability of the simulator, and we are not yet investigating efficient solutions to robot problems.

The key feature of this aspect is that the software is developed by the VLAB user using a standard high level language like Java. This means that the developer is not required to learn a new specialized language, and more resources may be devoted to the study of robotic issues.

**Fig. 4.** VRML rendering of a simulation

Figure 4 shows a snapshot taken from a simulation; we may see some items inside the home area and two robots which are working to accomplish the given task.

In the second environment we have developed, the team is again engaged in foraging, but the operating conditions are quite different, because two robots are necessary to carry an item in home area. This task is actually been developed to study cooperative behaviors. In this case robots communicate explicitly because they have to seek items but also to ask for help when they need a mate to back home. Communication systems may also be used to develop partially centralized architectures, with an agent which plans and then communicates decisions to other agents which act.

## 6 Conclusions

In this paper we have presented a distributed simulation software for multi-robot systems This simulator allows many users to instantiate robots which will live and interact in a shared virtual environment; furthermore users may specify robots morphology and study how performance is influenced by different robot capabilities. Comparisons with real robot systems have shown that the simulations actually performed are realistic, and may be used to set up robotic algorithms for a later migration toward real platforms.

In the near future we plan to develop new simulation environments to allow the study of a broad class of multirobot tasks; we will implement new modules to simulate more sensing and acting capabilities. Then we will move toward a distributed architecture for the server; we are evaluating the feasibility of setting up a network of computers cooperating one another to carry out server services. Having developed the software in Java, the natural way is to move toward RMI, but also CORBA is an interesting alternative which opens the possibility to realize the server using different languages. Finally, we will investigate the impact of net delays on the performance of the system; it is our intention to move from a LAN based configuration to a WEB based, with clients and server situated in different geographical location.

We do believe that the methodological framework realized through the VLAB experience is very useful to give the robots prediction capabilities that are necessary to achieve intelligent behaviors in real-time applications in the real world.

**Acknowledgements**
This research has been partially supported by the Ministry for Education and Research (MURST Certamen project), by the Italian National Council of Research (CNR special project), and by the Italian Energy Agency (ENEA Parallel Computing Project). We would like to thank Pierluigi Patuelli, for developing some simulation experiments using VLAB.

## References

1. T. Arai, J. Ota. (1995) Let-us Work Together - Task Planning of Multiple Mobile Robots, Proceedings of IEEE/RSJ Int. Conf. on Intelligent Robot and Systems (IROS'95), 298-303
2. L. E. Parker. (1998) Toward the Automated Synthesis of Cooperative Mobile Robot Teams, Proceedings of SPIE Mobile Robots XIII, 3525, 82-93.
3. E.Corten, L.Dorst, B.Krose. (1998) OASIS: Open Architecture for Simulation with Intelligent Systems, Proceedings of the Fifth Intelligent Autonomous Systems Conference, 6-12.
4. Y.Mashimoto, T. Miyazaki, M. Inaba, H. Inoue. (1999) View Simulation System: A mobile Robot Simulator using VR Technology, Proceedings of Int. Conf. on Robotics and Automation (ICRA 1999), 936-941.
5. J.J.G.Ramos, S.Maeta, M. Bergerman, S. Bueno, L. Mirisola, A. Bruciapaglia. (1999) Development of a VRML/Java Unmanned Airship Simulating Environment, Proceedings of the Int. Conf. on Intelligent Robots and Systems (IROS 1999), 1354-1359.
6. R.Brooks. (1991) Intelligence without reason. AI memo NO.1293, MIT Lab.
7. R. Brooks. (1986) A Robust Layered Control System For A Mobile Robot. IEEE Journal of Robotics and Automation, 2 (1), 14-23
8. Y.U. Cao, A.S. Fukunaga, A.B. Khang. (1997) Cooperative Mobile Robotics: Antecedents and Directions. Autonomous Robots, 4 (1), 7-27.
9. E. Pagello, A. D'Angelo, F.Montesello, F. Garelli, C. Ferrari. (1999) Cooperative behaviors in multi-robot systems through implicit communication. Robotics and Autonomous Systems, 29 (1), 65-77.
10. C.Waters, W.Burrus. (1997) The rise of shared virtual environments. IEEE Spectrum, 34 (3), 20-25.
11. T. Hudson, M. Lin, J. Cohen, S. Gottschalk, D. Manocha. (1997) V-COLLIDE: Accelerated collision detection for VRML. Proceedings of VRML '97, Monterey, CA, 24-26.
12. Java HotSpot(Tm) Technology homepage, http://www.javasoft.com/products/hotspot/

# Robots that cooperatively enhance their plans

Silvia Botelho[1] and Rachid Alami[2]

LAAS-CNRS - 7, Avenue du Colonel Roche, 31077 Toulouse Cedex 4 - France

**Abstract.** This paper presents a general architecture for multi-robot cooperation and then focuses on a scheme called "M+ Cooperative task achievement". Its originality comes from the robots ability to detect and treat - in a distributed and cooperative manner - resource conflict situations as well as sources of inefficiency. We illustrate its use through a simulated system, which allows a number of robots to plan and perform cooperatively a set of servicing tasks in a hospital environment.

## 1 Introduction

Starting from the **Plan-Merging** Paradigm [1] for coordinated resource utilization, and the **M+ protocol** [3,2] for distributed task allocation, we have developed a generic architecture for multi-robot cooperation. This architecture is based on a combination of local individual planning and coordinated decision for incremental plan adaptation to the multi-robot context. We present and discuss here a set of cooperation issues which allow a set of autonomous robots not only to perform their tasks in a coherent and non-conflict manner but also to cooperatively enhance their task performance.

After a brief analysis of related work, we present our general architecture for multi-robot cooperation. We define informally our cooperative task achievement scheme and focus on the cooperative plan enhancement components. Finally, we present an implemented system which illustrates, in simulation, the key aspects of our contribution.

## 2 Related work

The field of multi-robot systems has become very large[6]. We restrict our analysis here to contributions proposing cooperative schemes at the architectural and/or decisional level. In such stream, *behavior-based* and similar approaches[17], [16], propose to build sophisticated multi-robot cooperation through the combination of simple (but robust) interaction behaviors. ALLIANCE [18] is a distributed behavior based architecture, which uses mathematically modelled motivations that enable/inhibit behaviors, resulting in tasks (re)allocation and (re)decomposition.

AI-based cooperative systems have proposed to provide models for the agents interaction which are domain independent. For example, Brafman [4]/Ephrati [12] enrich the STRIPS formalism, aiming to build centralized/decentralized conflict-free plans. Clement [7] develops specialized agents which are responsible for HTN individual plans coordination.

Several generic approaches have been proposed concerning goal decomposition, task allocation and negotiation [9]. PGP [11] (and later GPGP [8]) is a specialized mission representation that allows exchanges of plans among the agents. DIPART [19] is a scheme for task (re)allocation based on load balancing. Cooperation has also been treated through negotiation strategies [21] like CNP-based protocols[23], or BDI approaches where agents compromise to achieve the individual/collective goals ([13],[14],[24]). Another perspective is based on the elaboration of conventions and/or rules. Shoham [22] proposed "social behaviors" as a way to program multi-agent systems. In STEAM [25], coordination rules are designed in order to facilitate the cohesion of the group.

Cooperation for achieving independent goals has been mostly addressed in the framework of application-specific techniques such as multi-robot cooperative navigation [27,5].

## 3 Cooperation for Plan Enhancement

In the context of autonomous multi-robot systems, we identify three main steps that can often be treated separately: the *decomposition* of a mission into tasks (mission planning), the *allocation* of tasks among the available robots and the *tasks achievement* in a multi-robot context (Fig. 1). In this paper, we limit ourselves to this last aspect i.e. the concurrent achievement of a set of tasks. Indeed, we assume a set of autonomous robots which have been given a set of partially ordered tasks. This could be the output of a central planner [26], or the result of a collaborative planning and task allocation process [3]. One can consider this plan elaboration process is finished when the obtained tasks have a sufficient range and are sufficiently independent to allow a substantial "selfish" robot activity.

However, and this is a key aspect in robotics, the allocated tasks cannot be directly "executed" but require further refinement. Since each robot synthesizes its own detailed plan, we identify two classes of problems related to the distributed nature of the system: 1. coordination to avoid and/or solve conflicts and 2. cooperation to enhance the efficiency of the system. The first class has been often treated in the literature. The second class is newer and raises some interesting cooperative issues linked to the improvement of the global performance by detecting possible enhancements. We have developed a scheme, called *M+ cooperative task achievement*, which partially answers these questions by considering three features:

1. **opportunistic action re-allocation**: one robot can opportunistically detect that it will be beneficial for the global performance if it could perform an action that was originally planned by another robot;
2. **suppression of redundancy**: it may happen that various robots have planned actions which achieve the same world state. This feature provides the reasoning capabilities that allow them to decide when and which robot will achieve them, avoiding redundant executions;

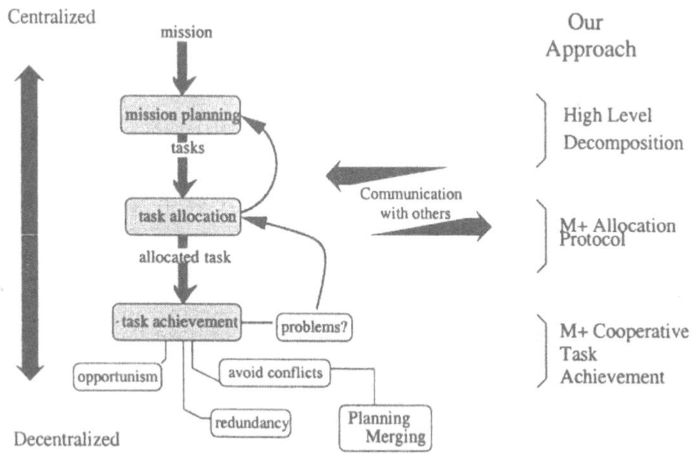


**Fig. 1.** An Architecture for multi-robot cooperation

3. **incremental/additive actions**: the robots detect that an action originally planned by one robot can be incrementally achieved by several robots and that this could be beneficial to the global performance.

## 4 Mechanisms for Plan Enhancement

The *M+ task achievement* scheme is based on independent planning capabilities together with a set of cooperative mechanisms that implement an incremental plan validation provess. Starting from the task that have been allocated to it, a robot produces its own plans called *individual plans*. It must then negotiate with the other robots in order to incrementally adapt its actions in the multi-robot context.

*A state description:* The world model we use has been specially devised to allow reasoning on the cooperative issues mentioned above while maintaining a STRIPS-like representation in order to allow the robots to invoke efficient practical planners (in our implementation, we use Propice[10] with a STRIPS-like IPP [15] planner). The world state is described through a set of predicates. While *stable* predicates represent constant environment features (e.g. `CONNECTED(A1):A2`, `CONNECTED(A1):A3`), *evolutive* predicates represent features that can be changed and whose modification can be planned (e.g. `GRIPPER(R1): EMPTY`). Besides, for a given robot, we define a subset of *evolutive* predicates - called *exclusive* predicates - that can only be changed by the robot itself (e.g. `POSITION-ROBOT`).

*Social Rules:* This cooperative activity is based on the satisfaction of a set of constraints expressed in terms of what we have called ***social rules***. *Social* *rules* have been introduced in order to produce easily *merge-able* plans. They impose constraints that must be taken into account during the planning and also during the validation process[1]. We have defined three classes of rules[2]:

1. `end`: ($END\text{-}RULE$ *pred*), where *pred* must be satisfied at the end of each robot activity. This class guarantees a known final state, allowing the planner to predict the state of an attribute (initial state for the next plan). For example, the rule: *D1 door must be maintained closed* can be represented as ($END\text{-}RULE$ ($STATE\text{-}DOOR(D1,CLOSED)$))
2. `amount`: ($AMOUNT\text{-}RULE$ $(a : v), u, (s_a : s_v)$) where the "resource" $(a : v)$ represented by an attribute $a$ and a value $v$ is limited to a maximum of $u$ user entities. The $s$ field (attribute and value) is a *proposed state* to avoid the rule violation. Note that such rules allow to describe the resource constraints of the system. For instance *a limitation of 2 robots at desk D1* can be represented by ($AMOUNT\text{-}RULE$ ($POS\text{-}ROBOT$:$D1$),2,($POS\text{-}ROBOT$: $OPEN\text{-}AREA$), where it is proposed to send the robot to an $OPEN\text{-}AREA$, in order to satisfy the rule.
3. `time`: ($TIME\text{-}RULE$ $pred, u, s$), where *pred* can be maintained true only during a given amount of time $u$. The $s$ field is a *proposed state* which can be used by the planner in order to avoid the violation of the rule. For instance the rule: *Machine M1 must be ON at most 10 minutes* can be represented as ($TIME\text{-}RULE$ ($STATE(M1,ON)$),10,($STATE(M1,OFF)$)), where it is proposed to turn off $M1$ to avoid the rule violation.

*The use of social rules in the planning phase:* We associate to the social rules a value called ***obligation level*** in order to distinguish between rules that must be systematically respected in order to obtain merge-able plans and other rules whose satisfaction can be deferred i.e planned but not necessarily executed.

Whenever a robot plans, it considers the ***proposed final states*** of the rules as mandatory goals that will be added to its current list of goals. However, depending on the obligation level, goals will be posted 1. as a conjunction with the current robot goals or 2. as additional goals that the robot will try to satisfy in subsequent planning steps. In such case, the planner will produce ***additional plans*** that will achieve each low-level obligation social rule.

During the execution of a plan, the robot may or may not execute these additional plans, thus neglecting the ***proposed state*** and "violating" a social rule. Note that if another agent asks the robot to fulfill the ***rule proposed state***, it will then (an only then) perform the associated additional plan. The ***obligation level*** may also change depending on the context.

[1] Note that this notion is different, or even complementary, from the social behaviors proposed by [22]. While *social behaviors* are explicitly coded in its reactive task execution, the *social rules* are used at the robot decision level as constraints in its planning and negotiation activity.

[2] One can imagine other classes of rules related to the application domain.

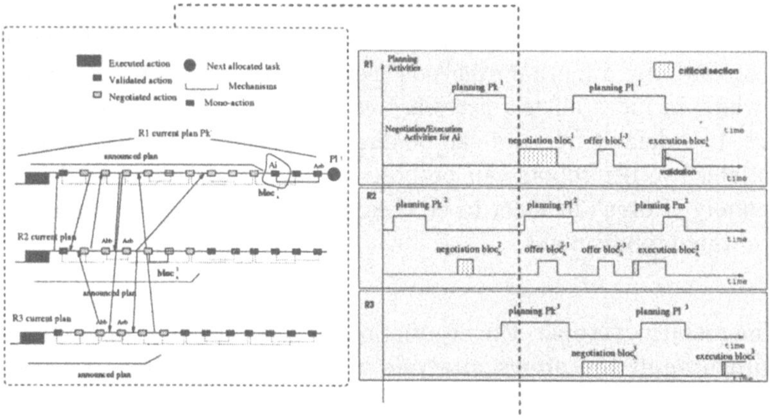


**Fig. 2.** The task achievement process and its protected activities

*Operations on plans:* Potential conflicts and cooperation opportunities between robots will be managed through actions that modify *interference predicates.* Such predicates are composed of *non-exclusive* predicates and of all predicates that belong to *social rules.* The cooperation activity will be based on the notion of *Block of actions.* A block is a sub-plan which begins with an action that changes an *interference* predicate and finishes with an action that changes again the same predicate.

We have defined the following *mechanisms* for plan modification: `insert_message_wait` that introduces a temporal order constraint between two actions belonging two robots, and `insert/delete`, when an action is re-assigned to another robot or when an action execution is neglected.

## 5 The task achievement process

The M+ task achievement process involves three activities that correspond to different temporal horizons and may run in parallel (Fig. 2): 1. task planning which produces a mono robot plan; 2. the plan negotiation activity which adapts the plan to the multi-robot context; and 3. the effective plan execution. While task planning is purely internal, the other activities are performed in a critical section in order to ensure a coherent distributed multi-robot plan management and execution.

**The task planning activity:** This is a standard task planning activity. The robot invokes its own planner and it incrementally appends new sequences of actions to its current plan. This new plan does not consider explicitly the other robots' plans. We call it a *mono-plan.* However, the obtained plan satisfies the social rules and is consequently easily *merge-able.*

**The plan negotiation:** Let us assume that $R_1$ begins a negotiation process in order to introduce its action $A_i$ in the multi-robot context. This operation is "protected" by a mutual exclusion mechanism[3]. The result is a coherent plan which includes all the necessary coordinations and some cooperative actions. It is default free and can be directly executed. However, it remains "negotiable" (other robots can propose a plan modification) until it is incrementally "frozen" in order to be executed. Fig. 2 shows an instantaneous state of three robots plans.

**The negotiation steps:** The negotiation process comprises three steps: the **announcement**, the **offers analysis** and the **deliberation.** During this process, a robot negotiates all the *blocks of actions* of its current plan which are not yet announced.

**Step 1: the announcement.** Whenever a robot (e.g $R_p$) wants to negotiate an action $A_i^p$ in the multi-robot context, it announces it by providing the associated block of actions. After having received an action announcement, the other robots search for possible coordinated and cooperative blocks in their announced/validated plans, and send back their offers to $R_p$.

**Step 2: $R_p$ analyzes the offers.** *Coordination:* this analysis is directly derived from the Plan-merging paradigm[1]. Block insertion is performed incrementally by adding temporal constraints to the robots plans[4].

*Cooperation:* in a cooperative scenario, $R_p$ verifies which candidates are able to participate and builds a *cooperative final block*, choosing the agent(s) that will achieve the actions. Due to the need of respecting the *social rules*, it may occur that some robots can not participate in the *cooperative final block.*

**Step 3: Deliberation.** $R_p$ sends to the other robots the result of the announcement process.

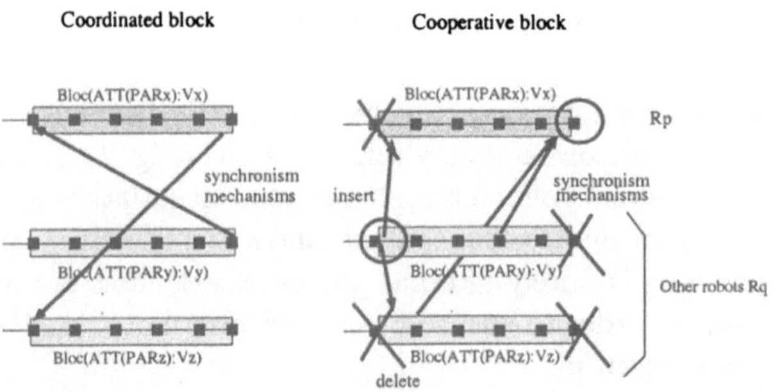


**Fig. 3.** Plan Modification mechanisms

---

[3] We assume that the robots are equipped with a reliable inter-robot communication device.

[4] We use the mechanisms described in [20] for detecting and treating in a distributed way the deadlocks that may occur

Note that such a negotiation process involves only communication and computation and concerns future (short term) robot actions. It can run in parallel with execution of the current coordination plan.

**Execution process:** Before executing an action, the robot **validates** the blocks pf actions associated with it. Indeed, a block is "negotiable" until its **validation.** Once validated, it is "frozen" and the other robots can only perform insertions after a validated block. Action execution causes the evolution of the system, resulting in events that will entail new planning, negotiation and execution steps for the robot itself and for the other robots.

## 6 Illustration and future work

A first version of the scheme has been implemented. We describe here below an illustrative example of its use. There are three mobile robots in a hospital environment. Servicing tasks are items delivery to beds as well as bed cleaning[5]. Fig. 4 shows the tasks goals (there are 5 partially ordered tasks: `T0,...T4`) and the initial world state description[6].

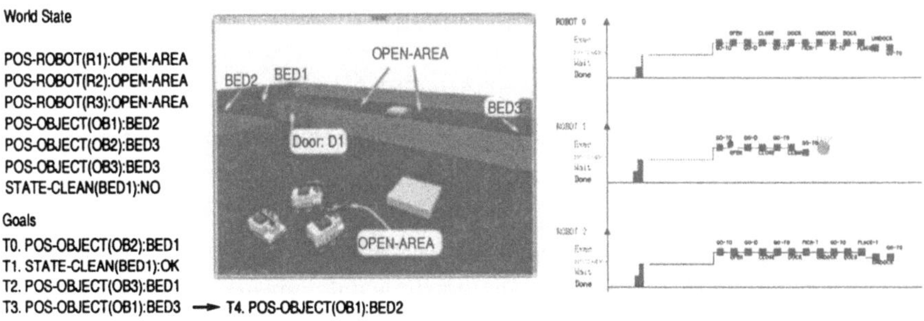


**Fig. 4.** Example 1: Transfer object and clean beds in a hospital area

The robots must respect the following *social rules*: 1. an `amount` rule (with low *obligation level*) that limits the number of robots near a bed to one, $(AMOUNT{-}RULE\ (POS{-}ROBOT{:}BED1),1,(POS{-}ROBOT{:}OPEN{-}AREA)$ and 2. an `end` rule (with high *obligation level*) $(END{-}RULE\ (STATE{-}DOOR(D1,CLOSED)))$ that requires to close the door. Besides, there are potentially the following coordination and cooperation issues: 1) coordination for resource conflict near the beds (rule 1); 2) `open/close` door actions can be potentially redundant; and 3) `clean bed` is a cooperative incremental action (only *non-*

[5] Each robot control system runs on an independent Sun workstation which communicates with the other workstations through TCP/IP.

[6] Due to the lack of space, we exhibit here a simplified world state representation.

*exclusive/incremental* effects: STATE-CLEAN(<bed>):OK) that allows cumulative effects when executed several times or by several robots.

The set of tasks is transmitted to the three robots. After a first phase (not described here), they plan and incrementally allocate the tasks using *M+ protocol*[3]. In a first step, r0 allocates T2 (i.e. POS-OBJECT(OB3): BED1), r1 allocates the cleaning task T1 and T3. r2 is in charge of T0. T4 has not been allocated yet. Fig. 5 shows the individual plans before any negotiation while Fig. 6 shows their state after a number of negotiation processes. Note that r1 has elaborated a plan with six actions in order to achieve its main goal STATE-CLEAN(BED):OK and to satisfy the social rule requiring (STATE-DOOR(D1,CLOSED)) with a high obligation level. Besides, it has also produced an *additional plan* that satisfies rule 1 (with a low obligation level) by introducing a go-to(OPEN-AREA) action.

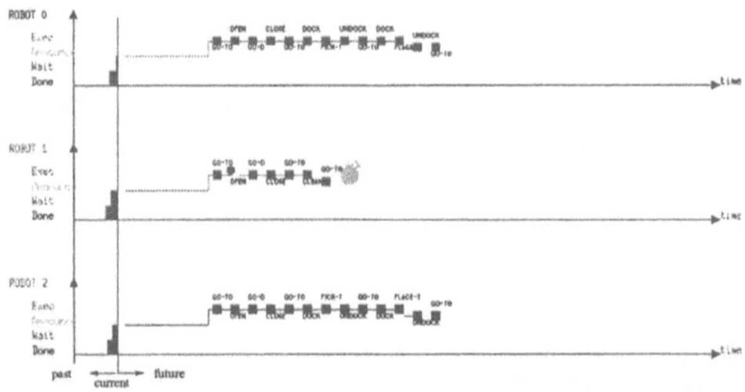


**Fig. 5.** M+ Allocation result: Individual plans before negotiation process

**Example 1:** The robots engage a negotiation process. They negotiate their open-close blocks. Finally, there will be only one open-close sequence instead of three. The robots decide that (r0) will open the door and that (r1) will close it and how this will constrain temporally their plans. Fig. 6 shows the result of this negotiation with the deleted actions represented as circles. One can also notice, that the robots have satisfied the *social rule* associated to the robot position near the beds.

Continuing the incremental negotiation, it is time for r1 to negotiate its clean action. It is an *incremental* action. r1 announces it, but, in this first attempt, no robot is a priori concerned. After a while r0 and r2 plan to be next to BED1 (for delivering an object) and send opportunistic offers to r1. The added clean actions to r0 and r2 plans are represented by a different filling pattern in Fig. 6.

The overall process continues; the tasks are incrementally planned, negotiated and executed. Fig. 7 shows the final result of this run. One can observe that the tasks have been achieved without conflict: r0 and r2 wait until r1 leaves BED1 (1), r0 waits until r2 leaves BED3 (2) and BED1 (3). Moreover,

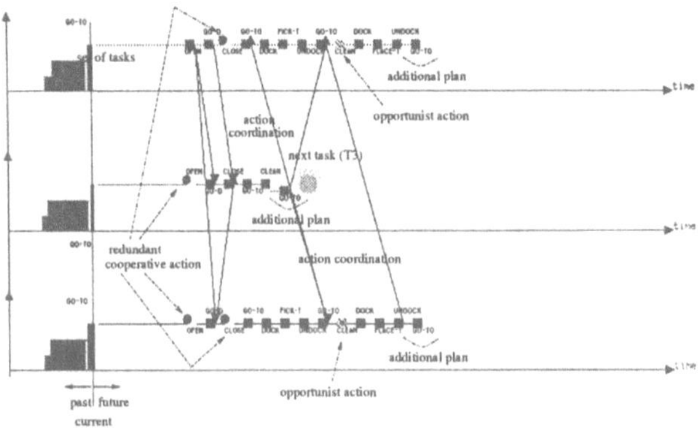


Fig. 6. M+ negotiation

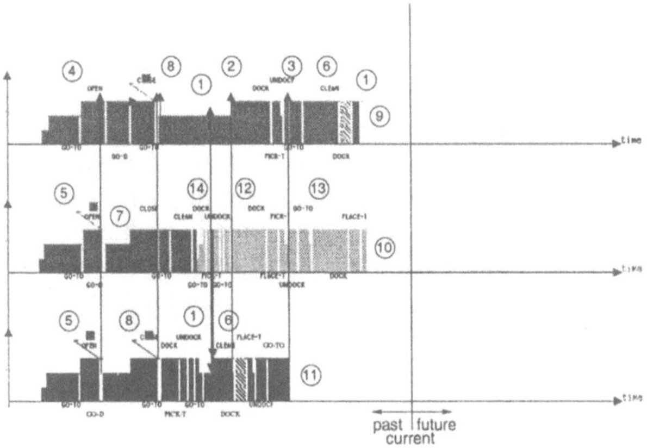


Fig. 7. Example 1: Final result of M+ task achievement

they have also exhibited several cooperative interactions. Indeed, r0 opens the door (4) and *opportunistically* r1 and r2 take advantage of this, deleting their open action from their current plans (5). Besides, r0 and r2 also help r1 to clean a bed (6). Finally, r1 closes the door for all (7) (r0 and r2 delete their close action (8)). Note also that r0 (9) and r1 (10) have neglected to execute their additional plans that make them go to an OPEN-AREA because no robot has requested them to leave the beds. This was not the case for r2 because r0 has requested it to free BED1 (11).

**Examples 2 and 3:** Due to lack of space we do not give details here on other runs of the same example but with different conditions. In example 2, r0 and r2 are slower and they cannot help (Fig. 8) r1 to clean BED1. In example 3, we have inhibited the cooperative mechanisms and allowed only coordination. The global performance clearly suffers from this.

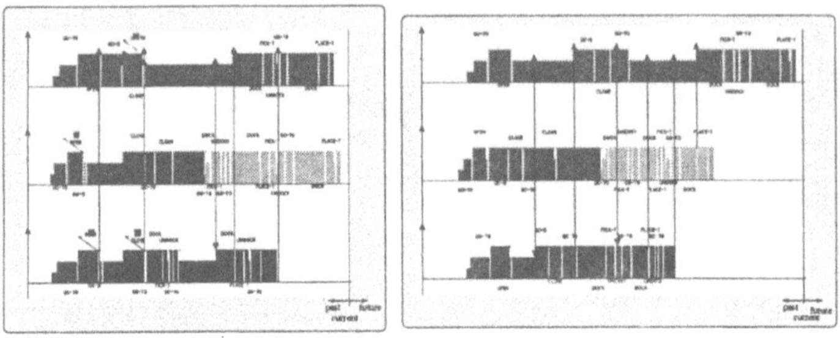

**Fig. 8.** Examples 2 and 3

**Conclusion:** We have proposed and discusses a scheme for cooperative multi-robot task achievement. This scheme is a key component of a general architecture for multi-robot cooperation. Its main originality comes from its ability to allow the robots to detect and treat - in a distributed and cooperative manner - resource conflict situations as well as sources of inefficiency among the robots. We have presented its main ingredients and mechanisms, and illustrated its use through a simulated system.

Our future work is twofold. First, We envisage to validate our approach through a number of significant application domains and to implement it on real laboratory robots. Besides, we would like to extend and further formalize the overall system and its representational and algorithmic ingredients. Another interesting aspect is the fact that such multi-robot architectures, raise new complementary issues and constraints which are not correctly treated by the existing task planners.

## References


1. R. Alami, S. Fleury, M. Herrb, F. Ingrand, and F. Robert. Multi-robot cooperation in the martha project. *IEEE Robotics and Automation Magazine, Special Issues: Robotics and Automation in Europe*, 1997.
2. S. S. C. Botelho. A distributed scheme for task planning and negotiation in multi-robot systems. In *ECAI'98*, 1998.
3. S. S. C. Botelho and R. Alami. M+: a scheme for multi-robot cooperation through negotiated task allocation and achievement. In *IEEE ICRA'99*, 1999.
4. C. Boutilier and Brafman R. Planning with concurren t interaction actions. In *AAAI'97*, 1997.
5. A. Brumitt, B. Stentz. Dynamic mission planning for multiple mobile robots. In *IEEE ICRA'96*, 1996.
6. Y. Cao, A. Fukuna, and A. Kahng. Cooperative mobile robotics: Antecedents and directions. *Autonomous Robots*, 4:7–27, 1997.
7. B. Clement and E. Durfee. Top-down search for coordinating the hierarchical plans of multiple agents. In *Third International Conference on Autonomous Agents*, pages 252–259. Association of Computing Machinery, 1999.
8. K. Decker and V. Lesser. Generalizing the partial global planning algorithm. In *Int Journal of Cooperative Information Systems 92*, 1992.

9. M. DesJardins, E. Durfee, Ortiz C., and M. Wolverton. A survey of research in distributed, continual planning. *AI Magazine*, pages 13–22, 1999.
10. O. Despouys and F Ingrand. Propice-plan:toward a unified framework for planning and execution. In *ECP'99*, 1999.
11. E. Durfee and V. Lesser. Using partial global plans to coordinate distributed problem solvers. In *IJCAI87*, 1987.
12. E. Ephrati, M. Perry, and J. S. Rosenschein. Plan execution motivation in multi-agent systems. In *AIPS*, 1994.
13. T. Fukuda, Y. Kawauchi, M. Buss, and H. Asama. A study on dynamically reconfigurable robotic systems. JSME *International Journal*, 34(2):295–302, 1991.
14. N. Jennings. Controlling cooperative problem solving in industrial multi-agent systems using joint intentions. *Artificial Intelligence*, 75, 1995.
15. J. Koehler, B. Nebel, J. Hoffmann, and Dimopoulos Y. Extending planning graphs to an adl subset. In *ECP97*, 1997.
16. R. Mackenzie, D. Arkin. Multiagent mission and execution. *Autonomous Robots*, 4:29–52, 1997.
17. M. Mataric. *Interaction and Intelligent Behaviour*. PhD thesis, Massachusetts Institute of Technology, 1994.
18. L. Parker. Alliance: An architecture for fault tolerant multirobot cooperation. *IEEE Trans. on Robotics and Automation*, 14(2):220–239, 1998.
19. M.E. Pollack. Planning in dynamic enviroments: the dipart system. *Advanced Planning Thechnology: Technology Achievements of the ARPA/Rome Laboratory Planning Initiative*, 1996.
20. S. Qutub, R. Alami, and F Ingrand. How to solve deadlock situations within the plan-merging paradigm for multi-robot cooperation. In *IEEE IROS'97*, 1997.
21. J. S. Roseschein and G Zlotkin. Rules of and encounter: Designing convention for automated negotiation among computers. *Artificial Intelligence - MIT press*, 1994.
22. Y. Shoham and M. Tennenholtz. On social laws for artificial agent societies: off-line design. *Artificial Intelligence*, (75):231–252, 1995.
23. R. Smith. The contract net protocol: High-level communication and control in a distributed problem solver. *IEEE Transactions on Computers*, c-29(12), 1980.
24. G Sullivan, A. Glass, B. Grosz, and S. Kraus. Intention reconciliation in the context of teamwork: an initial empirical investigation. *Cooperative Information Agents III, Lecture Notes in Artificial Intelligence*, 1652:138–151, 1999.
25. M. Tambe. Agent architectures for flexible, practical teamwork. In *First International Conference on Autonomous Agents*, 1998.
26. T. Vidal, M. Ghallab, and R. Alami. Incremental mission allocation to a large team of robots. In *IEEE International Conference on Robotics and Automation*, April 1996.
27. S. Yuta and S. Premvuti. Coordination autonomous and centralized decision making to achieve cooperative behaviors between multiple mobile robots. In *IEEE IROS'92*, 1992.

# Chapter 3

# Communication and Knowledge Sharing

# Performance Evaluation of Autonomous Knowledge Acquisition and Sharing by Intelligent Data Carriers

Daisuke Kurabayashi[1], Katsumi Konishi[2], and Hajime Asama[1]

[1]RIKEN: The Institute of Physical and Chemical Research, 2-1 Hirosawa, Wako-shi, Saitama, 351-0198 Japan.
[2]The University of Tokyo, 7-3-1 Hongo, Bunkyo-ku, Tokyo, 113-8656, Japan

**Abstract.** In this paper, we built a device and algorithm to implement autonomous robots that can enhance the efficiency through autonomous knowledge acquisition and sharing. We also propose a quantitative evaluation algorithm of task execution by autonomous mobile robots. In the robotic system, the "Intelligent Data Carrier" provides knowledge of navigation for autonomous mobile robots. An IDC summarizes fragments of knowledge from individual robots and reports the best direction to a destination that a robot wants to go. Based on a stochastic model, we establish a function to evaluate effectiveness of knowledge management by the IDCs. We verify effectiveness of the proposed algorithm by simulations.

**Keywords.** Autonomous knowledge acquisition and sharing, Intelligent data carrier, Stochastic model.

# 1 Introduction

Currently, researchers are trying to create a robotic system that can function in any general environment. Realization of autonomous task execution would be especially advantageous in hazardous environment. Most of robotic systems require a model environment in order to execute tasks effectively. Autonomous robotic systems should create models of the environment by themselves. However, such tasks are not easy for current autonomous robots because they have only limited ability to sense and thus survey the environment. In such cases, the method of knowledge acquisition and sharing becomes very important.

Some researchers have proposed local communication methods for enhancing communication between robots[1]-[4]. In previous related studies, only robots kept the knowledge. Let us consider what takes place on the ground, ants forage effectively by pheromone trails, dogs claim territory by smell. Social creatures improve the efficiency of their actions by storing information in the environment. Simulation studies by Drogoul and Feber[5] have suggested that the pheromone trails are very effective for completion of iterative transportation tasks to be performed by autonomous agents. This

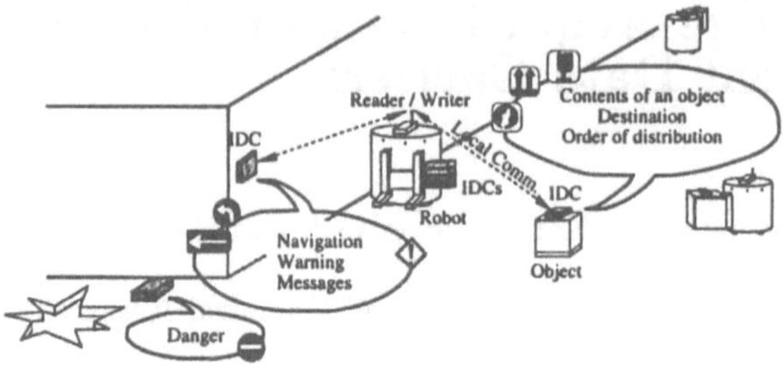


Figure 1: An overview of local communication by the IDC

kind of data storage and communication system can be applied by creating model environments for robotic systems for the sharing of knowledge and for cooperation.

In this paper, we propose a device that enables local communication, we refer to the device as an "Intelligent Data Carrier (IDC)" (Fig. 1). We not only develop such a device but also propose in this study an algorithm to enhance the efficiency of task execution performed by autonomous robots via knowledge sharing and acquisition through the intelligent data carrier system in particular environments.

The layout of this paper is as follows. In section 2, we describe the intelligent data carrier, and settlement of the problem. In section 3, we propose an algorithm for autonomous robots to acquire and share knowledge of navigation. We created a mathematical model to analyze the behaviors of robots and to evaluate the efficiency of the proposed algorithm in section 4. Section 5 provides a summary of the paper.

# 2 Problem settlement

## 2.1 Intelligent Data Carrier System

The authors have developed an Intelligent Data Carrier (IDC)[6][7] to reduce the traffic of global communication by providing local communication links and local information management functions. By reading information from and writing it into the IDCs, robots can use them as media for inter-robotic communication.

The IDC system consists of portable information storage (tags) and read-write devices carried by the robots (Fig. 2). Tags are usually refereed to as an "IDC". A tag has its own CPU, memory, and battery. A user can download and execute original program into the tags. The specifications of the IDC are shown in table 1. By placing the IDCs in specific locations in a particular environment, robots can allocate functions to act as agents for

Table 1: Specifications of the IDC

| | |
|---|---|
| Media | Electromagnetic wave |
| Frequency | 290, 310 [MHz] |
| Memory | 32 Bytes |
| Modulation | ON/OFF Keying |
| Data rate | 1200 [bps] |
| Power Source | a Li-ON battery (3.6V) |
| Size | tag: 110 x 65 x 25, reader/writer: 195x130x50 [mm] |

(a) Tags (b) A reader/writer (c) Portable system

Figure 2: The prototype of the IDC system

information storage and management (Fig. 2(c)). The communication range is up to 3.0 [m].

## 2.2 Problem settlement

In this paper, we consider iterative transportation tasks (e.g. [8]). A robot has to carry objects to given destinations. We posit an environment in which several destinations are located. When a robot achieves desired destination, it receives instructions regarding the next destination and then continues with the task at hand.

We assume that robots do not have maps, because fixed map, made by humans, may decrease the flexibility of autonomous robotic systems. We consider adaptability to unknown environments in crucial. We assume that a robot consists of the following characteristics.

- A robot does not have a map and does not estimate its global position.
- A robot can sense wall and distinguish paths, junctions and destinations. It can also distinguish branches at junctions.
- A robot can remember the last visited destination and count, in steps, the duration of running time.

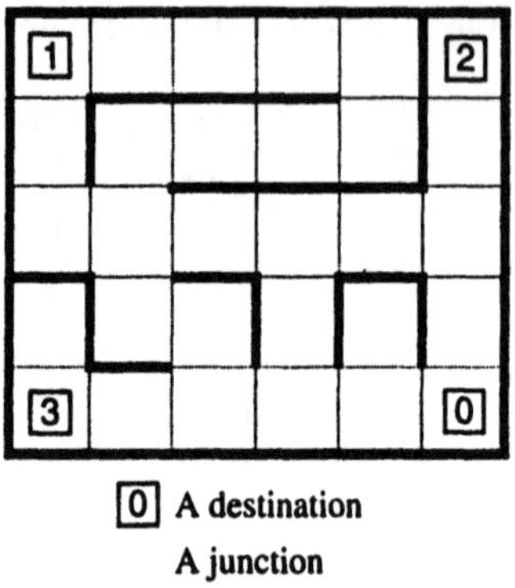


Figure 3: An example of a work area

A robot cannot understand its global position, but it can understand local situation around them.

In this example, we assume that the transportation task is as follows.

- A robot is given only the ID number of a destination. When a robot arrives at the destination, it receives another destination ID at random.
- The work area is a maze-like environment which consists of square cells and walls (Fig. 3).
- A robot can move to neighboring cells, but only one step at a time.
- We do not consider cases of collisions among robots.

# 3 Knowledge acquisition and sharing

## 3.1 Algorithm to acquire and share knowledge

In this section, we propose an algorithm to build knowledge for navigation autonomously. A robot should select feasible branches at junctions, such branches lead to the shortest path to its current destination, for effective transpiration is the goal. We set IDCs at junctions to facilitate autonomous robots' decisions. Robots can store and share their fragments experience by the IDCs. A robot does not need to estimate its global position if a robot can obtain sufficient knowledge from an IDC.

As we assumed, a robot does not know its global position. Thus we propose an algorithm to locate where branches connect at junctions to expected destinations. This proceeds according to the last visited destination and entered branch. For example, when a robot, which has started destination 1,

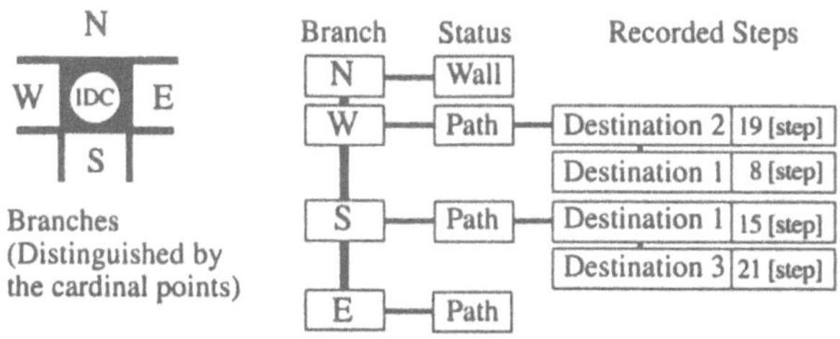


Figure 4: Data structure in an IDC

enters a junction through a southern branch, we can expect the southern branch may lead us to destination 1.

We describe data structure of an IDC in Fig. 4. At the initial state, no valid data is recorded. When a robot enters a communication area of an IDC at a junction, it reports the branch of entry, the ID number of the most recently visited destination, and running step measured from the destination to the IDC.

When the IDC has already received data about the same destination in the same branch, it compares the current running step with a former one. If the new one is shorter than the former, the record is renewed.

When a robot wants to go to destination 1, it should choose a branch leading to the fewest numbers of steps from the destination. In the example of Fig. 4, the robot should choose the W (western) branch. We implemented algorithm of probabilistic blanch selection algorithm. The procedure of the algorithm is as follows.

1. When a robot can not communicate with an IDC at a junction, it chooses a branch at random.

2. A robot whose destination is $j$ comes a junction which has $m$ branches. We describe recorded steps from destination $j$ in branch $i \in m$ as $t_{ij}$. We can find four data, $t_{W1} = 8$,$t_{W2} = 19$,$t_{S1} = 15$,$t_{S3} = 21$ in the example shown in Fig. 4.

3. When the IDC has no record about destination $j$, the robot chooses a branch at random.

4. When the IDC has one or more branches which contains record about destination $j$, it calculates $s_{ij} = \frac{1}{t_{ij}}$. If branch $i$ has no record about destination $j$, set $s_{ij} = 0$.

5. Calculate probability $p_{ij}$ to choose branch $i$ as equation (1). $P_{min}$ denotes a fixed probability that a robot chooses a branch randomly.

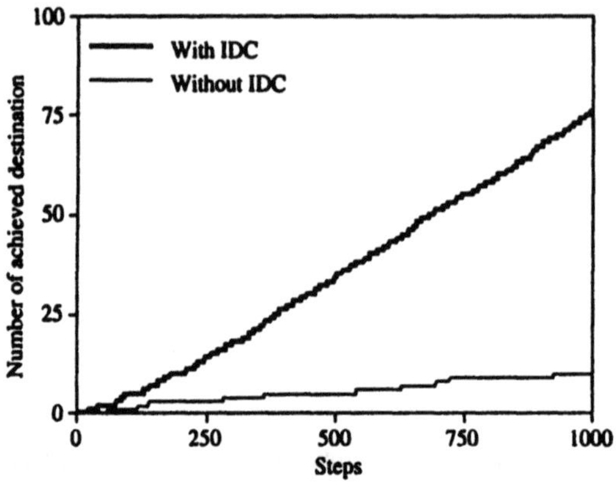


Figure 5: Comparison of task execution

$$p_{ij} = (1 - p_{min})\frac{s_{ij}}{\sum_j^m s_{ij}} + \frac{p_{min}}{m} \tag{1}$$

## 3.2 Simulation results

We verified the effectiveness of the proposed algorithm by performing simulations. We evaluated the number of achieved destinations by counting a constant number of steps. We assume an environment like the one given in Fig. 3 and performed simulations with or without IDCs at the junctions. Each robot worked for 1000 steps. We set $P_{min} = 0.10$. We set IDCs in all junctions when applying the IDC and the proposed algorithm.

Figure 5 shows results by single robot. We can see that the proposed algorithm with IDCs achieved about 600% more destinations than the algorithm without IDCs. We did not give the robot any knowledge about the environment in advance of performing simulations. Other robots can share the knowledge stored in IDCs by just the same algorithm. Those results suggest that the proposed algorithm and IDC system realized implicit cooperation among autonomous robots without explicit communication and without a priori knowledge.

# 4 Performance evaluation

## 4.1 Formulation

In this section, we formulate the effects of knowledge sharing and establish method of evaluating autonomous robotic systems with the IDC system. We

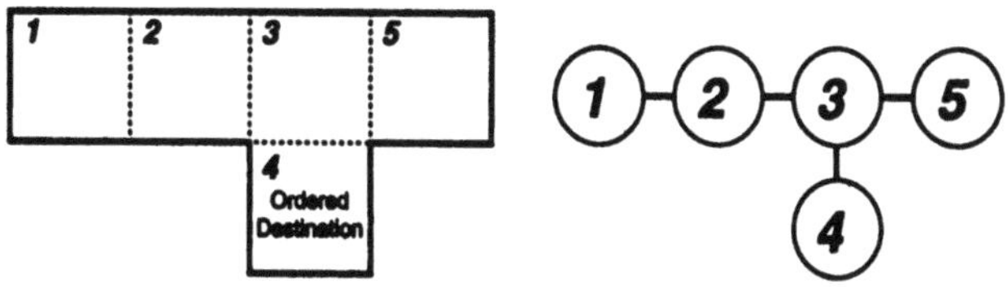


(a) Layout of a work area (b) Undirected graph

Figure 6: Example of evaluation model

estimate ideal performance of the knowledge sharing with IDCs as a performance index of the proposed algorithm. We calculated the expected number of steps needed to reach given destinations. In order to analyze the behavior of robots with IDCs, a stochastic model was introduced. The positions of robots were described by a state equation.

First, let us consider the cases lacking IDCs. We can assume that the work area of robots is divided by grids, as shown in Fig. 6(a). The work area itself is described by undirected graph, as shown in Fig. 6(b). The stochastic model is given based on the graph, as follows,

$$x(t+1) = Ax(t), \tag{2}$$

$$x(t) = [x_1(t)x_2(t)\cdots x_n(t)]^T. \tag{3}$$

In the above equations, $x_i$ denotes the probability that a robot is in the $i$-th node of the graph at step $t$. Therefore, $x_i$ must satisfy $0 \leq x \leq 1$. $A = (a_{ij})$ denotes the state transition matrix and $a_{ij}$ describes the probability that the robot in the $j$-th node moves into the $i$-th node. The robot that arrives at the given destination stops and stays there. We assume here that these robots disappear from the work area. Hence, if the given destination is $l$-th node,

$$a_{il} = 0 \qquad i = 0, 1, \cdots, n. \tag{4}$$

$$A = \begin{bmatrix} 0 & \frac{1}{2} & 0 & 0 & 0 \\ 1 & 0 & \frac{1}{3} & 0 & 0 \\ 0 & \frac{1}{2} & 0 & 0 & 1 \\ 0 & 0 & \frac{1}{3} & 0 & 0 \\ 0 & 0 & \frac{1}{3} & 0 & 0 \end{bmatrix} \tag{5}$$

In any case, the maximum eigenvalue of $A$ is smaller than 1, and it is clear that $x \to 0$ as $t \to \infty$.

Next, let us consider the case that employs IDCs. In the node with an IDC, a robot can obtain information about branch to a desired position, therefore, such robots move exactly into the node that is close to the ordered

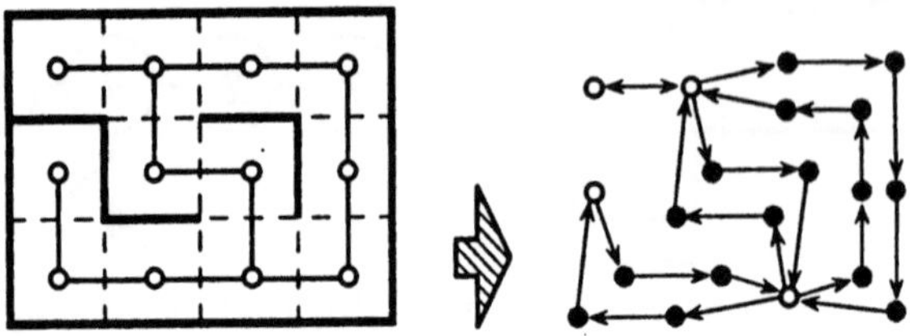

Figure 7: Arrangement of a work area

destination. In other words, such robots will move into the next node with a probability 1.

$$A = \begin{bmatrix} 0 & \frac{1}{2} & 0 & 0 & 0 \\ 1 & 0 & 0 & 0 & 0 \\ 0 & \frac{1}{2} & 0 & 0 & 1 \\ 0 & 0 & 1 & 0 & 0 \\ 0 & 0 & 0 & 0 & 0 \end{bmatrix} \tag{6}$$

## 4.2 Evaluation function

Here, $A_k$ is a transition matrix which indicates transition of a robot seeking destination $k$. We calculate expected steps $E_k$, it will take to reach destination $k$.

$$\begin{aligned} E_k &= \sum_{t=0}^{\infty} c^T x(t) \\ &= c^T \sum_{t=0}^{\infty} A_k^t x(0) \\ c &= [11 \cdots 1] \end{aligned} \tag{7}$$

, where $x(0)$ indicates initial conditions, and $\sum_i^n x_i(0) = 1$.

As a single work area may have several destinations, we evaluated the average of expected steps when a work area has $m$ destinations:

$$E_{index} = \frac{\sum_{i=1}^{m} E_i}{m}. \tag{8}$$

We created a model for evaluation by employing a simple Malkov model. A robot could proceed back and forth on the model even when in the middle of a corridor. Actual robots do not perform useless movements. We introduce a directed graph to avoid the movements (Fig. 7).

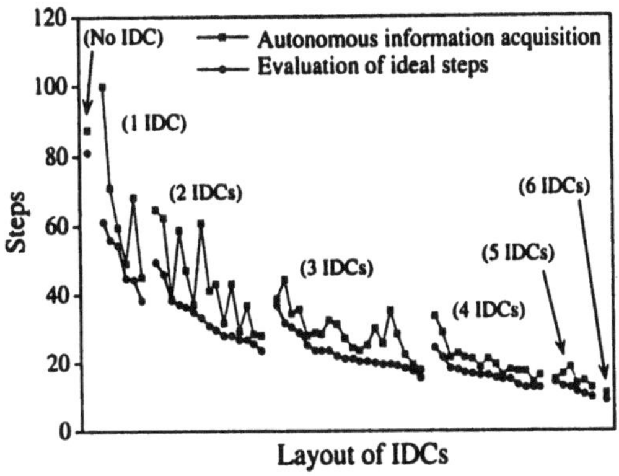


Figure 8: Evaluation of the algorithm to acquire navigation information by autonomous agents.

## 4.3 Performance evaluation

The performance of the proposed system depends on the IDC layout. We considered $2^6$ combinations for six junctions in the map (Fig. 3). We evaluate performance of transportation in each case by the use of simulations. We compare the averaged number of steps to reach given destinations. Figure 8 offers comparison of the ideal number of steps calculated by the proposed method with the actual number of steps obtained by simulation during 10000 steps. The vertical axis shows averaged steps needed to reach given destinations. The horizontal axis indicates combinations of IDC layout possibilities. We sorted the combinations according to ideal steps.

A robot driven by the proposed algorithm, described in section 4, had to acquire knowledge for navigation autonomously because IDCs have no data at the beginning of the task. The comparison shows how much the overhead of autonomous knowledge acquisition is.

The robot in simulations required 128% steps on average to reach a desired destination. The standard deviation, $\sigma$, was 0.173. We evaluated that the proposed robotic system with IDCs pays +28% steps for autonomous knowledge acquisition and sharing.

# 5 Conclusion

In this paper, we proposed a device and algorithm to implement autonomous robots that can enhance the efficiency through autonomous knowledge sharing and acquisition system, refereed to as an "intelligent data carrier" system. In conclusion, we accomplish the following:

- We developed a device that enables local communication, and we refer to this device as an "Intelligent Data Carrier (IDC)".
- We implemented a system of autonomous navigation knowledge acquisition and sharing, carried out by autonomous mobile robots.
- We formulated behaviors of autonomous robots and proposed method to evaluate performance of the proposed algorithm.

When some robots are broken, we need only to replace them with new ones. The new robots can perform as effectively as former robots. This is possible because not the individual robots but rather the environment has the knowledge, which the robots can share easily. We conclude that the proposed system successfully realized a robust autonomous robotic system.

# References


1. Yoshida, E., et al. (1995) A Design Method of Local Communication Area in Multiple Mobile Robot System, Proc. IEEE Int. Conf. on Robotics and Automation, 2567-2562.
2. Yoshida, E., et al. (1996) Evaluating the Efficiency of Local and Global Communication in Distributed Mobile Robotic Systems, Proc. IEEE/RSJ Int. Conf. on Intelligent Robots and Systems, 1661-1666.
3. Balch, T., Arkin, R.C. (1994) Communication in Reactive Multiagent Robotic Systems, Autonomous Robots, No. 1, 27-52.
4. Ichikawa, S., Hara, F. (1996) Experimental Characteristics of Multiple-Robots Behaviors in Communication Network Expansion and Object Fetching, Distributed Autonomous Robotic Systems 2, 183-194.
5. Drogoul, A., Feber, J. (1992) From Tom Thumb to the Dockers: Some Experiments with Foraging Robots, From Animals to Animats 2, 451-459.
6. Fujii, T., et al. (1996) Co-evolution of a Multiple Autonomous Robot System and its Working Environment via Intelligent Local Information Storage, Robotics and Autonomous Systems, Elsevier, 19, 1-13.
7. Kurabayshi, D., Asama, H. (2000) Knowledge Sharing and Cooperation of Autonomous Robots by Intelligent Data Carrier System, Proc. IEEE Int. Conf. on Robotics and Automation, 464-469.
8. Yoshimura, Y., et al. (1996) Iterative Transportation Planning of Multiple Objects by Cooperative Mobile Robots, Distributed Autonomous Robotics Systems 2, Springer, 171-182.

# Collective Grounded Representations for Robots

Louis Hugues

Université Pierre et Marie Curie, LIP6
4 place Jussieu, 75252 Paris Cedex 05, France.
email: louis.hugues@lip6.fr

**Abstract.** Realistic applications of autonomous Robotics face a lot of difficulties in real environments. To navigate, self-localize and cooperate in such contexts, a multi-robot system has to find a way to represent and share knowledge about the world it is living in. In this paper we discuss the necessity and the advantages of grounding representations and communications of a robots team in real perceptions. We describe a technical framework designed for this purpose, which is based on the classification of the perceptions of a robots population. Percepts must have specific properties to enable grounding, and we present particular percepts well suited to this process. Finally we discuss the possible usages of grounded classes and propose a grid representation for fusion and exchange.

*Keywords:* Multi-Robots Systems, collective representations, grounding.

## 1 Introduction

Realistic applications of autonomous robotics face the difficulties that arise in most indoor environments. Real environments are usually partially unknown, not prepared for robots, and they are subject to dynamical changes. In those environments, perceptions of robots are uncertain, robots build and follow imprecise maps, robots get lost, they are starving while looking for energy, their knowledge becomes blurred during long term operation. Beside this, most realistic applications require the deployment of several robots in the same environment, for cleaning, surveillance, exploration, guided tour, delivery, animation... This is needed at least to scale to normal environments size and often needed by the nature of the tasks.

We believe that concrete problems such as navigation, localization and map building benefit of being envisaged from the outset in the multi-robot perspective. If this is envisaged later the potential solutions may not be adaptable to a multi-robot context and will not take advantage of the robot team's distributed point of view and multi-sensor capacities. In order to face those problems, we argue in the first section of the paper, that a shared medium for representation and communication between robots has to be discovered in the available set of real perceptions. In the second section a technical framework for collective representations is described, based on classification of the robots' population perceptions. The third section presents particular perceptions, well suited to grounded representations. Preliminary results of percept classification are given to illustrate the approach. In the last chapter some usages of grounded collective representations are discussed and Percept Distribution Grid are proposed to support cooperation in dynamic environments.

This work is part of the larger project Microbes headed by Alexis Drogoul at the LIP6 in Paris(Computer Science lab. of university Paris 6 ). The aims of the Microbes project are to enable the long-term adaptation of a robots colony in human standard environments and to study human/robots interactions and arising problems apart from a specific task context. The robots used in the project are Pioneer 2DX equipped with 16 ultrasonic sensors, a ccd color camera, a compass and a radio modem. All computations are performed on-board on the embedded Linux-based PC. The project takes place in the laboratory, a typical office area of 4000 square meters.

## 2 The need for grounded collective representations

A lot of activities performed by a robot such as mapping o learning pathways, involve the use of some kind of representations. These representations do not necessarily refer to a human-like abstract model. Beside this, a multi robot system benefits to mix up and collate robots partial informations and therefore use some communication scheme, for instance when specifying a meeting point, or building a global map from partial ones. It is clear that cooperative activities require the use of a shared medium where the communication can take place. In simulated robotic systems a shared medium can be easily found using the underlying precise model of the world and robots can communicate precise locationsand precise perceptions. However, in a real multi-robots team, deployed in an unknown environment, this is not the case. Robots' locations are imprecise and sometimes unknown, sensors are noisy, not necessarily following Gaussian models and due to the dynamicity of the environment, unpredictable events may happen. Also in a world that resists to an immediate symbolic transcription, the robots may have difficulties to communicate initially with symbolic tokens. So, basically, what can the robots use to communicate ?

Our hypothesis is that, in a real multi-robot system involved in concrete tasks, there should be a tight relation between the perceived world, the internal representations and the communication language elements, and that this bottom-up chain should be anchored in the real perceptions. As an example, the problem of self-localization and the problem of sharing the description of a particular location between robots are similar, since they both rely on finding sufficiently stable features in the environments. Generally we could say that representing the world and talking to others about the world are two facets of the same problem. Both require an initial material, strongly related to what it is possible to do and to perceive in the environments. Grounding in robotic systems is a promising direction, as pointed out by Brooks [1], and it is the natural emanation of today's ideas, that progressively blur the mind-body dualism. In Collective Robotics somehow similar orientations are proposed in papers such as [2,3]. However these papers do not address the problem of long term adaptation in large environments.

Of course the internal representations used by a robot can categorized by human beeing categorized landmarks (doors, corners, crosspoint, lanes ..) and the communication chunks be list, graphs,... constituted with those elements. But, in an unknown, dynamical environment, trying to ground representation and communication in real perceptions has the following advantages:

- The robots elaborate and communicate about what they are 'used' to perceive and not what we guess they should perceive.
- This prevents us from developing feature recognition mechanisms that could be too much specific to a given environment.
- This prevents us to follow a misleading human 'a priori', allowing a better adaptation of the robots to their environment.

For this purpose we favor the emergence of a collective substratum that can be used as an elementary material for representation and communication.

## 3 Percept class language

The proposed substratum is a rudimentary language: a finite set of classes whose 'meaning' is well known in the robot team. The set of classes is obtained by classifying the percepts of the robot population with a self organizing classifier. Several authors [4,5] have used unsupervised classifiers to classify perceptions of a single robot (ie, sets of sonar values) in a robust way. After sufficient learning, percepts are input in the classifier and the output classes are placed as landmarks indicators in a topological map. Here, in addition to this method we broadcast the percepts over the robot team to obtain homogeneity and use rather different percepts (see next chapter). We also choose to consider only percepts that are sufficiently singular: we do not take too frequent percepts (not informative) or too rare (spurious) ones into account. The class set is built in four steps (see fig 1):

- *Perception*: each individual regularly produces percepts while exploring its environment.
- *Broadcast*: the percepts produced are broadcasted (through the radio modem) to all other members of the team. Each of them constitutes a set of its own percepts plus the received ones.

- *Singularity Filtering*: the set of percepts is filtered so as to retain only the percepts which are sufficiently singular with respect to the whole set.
- *Classification*: each robot classifies the set of percepts. After classification, the resulting class set is the same for each individual.

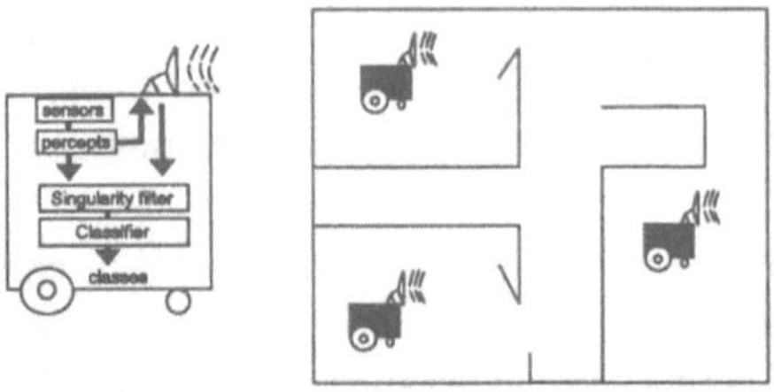


**Fig. 1.** Each robots broadcasts its percepts to the other ones. Individual percepts and received ones are input to an unsupervised classifier

After a sufficient maturation phase each robot can use the classifier to categorize its own percepts. The classes then constitutes an elementary language whose signification is well known among the robots and anchored in the possible perceptions. More elaborated representations (maps) can later be built using those elementary bricks.

The filtering and the classification may be performed with a Kohonen Self-Organizing Map (SOM)[6]. A SOM classifier is constituted by a 2D lattice of nodes. Each node $i$ corresponds to a class and stores a reference vector $\boldsymbol{m}_i$, a representative of the class. The class $c$ of a new input vector $\boldsymbol{x}$ can be obtained by comparison to reference vectors using appropriate distance:

$$c = argmin_i \, \|\boldsymbol{x} - \boldsymbol{m}_i\|$$

During learning, the reference vectors are updated to reflect the distribution of the input vectors, for each new input vector, the winning node and neighbors are moved towards the input vector value. At the end of the learning process the reference vectors are topologically ordered.

A SOM classifier has the following interesting properties:

- *Topology preservation* : the similarity relationshipss that exist between the percepts is maintained between their corresponding classes. It allows to use a distance between classes.
- *Unsupervised learning* : autonomous exploration is possible.
- *Error quantification* : the distance between an input vector and its class reference vector : $\|\boldsymbol{x} - \boldsymbol{m}_c\|$ , can be used to reject irrelevant percepts (ie: caused by temporary obstacles).

In the above process, the filtering step is required because we want to focus on the singular aspects of the environment, and do not want to take percepts that are too frequent into account. The SOM classifier over-fits the most frequent percepts, and thus the less frequent ones have a higher quantification error. We use this property for filtering, with a first SOM classifier. The percepts are classified a first time and the average quantification error for all percepts is computed. Then the percepts with a quantification smaller the average error (over-fitted) are eliminated. In the classification step remaining percepts are simply classified again using SOM.
In order to be anchored in the complete (eventually changing) environment the building of the substratum has to be coupled with an appropriate exploration behavior and a regular update of the classifier.

# 4 Choosing appropriate percepts

The word 'percept', here, refers to the result of some transformation applied to the raw sensors. This can be vertical lines extracted from video images, derivative of sonar values over the time to detect sudden variations, precise landmark detections such as doors.... The transformations can involve several sensors, (sonar and odometry for instance), they can also incorporate internal states values, or they can come from an active perception behavior. While raw sensors are most of the time immediate (ie, sonar range), percepts can integrate information over time and can contain informations that are currently out of reach (ie, after passing a door).

In a grounded system, the choice of an appropriate transformation from raw sensors to percept is necessarily crucial because it orientates the whole system. As an indication, we can refer to the importance of this transformation in animal nervous systems [7]. We suggest that the transformation is not only to obtain more accurate and precise information about the environment, but also to obtain an ambient feeling that captures invariant structures of the environment on sufficiently wide areas. Ambient here means invariant relatively to specific poses, invariant relatively to particular trajectories and possibly invariant to local sensing sequences. To illustrate this, one may think of sauntering in a town and being successively under the influence of a narrow street ambiance, then an open place ambiance, and so on. This kind of property is desirable in a mapping system, since it should help stable disambiguation of sufficiently wide areas. We are therfore looking for percepts with following properties:

- Percept that are not instantaneous, integrated over a period of time, containing informations not accessible to immediate senses (out of reach).
- Percepts that are ambient, as much as possible invariant to particular poses and that do not rely on detection of details.

And to fulfill the requirements of the SOM learning :

- Synthetic percepts, especially to shorten SOM processing and broadcast.
- Comparable percepts because SOM requires a distance between input vectors. Also we want to know if two percepts are sufficiently different.

## 4.1 Percepts obtained from an occupancy grid

Sonar sensors are widely used in the robotic systems and their disadvantages are also well known, since they are subject to noise and specular reflections. Moreover they give a poorly informed and imprecise vision of obstacles surrounding a robot (in our case 16 sonar ranges around the robot), and this at a rather low rate ( 1/3 of a second). H.P. Moravec and A. Elfes [8,9] have proposed a statistic method to improve sensors information and to fuse heterogenous sensors. The occupancy grid have been succesfully used in a number of works [10,11]. In the occupancy grids, the space around the robot is tessellated into small square cells (env. 15x15 cm). A probability value associated to each cell represents the occupancy probability of the corresponding location. Initially the probabilities are unknown (0.5), and as the robot progresses the probabilities are refined (using Bayes' rule), the cells receiving more sonar hits having their value increased. The moves of the robot in the grid reflect the moves of the robot according to the internal odometer. Figure

**Fig. 2.** Occupancy grid obtained in a part of our office environment (here a large corridor and a narrow lane - env. 20 m long). Grey, black and white cells correspond respectively to unknown, occupied and free space.

**Fig. 3.** Local occupancy grid (using 16 sonars and odometer, no position correction) - size 6x3 meters

2 shows an occupancy grid obtained in our office environment. The sonars have been modeled by single rays. Here the odometry drift and slippage have been partially compensated by using a mechanism based on the detection of wall orientation. However the local odometry precision is sufficient to build local grids (several meters wide) without any position correction, thus without presupposition on the conformation of the environment (Fig.3).

A percept with the properties stated in precedent paragraph (Fig.4) can be obtained from this local grid: The mass center of the grid constitutes a virtual point of view, independent of the trajectory. Drawing n rays (n=36) from this point, we can build a vector of n distances to the nearest obstacles. The orientation of the first ray is given by the north of the filtered magnetic compass. Such vectors can be compared easily using an Euclidean distance. This percept canconstitute an input in the SOM classifier. Percept with similar shape and orientation have similar resulting classes.

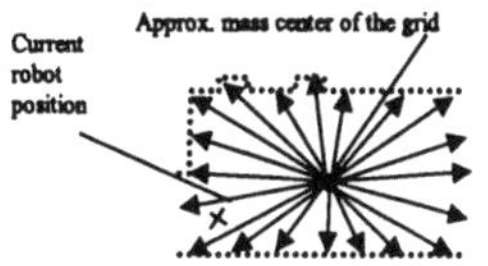


**Fig. 4.** The percept, obtained from the local occupancy grid matured during a short period, is a vector of n distances measured from the current mass center of the grid to the nearest obstacles

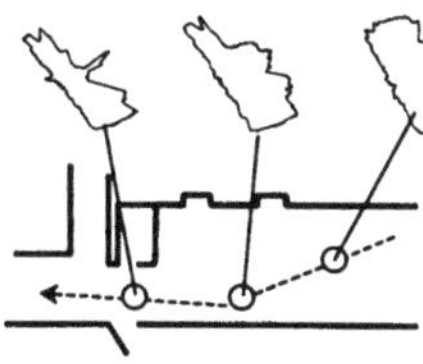

**Fig. 5.** Successive percepts obtained along a path in the office environment (robot experiment). They are rotated according to the internal compass

### 4.2 Example of Percept classification

Figure 6 shows the class reference vectors obtained by a single robot during the exploration of a part of the environment. This gives a short vocabulary of 6x10 classes. The classes around class (5,4) correspond to narrow North-West/South-East corridors with opening at NE. Around (5,6) are larger corridors with larger opening. Classes around (2,4) corresponds to transition between large and small corridor. Corridors with different orientation have classes around (6,10). Wide areas are around (1,8). This classification is solely intended as an example of the proposed approach.

## 5 Usage of the Percept classes

### 5.1 Support for traces and maps

During its exploration the robot can produce single traces having the following conformation: a trace is a succession of linked nodes; each node is labelled with the corresponding percept class, each link is labelled with geometric distance and compass sense orientation. A new node can be added to the trace when sufficient distance has been traveled through (distance may be obtained by integrating the odometry value) and when the distance between successive classes exceeds a given threshold (exploiting the topology preservation property of SOM classifier).
The traces can be combined into complex topological maps. The difficult

**Fig. 6.** SOM reference vectors obtained during the exploration of a part of our office by a single robot.

problem of maintaining a consistent map during long term operation, as well as the localization/re-localization problems can be eased by using ambient percept and percept class comparison. In a populated environment such as ours, invalid percepts can be rejected using quantification error.

### 5.2 Fusion and exchange-Percept Distribution Grids

The fact that a representation can use classes that are significant for all the robots helps the design of various exchange schemes. Primarily robots can exchange useful pathways that they have discovered, for instance exploration pathways starting at the power stations.

Topological representations are rather difficult to match and are not easily tractable. Alternatively we propose that the robots exchange discretized representations of the environment. Each robot can project its traces, and topological maps into a tessellated representation of the environment, a grid made of large cells (ie: 2x2 meter wide) and where each cell can store the probability distribution (histogram) of the encountered percept classes (fig 7). The cells can be labelled with utility informations (ie : visited or not). Arbitrarily the fixed point constituted by the power station is set in the center of the world. The histogram can reflect variations in the environment such as doors opened, closed , partially opened... The possibility of a grid description of the world with large cells directly rely on the ability to distinguis stable ambient percepts in the environment, each cell being under the influence of a small set of percepts. Of course the reliability of such grids depends on the correctness of the topological maps. However, even imprecise, they can provide a basis for cooperative behaviors.

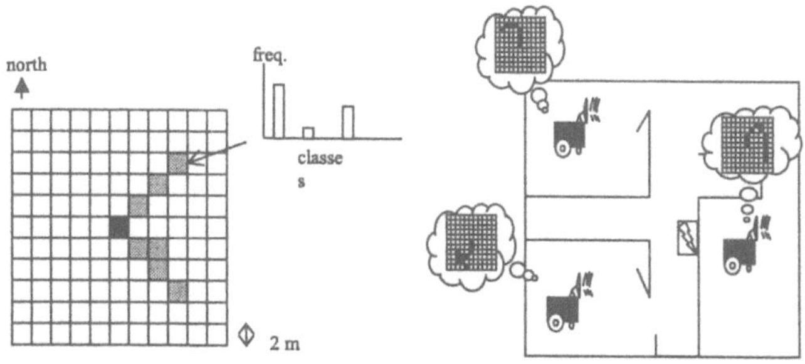


**Fig. 7.** Percept Distribution Grids: The environment (here hypothetical) space is tessellated into 2x2m cells. The grid keeps memory of the percept classes that have been observed in each cell. For a given location, several percepts may be observed (door closed, door semi-opened ..), therefore each cell keeps the histogram of observed classes.

The robots can exchange/fuse their maps cyclically to obtain a collective global map. The detection of overlapping zones is eased by the small amount of grid cells and by the fact that partial maps follow the same orientation (compass sense), and have a common fixed point (power station) then the size of the search space is reduced. A territorial division behavior, useful in most office environment applications, can be obtained by minimizing the overlapping areas between individual maps.
The grid provides the robot with a shared medium where synchronization, resources sharing, etc.. can take place more easily than in the physical world. The detection of Robots encounters, difficult to realize in the physical world, can be made easier by traces and grids confrontation. By comparing recent traces robots can confirm if they have encountered a colleague.

### 5.3 Meeting human models and representations

At the end, in most of multi-robot applications, it is necessary to meet the models and representations commonly used by humans. This is required for monitoring purpose or control purpose, for the design of behaviors that are somehow related to human activities. A simplistic but realistic approach can be to label subsequently the set of SOM classes. A user can visualize the percept reference vectors, and associate a description, such as 'large corridor' , 'narrow lane', 'opened door at south ' etc... The labelling can also be applied to the grids, by specifying points of interest (ie, forbidden places, etc ...).

## 6 Conclusions and future work

In this paper we have presented a method for grounding a robot team's representations and communications in real perceptions. We have argued that this is an essential feature to deploy a group of robot in real environments. A technical framework has been described, relying on broadcast and classification of adequate perceptions. Preliminary results have been presented to illustrate the approach. In future work we plan to perform larger experimentations with a team of six robots, studying particularly : the classification process, the relevance of discovered classes, the building of reliable maps with those classes and the confrontation of individuals' maps. A second set of experiments will then be carried in which we will extend the ambient percept with visual information (ie, color histogram ).
We believe that the classification of the population's percepts provides a material for various learning experiments. Implicitly, the sharing of percepts in a robot population is an attempt to increase the initial amount of common experience, suggesting that the possibility of any communication act depends on such common experience, a 'common ground'.

## References


1. R. A. Brooks. (1991) Intelligence Without Reason, MIT AI Lab Memo N0 1293.
2. A. Billard, K. Dautenhahn. (1997) Grounding communication in situated, social robots, proceedings 1997 of Towards Intelligent Mobile Robots TIMR.
3. D. Jung, A. Zelinsky. (2000) Grounded Symbolic Communication between Heterogeneous Cooperating Robots, Autonomous Robots Journal, special issue on Heterogeneous Multi-robot Systems, Kluwer Acadamic Publishers, Balch, Tucker, Parker Lynne (eds), Vol 8, No 3.
4. A. Kurz. (1996) Constructing Maps for Robots Navigation Based on Ultrasonic Range Data, Transactions on Systems, Man and Cybernetics, Vol 26, No 2, 233-242.
5. U. Nehmzow, T. Smithers. (1991) Mapbuilding using Self-organizing Networks, From Animals to Animats, J.A. Meyer and S. Wilson (eds.), MIT Press, 1991, 152-159.
6. T. Kohonen. (1997) Self-Organizing Maps. Springer-Verlag, New York.
7. D. Hubel. (1995) Eye, Brain and Vision. Scientific American Library No 22, W.H. Freeman, New York.
8. H .P. Moravec. (1988) Sensor Fusion in Certainty Grids for Mobile Robots, AI Magazine, Summer 1988 ,61-74.
9. A. Elfes. (1989) Using Occupancy Grids for Mobile Robot Perception and Navigation. IEEE Computer, june 1989 , 46-57.
10. B. Yamauchi, P. Langley. ( 1997) Place Recognition in Dynamics Environnements, Journal of Robotic Systems, Vol 14 , No 2 , 107-120.
11. S.Thrun, A. Bucken, W. Burgard, D. Fox, T. Frohlinghaus, D. Hennig, T. Hofmann, M. Krell, T. Schmidt. (1997) Map Learning and High-Speed Navigation in RHINO, AI-based Mobile Robots : Case studies of successful robot systems. D. Kortenkamp, R.P. Bonnasso, R.R Murphy (eds) MIT Press.

# Effects of Static and Dynamic Variety in the Character of Robots on Group Intelligence of Multi-Robot System

Sumiaki Ichikawa* and Fumio Hara**
Science University of Tokyo, Japan

1-3, Kagurazaka, Shinjukuku, Tokyo, JAPAN 163-8601
* ichikawa@hafu0103.me.kagu.sut.ac.jp
** hara@hafu0103.me.kagu.sut.ac.jp

**Abstract:** We focus on the effects of variety in robot characteristics in a multi-robot system, with variety possibly contributing toward improving robotic capability. The multi-robot system was assigned the task of building up a communication network within a simple maze-like area by use of each robot's local communication capability. The multi-robot system is composed of autonomous robots each assigned simple functions and simple action principles. A communication network is formed by a series of node robots, which relay communication signals. Dense allocation of the node robots within the work area degrades the system's capability to build a communication network. In order to eliminate this drawback, we add another action principle so as to realize reallocation of the node robots according to how often they communicate with other robots (called "communication frequency" herein). The condition to trigger the action principle is the threshold value of communication frequency assigned to each robot. In the present study, we often assign different threshold values to different robots. The differential is important for generating variety in a multi-robot system. We introduce two types of threshold values, static threshold values and dynamically variable threshold values. Computer simulation studies show that variety can improve robotic capabilities. Moreover, dynamically variable threshold values are more effective than are static threshold values. We thus point out the importance of dynamic variety in a multi-robot system for attaining higher performance.

Keywords: multi-robot, emergent intelligence, local communication, static variety, dynamic variety

## 1. Introduction

When a number of robots move together within a single area, physical and informational interactions occur among the robots, generating a new capability in the robot group. We call such a situation 'emergence of group intelligence'. We posit that this phenomenon is caused by complex interaction among the robots in the group, and therefore, 'emergence of group intelligence' is expected to occur not in a single-robot system, but only in a multi-robot system. We believe that emergence of group intelligence leads to realization of a particularly useful multi-robot system, even if the robots involved are equipped with only simple mechanical functions and action principles. This belief is based on the premise that generating complex behaviors in a group of robots does not always require advanced functions or sophisticated action principles. We expect to attain simplicity in the design of mechanical function and action principle of a single robot, and also expect such a multi-robot system to attain not only a fundamental capability to achieve a certain task, but also adaptability, fault tolerance, ease of maintenance, size reduction, and other desirable characteristics.
A group of engineers and scientists have recently focused on 'embodiment' of a robot

system. The concept of embodiment is based on dynamics and morphology of a robot that interacts with its environment. This idea is important in robot system design, and expected applications include generating a new robotic function, intelligence, or capability without explicit design consideration [1][2]. Embodiment and emergence of group intelligence seem to be analogous to some degree, because in both cases complex interaction occurs between a robot and its environment.

From this viewpoint, we have investigated a multi-robot system composed of mobile robots each equipped with a function of local communication, and found a certain emergent group intelligence, such as searching the given space, building a communication network, and returning to the origin point. However, we have found that we still need to improve on the multi-robot capability of building a communication network [3]. When the communication node robots, which play a key role in building a communication network, happen to be densely allocated, they come to obstruct each other's behavior. This becomes an unfavorable factor which degrades the capability of our multi-robot system to build a communication network.

In order to improve the communication network capability of the multi-robot system, in this paper, we introduce a new function and a new action principle assigned to each robot of our multi-robot system. The new function is to calculate frequency of communication signal reception and classify the frequency into one of two levels, 'high frequency' or 'low frequency,' by reference to a threshold value. The new action principle is to revise (refresh) the role of a robot from relaying a communication signal to searching space when a 'high frequency' signal is detected. We call this 'refresh action'. We focus on the two manners of setting threshold values for the refresh action; one employs a fixed threshold value, and the other employs a variable threshold value. Note that, in both methods, different robots are generally assigned different initial threshold values. Here, we point out that the difference in threshold value is for representing one type of variety in the multi-robot system. Thus, we are interested in the effects of variety in the characteristics of robots composing the multi-robot system on the system's capability to build a communication network.

We expect this kind of variety in a multi-robot system to yield a higher-performance group intelligence that elicits cooperative behaviors or adaptive behaviors in the multi-robot system.

## 2. Action Principle for Building Communication Network

We consider a multi-robot system consisting of autonomous robots equipped with two functions; moving on the ground with collision avoidance behavior and communicating with other robots. A communication capability is characterized by a specific range of communication signal transmission. Each robot is assigned action principles, such as

Mode 0: move forward and avoid collision
Mode 1: reverse direction
Mode 2: move forward
Mode 3: remain in place and relay a communication signal

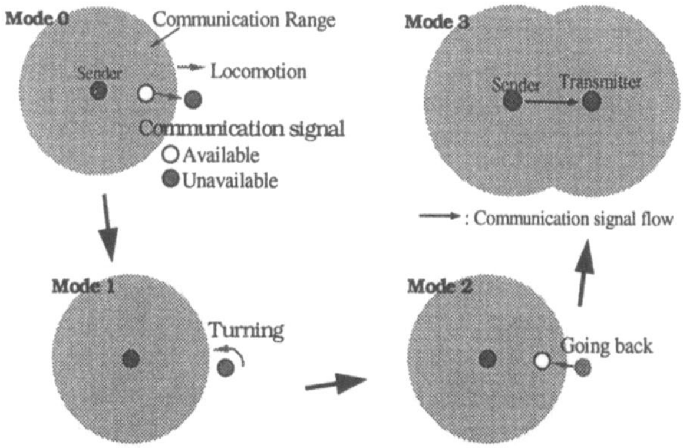


Fig. 1 Illustration of communication expansion behavior

'Mode 0' and 'Mode 3' are main action modes, and the other action modes are merely transient action modes between the two main action modes. Each robot starts at 'Mode 0' in the initial situation. A communication network is then expanded step by step by employment of 'Mode 3'. The manner of expanding the communication network is as follows: A specific robot within the multi-robot system is given priority. The specific robot remains at its initial position and transmits a communication signal at regular intervals time steps). Here, the specific robot is called a 'Sender'. Other robots are initially positioned within a communication range centered on the specific robot, and the robots move about with collision avoidance while they receive a communication signal at regular time steps. When a robot moves outside the communication range, it changes its action mode to 'Mode 1'. A 'Mode 1' robot is to reverse its direction of movement. Upon completing the reversal, the robot changes its action mode to 'Mode 2'. A 'Mode 2' robot is to move forward until a communication signal is received again, at which point the robot changes its action mode to 'Mode 3'. The 'Mode 3' robot reenters the previous communication range, and is to relay the communication signal. Thus, the communication range of the multi-robot system is expanded by the new 'Mode 3' robot.

## 3. Refresh Action

3.1 Communication frequency and population density of robots

We define communication frequency of a robot to be the average number of communication signals received by a single robot over the previous 100 time steps. When several communication signals are received within the same time step, they are added to the total. The communication frequency of a robot shows correspondence with the number of 'Mode 3' robots within the communication range of the robot. For example, when the time step of the regular communication transmission is 14 and seven 'Mode 3' robots are present within the communication range of the robot, the robot approximately calculates a communication frequency of 0.5 (= 7 / 14). A robot recognizes the status of communication frequency, high or low, by reference to a threshold value FTH.

3.2 Action principle for refreshing

The following two principles must be considered in our method of building a communication network [3]: One principle is that the multi-robot system cannot avoid dense allocation of 'Mode 3' robots in advance, because robots do not consider other robots' action modes. The other principle is that the multi-robot system cannot recover from such a crowded situation, because in our method 'Mode 3' is the terminal action mode. These two principles will constitute a drawback that will degrade the capability of building a communication network. When many 'Mode 3' robots are allocated at excessive density within a portion of the working area, they become obstacles for other moving robots. Excessive allocation of 'Mode 3' robots also means a decrease in the number of 'Mode 0' robots; that is, the multi-robot system will be degraded in its capability to explore the working area.

In order to eliminate the unfavorable situation described above, we introduce a new action principle assigned to each robot. That is, when a 'Mode 3' robot detects a high communication frequency, the robot changes its action mode to 'Mode 0' and can move away from the unfavorable situation. We call this new action a refresh action.

## 4. Effects of Distribution of Static Threshold Values in Multi-Robot System

4.1 Distribution of Threshold Values

Each robot of the multi-robot system is assigned a unique, fixed threshold value. To investigate the effects of the distribution of the static threshold values in the multi-robot system on the capability to build a communication network, we carried out the following three types of computer simulation experiments:

Sim-A: study of distribution effects

In one case, refresh action is not used. In three cases, identical threshold values are assigned to all robots; 0.5, 0.75 and 1.0, respectively. In the last case, static threshold values are assigned within the range of 0.5 to 1.0.

Sim-B: study of distribution range effects

The lower range of the threshold value is fixed at 0.5. Eight cases employ different upper ranges; from 0.6 to 1.3 at intervals of 0.1.

Sim-C: study of distribution position effects

The width of the distribution range is fixed at 0.5. Four cases employ different distribution ranges; 0.4 - 0.9, 0.5 - 1.0, 0.6 - 1.1 and 0.7-1.2, respectively.

4.2 Conditions for computer simulation experiment

The multi-robot system operates in a maze-like area shown in Fig. 2. The area consists of segmented corridors of 1.0 m in width. A single robot is 10 cm in diameter and proceeds forward at the rate of 0.75 cm per time step. The communication range of a single robot is 1.0 m within the robot position.

A computer simulation study is carried out in the following manner: A sender robot is allocated at the center of area S in Fig. 2. Only the sender robot sends a communication signal at a regular time step. Area S indicates the region where the other robots are allocated and which is within the communication range of the sender robot. The action mode of each allocated robot is initially set to 'Mode 0.' Area A and area B indicate the target boundaries of the multi-robot system. Forty robots are employed in the given task for 5,000 time steps per single trial of the experiment. The 5,000 time steps correspond to about five times the minimum time cost for a robot to move from area S to B. In each of the three types of simulation experiments, twenty trials of simulation are carried out. When any robot reaches one of the goal areas within the 5,000 time steps, we consider the task of building a communication network to the respective goal area to have been accomplished.

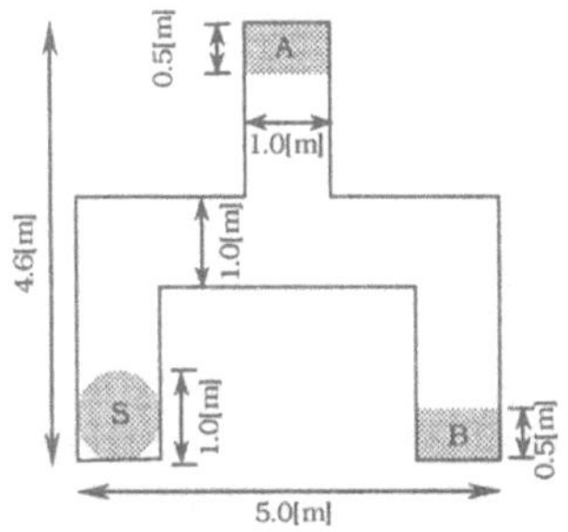


Fig. 2 Work area for simulation experiment

4.3 Results of computer simulation experiments

Fig. 3 shows the results of three simulation experiments; Sim-A, Sim-B, and Sim-C. Overall, the simulation studies show that area A is reached more frequently than is area B. This is due to the distance from S to A being shorter than the distance from S to B. In Sim-A, the results of the three cases employing identical threshold values (0.5, 0.75, 1.0) closely correlate with the results of the case where refresh action is not used (none). Task achievement is improved only in the case where threshold values are distributed (0.5 - 1.0). Thus, distribution of threshold values is said to show an effect on the capability to build a communication network. In Sim-B, the five cases appearing in the right-hand side of Fig. 3 b) are superior to the other three cases; when the range width is 0.4 or larger, the capability is superior to the case where range width is narrower. In Sim-C, the capability is higher when the distribution range is high. As shown in Fig. 3c), the capability is the highest for the distribution range of 0.5 - 1.0. These results show that imparting a distribution to the threshold values is useful for improving the capability of the multi-robot system to build a communication network. In these experiments, the multi-robot system shows optimal performance in the case where distribution range is 0.5 - 1.0.

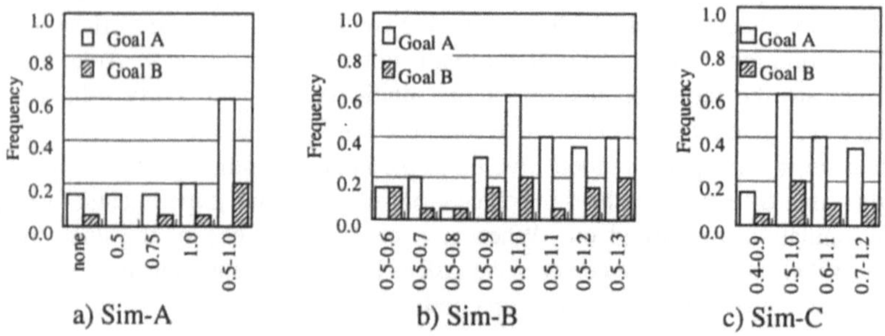


Fig. 3 Frequency of task accomplishment for 5,000 time steps

4.4 Characteristics of stimulating refresh action

We examine the characteristics of multi-robot behaviors by employment of the refresh action. In Sim-A, we calculate the average of the frequency of refresh action for twenty trials in each case. The results are shown in Fig. 4 a). The highest frequency, about 450 times, is observed for the case where FTH = 0.5. As shown in Fig. 4 a), when the threshold value FTH is increased, the frequency of refresh action decreases. In contrast, in the case where the threshold value shows a distribution of FTH = 0.5 - 1.0, the frequency of refresh action is an intermediate value between the frequency for FTH = 0.75 and the frequency for FTH = 1.0.

In a single trial for each case of Sim-A, we calculate the average numbers of 'Mode 3' robots and the average population densities of 'Mode 3' robots. The results are shown in Fig. 4 b). The average population density is evaluated by dividing the number of 'Mode 3' robots by the distance between the Sender robot and most distant 'Mode 3' robot. When a multi-robot system does not use the refresh action, indicated as FTH = none in Fig. 4 b), about twenty-five 'Mode 3' robots are allocated and the population density is about fourteen. However, when the system uses the refresh action, in all cases the population density of 'Mode 3' robot is reduced.

The average of the threshold values in the case of distributed threshold values, where FTH = 0.5 - 1.0, correlates closely with the average in the case of identical threshold values, FTH = 0.75. Thus, the two population densities are almost equal, as shown in Fig. 4 b). However, as shown in Fig. 4 a), the frequency of the refresh action is noticeably different. This accounts for the difference in frequencies of task achievement, as shown in Fig. 3.

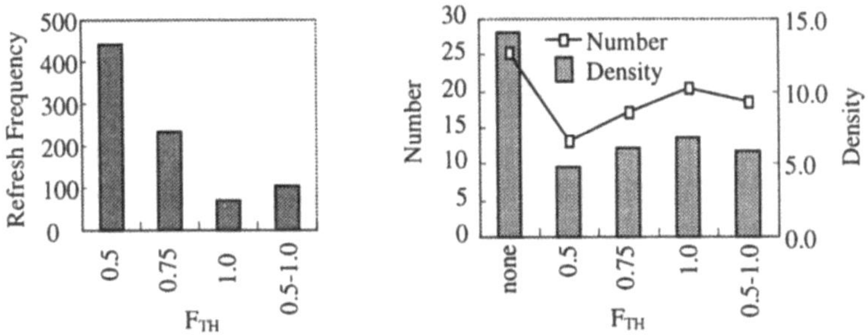

a) Frequency of refreshes b) Number and density of 'Mode 3' robots

Fig. 4 Characteristics of Sim-A

4.5 Characteristics of multi-robot behaviors using refresh action

From the results mentioned above, we found the following: Employment of the refresh action with a well-adjusted threshold value can eliminate excessive allocation of 'Mode 3' robots. However, when identical threshold values are assigned, the lower the threshold value, the higher the frequency of refresh action. Frequent refresh action costs additional time. Therefore, frequent refresh action is one factor which degrades the capability to build a communication network. However, imparting a distribution to threshold values assigned to the robots can prevent such a frequent refresh action, thereby improving the capability to build a communication network.

## 5. Effects of Refresh Action with Dynamic Threshold Value

5.1 Conditions for computer simulation experiment

We introduce a dynamically variable threshold value for refresh action. In this case, sixty robots are employed in computer simulation experiments. Work area and task are the same as those in the paragraph 4. We perform the following three types of simulation experiments.

Sim-1: An identical fixed threshold value, FTH = 0.75, is assigned to all robots.

Sim-2: The robots are assigned different fixed threshold values, FTH ranging from 0.5 to 1.0.

Sim-3: The robots are assigned different variable threshold values, FTH ranging from 0.5 to 1.0.

5.2 Change in variable threshold value for communication frequency

When a robot changes its action mode to 'Mode 3,' the variable threshold value of the robot changes in the following manner. If the communication frequency is high, the variable threshold value decreases at a specific rate. In contrast, if the communication frequency is low, the variable threshold value increases at the specific rate. In the present study, the specific rate is set at 0.066. The variable threshold value FTH ranges from 0.25 to 1.25. This range corresponds to 4 to 20 robots being 'Mode 3' robots. When the changed threshold value falls outside the range 0.25 - 1.25, the threshold value is reset at the center value of the range, 0.75.

5.3 Results of computer simulation experiments

Fig. 5 shows the frequencies of task achievement and the frequencies of refresh action for a single trial, representing the average among twenty trials in each of the three types of experiments.

Comparing the results of Sim-1 with the results for the same case, FTH = 0.75, in Sim-A (refer to Fig. 3), we find that the results (task achievement) of Sim-1 are generally superior to those of Sim-A. Comparing the results of Sim-2 with those of the same distribution range, FTH = 0.5 - 1.0, in Sim-A, we find that, in only the case of 10,000 time steps, the results of Sim-2 are superior to those of Sim-A. The former comparison shows that a large population size in a multi-robot system generally increases the capability to build a communication network, but the latter comparison case shows that this effect of increasing the capability may be subjected to a certain limitation.

In Fig. 5 a), Sim-3 shows the highest capability to build a communication network among the three types of simulation experiments. With respect to the frequency of refresh action, in Fig. 5 b), Sim-3 shows the lowest frequency among the three types of simulation experiments.

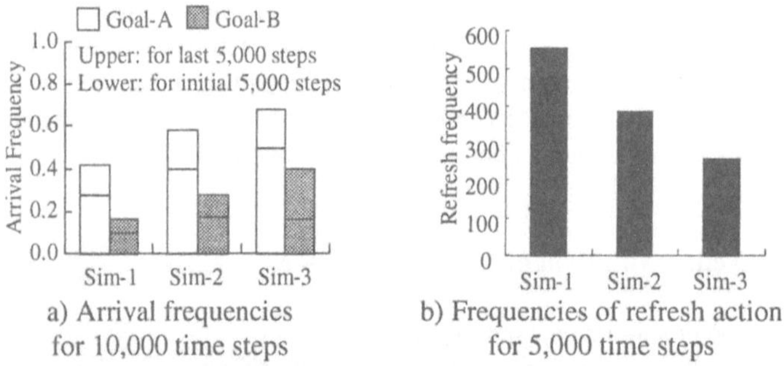


Fig. 5 Results of Simulations, Sim-1, 2, 3

5.4 Characteristics of changeable threshold values

Although the initial distributions of the threshold values in Sim-2 and Sim-3 are identical, the two simulations show different capabilities. Thus, in order to examine the characteristics of the effects caused by the fixed threshold value and those caused by the variable threshold value, we focus on the distribution of threshold values of 'Mode 0' robots at the end of each simulation trial. As 'Mode 0' robots are expected to become 'Mode 3' robots in the near future, the distribution of the threshold values of the 'Mode 0' robots can be said to represent the capability that the multi-robot system will attain in the near future.

We compare the three statistics of threshold values for following points: One is for common initial threshold value in Sim-2 and Sim-3, another is for the threshold values in Sim-2 at the end of the simulation trial, and the remaining is for the threshold values in Sim-3 at the end of the simulation trial. First, we calculate the average of the threshold values for each of twenty trials, and then calculate the average of the twenty average values for each of the three types. Fig. 6 a) shows the averages of the statistics. Fig. 6 b) shows the standard deviations of the statistics. Averages for Sim-2 and Sim-3 are lower than the average for initial value and they are almost equal, at about 0.7. However, the standard deviation for Sim-2 decreases slightly from the initial value, but that of Sim-3 increases from the initial value.

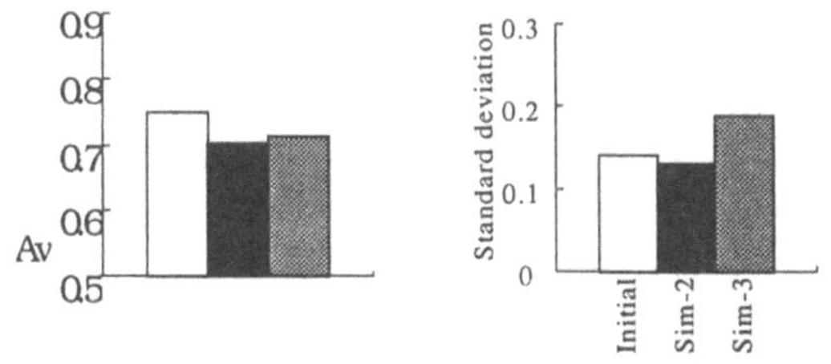


a) Averages b) Standard deviations

Fig. 6 Statistics of threshold values

Because the threshold value of each robot in Sim-2 is fixed, the decrease in the average of threshold values in 'Mode 0' robots indicates an increase in the average of threshold values in 'Mode 3' robots. The higher the threshold value of a robot, the higher the expectation that the robot will become a 'Mode 3' robot. Thus, in Sim-2, the average of threshold values of 'Mode 0' robots becomes smaller as shown in Fig. 6 a), and the distribution of the threshold value becomes narrower as shown in Fig. 6 b). Here, note that the narrowing in the distribution of the given threshold values indicates that the capability to build a communication network approaches that attained in the case where identical threshold values are employed. Thus, even with distribution of the threshold value, the multi-robot system may be degraded in capability for a long time when each robot is assigned a fixed threshold value. In contrast, in Sim-3 employing variable threshold values, the distribution; i.e., standard deviation, of the threshold values increases from initial standard deviation.

The foregoing discussion indicates that maintaining the distribution of threshold values in a multi-robot system markedly contributes to maintaining the capability to build a communication network of high performance. This feature is evident in the result of a comparatively difficult task, reaching area B, as represented by the higher frequency of reaching B for the last 5,000 time steps in Sim-3 than in the case of Sim-2, as shown in Fig. 5 a).

## 6. Conclusions

We focused on the capability of a multi-robot system to build a communication network using refresh action, which is expected to alleviate the dense allocation of 'Mode 3' robots. In our multi-robot system, population density of 'Mode 3' robots can be represented by communication frequency. Each robot can trigger the refresh action according to the threshold value of communication frequency. We focused on three manners of assigning threshold values. In one type, the robots are assigned identical, fixed threshold values; in another type the robots are assigned different, fixed threshold values distributed within a range; and in the remaining type the robots are assigned different, variable threshold values that were distributed within a range and were varied according to the statuses of the respective robots. Both types using different threshold values improved the capability of the multi-robot system, but employing identical threshold values did not improve the capability. However, when different threshold values were set for different robots, refresh actions were not triggered substantially simultaneously, which means that the occasions of triggering the refresh action were distributed within a certain range of number of time steps. In our simulation studies, we found the most effective range of the distribution of threshold values. According to our investigation, the range of the distribution of the fixed threshold values became smaller, but that of the variable threshold values became larger.

We believe that further investigation is necessary in order to obtain more general conclusions, because our stimulation studies examined the parameters heuristically. However, we can point out the following as an important feature of a multi-robot system: In our multi-robot system, the range of the distribution of the threshold values represents variety in the multi-robot system. Generally, variety in a multi-robot system improves the system's capabilities, but if variety in a multi-robot system it fixed or static, the improvement is not maintained for a long time. In contrast, variable or dynamic variety is expected to elicit more effective performance.

## Acknowledgement

This research work was partially supported by the JSPS project "Research for the Future" 96P00803. The authors express their heartfelt gratitude to those who involved in the project.

## References

[1] M. Asada (1998). Emergence of Intelligence Based on Physical Embodiment: Journal of Japanese Society for Artificial Intelligence 13-1, 14-15

[2] Y. Kuniyoshi and L. Berthouze (1999). Towards Emergence of Embodied Interaction Dynamics: Journal of the Robotics Society of Japan 17-1, 29-33

[3] S. Ichikawa and Fumio Hara (1996) "Experimental Characteristics of Multiple-Robots Behaviors in Communication Network Expansion and Object-Fetching" Distributed Autonomous Robotics System 2, Springer-Verlag, pp. 183-194

# Communication Fault Tolerance in Distributed Robotic Systems

Péter Molnár[1] and Jens Starke[2]

[1] Center for Theoretical Studies of Physical Systems, Clark Atlanta University, USA, pmolnar@cau.edu
[2] Institute of Applied Mathematics, University of Heidelberg, Germany, starke@iwr.uni-heidelberg.de

**Abstract.** The task of assigning a team of mobile robotic systems to individual job-locations has many challenges. We use the dynamical systems approach of coupled selection equations to achieve this problem. This intrinsically distributed algorithm has several advantages over traditional integer programs and other distributed approaches: 1) no backtracking is needed, 2) it can be used for NP-hard problems, such as assigning multiple robots with different capabilities to a certain job (three- or higher-index assignment problems), and 3) feasibility of the obtained solutions can be guaranteed.

The key point of the applicability in real distributed environments is a distinctive communication fault tolerance so that the necessary data communication between the different processes does not alter to an Achilles heel of the system. Therefore, the present paper addresses the loss of messages in a distributed robotic system based on coupled selection equations, and demonstrates the remarkable communication fault tolerance of this specific dynamical system approach by computer simulations.

## 1 Introduction

The problem addressed here is to control a group of robots so that they move to certain locations or targets in order to perform a task at that target. Applications of that type can be found in first incidents response, taxi on demand, and flexible manufacturing environments, such as those in the airplane industry.

In order to operate efficiently, explicit group behavior is required to assign only one robot to each target. Additional constraints are usually given by goal functions, such as minimizing the travel distance or working costs of the robots at these targets. Problems of this kind can serve as an evaluation platform for different approaches to flexible manufacturing systems. An overview of the various concepts and directions in this field is given in [2].

Navigation of the robots is achieved by using artificial force fields, that represent obstacles and other robots. Such force fields have been widely used in motion planning with various representations of obstacles (concentric, rotational etc.) [1].

Target assignment and motion planning are two aspects that often have been treated independently. The proposed self-organized control system combines these two aspects, in order to allow the robots to start moving before their destination is finalized.

## 2 Self-organized Control by Coupled Selection Equations and Behaviour Based Navigation

In the following, we use the concept of preferences and negotiation represented by a well defined and analysed dynamical system, the so-called coupled selection equations [5]. Hereby each robot will determine a preference value for each of the offered targets. The value lays between zero and one, whereby a larger value indicates a stronger preference. These preference values can be determined based on a cost function that may include, the proximity of the robot to a particular target, and the suitability of the robot's capability for the required task.

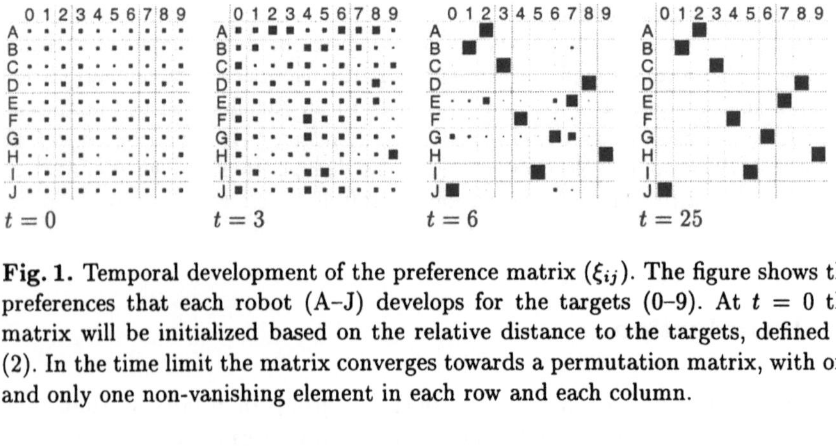


**Fig. 1.** Temporal development of the preference matrix $(\xi_{ij})$. The figure shows the preferences that each robot (A–J) develops for the targets (0–9). At $t = 0$ the matrix will be initialized based on the relative distance to the targets, defined in (2). In the time limit the matrix converges towards a permutation matrix, with one and only one non-vanishing element in each row and each column.

Our simulation uses a concentric repulsive potential to guide the robots away from obstacles. This approach does not produce the most efficient path around an obstacle, but can be easily created by the data of proximity sensors on the perimeter of a robotic unit. Hereby, the robot does not have to recognize the shape and size of the obstacle. In contrast to methods described in [1] we do not use an attracting force field to represent the goal but rather use the formalism of behavioural force models, which has been developed and used to model the dynamic of pedestrian crowds [3]. Here, each robot is guided by a destination vector $\mathbf{e}^0$ which finally points directly to the selected goal after the dynamic selection process. In our case, each robot has a number of potential goals. We define the direction vector $\mathbf{e}^0$ as a kind of linear combination of the direction vectors to all targets weighted by the preference coefficient $\xi_{ij}$ for these targets, as illustrated in Fig. 2. The destination of the robotic

**Fig. 2.** The robotic unit "x" is located in the neighborhood of two targets. Outlined arrows depict the unit vectors $\mathbf{e}_1^0$ to target "T1" and $\mathbf{e}_2^0$ to target "T2". Solid arrows represent the unit vectors multiplied by the coefficients $\xi_{xj}\mathbf{e}_j^0$ and the linear combination $\xi_{x1}\mathbf{e}_1^0 + \xi_{x2}\mathbf{e}_2^0$ of these vectors. The figures demonstrate the process of making a decision (here, without having the robotic unit move): initially both coefficients have about the same value ($\xi_{x1} = 0.5$, $\xi_{x2} = 0.5$). In the dynamic process $\xi_{x1}$ increases after some time to 0.7 while $\xi_{x2}$ diminishes. Finally, only one of the coefficients ($\xi_{x1} = 1$) increases to 1 while all others converge to zero ($\xi_{x2} = 0$). This results in a destination vector $\mathbf{e}_i^0$ pointing to target T1.

units is therefore determined by the preference coefficient $\xi_{ij}$ where $i$ denotes the robot and $j$ the target. The dynamics of these preferences $\xi_{ij}$ is given by coupled selection equations [5] which assign one and only one robotic unit to each of the targets in a dynamic process. Hereby, the assignment emerges in the limit of large times as an dynamics and can be guaranteed to be feasible [5]. Fig. 1 shows an example of this behaviour. In the case of a surplus on robots, some of the units will move to an idle position and wait unless one of the other robots breaks down, or a new target emerges [4].

In a distributed case each robot $i$ has its own instance of the preference matrix ($\xi_{ij}$). Though, it will only compute the coupled selection equation defined later in Eq. (1) for it's own preferences, i.e. only the row $i$ of the matrix. In addition the device has to send data packages including its recent preference vector (i.e., the corresponding row of the preference matrix) to the other units. We assume a non-confirmed protocol, such as multicast UDP.

The robot sends its preferences with a certain periodicity to all units in the range of reception. The fact that robots far away won't be able to receive the message, does not create a problem for many applications where only robots within a certain range are suitable to serve the same targets.

All remaining rows of the matrix will be updated when a message of the corresponding robot arrives. Updates will occur with a certain frequency. In between updates, the matrix elements remain constant. The length $\Theta$ of the interval between each update will affect the quality of the assignment solution. Communication faults that lose messages with a certain probability $p$ will even increase the time until matrix elements get updated.

We use computer simulations of the self-organized and distributed control system and the communication of data packages to study to which degree communication may deteriorate, and still produce feasible assignments.

## 3 Mathematical Representation of the Distributed Control System

This section presents the mathematical description of the control system in its two parts: the assignment problem solver and the behaviour based navigation model.

### 3.1 Selection Process and Assignment of Robotic Units to Targets

The dynamics of the coupled selection equations [5] is given here by

$$\dot{\xi}_{ij} = \kappa \xi_{ij} \left( 1 - \xi_{ij}^2 - \beta \sum_{i' \neq i} \xi_{i'j}^2 - \beta \sum_{j' \neq j} \xi_{ij'}^2 \right) \tag{1}$$

with $\beta > 1/2$. The time scaling $\kappa$ adjusts the change per time compared to the later defined equation of motion (5) for the robot's velocities. Non-negative initial values ensure for square matrices $(\xi_{ij})$ that the system will always asymptotically end in a stable solution of permutation matrices, i.e., there is one and only one non-vanishing element which is equal to 1 in each row and in each column. A proof is given in [5]. Due to this fact, in the selection of the destination there is a one to one correspondence of robotic units to targets. In the case of a surplus of robotic units the coupled selection equations (1) ensure that there is not more than one target as destination for each of the robotic units. Because of the property of selection equations to select the largest initial value in the long time limit, these coupled selection equations can be used for assignment problems in combinatorial optimization [5]. Even though the global optimum is not necessarily reached, the outcome compares very well to other optimization algorithms [6].

For the initial values of $\xi_{ij}$ we use the linear transformed EUCLIDean distance of the robotic units at $\mathbf{r}_i(0)$ to the targets at $\mathbf{g}_j$

$$\xi_{ij}(0) = 1 - \frac{||\mathbf{r}_i(0) - \mathbf{g}_j||}{\max\limits_{i',j'} (||\mathbf{r}_{i'}(0) - \mathbf{g}_{j'}||)}. \tag{2}$$

Using this transformation, targets which are located closer to the robotic units than others obtain larger initial values, i.e., a larger preference for the final selection. From the definition follows directly that $\xi_{ij}(0) \in [0,1]$ for all $i, j$. However, more sophisticated cost functions including manufacturing costs at the target's location, could be considered as well.

The time dependent matrix $(\xi_{ij})(t)$ depicts the preferences that each of the robotic units has for a certain target, or the preferences the targets have for being served by a specific robot. It can be directly linked to the mobile units by choosing the destination vector

$$\mathbf{e}_i^0(t) = \mathbf{\Phi}_{\gamma\delta} \left( \sum_j \xi_{ij}(t) \mathbf{\Phi}_{\gamma'\delta'} (\mathbf{g}_j - \mathbf{r}_i(t)) \right) \tag{3}$$

as a kind of linear combination of the difference vectors of the robotic unit $i$ to all targets. This linear combination of the with $\xi_{ij}(t)$ weighted normalized difference vectors $\mathbf{\Phi}_{\gamma'\delta'} (\mathbf{g}_j - \mathbf{r}_i)$ is demonstrated in Fig. 2. Hereby, the function

$$\mathbf{\Phi}_{\gamma\delta}(\mathbf{x}) = \frac{1}{||\mathbf{x}|| + 1/(\gamma||\mathbf{x}|| + \delta)} \cdot \mathbf{x} \tag{4}$$

with $\gamma, \delta > 0$ mainly normalizes the vector $\mathbf{x}$ but avoids a singularity at $\mathbf{x} = \mathbf{0}$.

Once the dynamical system of coupled selection equations has reached a stable point, there will be exactly one destination vector left for each of as many robotic units as there are targets. All remaining units will stop moving as soon as all of their corresponding coefficients converge close enough to zero.

## 3.2 Equation of Motion to Control the Robotic Units

The formalism of behavioural force models is used here to guide autonomous mobile robotic units through their environment. Even though no external forces are exerted onto the robotic units, the intended change of the robots' velocity vectors $\mathbf{v}_i(t) \in \mathbf{R}^2$ (or $\mathbf{R}^3$) is formulated in analogy to physical systems as equation of motion

$$\frac{d}{dt}\mathbf{v}_i(t) = \frac{1}{\tau}\left(v_i^0 \mathbf{e}_i^0(t) - \mathbf{v}_i(t)\right) + \sum_{i' \neq i} \mathbf{f}_{ii'}^{\mathrm{r}}(\mathbf{r}_{i'} - \mathbf{r}_i) + \sum_k \mathbf{f}_{ik}^{\mathrm{o}}(\mathbf{x}_k - \mathbf{r}_i). \quad (5)$$

Hereby, the direction to the destination of a robotic unit is given by the normalized vector $\mathbf{e}_i^0(t) \in \mathbf{R}^2$ (or $\mathbf{R}^3$) which results out of a set of coupled selection equations, that may assign the unit to a target in the course of its tour. Usually, the robotic devices operate best at a certain speed $v^0 \in \mathbf{R}$ which is determined by the nature of the robotic units, similar to the constant walking speed we observe with most pedestrians. Within a certain relaxation time $\tau$ the units velocities tend to the desired ones.

In addition, short range force fields $\mathbf{f}_{ii'}^{\mathrm{r,o}}(\mathbf{r}) \in \mathbf{R}^2$ (or $\mathbf{R}^3$) change the robots' directions and speeds in order to avoid collisions with other robotic units and obstacles. They are defined by

$$\mathbf{f}_{ii'}^{\mathrm{r,o}}(\mathbf{r}) = \begin{cases} (\tan g(\tilde{r}) - g(\tilde{r})) \frac{\mathbf{r}}{\|\mathbf{r}\|} & \text{for } 0 < \tilde{r} \leq \sigma^{\mathrm{r,o}} \\ 0 & \text{for } \tilde{r} > \sigma^{\mathrm{r,o}} \end{cases} \quad (6)$$

with $\tilde{r} = \|\mathbf{r}\| - d_i^{\mathrm{r}}/2 - d_{i'}^{\mathrm{r,o}}/2$, $g(\tilde{r}) = \frac{\pi}{2}\left(\frac{\tilde{r}}{\sigma^{\mathrm{r,o}}} - 1\right)$ and $\|\cdot\|$ is the EUCLIDean norm, around each of the other units $i'$ at the locations $\mathbf{r}_{i'}$ with diameter $d_{i'}^{\mathrm{r}}$ and obstacles at the locations $\mathbf{x}_{i'}$ with diameter $d_{i'}^{\mathrm{o}}$. The distances maintained between robotic units and obstacles can be chosen with appropriate range parameters $\sigma^{\mathrm{r}}$ and $\sigma^{\mathrm{o}}$. Using this short range potential force avoids unwanted local minima or spurious states where the system could stuck in. These repulsive forces can be calculated with the information from infrared or ultrasonic sensors that detect the distance to an obstacle within a given range. Many mobile robots are equipped with a number of such sensors around their perimeter.

# 4 Simulation Results Demonstrating the Fault Tolerance

The simulation program is based on the equations (1) and (5) as described in section 3. The differential equations have been integrated by using the EULER forward method with time steps $\Delta t = 0.05$ for the navigation part (5). The time scale of the coupled selection equations (1) is accelerated by factor $\kappa = 3$.

Every robot receives an update of the other robots' preferences in intervals of $\Theta = 1.00$. After $n = \frac{\Theta}{\Delta t}$ iterations the simulation program will copy the each robot's current preferences into the matrix $(\xi_{ij})$ of any other unit. Based on a random number, that will be generated for each transaction, and the package loss probability $p$, the program will decide whether the matrix elements will actually be copied. It is assumed that the robots communicate by sending data-packages using a non-confirmed protocol, such as UDP. Therefore, a data-package containing the recent preferences of one robot, might be received by some but not all devices.

Ten simulations have been run for each of ten values for $p$. Table 1 shows the elapsed time until every robot of the system reached a target and the final target assignment of the robots (A–J) for each of the simulation runs.

The collision avoidance maneuver of the robots extends the time they need to reach their destination. If a certain number of messages got lost, robots might choose a slightly different assignment which can actually result in shorter travel times, because the new path collides less often with obstacles. This explains why in the presented simulations the average elapsed time of ten simulation runs with 20% package loss probability is lower than the value in the 0% case (Fig. 3).

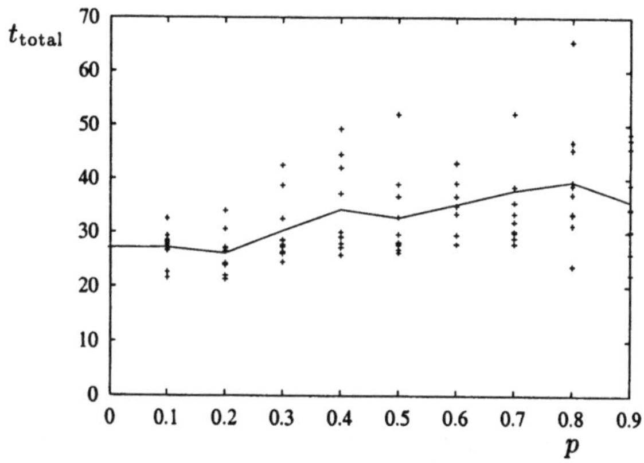


**Fig. 3.** Elapsed times $t_{\text{total}}$ until every robot reached a target. The solid line graph shows the average time over the package loss probability $p$.

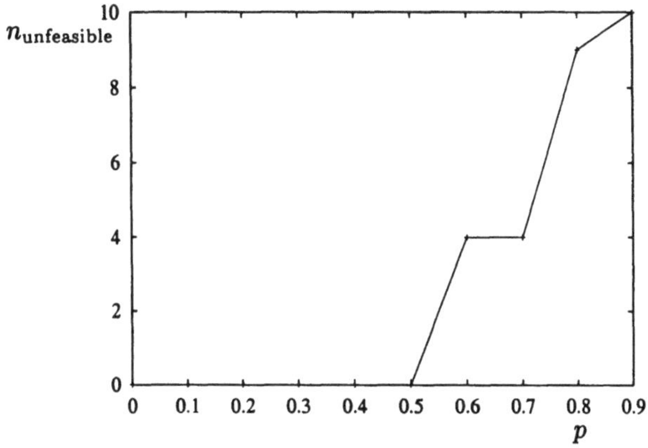


**Fig. 4.** Number $n_{unfeasible}$ of unfeasible solutions in dependence of the package loss probability $p$. Unfeasible solutions are assignments that allow more than one robots to select the same target. Due to the fault tolerance, there are feasible solutions only ($n_{unfeasible}$=0) up to a package loss of 50%.

Higher rates of package losses will violate the constraint of assigning exactly one robot to every target. Simulations with package loss probability $p \geq 0.6$ may result in unfeasible assignments, where more than one robot approaches the same target (Figure 5.d).

The way the number of these incidents increases with more frequent communication drop-outs is depicted in Figure 4. Even with no communication at all, robots still would approach a target. The target of their choice would depend on the initialization value $\xi_{ij}(0)$. However, several devices may select the same target, and some targets might be left out.

Our results show that the used distributed control [4], [7] is extremely stable and robust against communication drop outs up to 50%.

## 5 Conclusion and Outlook

We have shown, how the self-organized control system for distributed autonomous mobile robotic units behaves under (simulated) communication faults. The results demonstrate the tolerance of the task assignment system under communication faults, which all distributed systems are exposed to.

The success of the control system depends on the intervals between updates of the elements of the preference matrix $(\xi_{ij})$. Both factors, the frequency of data messages and the probability $p$ of package losses contribute to the quality of the task assignment.

Frequent package losses in the communication channel can be compensated by shorter intervals between the broadcast of updates. In a technical implementation, robots should not send messages at a fixed rate, but rather randomly change the time between two transmissions. This can avoid periodical peeks in the data traffic, which yield to increased numbers of package

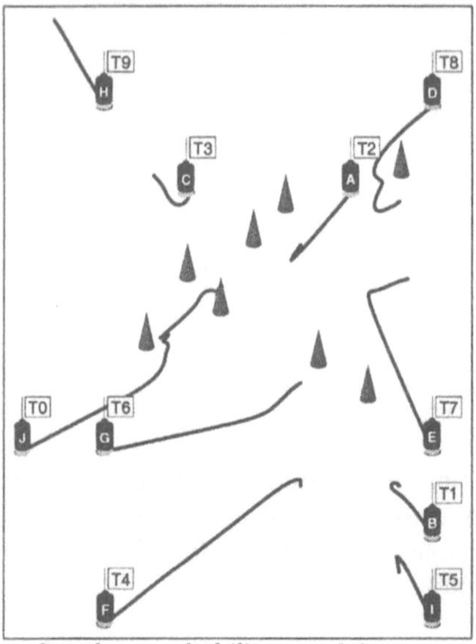

a. data loss probability $p = 0.0$

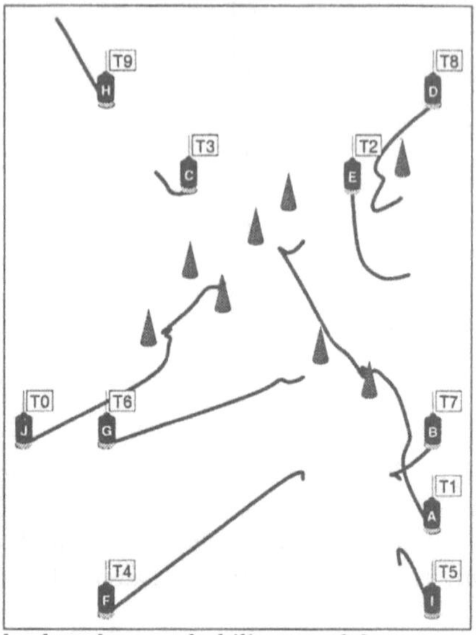

b. data loss probability $p = 0.2$

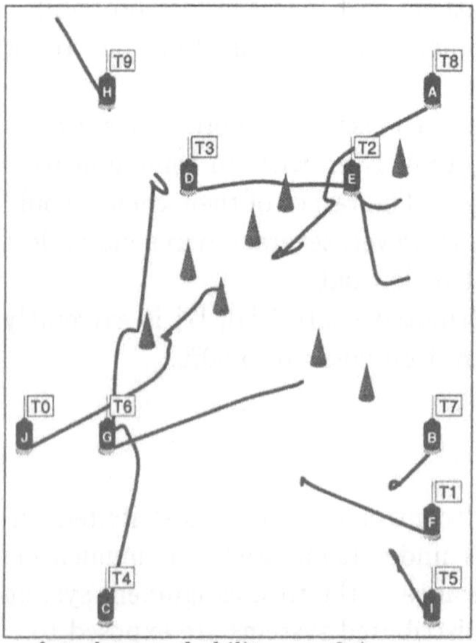

c. data loss probability $p = 0.4$

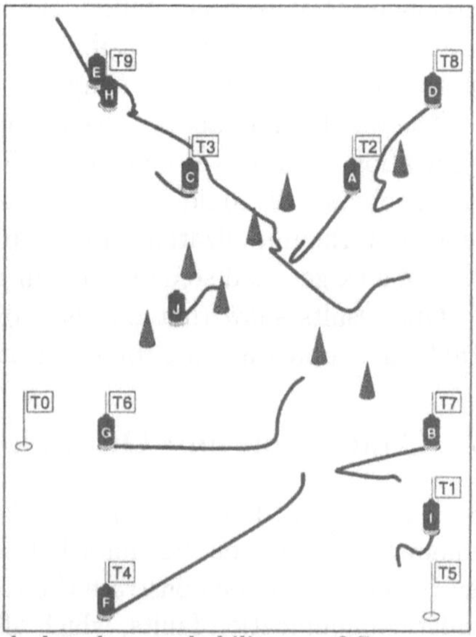

d. data loss probability $p = 0.7$

**Fig. 5.** Simulation runs for different values of the data loss probability $p$: a. no message loss, the system finds its optimal assignment, at $t = 27.1$ each robot has reached a target; b. 20% of the messages between robots got lost, the elapsed time is 34.0; c. 40% package loss, $t = 42.0$ and d. 70%, $t = 69.9$.

**Table 1.** Simulation results for different probabilities $p$ for message loss. The table shows the elapsed time $t_{total}$ until every robot of the system reached a target, and the final target assignment. Non-feasible assignments are written in brackets.

| $p$ | # | $t_{total}$ | Assignment A–J | $p$ | # | $t_{total}$ | Assignment A–J |
|---|---|---|---|---|---|---|---|
| 0 | 1 | 27.1 | 2 1 3 8 7 4 6 9 5 0 | 0.5 | 1 | 36.7 | 1 4 3 2 8 6 7 9 5 0 |
| | 2 | 27.1 | 2 1 3 8 7 4 6 9 5 0 | | 2 | 27.6 | 2 6 3 8 7 4 1 9 5 0 |
| | 3 | 27.1 | 2 1 3 8 7 4 6 9 5 0 | | 3 | 28.1 | 7 1 3 8 6 4 2 9 5 0 |
| | 4 | 27.1 | 2 1 3 8 7 4 6 9 5 0 | | 4 | 26.8 | 6 5 3 8 7 4 2 9 1 0 |
| | 5 | 27.1 | 2 1 3 8 7 4 6 9 5 0 | | 5 | 39.0 | 2 7 0 8 4 1 6 9 5 3 |
| | 6 | 27.1 | 2 1 3 8 7 4 6 9 5 0 | | 6 | 28.0 | 3 1 0 8 2 5 6 9 7 4 |
| | 7 | 27.1 | 2 1 3 8 7 4 6 9 5 0 | | 7 | 26.2 | 3 1 9 8 7 4 6 2 5 0 |
| | 8 | 27.1 | 2 1 3 8 7 4 6 9 5 0 | | 8 | 32.8 | 2 0 3 8 7 4 6 9 5 1 |
| | 9 | 27.1 | 2 1 3 8 7 4 6 9 5 0 | | 9 | 52.0 | 8 7 3 0 2 4 9 1 5 6 |
| | 10 | 27.1 | 2 1 3 8 7 4 6 9 5 0 | | 10 | 29.6 | 9 1 0 8 2 4 7 6 5 3 |
| 0.1 | 1 | 29.3 | 2 1 3 8 7 4 6 9 5 0 | 0.6 | 1 | 42.9 | 2 1 9 8 4 5 3 7 6 0 |
| | 2 | 26.8 | 2 7 3 8 1 4 6 9 5 0 | | 2 | 29.5 | 7 1 2 8 6 4 0 9 5 3 |
| | 3 | 28.0 | 2 1 3 8 7 4 6 9 5 0 | | 3 | 27.8 | [7 1 0 8 9 4 2 8 5 3] |
| | 4 | 32.5 | 1 7 3 8 2 4 6 9 5 0 | | 4 | 36.7 | 1 4 3 8 2 0 7 9 5 6 |
| | 5 | 26.6 | 2 1 9 8 7 4 6 3 5 0 | | 5 | 43.0 | [7 5 4 2 8 3 6 9 8 0] |
| | 6 | 21.6 | 2 1 3 8 7 4 0 9 5 6 | | 6 | 33.4 | [8 7 3 2 1 4 4 9 5 0] |
| | 7 | 28.5 | 2 1 3 8 7 4 6 9 5 0 | | 7 | 39.2 | 2 7 6 8 4 5 3 9 1 0 |
| | 8 | 27.6 | 2 1 3 8 7 4 0 9 5 6 | | 8 | 29.5 | [2 1 3 8 7 4 7 0 5 9] |
| | 9 | 28.2 | 2 1 3 8 7 4 6 9 5 0 | | 9 | 34.9 | 2 1 3 2 5 4 0 9 7 6 |
| | 10 | 22.7 | 2 1 3 8 7 4 0 9 5 6 | | 10 | 34.9 | 9 7 3 8 2 1 4 6 5 0 |
| 0.2 | 1 | 24.3 | 2 1 0 8 7 4 6 9 5 3 | 0.7 | 1 | 52.1 | 7 4 9 8 3 1 0 6 5 2 |
| | 2 | 23.9 | 2 1 3 8 7 4 0 9 5 6 | | 2 | 28.9 | 1 2 3 8 7 6 4 9 5 0 |
| | 3 | 27.2 | 2 1 3 8 7 4 6 9 5 0 | | 3 | 31.9 | 8 1 0 6 7 4 2 9 5 3 |
| | 4 | 30.6 | 7 1 9 8 2 4 6 3 5 0 | | 4 | 30.2 | [7 8 8 3 2 4 1 9 5 6] |
| | 5 | 22.0 | 2 1 3 8 7 4 0 9 5 6 | | 5 | 38.5 | 3 7 0 8 2 6 4 9 5 1 |
| | 6 | 27.0 | 2 1 3 8 7 4 6 9 5 0 | | 6 | 35.5 | 1 7 3 2 8 0 4 9 5 6 |
| | 7 | 34.0 | 1 7 3 8 2 4 6 9 5 0 | | 7 | 33.4 | [1 7 3 8 2 4 0 9 5 9] |
| | 8 | 21.3 | 2 1 3 8 7 4 0 9 5 6 | | 8 | 27.8 | 0 7 9 8 1 4 6 9 5 3 |
| | 9 | 26.5 | 2 6 0 8 7 4 1 9 5 3 | | 9 | 69.9 | [2 7 3 8 9 4 6 9 1 0] |
| | 10 | 24.1 | 2 7 3 8 1 4 0 9 5 6 | | 10 | 29.9 | [8 1 9 3 7 4 6 0 5 3] |
| 0.3 | 1 | 26.1 | 0 1 3 8 2 4 7 9 5 6 | 0.8 | 1 | 37.1 | [3 0 9 1 2 4 8 6 5 9] |
| | 2 | 26.4 | 6 7 0 8 1 4 2 9 5 3 | | 2 | 23.7 | [3 7 9 8 2 4 1 9 5 9] |
| | 3 | 27.3 | 2 1 9 8 7 4 6 3 5 0 | | 3 | 31.3 | [2 1 3 8 1 4 0 9 5 6] |
| | 4 | 42.5 | 2 0 3 1 8 4 7 9 5 6 | | 4 | 38.8 | [6 5 0 8 7 4 9 9 1 3] |
| | 5 | 24.5 | 4 1 3 8 7 6 5 9 0 2 | | 5 | 33.3 | 2 1 9 7 0 4 6 3 5 8 |
| | 6 | 28.6 | 2 1 3 8 9 4 7 6 5 0 | | 6 | 39.4 | [3 7 4 9 8 5 0 9 5 6] |
| | 7 | 38.8 | 2 1 3 8 0 4 7 9 5 6 | | 7 | 45.4 | [5 1 8 8 2 3 0 9 2 9] |
| | 8 | 27.7 | 3 1 6 8 7 4 2 9 5 0 | | 8 | 65.5 | [9 5 3 8 8 0 7 9 4 6] |
| | 9 | 28.5 | 3 1 0 8 2 4 7 9 5 6 | | 9 | 46.8 | [5 1 3 9 8 7 6 9 5 4] |
| | 10 | 32.5 | 2 7 3 8 0 1 4 9 5 6 | | 10 | 33.5 | [7 1 3 8 9 4 6 0 5 3] |
| 0.4 | 1 | 30.0 | 2 1 3 8 5 4 6 9 7 0 | 0.9 | 1 | 34.2 | [2 1 9 8 0 4 0 6 5 3] |
| | 2 | 42.0 | 8 7 4 3 2 1 6 9 5 0 | | 2 | 30.1 | [2 1 8 8 7 4 4 9 5 0] |
| | 3 | 29.1 | 2 1 3 8 7 4 6 9 5 0 | | 3 | 48.4 | [2 7 9 8 5 1 7 9 5 4] |
| | 4 | 37.2 | 3 1 0 8 4 2 6 9 5 7 | | 4 | 45.6 | [2 1 9 7 8 4 7 9 5 0] |
| | 5 | 25.8 | 2 1 3 8 1 7 6 9 5 0 | | 5 | 34.4 | [4 1 9 9 8 5 2 9 5 3] |
| | 6 | 27.9 | 2 1 0 8 6 4 7 9 5 3 | | 6 | 38.9 | [2 7 9 8 5 6 4 7 5 0] |
| | 7 | 44.5 | 6 1 9 8 2 4 0 7 5 3 | | 7 | 30.2 | [2 1 9 8 8 4 0 9 5 3] |
| | 8 | 29.2 | 0 7 3 8 1 4 6 9 5 2 | | 8 | 26.0 | [2 4 9 8 2 4 7 9 5 3] |
| | 9 | 27.2 | 2 1 3 8 7 4 6 9 5 0 | | 9 | 22.1 | [9 1 3 8 7 4 4 9 5 6] |
| | 10 | 49.3 | 9 1 3 8 2 4 6 7 5 0 | | 10 | 47.3 | [4 1 2 8 8 5 7 9 7 3] |

collisions (on an wireless ethernet). Further studies will include an experimental setting that uses wireless ethernet.

For many technical implementations, data communication is the most vulnerable part of distributed systems. In this paper we demonstrated how our control system based on coupled selection equations can overcome this vulnerability.

## References

1. R. Arkin. *Behavior-Based Robotics.* The MIT Press, Cambridge, London, 1998.
2. George A. Bekey and Arvin Agah. Group behavior of robots. In Shimon J. Nof, editor, *Handbook of Industrial Robotics*, chapter 23. John Wiley & Sons, Inc., second edition, 1999.
3. D. Helbing and P. Molnár. Social force model for pedestrian dynamics. *Physical Review E*, 51:4282–4286, 1995.
4. P. Molnár and J. Starke. Control of distributed autonomous robotic systems using principles of pattern formation in nature and pedestrian behaviour. submitted, 1999.
5. J. Starke. *Kombinatorische Optimierung auf der Basis gekoppelter Selektionsgleichungen.* PhD thesis, Universität Stuttgart, Verlag Shaker, Aachen, 1997.
6. J. Starke and M. Schanz. Dynamical system approaches to combinatorial optimization. In D.-Z. Du and P. Pardalos, editors, *Handbook of Combinatorial Optimization*, volume 2, pages 471 – 524. Kluwer Academic Publisher, Dordrecht, Boston, London, 1998.
7. J. Starke, M. Schanz, and H. Haken. Self-organized behaviour of distributed autonomous mobile robotic systems by pattern formation principles. In T. Lueth, R. Dillmann, P. Dario, and H. Wörn, editors, *Distributed Autonomous Robotic Systems 3*, pages 89 – 100. Springer Verlag, Heidelberg, Berlin, New York, 1998.

# Chapter 4

# Biological Inspirations

# Blazing a trail: Insect-inspired resource transportation by a robot team

Richard T. Vaughan, Kasper Støy, Gaurav S. Sukhatme, and Maja J. Matarić

Robotics Research Laboratories, University of Southern California,
Los Angeles, CA 90089-0781

vaughan|kaspers|gaurav|maja@robotics.usc.edu

**Abstract.** We demonstrate a team of real robots that cooperate to robustly transport resource between two locations in an unknown environment. The robots use a trail laying and following algorithm inspired by the trail following of ants and the waggle dance of honey bees. Rather than directly marking their environment, the robots announce landmarks in their odometric localization space. The system tolerates significant odometric drift before it breaks down.

## 1 Introduction

In a recent paper we described a method for establishing a robot 'supply column' inspired by the trail-laying of ants and the waggle dance communication of bees. We demonstrated a group of simulated mobile robots that communicate by leaving landmarks in their shared localization space. By laying and following trails of such landmarks, the robots could robustly transport resources between two areas in an unknown environment [1].

The algorithm's robust performance in simulation, particularly with respect to localization errors, indicated that the method should be suitable for use in real robots. In this paper we present our first results with real robots. We briefly reiterate our motivation for tackling this task, then describe the trail laying algorithm and its implementation on a team of four Pioneer 2 DX robots. The results from an initial experiment are presented, showing that the method transferred successfully into the real world.

### 1.1 Ants, trails and localization

Ants form supply columns to relocate valuable items through complex, dynamic environments. Their remarkable effectiveness is due to a highly robust strategy of local interactions among a large number of autonomous agents. The chemical trails formed by ants along their supply routes are robust with respect to changes in the environment and to the 'failure' of many individual ants [2,3]. These properties are attractive to agent designers in general and in particular to robot builders who can see immediate applications for robot supply columns in hazardous or tedious environments.

Ant-inspired solutions to various search problems have been demonstrated [4–8], and chemical trail laying and following has been demonstrated in robots [9,10]. Holland *et al* exploit stigmergy in a foraging task with real robots [11]. However it is often impractical and sometimes undesirable for robots to physically mark their environment, so in our method a group of robots deposits landmarks in a shared *localization space.*

We define localization space as any consistent spatial or topological representation of position. Such a space is *shared* if there is some (probably imperfect) correlation between the representations maintained by two or more individuals. A prime example is the Global Positioning System (GPS). Two systems equipped with GPS share a metric localization space in planetary coordinates. Similarly two robots that start out with known positions in the same coordinate system and maintain a position estimate via odometry share a localization space. In both examples each robot has only an *estimate* of its position in the true space, but the true space is common to both.

The robots in these experiments are localized by integrating their odometry to maintain a position and heading estimate. All robots are started from the same position so their coordinate systems are initially aligned. They share a localization space, but the correlation between the spaces will deteriorate over time as the error in each individual's localization estimate increases.

## 2 Task and approach

We examine the ant-like task of having multiple autonomous robots find and repeatedly traverse a path in an unknown environment between a known 'home' and a supply of resource at an initially unknown position.

Achieving this task reliably with robots would meet real current and future needs: a factory may require a supply of widgets manufactured at position A to be transferred by robot to position B; an army may require supplies (medicine, food, ammunition) to be transported over hazardous and uncertain terrain; a colony on Mars might want to to retrieve resources from a distant autonomous mining robot. The start location and the existence of one or more resource locations may be the only known features, and these may be a few meters or several kilometers apart. Establishing a reliable automated supply column robust with respect to loss of individual robots could be very valuable.

The experiments described in this paper demonstrate a simple algorithm for resource transportation by robot teams using mechanisms loosely analogous to the ant trail following and the honey bee 'waggle dance' [12]. Instead of directly modifying their physical environment like ants, or performing an information-transmitting dance like bees, our robots communicate localization space landmarks over a wireless network. This communication requires very little bandwidth and is well within the capabilities of current networks. Our communication strategies are informed by the work of [13] but are more

**Fig. 1.** The Pioneer 2DX robots used in the experiments.

complex because we are tackling a more realistic task in a more complex environment.

# 3 Robots

The experimental platform for these experiments was four identical ActivMedia Pioneer 2DX robots. These are fairly small (40x50cm), differential-steering and fitted with PC104+ Pentium computers running Linux. Each robot has front and rear sonar rings and wheel encoders as its basic sensors. For these experiments a SICK LMS scanning laser rangefinder was fitted to the front of each robot, providing excellent quality range readings over $180^o$. The SICK gives two samples per degree and an range accuracy of a few millimeters to most surfaces. BreezeNet 802.11 wireless Ethernet connected the robots and our workstations with a nominal bandwidth of 3Mbits/sec. The robots are shown in Figure 1, named Ant, Bee, Punky and Tanis.

# 4 Robot controller

## 4.1 Trail laying algorithm

Robots start from a home position and search for the resource. On reaching the resource, they receive a unit of resource and must return home with it, then return to fetch more resource repeatedly for the length of the trial. Each robot records its movements as a sequence of landmarks in localization space and shares the path with its team mates after each successful traverse. The algorithm consists of three decision processes running in parallel. A schematic is shown in Figure 2.

The trail laying algorithm is independent of the method used to drive the robot's wheels; rather it can provide a hint of a 'good' heading to go in from the robot's current location. In this implementation the robots initially search the environment by random exploration: following walls and turning at random at intersections. This was chosen because it is simple, requires

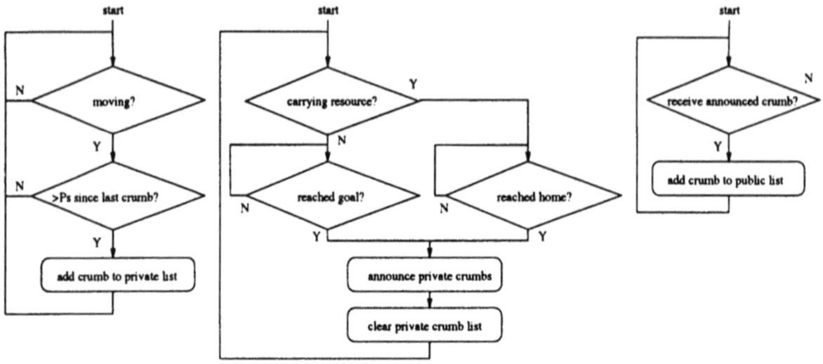


**Fig. 2.** Schematic of the trail laying algorithm. Three decision processes run in parallel to manipulate the private and public crumb lists.

no *a priori* knowledge of the environment and will always find a path if one exists, given sufficient time. As it moves through the environment each robot records its current position and heading estimate at regular time intervals. We refer to this as 'crumb dropping'. As time goes by, a robot stores many such records on its initially empty *private crumb list.*

If a path to the resource exists, a robot will eventually reach it. Whenever a robot reaches the resource it announces the contents of its private crumb list on the broadcast channel of the shared network. All the robots (including the sender) receive the crumbs and add them to their initially empty *global crumb list.* The sender's private crumb list is cleared. Communicating the route only upon completing a successful trip corresponds loosely to the honey bee's waggle dance, as compared to the ant which leaves a trail as it goes along.

The first robot to reach the resource must have used the most time efficient path yet discovered and all robots now have a list of waypoints describing this path. At each timestep each robot examines its global crumb list and computes the average heading of all those crumbs that lie within a threshold radius of its current position estimate. If there are no such crumbs, the robot performs random exploration, otherwise it does wallfollowing and turns in the direction of the average heading at intersections. If a robot is carrying resource, i.e. it has reached the supply of resource, it will follow the opposite heading back to the home position. On reaching home a robot also broadcasts and clears its private list.

Thus a robot whose position estimate is similar to the position of a dropped crumb will tend to move in the same direction as the successful robot that passed that way earlier. By following a trail of crumbs each robot can find the resource much faster than by random exploration.

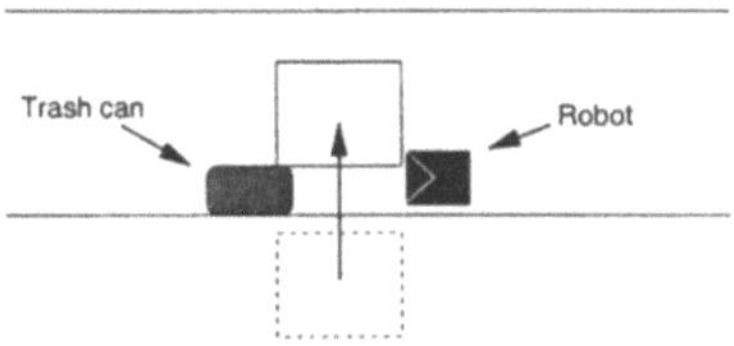


**Fig. 3.** Principle of the sliding box algorithm. The free-space box is initially placed to the left of the robot and is slid in the direction of the arrow until there are no obstacles detected inside the box.

### 4.2 Navigation

Navigation and obstacle avoidance is provided by two simple high-level behaviours: *Navigate* and *panic. Navigate* drives the robot around the environment following walls, turning down corridors and following crumb trails. Navigate is described in detail below. An emergency-stop mechanism runs continuously in parallel, monitoring the laser sensor; if the robot gets too close to any obstacle it will stop and switch to the *panic* behaviour.

*Panic* is designed to cope with situations that confound *navigate.* It 'unsticks' the robot from obstacles, dead ends and traffic jams by turning on the spot in a random direction until it detects free space and moves forward. After a few seconds the behaviour times-out and switches to the *navigate* behaviour.

The navigate behaviour is fairly sophisticated and contains several sub behaviours. It exploits the accuracy of the SICK laser scanner to make precise movements, turning tight corners and closely following walls.

At each time step the behaviour detects corridors to the left and right of the robot using the laser scanner. A corridor is identified if the extreme left and rightmost laser samples show a nearby obstacle followed by a large discontinuity and an opening wide enough to accommodate the robot. If the robot detects a corridor and it is not already turning, there is a 50% chance that the robot will decide to take the new corridor and make a sharp $90^o$ turn into the corridor. The corridor-seeking behaviour is overridden by a trail-following component if the trail points away from the opening.

If the robot is heading in approximately the opposite direction to the trail it makes a sharp $180^o$ right turn. If there is no reason to make sharp turns the behaviour defaults to wall following.

The wall following behaviour is implemented using a sliding box algorithm. A virtual box a little bigger than the robot is placed to the left in the laser scan (see Figure 3). The box is moved left to right in front of the robot until the laser scan shows no obstacles within the box. The robot moves forwards at a constant speed, turning towards the center of this box . If there are no obstacles in the first box that is placed it is assumed that the robot is in a big open space and therefore moves forward.

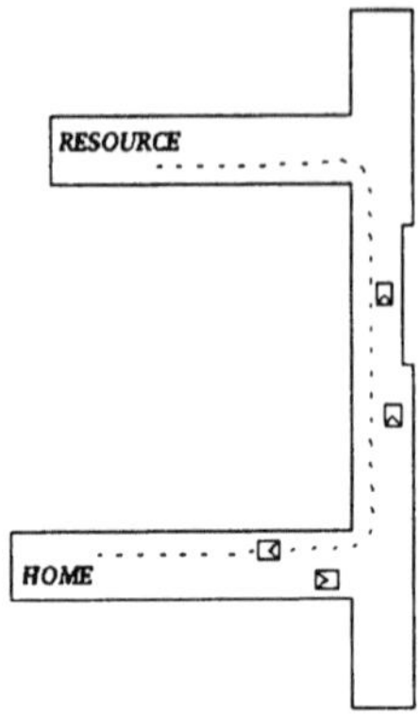


**Fig. 4.** Diagram of the experimental environment; part of our office building. The dashed line is an example crumb trail. Robots (small rectangles ) are drawn to indicate scale.

The advantage of this approach compared with approaches that try to keep a fixed distance to the wall is that the size of the box can be tuned so that the robot keeps minimum distance to the wall but still has time to move away from the wall to avoid obstacles. If no open space is found the control system turns the robot in the opposite direction of the nearest obstacle.

This navigation controller is carefully designed to make the most of the limited space available in our environment. Navigate will keep control of the robot and follow the trail unless the emergency stop mechanism detects an obstacle critically close to the front of the robot. These behaviours make it possible for the robots to perform their task in small rooms and narrow corridors, passing within as little as 40cm of each other before interference occurs.

## 5 Experiment

The system was tested in our office building, with the resource separated from the home position by two corners and approximately 33m of corridors. To follow the optimal route to the resource, the robots must make two correct turning decisions based on the hint govern by the crumb trail. The layout is shown in Figure 4 and is similar to that used in our recent simulation experiments [14].

The robots are started one at a time from the same spot at the home position, with the same orientation. This position is taken as the the origin of their odometric localization space. The controller is run and the robot starts to explore the environment. When a robot comes within 1m of the the home or resource position the experimenter notifies it by sending a message over the network. The trial runs for 30 minutes and the the robots are fully autonomous for the whole trial.

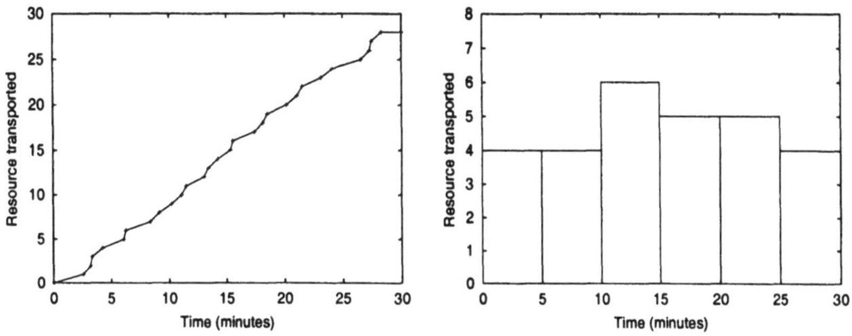


**Fig. 5.** Total units transported (left) and frequency of transportation (right).

Every time a robot has completed a home-resource-home round trip (and therefore notionally transported a unit of resource) the time and the name of the robot is noted. Additionally the crumb trail broadcast by each robot is logged.

### 5.1 Results

Ant was the first robot to reach the resource, at 1.25 minutes. It broadcasts the first crumb trail to its colleagues. Tanis and Bee were already heading towards the resource, but Punky was heading in the wrong direction, back towards the home position. It was observed to immediately turn $180^o$ and follow the wall towards the resource. Ant turned around at the resource position and followed its trail back to the home position, correctly turning right at both corners. The other three robots also turned correctly and reached the resource by the optimal route.

All four robots continued to shuttle between home and resource for the next few minutes, until Punky and Bee came too close together as they negotiated the same corner in opposite directions. Their controllers went into *Panic* mode and they moved into open space, leaving the trail. Both then switched back into *Navigate* and quickly recovered the trail and followed it in the correct direction. Such near-collisions and recoveries occurred 5 times during the trial.

Until 20 minutes all robots made consistently correct turning decisions, but after 20 minutes they would occasionally make an incorrect choice, indicating that their localization estimates were diverging. After making one or two incorrect turns, the robots would apparently recover and follow the trail again. It may be that the robot wandered by chance into a region where the trail was better-formed than elsewhere, but we have insufficient data to analyse this at the time of writing.

Figure 5(left) shows the units of resource transported over time. The slope of the curve levels out at around 27 minutes; no more units were transported after this time. The robots seemed to suddenly lose any coherence in their

**Fig. 6.** The trails announced by each robot over the 30 minutes trial, illustrating the the accumulating odometry errors and the large variation between robots.

movements. They appeared to turn at random and ended up jammed together in one spot until the end of the trial at 30 minutes.

The amount of resource transported by each robot was:

| Tanis | Ant | Bee | Punky | Total |
|---|---|---|---|---|
| 7 | 7 | 6 | 8 | 28 |

Since the time for a round-trip from home to resource is approximately 3 minutes the robot group performed three times better than a single robot performing under optimal conditions.

Figure 5(right) shows how the frequency of transported resource changes over time. It can be seen that the amount of resource transported per time unit decreases over time due to the increasing odometry error. The unbounded odometry error causes the system to break down completely at around 27 minutes.

In figure 6 the trails broadcasted by each robot is shown. It can be seen that the odometry of Ant and Bee is worse than the odometry of Tanis and Punky whose odometry only rotates slightly during the entire trial.

Despite the bad odometry of Ant and Bee these robots still perform well, indeed better than the authors expected. This is due to the robustness of the algorithm with respect to rotation in the environment; the average crumb heading does not have to point the robot in exactly the correct direction. When the robot detects an opening it turns into the opening if the average heading of the crumbs points to the side where the opening is detected. This means that the robot will make turns correctly as long as the average crumb heading lies within $90^o$ of the direction of the opening. This provides a very high tolerance to localization error.

The robots also get a little help from the particular environment we used. The correct turn is the same at each of the two corners, so if the localization error is such that the two corners are confused, perhaps by a $90^o$ rotation, the 'correct' turn is taken. Future experiments will be design to avoid this advantage.

## 6 Conclusion and further work

We have shown that our insect-inspired trail laying algorithm is effective for a real robot team. Sharing landmarks in localization space instead of the real world allows us to achieve some of the benefits of stigmergic behaviour while avoiding the difficulties of laying and detecting physical trail-markers.

The trail-laying algorithm depends on the robots having a common localization space, but can tolerate a significant amount of error before it fails [1]. The system in this paper eventually breaks down due to the unbounded accumulation of odometry errors. The working life of the system could be extended by reducing the rate of drift, or if the localization error is kept within some relatively high bounds it will work indefinitely.

Methods for slowing odometric drift have been described by [15], and bounding drift by [16]. We aim to enhance our localization in the near future, likely using real-world landmarks to maintain the commonality of the robots' localization spaces.

Meanwhile we are investigating some immediate extensions of this work, such as moving home and resource positions, multiple sources of resource, and restricting the communication range of the robots.

The system's performance was degraded by near-collisions between robots. The phenomenon of interference is an important problem for multi-robot systems, particularly in constrained spaces such as home and office environments. In these experiments we deliberately worked only in the corridors, where robots could always pass each other. We are extending our controllers to cope with co-localization conflicts, such as two robots passing in opposite directions through a narrow doorway. Results of our initial simulation work have been submitted elsewhere [14] and a similar approach will be implemented on the real robots in the near future.

## Acknowledgements

Thanks to Brian Gerkey for endless and uncomplaining help with the Pioneer robots and Barry B. Werger for discovering a disconnected serial cable inside a robot. This work is supported by DARPA grant DABT63-99-1-0015, NSF grant ANI-9979457 and DARPA contract DAAE07-98-C-L028.

## References


1. Richard T. Vaughan, Kasper Stoy, Gaurav S. Sukhatme, and Maja J. Mataric. Whistling in the dark: cooperative trail following in uncertain localization space. In *Proc. Fourth Int. Conf. Autonomous Agents*, to appear 2000.
2. B. Holldopler and E.O. Wilson. *The Ants.* Springer-Verlag, 1990.
3. O.W. Richards. *The Social Insects*, page 113. Harper Torchbook, 1970.
4. M. Dorigo, V. Maniezzo, and A. Colorni. The ant system: Optimization by a colony of cooperating agents. *IEEE Transactions on Systems, Man, and Cybernetics-Part B*, pages 26(1):29–41, 1996.
5. J. L. Deneubourg, S. Aron, S. Goss, J. Pasteels, and G. Duerinck. The dynamics of collective sorting: Robot-like ants and ant-like robots. In *Proc. 1st Int. Conf. Simulation of Adaptive Behaviour*, pages 356–363, 1990.
6. Ralph Beckers, Owen E. Holland, and Jean Louis Deneubourg. From local actions to global tasks: Stigmergy and collective robotics. In *Proc. Artificial Life IV.* MIT Press, 1994.
7. Simon Goss, Jean Louis Deneubourg, R. Beckers, and J. Henrotte. Recipes for collective movement. In *Proc. European Conference on Artificial Life*, pages 400–410, 1993.
8. Jean L. Deneubourg, G. Theraulaz, and R. Beckers. Swarm-made architectures. In F. Varela and P. Bourgine, editors, *Proc. European Conference on Artificial Life*, pages 123–133. MIT Press, 1992.
9. Titus Sharpe and Barbara Webb. Simulated and situated models of chemical trail following in ants. In *Proc. 5th Int. Conf. Simulation of Adaptive Behavior*, pages 195–204, 1998.
10. R.A. Russell. Ant trails - an example for robots to follow? In *Proc. 1999 IEEE International Conference on Robotics and Automation*, pages 2698–2703, 1999.
11. C. Melhuish, O. Holland, and S. Hoddell. Collective sorting and segregation in robots with minimal sensing. In *Proc. 5th Int. Conf. Simulation of Adaptive Behaviour*, 1998.
12. K. von Frisch. *Bees. Their Vision, Chemical Senses, and Language.* Cornell University Press, Ithaca N.Y., 1950.
13. C. Melhuish, O. Holland, and S. Hoddell. Using chorusing for the formation of travelling groups of minimal agents. In *Proc. 5th Int. Conf. Intelligent Autonomous Systems*, 1998.
14. Richard T. Vaughan, Kasper Stoy, Gaurav S. Sukhatme, and Maja J. Mataric. Go ahead, make my day: Robot conflict resolution by aggressive competition. In *Proc. 6th Int. Conf. Simulation of Adaptive Behaviour*, 2000 (submitted).
15. S.I. Roumeliotis, G.S. Sukhatme, and G.A. Bekey. Smoother based 3-d attitude estimation for mobile robot localization. In *Proc. 1999 IEEE International Conference in Robotics and Automation*, May 1999.
16. Sebastian Thrun. Learning maps for indoor mobile robopt navigation. *Artificial Intelligence*, 1:21 to 71, 1999.

# Artificial Emotion and Social Robotics

François Michaud[1], Paolo Pirjanian[2], Jonathan Audet[1], and Dominic Létourneau[1]

[1] Université de Sherbrooke, LABORIUS - Mobile Robotics and Intelligent Systems Laboratory, Sherbrooke, Sherbrooke (Québec Canada) J1K 2R1
[2] Mechanical and Robotic Technologies Group, Jet Propulsion Laboratory, Pasadena CA 91109-8099, USA

**Abstract.** Emotion is a concept increasingly used in designing autonomous robotic agents, especially for allowing the satisfaction of basic conditions crucial to the robot survival, for responding emotionally to situations experienced in the world and to humans interacting with it. But psychological evidence also suggests that emotion plays an important role in managing social interactions in groups. This document is a position paper explaining why emotion can benefit the field of distributed autonomous robotic systems. It also outlines research issues that need to be addressed to validate the usefulness of the concept of artificial emotion in social robotics.
**Keywords.** Decentralized Autonomous Systems, Group Behaviors, Cooperation of Multiple Robots.

## 1 Introduction

The idea of using emotion in the design of autonomous agents is not new, and the last three years have shown a great increase in the acceptability and usability of the concept of emotion, especially in designing software agents. A small number of researchers are also using ideas related to emotion in designing robotic agents. The concept of *Artificial Emotion* is principally used in making robots respond emotionally to situations experienced in the world or to interactions with humans. One goal is to help establish the believability that the robot is actually interacting purposefully with the user and not simply executing a program. Another goal is to design computational models of emotion, derived from psychological studies or neuroscience research. How emotion can influence the learning process of an autonomous agent is also considered.

Theories in psychology indicate that emotion plays an important role in such abilities, and also point to other uses that could greatly benefit the field of distributed autonomous robotic systems. The purpose of this paper is to present new research directions involving the concept of artificial emotion in social robotics. The paper starts by reviewing research done in 'emotional' robotics, followed by summarizing indications taken from psychological studies concerning the roles of emotion. It then explains how artificial emotion can be used in social robotics, and outlines research issues that need to be addressed to validate the usefulness of the concept.

## 2 Emotional Robotics

The concept of emotion has just recently started to be used in mobile robotic research, mostly by emotive expression in the behavior of the robot for human-machine interactions. For instance, Velasquez's work [24] is oriented toward computational neural models of emotions necessary for satisfying basic conditions crucial to survival (e.g., escape from danger, obtain food and shelter), and integrate them in complete systems involving drives, perception, behavior and motor control. For emotions, Velásquez identified and created explicit models for six different emotion families: *Anger, Fear, Distress/Sadness, Enjoyment/Happiness, Disgust, Surprise.* The basic computational unit is made of *Releasers* which filter sensory data and identify special conditions before being sent to a neuron. Releasers can be innate (called *Natural*, creating primary emotions) or learned (from predicting the occurrence of natural releasers and used in generating secondary emotions). Each releaser has a short-term memory and can habituate to stimuli using a model of rate-sensitive habituation process. The activation of an emotional neuron, in addition to being influenced by releasers, considers temporal decay and direct influences from other neurons. Emotional learning is done using a modified Hebbian rule. In his model, mood are explained as low levels of arousal of emotion systems, and temperament are modeled through the different values of parameters (thresholds, gains and decay rate) of emotions. His work is used to control Yuppy, an emotional pet robot of the MIT AI Lab. The robot has nineteen self-interested behaviors directed in most part toward satisfying its needs and interacting with humans. It is programmed to respond emotionally to situations experienced in the world and to human interacting with it.

Another example is Kismet [1,2], designed to interact socially with human "parents". Kismet must generate meaningful interactions with the caretaker, regulating these interactions to maintain an environment suitable for the learning process, and assisting the caretaker in satisfying the robot's drive. The work focusses on the role motivations and facial expressions play in maintaining an appropriate level of stimulation during social interaction. The robot responds with expressive displays which reflects an ever-changing motivational state and which give the human cues on how to satisfy the robot's drive while neither over-whelming nor under-stimulating the robot. These emotive expressions then serve as communicative acts. The framework proposed is made of a perception system, a motivational system, an attention system, a behavior system and a motor system. An emotion subsystem is part of the motivational system. A given emotion's level of activation (ranging between [0, max] and determined by a threshold) is influenced by drives, pain and other emotions. Each drive is partitioned into three regimes (homeostatic - the operation setpoint of the drive, over-whelmed or under-whelmed) and for a given drive, each regime potentiates a different emotion and hence a different facial expression. Influences from other emotions serve to prevent conflicting emotions from becoming active at the same time, using mutually inhibitory connections. The experiments done with the Kismet robotic face involved having up to four drives (fatigue, social, security and stimulation), three consummatory behaviors (sleep, socialize and play), two visually-based percepts (face and non-face), five emotions (anger, disgust, fear, happiness and sadness) and up to ten expressive states. Fatigue and sleep create a self-regulation method used for restoring all the drives, to let the robot have some time to regulate its interaction with the external world (especially when they are inappropriate), to consolidate its learned anticipatory models and integrate them with the rest of the internal control structure. The security drive is to determine when learned anticipatory models of the effects of its actions on the world are adequate.

Emotional robotics is also concerned with learning, but only in simulation yet:

- Gadanho and Hallam [8] experimented with an emotion-based evaluation of the context to see if it can be adequately used as a reinforcement signal for policy acquisition by Q-learning. Their conclusion is that emotions do not provide a good evaluation of what is going on at any one moment, but are a sort of mixed evaluation the robot acquired from its past experiences. Emotions are more suited to modulate learning parameters and the exploration versus exploitation ratio. They also studied whether emotions can successfully fulfill the rôle of determining state transition in order to decide when to adapt a controller in a continuous learning experiment [9]. It resulted in drastic cuts in the number of triggerings of the learning controller while maintaining overall performance.
- Foliot and Michel [5] present an approach based on the assumption that emotion can be seen as the basis of cognition because it provides a default functional model. Their goal is to show how emotion based structures could contribute to the emergence of cognition by creating suitable learning conditions.

In regards to social robotics, the only reference found involving emotion is about an approach that uses a frustration variable to change the behavior-selection strategy, making the robots specialize in doing one of two tasks [22]. However, the approach was only validated in simulation.

## 3 Psychological Foundations Concerning the Roles of Emotion

Psychological research on emotion is focused primarily on three aspects: biology of emotion, emotion and interpersonal communication, and emotional development [23]. From an engineering point of view, the second aspect is more interesting. Our research suggests that emotion can serve three important roles in designing autonomous robots:

- *Emotion to Adapt to Limitations.* Emotion plays a role in determining control precedence between different behavior modes, coordinating plans and multiple goals to adapt to the contingencies of the world (under constraints of time and other limited resources), especially in imperfectly predictable environments [7,17,20]. Uncertainty prevents any complete dependence on predictive models in human planning – people typically think only a step or two ahead and they respond, moment-by-moment, to the new arrangements of the environment that their actions help to create [17]. The adaptation problems that emotion help solve are finding an equilibrium between the subject's concerns and the environment by signalling the occurrence of concern-relevant events, by recognizing the plan junctures that these events imply, and by instigating control shifts accordingly [7].
- *Emotion for Managing Social Behavior.* In relation to social behavior, Plutchik [20] interestingly points out that emotions are in direct association with the **universal problems of adaptation**, which are:
  - **Hierarchy**: dominance hierarchies are connected primarily with certain types of basic emotions, namely anger and fear. They are also connected to the personality derivatives of these emotions, which are dominance and submission.
  - **Territoriality**: the basic emotions related to territoriality are exploration and its opposite, surprise (control or dyscontrol). They are developed through exploration of the environment.
  - **Identity**: this is related to questions like who we are, what group we belong to. It is used for cooperative hunting, group defense, social signaling, social communication, built in through genetic coding mechanisms to recognize other organisms (size, shape, color, markings, sound patterns, chemical and olfactory cues). Emotions associated with identity are acceptance and rejection.
  - **Temporality**: this allows to take into consideration the limited duration of an individual's life. Distress signals for social support, nurturing responses in other members of the social group are also considered. Sadness, distress and joy contributes in solving this adaptation problem.

  These universal problems of adaptation have important implications. They provide a general way of looking at life problems, and a set of implicit dimensions that one can use to assess the environmental and/or social demands on a given species. For any group of agents, one can ask: How are problems of hierarchy, territoriality, identity and temporality expressed for that group? What kind of adaptations has the group made for dealing with each of these problems? Another implication is that they provide "another kind of rationale for the existence of certain emotions, since emotions are reflections of the adaptations that animals make to the universal problems. Since these problems are universal, the emotions that are derived from them may be thought of as universal, basic, or primary"

[20]. Plutchik's theory also suggests the possibility that emotions "are functional adaptations for establishing a kind of social equilibrium. This would imply that emotions enter into every social transaction and help to establish a balance of opposing forces. These balances are always temporary and frequently change as we move through life from one conflict to another" [20]. This is also suggested by Oatley and Johnson-Laird [17], especially in the context of creating mutual plans: "Mutual plans cannot be innately wired into the cognitive systems; they must be created in the minds of more than one individual by implicit or explicit agreement." These plans are partly under the control of both participants and partly governed by conventions of their society. One way to set up a mutual plan is to make a promise to somebody, creating an obligation and a corresponding expectation. Emotions are then used to communicate junctures in mutual plans among individuals in social groups.

- *Emotion for Interpersonal Communication.* In order for emotions to regulate behavior in social interaction, emotion also has a communicative role, as suggested in the previous paragraph. Emotion plays important functions in social signalling. Ethologists believe that emotional expression have a communicative function and act as releasers for the coordination of social behavior. There are signals that promote group cohesion, signals to communicate about external environment, and intraspecific threat signals. It is to an animal's advantage to communication its intentions, and to be sensitive to such messages from others [15]. For animals, Darwin believes that joy and anger occupy opposite ends of a single dimension of sociability, while fear would lie on a different dimension [12]. Display, the expression of emotion, has evolved as a mean of communication, generating signals to indicate how they would react to a social encounter. Emotional expression promotes individual isolation (as it may be necessary in defending something) or to promote group (as different social circumstances might require). In fact, the role of expression in emotion can be seen from three different views: the situation is evaluated by emotion that lead to an expression; expression may be a reaction to the situation that also produces the emotion; the expression may affect the emotion rather than the other way around [15]. Emotion then serves a dual purpose: it is a communication act and it is a sensed state.

The concept of emotional communication also helps explain why the great variety of human emotions depends so much upon the social context: it is because of the richness and subtlety of human social relationships [12,17]. While humans have emotions to fit virtually any kinds of social situations, animals probably only have emotions to deal with certain kinds of survival problems, and for which there are some strong adaptive pressure [12]. Complex emotion then evolved for organisms living in society and having higher cognitive ability and a flexible behavior repertoire [21]. This leads us to believe that the concept of artificial emotion is required in order to design autonomous robots showing higher level of intelligence.

## 4 Artificial Emotion for Social Robotics

As described in Section 2, the use of emotion for interpersonal communication is the primary focus of research involving artificial emotion. The role of emotion for adapting to limitations is partly addressed, and in spite of the psychological evidence, no research has directly addressed the use of emotions to regulate social behavior in a group of robots. The ALLIANCE architecture [18], without explicitly using the term, is the one that comes closest to the notion of artificial emotion to regulate group behavior. In a foraging task, the approach allows robots to broadcast information about their goals to affect their motivations. *Impatience* and *Acquiescence* are internal states used to coordinate the robots. By using internal states, communication and learning, ALLIANCE exploits some of the characteristics that can be associated with artificial emotion. However, we still have to make significant progress in order to design autonomous robots capable of working in real life settings (like our homes, offices, market places, etc.). Dynamic and unpredictable conditions occur constantly in everyday situations, and a robot has to deal with them with limited energy, perception, action and processing capabilities. Also, a robot operating in such conditions requires the ability to interact with different individuals (robots, humans or other types of agents). Based on Section 3, more emphasis should be put on the concept of artificial emotion in order to achieve these goals, which are related to social robotics. Here are the characteristics associated with social robotics, taken from [4], with some comments on how emotion can be associated with them:

1. *Agents are embodied.* This characteristic includes robots, humans and animals. Emotions are something that is shared (at various levels) by humans and animals, and it can be interesting to add them to robots.
2. *Agents are individuals, part of a heterogeneous group (the members are not identical and have individual features, like different sensors, different shapes and mechanics, etc.).* The number of works involving this aspect is growing [4,13,16,19]. The role of emotion in adapting to limitations of the robot (in operating and in modeling the environment) can certainly help manage the heterogeneity in a group.
3. *Agents can recognize and interact with each other and engage in social interactions as a prerequisite to developing social relationships.* "The ability to distinguish the agents with whom one is interacting from everything else in the environment is a necessary condition for intelligent interaction and group behavior" [11]. This has up to now been mostly done using IR, explicit radio communication of the positions of the robot obtained from a positioning system, and vision [3].

4. *Agents have 'histories'; they perceive and interpret the world in terms of their own experiences.* This is associated with specialization. An artificial emotion can have a level of activation affected by different factors over time, representing this way the history of past experiences.
5. *Agents can explicitly communicate with each other. Communication is grounded in imitation and interactions between agents, and meaning is transferred between two agents by sharing the same context.* Current approaches involving interaction via communication are mostly oriented toward implementing electronic media access protocol [3]. Artificial emotion can serve as a rich abstraction of the current state and goals of a robot, a kind of implicit model of its intentions. This constitutes a simple way of establishing a shared meaning and create a basis for communication, without having to transmit large amount of data.
6. *The individual agent contributes to the dynamics of the whole group (society) as well as the society contributing to the individual.* Redundancy, fault tolerance and cooperation are then important.

Also related to social robotics is the notion of cooperative robotics [3], which involves three fundamental seeds: the mechanism of cooperation, the task and system performance. Regarding the mechanism of cooperation, Cao et al. [3] indicate that the realization of cooperative behavior must rely on group architecture (homogeneity/heterogeneity, ability to recognize and model others, and communication structure), resource conflicts (to share the environment, manipulate objects and communication), how cooperative behavior is motivated and achieved, and learning. As indicated for social robotics, these issues are also considered in the role of emotions.

## 5 Research Issues

The arguments presented in the previous sections point toward the conclusion that the concept of artificial emotion could lead to significant contributions to the field of social robotics, much more than what has been done until now. However, no solutions are provided to clearly validate the use of such a concept in the design of autonomous robots. The following issues need to be addressed:

- What computational mechanisms are necessary for implementation of artificial emotions for social robotics? Will the approach be based on a theory of emotion, or will it be derived empirically? On what grounds can we state that the approach is actually related to the concept of artificial emotion? If we try to summarize what we found in our research, we would need the following characteristics: 1) emotions are motives – they are causes of action; 2) cognition can trigger emotions – emotions can trigger cognitive operations; 3) emotions can cause expressive actions; 4) emotions can affect purposive, goal-directed action; 5) emotions can become goals; 6) the behavior also affects emotion [15].

- How can we make robots recognize and interact with each other? This is an important problem to resolve, especially if vision is preferred over positioning systems (which require some engineering of the environment). This issue is also related to the question of how robots will communicate their emotions.
- How can we efficiently evaluate the performance of an approach based on artificial emotions? The task must be chosen so that possible comparisons with approaches not using artificial emotions can be done. It must also make the robots adapt to their limitations which can be related to their sensing and acting abilities, to their heterogeneity, to the requirement of making local decision that will affect the group (and contrarily to approach with a centralized decision mechanism), and to shared resources. For instance, the medium size RoboCup competition would make an interesting experimental setup.

## 6 Conclusion

Artificial emotions have up to now been mostly used for emotive expression and response in human-robot interactions. However, the concept can be used to do much more, as argued in the present paper. This paper outlines different research perspectives in social robotics that can be addressed using the concept of artificial emotions. Our goal is to encourage research in these directions in order to generate new contributions to the field of distributed autonomous robotic systems.

**Fig. 1.** Group of Pioneer 2 mobile robots.

For our part, based on the guidelines indicated in this paper, we are in the process of preparing experiments involving the use of artificial emotions. What we plan to do is to use our group of robots shown in Figure 1 in a foraging task. The group of robots is made of six Pioneer 2 robots, three indoor and three outdoor models. Each robot is equipped with 16 sonars, a compass, a gripper, a camera, an ethernet-modem connection and an onboard computer. Note that the gripper of the outdoor model is not able to pick up small objects directly put on the floor, and that the indoor model has more difficulty moving over small objects. The camera of one robot may also be replaced by stereo vision, to increase the heterogeneity of the group. The programming environment is Ayllu, a tool for development of behavior-based control systems for intelligent mobile robots. Foraging is an interesting task because it has been used in various group robotic research, and comparisons are then possible. Works on territoriality, dominance hierarchy and caste for group strategies [6,10] will be useful. Finding the ideal number of robots for the foraging task [6] by dynamic role selection will also be addressed. For inter-robot recognition we plan to use visual cues, generated for example using a light signaling device that has shown to be an efficient way of explicitly communicating simple information to others, and at the same time be a rich source of implicit information for recognizing and interacting with others [14]. We also plan to use a charging station as a shared resource for the group, to study how artificial emotions could help manage 'survival' of the individuals in long-lasting experiments.

**Acknowledgments**

Research conducted by LABORIUS is supported financially by the Natural Sciences and Engineering Research Council of Canada (NSERC), the Canadian Foundation for Innovation (CFI) and the Fonds pour la Formation de Chercheurs et l'Aide à la Recherche (FCAR) of Québec. Special thanks to Maja J Matarić for her collaboration on this research.

## References

1. Breazeal C. (1998) Infant-Like Social Interactions Between a Robot and a Human Caretaker. In: Adaptive Behavior, Special Issue on Simulation Models of Social Agents
2. Breazeal C. (1998) A Motivational System for Regulating Human-Robot Interaction. In: Proc. of AAAI, Madison WI
3. Cao Y. U., Fukunaga A. S., Kahng A. B. (1997) Cooperative Mobile Robotics: Antecedents and Directions. Autonomous Robots, 4:1–23
4. Dautenhahn K. (1999) Embodiment and Interaction in Socially Intelligent Life-Like Agents. In: Computation for Metaphors, Analogy and Agent. Lecture Notes in Artificial Intelligence, Springer-Verlag, 1562: 102–142
5. Foliot G., Michel O. (1998) Learning Object Significance with an Emotion Based Process. In: Workshop on Grounding Emotions in Adaptive Systems, Conf. on Simulation of Adaptive Behavior
6. Fontán M.S., Matarić M. J. (1996). A Study of Territoriality: The Role of Critical Mass in Adaptive Task Division. In: Proc. 4th Int'l Conf. on Simulation of Adaptive Behavior, Cape Cod, September, The MIT Press

7. Frijda N. H. (1987) Comment on Oatley and Johnson-Laird's 'Towards a Cognitive Theory of Emotions'. Cognition and Emotion, 1(1):51–58
8. Gadanho S. C., Hallam J. (1998) Emotion-Driven Learning for Animat Control. In: Proc. Int. Conf. on Simulation of Adaptive Behavior: 354–359
9. Gadanho S. C., Hallam J. (1998) Emotion-Triggered Learning for Autonomous Robots. In: Workshop on Grounding Emotions in Adaptive Systems, Conf. on Simulation of Adaptive Behavior
10. Goldberg D., Matarić M. J. (1997) Interference as a Tool for Designing and Evaluating Multi-Robot Controllers. In: Proc. National Conf. on Artificial Intelligence (AAAI), Providence, RI, July: 637–642
11. Matarić M. J. (1994) Interaction and Intelligent Behavior. MIT AI Lab AI-TR 1495, Massachusetts Institute of Technology, Cambridge, May
12. D. McFarland D. (1982) The Oxford Companion to Animal Behavior, Oxford University Press
13. Michaud F., Matarić M. J. (1998) Learning from History for Behavior-Based Mobile Robots in Non-Stationary Environments. Special Issue on Learning in Autonomous Robots, Machine Learning/Autonomous Robots, 31/5:141–167/335–354
14. Michaud F., Vu M. T. (1999) Managing Robot Autonomy and Interactivity Using Motives and Visual Communication. In: Proc. Conf. Autonomous Agents:160–167
15. Mook D. G. (1987) Motivation. The Organization of Action, W. W. Norton and Company
16. Murphy R. R., Ausmus M., Bugajska M., Ellis T., Johnson T., Kelley N., Kiefer J., Pollock L. (1999) Marsupial-Like Mobile Robot Societies. In: Proc. Conf. on Autonomous Agents:364–365
17. Oatley K., Johnson-Laird P. N. (1987) Towards a Cognitive Theory of Emotions. Cognition and Emotion, 1(1):29–50
18. Parker L. E. (1998) Alliance: An Architecture for Fault Tolerant Multirobot Cooperation. IEEE Transactions on Robotics and Automation, 14(2):220–240
19. Parker L. E. (1999) Adaptive Heterogeneous Multi-Robot Teams. Neurocomputing
20. Plutchik R. (1980) A General Psychoevolutionary Theory of Emotion. In: Emotion: Theory, Research, and Experience, Academic Press 1(1):3–33
21. Rimé B., Scherer K. R. (1989) Les Émotions. Delachaux et Niestlé, Paris
22. Shibata T. , Ohkawa K., Tanie K. (1996) Spontaneous Behavior of Robots for Cooperation - Emotionally Intelligent Robot System. In: Proc. IEEE Int. Conf. on Robotics and Automation, Minneapolis, Minnesota:2426–2431
23. Strongman K. T. (1987) The Psychology of Emotion, John Wiley & Sons, Third edition
24. Velásquez J. D. (1998) A Computational Framework for Emotion-Based Control. In: Workshop on Grounding Emotions in Adaptive Systems, Conf. on Simulation of Adaptive Behavior
25. Velásquez J. D., Fujita M., Kitano H. (1998) An Open Architecture for Emotion and Behavior Control of Autonomous Agents. In Proc. Conf. on Autonomous Agents:473–474

# The Use of Biologically-Inspired Rules to Control a Collective Robotic Arm

Gene G. Korienek[1]*, Abigail B. Bautista[2], Tyson H. Harty[1], and Charles Leddon[1]

[1] Biological Control Laboratory, Oregon State University, Corvallis, OR 97331 USA
[2] Perceptual and Behavioral Adaptation Lab, NASA Ames Research Center, Moffett Field,
CA 94035 USA
* Corresponding Author: gene.korienek@orst.edu

**Abstract.** Robotic arms are generally not adaptable in unknown, dynamic environments. This research tested whether a robotic arm using a collective control architecture could exhibit adaptive behavior without centralized control. To simulate the physical properties of a collective robotic arm, we used a puppet arm with six identical elbow-shaped segments, each of which rotated orthogonally with respect to its adjacent neighbors. A control strategy of sensor-driven, distal leadership was applied to the collection of autonomous segments, each encapsulated with identical movement rules and without any central control or explicit communication. A goal of pointing to a target was chosen to assess the possible emergent behaviors of the system. The results indicated that the simple movement rules in a collective control architecture lead to coordinated pointing behavior.

**Keywords.** autonomous, collective, robotics, adaptive, emergence

## 1 Introduction

Managing movement and manipulation within a changing environment requires a form of control that exhibits a large degree of adaptability and represents a significant obstacle to designers of robotic arm control systems. Human arms tend to be very successful in this regard while robotic and prosthetic arms have yet to demonstrate a similar degree of flexibility. The issue appears to be the management of complexity, that is, successfully responding to changes in the environment during the course of completing a specified task.

In this investigation, we executed a control rule observed in human arms in a collective, highly-segmented, non-actuated puppet arm. This puppet was more productive than a software simulation in that the physical movement promoted an iterative control rule refinement process.

## 2 Biologically Inspired Control

A historical search of the human motor control literature reveals over 100 years of experimentation in the psychological [4] and biological [2] sciences addressing the issues of mammalian movement control. Sherrington [3], in particular, advanced the use of experimental observations of the mammalian and amphibian nervous system to uncover fundamental units of motor control, the basic unit of which he named the "reflex arc" [3]. Sherrington's reflex arc provides the foundation for several basic control constructs: feedback, reciprocal inhibition, reflexive facilitation, servomechanism, and facilitative chaining.

## 3 Robotic Puppet Arm

The puppet arm used in this study consisted of a collection of independent, physically constrained, mechanical segments. Communication between arm segments was limited to knowledge of the relative angles of neighboring segments. The movement decision rules of a given segment were derived from the direction and amplitude of rotation of the adjacent segment. This information was implicitly communicated through the environment, as there was no shared information between segments. Under these conditions, we examined how movement complexity in a 6-degree of freedom robotic arm could be managed using a biological control rule with a fully collective computational architecture.

Our robotic arm consisted of identical elbow joints and each segment had a 90-degree fixed bend, which allowed free rotation around the axes of the two adjacent segments to which it was connected. The "proximal segment" was at one end of the robotic arm attached to a stable-mounting base. The "distal segment" was the segment at the end not attached to the base and housed a pointing laser that facilitated the measurement of pointing performance (Fig. 1).

## 4 Control Rules

In this study, we used a conventional production rule format to model our movement rule, or movement decision. For example, in the human arm, there exists a relationship between the elbow flexors and elbow extensors called reciprocal inhibition [3]. If the elbow is flexing, then the activity of the elbow extensors will be inhibited. Sherrington [3] observed that a particular activation sequence of muscle groups was needed for a mammal (e.g., dog or horse) to move from a lying posture to an upright posture, which he called "facilitative chaining".

In this investigation, we used variations of the facilitative chaining rule to form the basis of control for each segment of a puppet arm. This means that a decision to move about a particular joint was made using the facilitative chaining rule. It also means that each segment of our robotic arm was making its own movement decision using its own facilitative chaining rule. This rule directed all but the distal segment's movements, which instead moved to minimize its target pointing error. A general form of the reflex-chaining movement rule is displayed below:

*IF:* *Adjacent segment rotation is x degrees Clockwise (or Counterclockwise)*
*THEN:* *Rotate Clockwise (or Counterclockwise) x plus (current parameter) degrees*

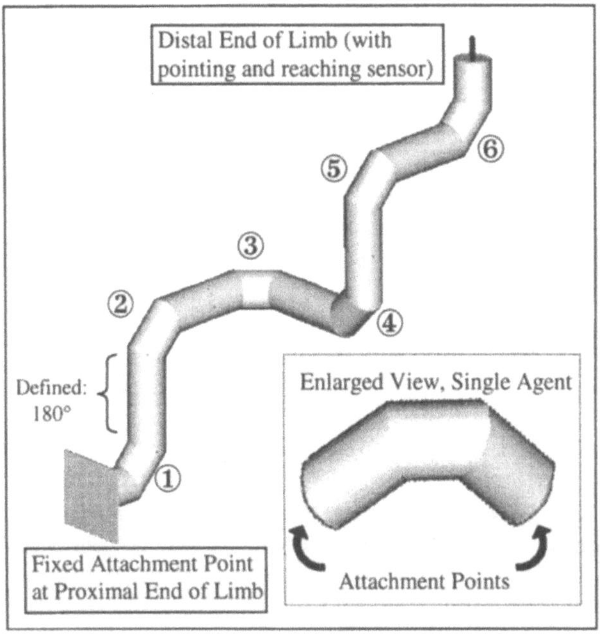


**Fig. 1:** The puppet robotic limb (6 segments) used in this study. Note the angular relationship between segments is based on the definition of 180° as shown.

# 5 Experimental Procedures

The data collection procedure began by manually rotating the distal arm segment to a point as close to the target as possible as indicated by a laser projecting from the end of the distal segment. The segment rotated through a fixed angle in either a clockwise (CW) or counter-clockwise (CC) direction. The values of direction and amplitude from the most recent segment rotation was input into a rule-based simulator and the calculated movement decision was applied to the next segment in the chain traveling from distal to proximal with reference to the fixed end of the arm. Limb movement was then initiated again by the distal segment making another best attempt to point at the target, and the process was repeated 10 times for each experiment. This setup allowed the puppet arm to perform as a physical simulation of an actuated robotic arm responding to the feedback from performance measurements made in three-dimensional space by a human experimenter (Fig. 2).

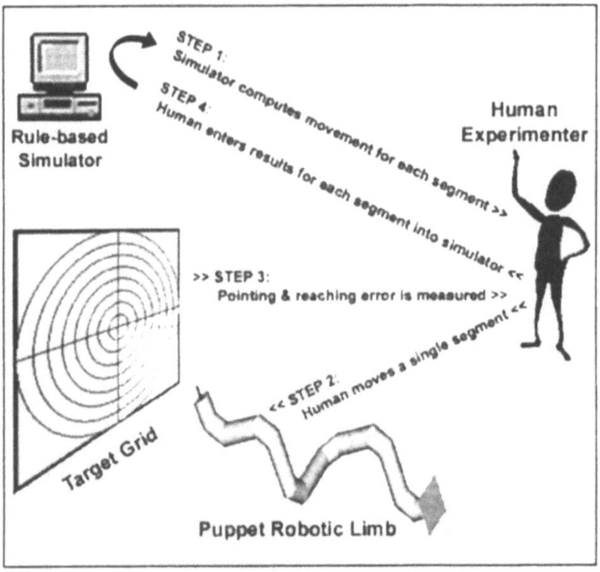


**Fig. 2.** Schematic of experimental procedures, showing the interaction between the human experimenter, rule-based computer simulator, and the puppet robot.

Three experiments were developed to test the facilitative chaining rules and identify factors that may interact with this robotic architecture. The goal of the rules was to optimize the distal segment's laser point to the target's location and was constant across all three experiments. The independent variables for each experiment were the two initial postures of the robotic arm (IP-1 and IP-2) and the particular movement rule being tested. IP-1 began with each segment fixed at 135 degrees relative to one another and IP-2 used segments initially at 90 degrees with respect to one another, where 180 degrees was defined as shown in Fig. 1.

The dependent measures chosen were anticipated to illustrate the development of an emergent pointing behavior. When mapped onto a two-dimensional Cartesian coordinate system, the constant error (CE) measures over successive trials displayed a path interpretable as searching behavior. From the absolute error (AE) measure, we were able to evaluate the overall effectiveness of the emergent pointing behavior. Thus by comparing AE for each movement rule in each condition, the pointing performance of the robotic arm could be directly assessed over the variety of applied rules. Additionally, a "distance-to-target" measure was used to determine if the rules produced preliminary emergent reaching behavior.

## 5.1 Experiment One: Ascending Facilitative Chaining

Experiment One compared the effectiveness of three different "ascending facilitative chaining rules" and determined their interactions with the two different initial postures (IP-1 and IP-2). The ascending movement rule for each segment prescribed the segment move the same amplitude plus an additional 2, 5, or 10 degrees in the same direction as its neighboring segment distal to it:

| | |
|---|---|
| *IF:* | *Adjacent segment rotation is x degrees Clockwise (or Counterclockwise)* |
| *THEN:* | *Rotate Clockwise (or Counterclockwise) x plus (2 or 5 or 10) degrees* |

As in all the experiments, the distal segment initiated a movement in an attempt to accurately point to the target, and the movement propagated via sensed environmental information only through inter-segment chaining to the proximal segment. This 2 (initial posture) X 3 (movement rule) experimental design was tested over a block of ten trials per condition.

## 5.2 Experiment Two: Descending Facilitative Chaining

While Experiment One used a movement rule that amplified the robotic arm's more proximal segments, Experiment Two tested descending facilitative chaining movement rules that concentrated the arm's movement capabilities in its distal end. The intermittent segments were again controlled to move the same distance as the previous segment but decreased by specified amplitudes:

| | |
|---|---|
| *IF:* | *Adjacent segment rotation is x degrees Clockwise (or Counterclockwise)* |
| *THEN:* | *Rotate Clockwise (or Counterclockwise ) x plus (-2 or -5) degrees* |

This created a 2 (initial posture: IP-1 and IP-2) X 2 (movement rule: previous -2, previous -5) experimental design tested over a block of ten trials per condition.

### 5.3 Experiment Three: Descending Facilitative Chaining with Power Rule

The delegation of power within human movement models has long been an area of research within motor control. Recent research has uncovered what appears to be a general power rule observed in human movement [1]. This work postulated the existence of a two-thirds power rule for angular velocity in human movement. To further extend the investigation of descending strategy, a movement rule that took into account the magnitude of the previous movement was employed using a two-thirds power format:

| | |
|---|---|
| *IF:* | *Adjacent segment rotation is x degrees Clockwise (or Counterclockwise)* |
| *THEN:* | *Rotate Clockwise (or Counterclockwise) x * 0.66 degrees* |

As in the two previous experiments two initial postures were examined; however, this experiment included only one descending movement rule across the two initial postures, creating a 1 (descending two-thirds movement rule) X 2 (IP-1 and IP-2) experimental design.

# 6 Results

### 6.1 Results for Experiment One: Ascending Facilitative Chaining

The ascending facilitative chaining rules of Experiment One generally produced broad searching patterns that were counterproductive for a pointing task due to an increasing movement amplitude, which propagated from the distal segment to the proximal end and hampered target attainment by continually driving the arm's position beyond beneficial limits. For this collective arrangement of arm segments to complete the cooperative task of target pointing for the arm as a whole, the behaviors of collective members should assist the distal segment in attaining its goal, and the descending rules failed in this respect.

Specifically, the +10 rule failed to show any systematic target-acquiring behavior, but the +2 and +5 ascending facilitative chaining rules showed preliminary target-acquiring behavior though were sensitive to the initial starting posture of the arm. In both initial postures, the arm displayed quadrant-specific behavior, where the distal segment pointed to locations in the quadrant in which it was initially oriented. Not surprisingly, the ascending movement rules failed to produce segmental facilitation for the distal segment in the pointing task. While the initial posture of the arm appeared to affect the arm's range, the ascending movement rules seem to have little effect on the overall range of the target distance.

As the results show, distal leadership paired with ascending rules seems an inappropriate combination for the emergence of pointing behavior in a collective arm. Whereas the ascending rules produced a large amplitude oscillatory searching behavior of the arm, an appropriate solution would attempt to attenuate this pattern with the descending movement rules of Experiment Two.

## 6.2 Results of Experiment Two: Descending Facilitative Chaining

The -2 and -5 descending facilitative chaining rules of Experiment Two proved to be effective in emerging target-acquiring behavior from the collective arm when initiated from the IP-2 posture. In the IP-2 posture, the arm was able to reach a final pointing solution, while in IP-1 it failed to show any systematic target-acquiring behavior, but instead created an oscillatory search pattern in the bottom two quadrants of the target area. The -5 rule was most successful, as the arm effectively immobilized its more proximal segments since the propagation of movement to more proximal segments was increasingly smaller. It is not surprising that this rule was most effective because of the larger subtractive parameter that dampened the oscillatory pattern quickly, focusing the movement solution to the target's location.

The arm's search pattern in IP-1 seemed overly constrained and appeared related to the starting orientation. When the arm started its movement with the distal segment in the bottom left quadrant, the search pattern was constrained to the lower two quadrants (Figure 3). In contrast, in the IP-2 posture, the arm's search pattern was not confined and covered all four quadrants of the target area. The success of these rules in IP-2 contrasted with the lack of success in IP-1 and indicates the important interaction of initial posture and control rules on pointing behavior.

A noteworthy result of using descending rather than ascending rules was the emergence of reaching in concert with the desired pointing behavior. As the robotic arm optimized its pointing position relative to the target, the distal segment's distance to the target was also optimized resulting in a reaching behavior that is preparatory to a grasping behavior for an end-effector. Both movement rules demonstrated an ability to reach for the target, and when combined with performance of the pointing task, the -5 rule proved to be most effective at reaching for the target.

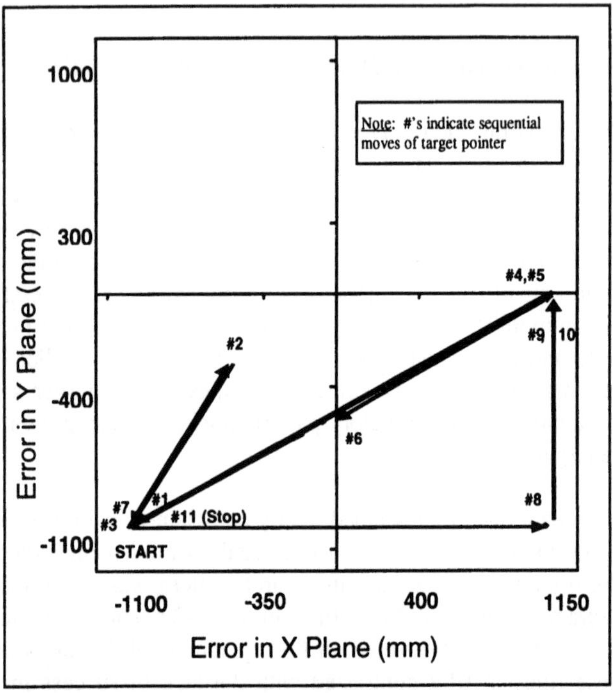


**Fig. 3.** Constant pointing error over trials for Experiment Two, showing constrained searching behavior of the robotic arm.

## 6.3 Results of Experiment Three: Descending Facilitative Chaining with Power Rule

With the descending facilitative chaining rule using a two-thirds power rule, the arm expanded its searching range with each trial until the target was acquired (Fig. 4). Overall, results were biased toward a search pattern where the arm spent time in three quadrants of the target area. As with the previous descending movement rules, the descending search patterns that emerged to facilitate pointing had excellent possibilities for control strategy applications in robotic arms.

The issue of performance, which is paramount in most robotic arms, was important for the two-thirds power rule executed in the IP-2 posture. Although the IP-1 variation produced results that were somewhat oscillatory in nature, the IP-2 trial block demonstrated a rapid decline in Absolute Error until a final solution was reached for the arm. Clearly, as experimental results presented previously would indicate, the IP-2 orientation has afforded broader search patterns that in turn produce more accurate solutions.

Interestingly, the resulting distance to target patterns for IP-2 did not show smoothly decreasing target distance that is a hallmark of human reaching behavior. Rather, the arm moved in a choppy inefficient fashion due to its 90° relationship between segments when performing its pointing task. Although the

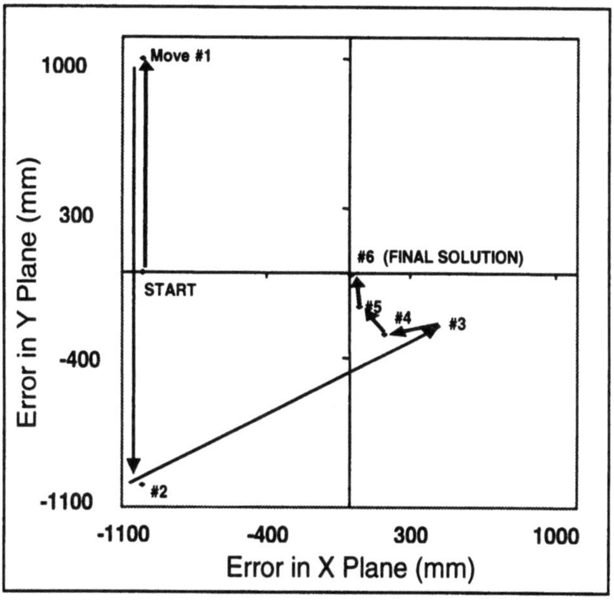


**Fig. 4.** Constant pointing error over trials for Experiment Three, showing a focusing-to-target pointing solution of the robotic arm.

arm produced target distances that decreased in IP-2, the resulting target distance from IP-1 were highly variable.

# 7 Conclusion

In this study, we used a puppet robotic arm comprised of six encapsulated robotic segments in a collective computational architecture to demonstrate the potential usefulness of biologically based movement principles. Our movement rules were derived from those identified by Sir Charles Sherrington [3] as the "Facilitative Chaining" of reflex arcs and included distal leadership to allow goal-directed behavior to self-organize and emerge among the collection of segments.

This control strategy enabled the robotic arm to search a space for a final solution to a pointing problem. The appearance of searching and reaching behaviors could not be directly determined by the simple rules nor from the implemented Facilitative Chaining strategy. Thus, it is apparent that these behaviors *emerged* from the collective interaction of sensors and movement rules of all six segments.

Although distal leadership of the robotic arm limited the order of movement execution, it enforced a goal on the arm's behavior. The laser pointer in the lead segment was allowed to reduce the distance between its projected point and the target. With this limited information, the leader had only Boolean decisions of closer or farther to the target. The problems due to this limitation were apparent when (1) the solutions closer or farther from the target were unavailable due to the arm's posture, causing a condition of paralysis; and (2) the arm came to a posture that obstructed the distal segment as well as the entire collective's movement.

Where ascending rules (Experiment One) only showed some success in searching for the target, application of descending rules (Experiments Two and Three) resulted in complete pointing behavior. As the ascending rules produced a high amplitude oscillatory search pattern in the arm, the descending rules produced a much attenuated search behavior and appropriate target pointing solution. Under descending rules, the segments of the collective moved in cooperation with the distal leadership in acquiring the target.

As pointing behavior emerged, it was important to assess whether our control rules facilitated behaviors (e.g. reaching) other than the one of interest, as such rules are likely to be fundamentally successful and result in generalizable control principles. Future work in this area will expand on this initial study in an attempt to produce robotic arms that can facilitate reaching and grasping. The beneficial applications for this work are self-evident for both humans (prostheses) and machines (robots). Context sensitive pointing, reaching, and grasping movements of the complexity and accuracy that we expect from human and robotic arms may be possible without explicit programming of the movement.

As robots were originally conceived as mechanical beings that simulated human behaviors, recent conceptualizations, though mostly fictional, strive to emulate human behavior in some fashion. At the practical level, the intent is to achieve movement that is as accurate, precise, and adaptable as that witnessed in highly skilled humans. With this goal in mind, it seems logical to base this approach on biological principles of movement while developing and implementing these control strategies.

For this initial series of experiments, we chose what was considered the basic unit of movement at the turn of the century, the reflex arc, to model our robot

[3]. Furthermore, we applied the fundamental principle--the control strategy of facilitative chaining--that was thought to organize these basic elements. As the results display, although not optimally, organized goal-directed behavior can emerge from these simple principles in mechanically as well as biologically based structures. On the other hand, the performance of this mechanical system will never be mistaken for that of the biological system it modeled.

# 8 References


[1] De Sperati, C. & Stucchi, N. (1995) Visual tuning to kinematics of biological motion: the role of eye movements. *Experimental Brain Research*, 105, 254-260.

[2] Sherrington, C. S. (1910) Flexion-reflex of the arm, crossed extension reflex, and reflex stepping and standing. *Journal of Physiology*, 40, 28-121.

[3] Sherrington, C. S. (1947) *The Integrative Activity of the Nervous System.* Yale University Press, New Haven.

[4] Woodworth, R. S. (1899) The accuracy of voluntary movement. Psychological Review, 3 (Suppl. 2).

# Chapter 5

# Reconfigurable Robots

# Micro Self-reconfigurable Robotic System using Shape Memory Alloy

Eiichi Yoshida, Satoshi Murata[1], Shigeru Kokaji[1], Kohji Tomita[1], and Haruhisa Kurokawa[1]

Mechanical Engineering Laboratory, AIST, MITI, 1-2 Namiki, Tsukuba-shi, Ibaraki 305-8564 Japan, eiichi@mel.go.jp

**Abstract.** This paper presents micro self-reconfigurable modular robotic systems using shape memory alloy (SMA). The system is designed so that various shapes can be autonomously formed by a group of identical mechanical units. The unit realizes rotational motion by using an actuator mechanism composed of two SMA torsion coil springs. We have realized micro-size prototype units and examined their fundamental functions by experiments. An extended 3D system and its self-reconfiguration algorithm are also discussed.
**Key words**: Modular Robotic System, Self-reconfiguration, Shape Memory Alloy Actuator, Micro-robot

## 1 Introduction

Recently, reconfigurable systems composed of robotic units have been investigated intensively for their versatility, flexibility, and fault-tolerance. This paper focuses on a micro-scale self-reconfigurable model aiming such applications as inspection robots in hazardous environments or micro-scale dextrous manipulators (Fig. 1).

Various self-reconfigurable robotic systems have so far been proposed that generate 3D shapes and motions [1]–[8] as well as 2D models [9]–[11]. Conventional electric motors used in these studies impose a limitation on miniaturization of the size of the system, due to their poor power/weight ratio in micro-scale. This has been major barriers to develop a micro modular robotic systems that can *actively* reconfigure themselves. Some micro-scale self-assembly systems have been reported [12,13]. These are developed by using microfabrication technology, but the driving force of the units is passive force such as surface tension that is not controllable.

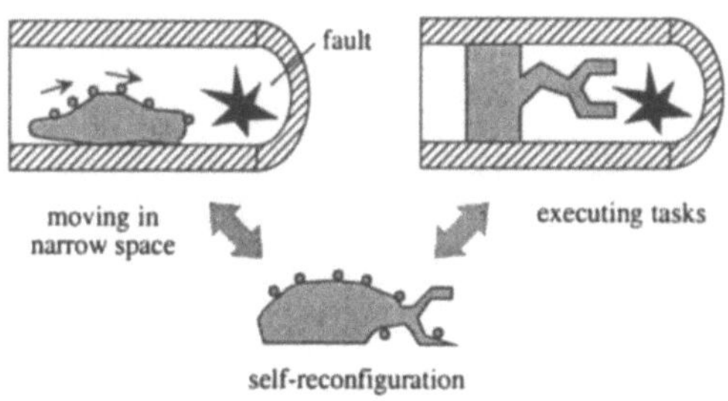


**Fig. 1.** Applications of a micro reconfigurable robotic system.

In this paper, we adopt an actuating mechanism driven by shape memory alloy (SMA) that has a high power/weight ratio even in micro-scale. Simple discrete end-to-end motion of SMA can overcome its major drawback, difficulty of precise control.

Although several micro SMA actuators have already been developed [14]–[16], it is still difficult to meet the requirements for self-reconfigurable systems. We have devised a rotational actuator mechanism using SMA torsion coil springs which satisfies both requirements of torque and motion area. So far we have confirmed the feasibility of the proposed actuator [17]. In this paper, we present 2D prototype units (5[cm] span and 80[g] weight) equipped with a microprocessor and inter-unit communication device. Their self-reconfiguration function will be verified through experiments. We will also report the development of the second prototype whose span and weight are reduced to 3[cm] and 15[g]. Finally, we will show that the unit system can be extended to three dimensions, and that the formerly developed distributed algorithm [18] can be applied to its self-reconfiguration.

## 2 Mechanism of Unit

This section outlines a 2D self-reconfigurable robotic system composed of identical units. We have designed the units' mechanism to be simple enough for miniaturization. The unit has a square shape, where two actuators at orthogonal vertices rotate male connecting parts that can be connected to female parts in another unit. Figure 2(a)~(c) shows the basic motion between two units. Unit U1 changes its relative position clockwise (CW) around U2 through appropriate operations of rotating actuators and connection mechanisms. By repeating the basic motion, a group of units can form various 2D shapes.

## 3 Actuator Mechanism using SMA

This section briefly presents an actuator mechanism dedicated to micro mechanical units. This actuator should have a rotation range of ±90° with

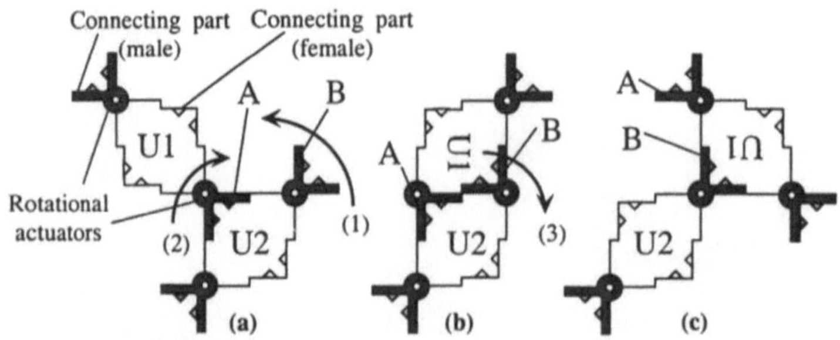


**Fig. 2.** Basic motion of two units.

enough torque to carry another unit. The reconfiguration operations require only three positions ($-90°$, $0°$, $90°$) realized by a simple end-to-end motion.

We have devised an actuator mechanism by using two SMA counter-torsion coil springs as shown in Fig. 3. The SMA torsion coil springs (memorizing the 0° rotation shape) are pre-loaded by twisting reversely by 180°. Without heating, the static torques balance and generate no torque. In this state, the connecting part is fixed at the original 0° position by a mechanical stopper (Fig. 4(a)).

The rotation takes place when one of the springs is heated, usually by electric current. Since Young's modulus of SMA increases drastically when the temperature exceeds its phase transformation temperature $A_f$, the heated spring generates a large torque in the direction to restore the memorized 0° rotation state.

Let us evaluate the performance of the actuator. The spring constant of a torsion coil spring $k_\tau$ [kgf·mm/deg] is in proportion to Young's modulus $E$ [17]:

$$k_\tau = \frac{\pi d^4 E}{11520 n D} = cE. \tag{1}$$

where $d$ and $D$ are the wire and spring diameters respectively, and $n$ is the number of turns. The relationship between its rotational angle $x°$ and the generated output torque $T$[kgf] is given as follows:

$$cE_2(B - x) = cE_1(B + x) + T \ . \tag{2}$$

where $E_1$, $E_2$ ($E_2 > E_1$) are Young's moduli below and above temperature $A_f$, and $B°$ is the angle by which both springs are twisted for pre-loading. By using Ti-Ni-Cu SMA that has a large difference between $E_1$ and $E_2$, we can diminish the reverse torque from the non-heated spring.

This actuator mechanism is appropriate to micro-size robotic systems since SMA keeps a relatively high power/weight ratio even in small scale compared to conventional electric motors [14]. The slow response in control (especially cooling) becomes less significant as the ratio of surface area to volume becomes large in small scale.

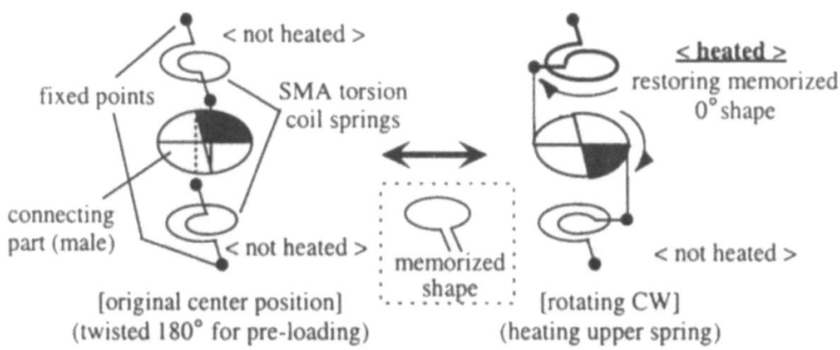


**Fig. 3.** Rotating actuator mechanism using SMA torsion coil springs.

## 4 Hardware Implementation and Experiments

We have developed prototype units and conducted experiments to verify their basic functions. Figure 4 shows the schematic view of the unit, which has a span of 5[cm] and weighs 80[g]. The unit is equipped with two SMA actuators at the orthogonal vertices which rotate the drums (male connecting parts). The original 0° position of a rotating drum is maintained rigidly by the stopper. The rotation becomes possible when the stopper is pulled back by heating the SMA spring. The stopper also limits the drum rotation within the range from −90° to 90° as illustrated in Fig. 4(a). The female connecting part has an auto-locking mechanism, also driven by SMA, which can hold and release the drum of male connecting part (Fig. 4(b)).

We adopt Ti-Ni-Cu SMA with $A_f$ = 70°C and $(E_1, E_2) = (500, 7000)$ [kgf/mm$^2$] for the prototype. The specifications of the actuator are $d$ = 0.6[mm], $D$ = 8[mm], $n = 2$ and $B = 270°$. For rotational angle $x = 0°$ and 90°, $T$ is calculated as 12.2[kgf·mm] and 7.68[kgf·mm] from (2), which is enough to move another unit.

Figure 5 shows a picture of units. The SMAs are driven by PWM (50[Hz], duty ratio 25%) through low-resistance MOS-FET (4[m$\Omega$]). The current was approximately 1[A] for input voltage 5[V] per SMA. The response of the actuator mechanism was prompt enough to complete 90° rotation within five seconds. Each unit is equipped with a controller using PIC microprocessor module BasicStamp II (Parallax Inc.) as shown in Fig. 6. It allows a unit to serially communicate with the wired host PC as well as other connecting units through connecting pins and electrodes in rotating drums as shown in Fig. 7.

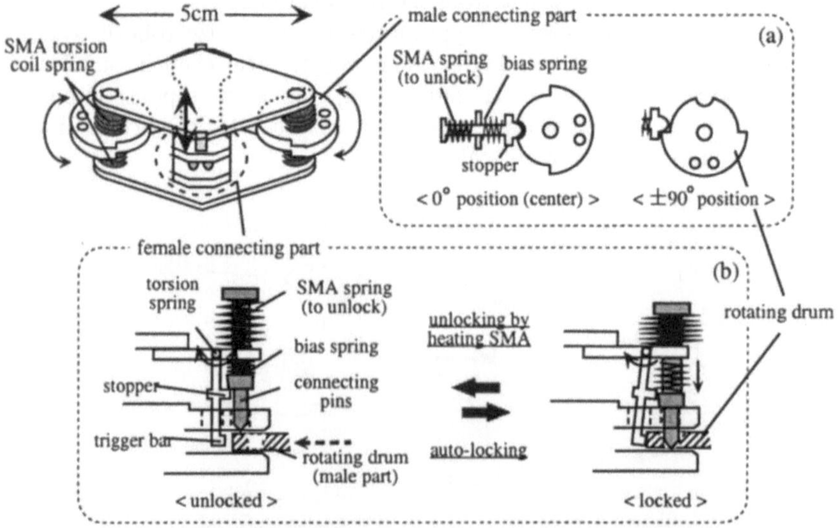


**Fig. 4.** Structure of the prototype unit.

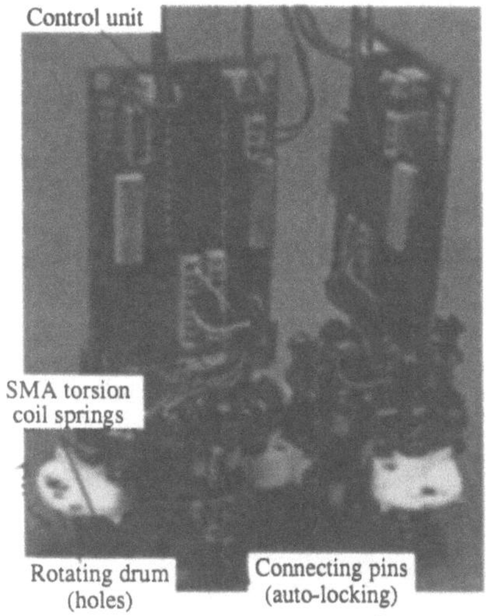


**Fig. 5.** Prototype unit (1st model).

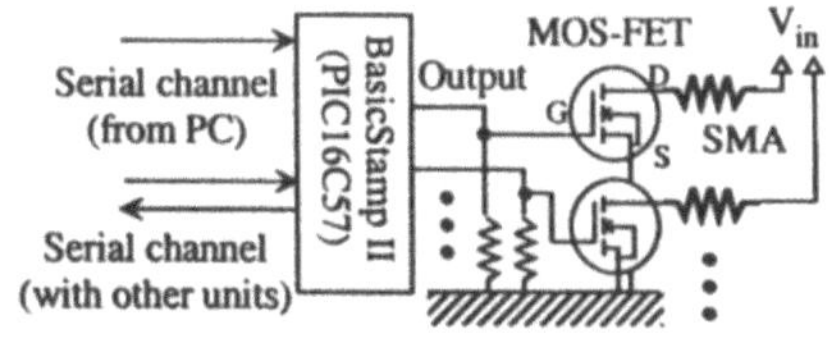


**Fig. 6.** Unit controller.

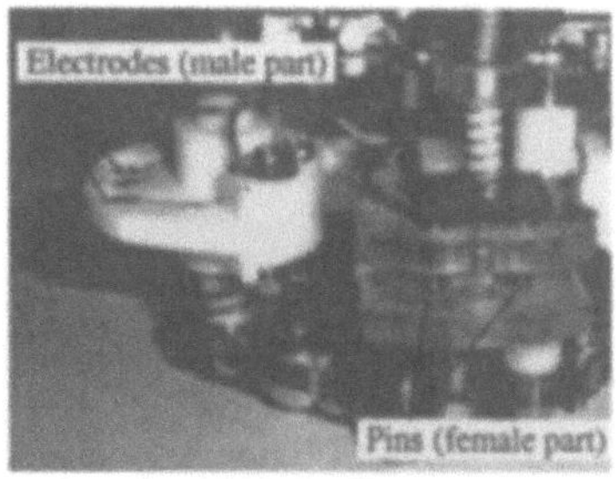


**Fig. 7.** Inter-unit connection.

In the current implementation, the motion sequence for the unit generated by a self-reconfiguration algorithm described later, is given from the host PC. A step motion of a unit requires locally coordinated motion of neighboring units as shown in Fig. 8. When a unit receives the command to make a step motion (clockwise or counterclockwise rotation), the unit asks neighboring units that the motion be supported by rotating and releasing of appropriate male and female connecting parts.

We have performed a self-reconfiguration experiment using six units illustrated in Fig. 9(a), where a group of units changes its shape from the initial to the final state on a horizontal plane. In this experiment, the units moves according to the commands from the host PC and the inter-unit communication device is used only as a sensor to detect the connection. As can be seen in Fig. 9(b)~(d), the desired inter-unit motion has been realized by repeating basic step motions.

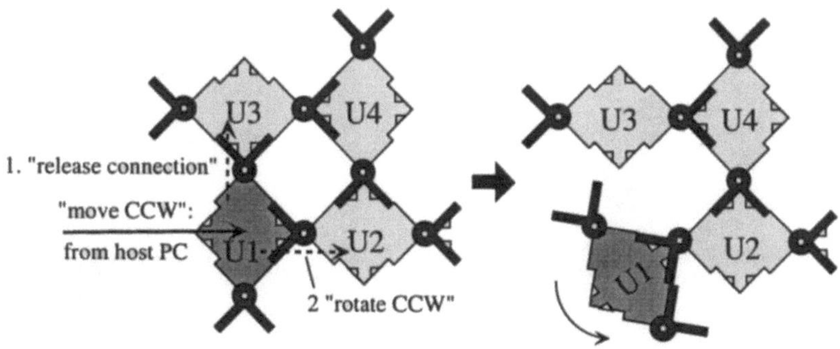


**Fig. 8.** Local motion coordination through inter-unit communication.

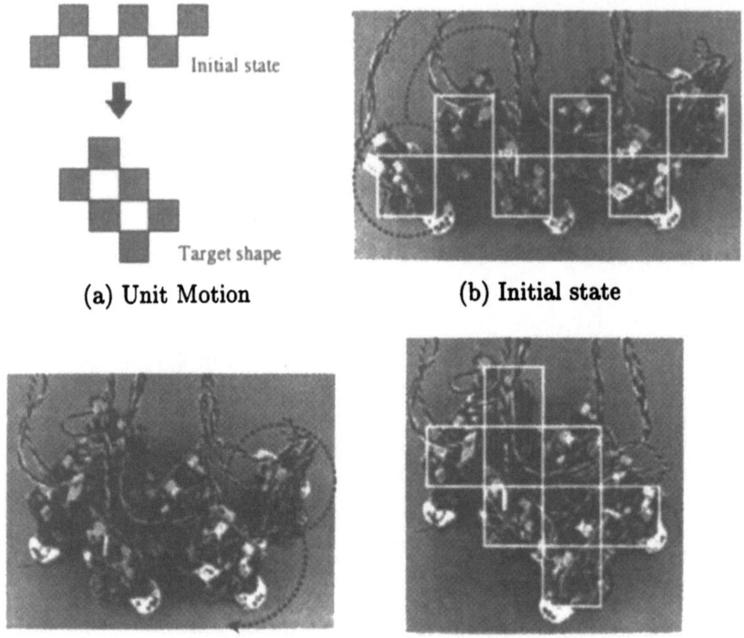


(a) Unit Motion (b) Initial state

(c) Transient state (d) Self-reconfiguration completed

**Fig. 9.** Reconfiguration experiment using six units.

## 5 Development of Micro-size Units

As mentioned in section 3, SMA actuator is advantageous both in power/weight ratio and response in micro-scale. In pursuit of wider applications requiring motion or tasks in narrow spaces, we are attempting further miniaturization.

Figure 10 describes a micro-size unit. Its structure is basically the same as the first model, but there are some differences in stopper mechanisms to lock and unlock rotating drums and connecting pins.

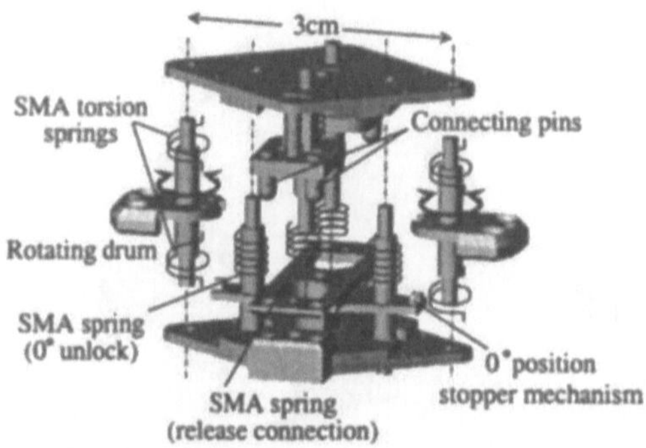


**Fig. 10.** Structure of micro-unit (2nd model).

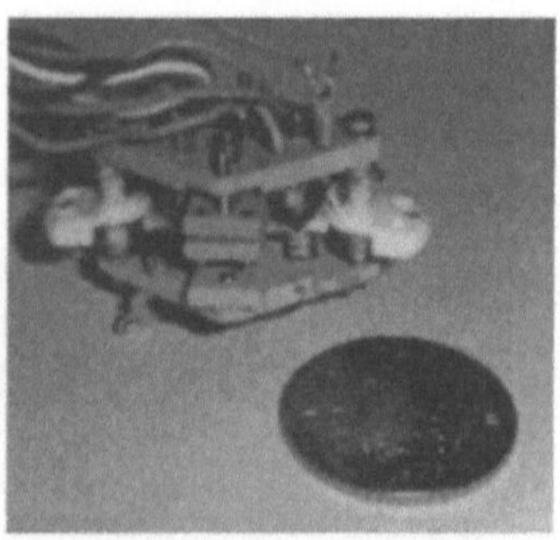

**Fig. 11.** Prototype of micro-unit.

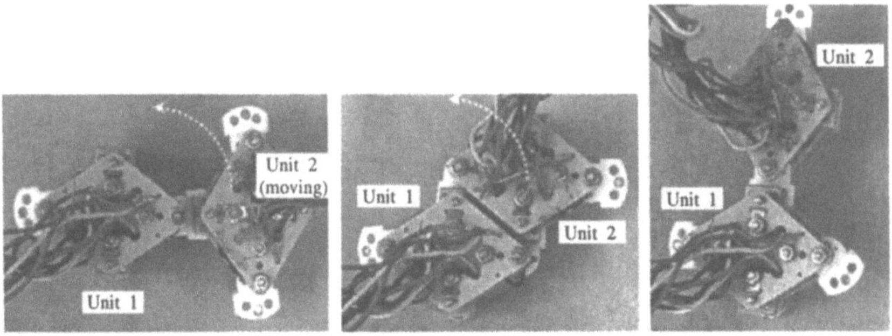


**Fig. 12.** Experiments of basic motion by two micro-units.

Figure 11 shows the prototype of the second model. It measures approximately 3[cm] span with 2[cm] height and weighs 15[g]. The dimensions of the torsion coil springs are $d = 0.6$[mm], $D = 3$[mm], $n = 2$ and $B = 270°$. The unit currently does not include the onboard microprocessor and is driven by an outside controller. The response was improved here as 90° rotation is made in approximately two seconds. Figure 12 demonstrates the units effectively realized the basic motion in Fig. 2. We are currently collecting experimental data to evaluate the performance of the actuator.

# 6 Extension to 3D Reconfigurable System

A 3D unit is designed by combining three 2D square units into a regular octahedron which can be connected at its vertices as shown in Fig. 13. This extension needs the current 2D connecting mechanism to be modified so that two male parts, meeting at the vertex with right angles, can interconnect each other. The SMA actuator should also be redesigned to generate sufficient torque.

The 3D units rotate in vertical and horizontal direction around the connecting vertex as shown in Fig. 14. A group of units can move on orthogonal-cubic lattice and form various 3D shapes as shown in Fig. 15. As this motion is compatible to that of the 3D self-reconfigurable structure [1,2], its distributed self-reconfiguration methods [18] can be applied to the octahedral 3D units.

We apply a self-assembly and self-repair method for large-scale systems (Fig. 16). It is based on a recursive description of the target shape by using a layered graph. Primitive description types are introduced that determine the geometrical relationship between "nodes", denoting a group of units here. By assigning another sub-structure to each nodes, various complex shapes can be described in a recursive manner [18]. A unit can belong to multiple levels in the course of self-assembly.

Given the description of a target shape, self-reconfiguration proceeds by assembling first top-level structure, then down to sub-structure and so forth, using inter-unit communication. Figure 17 shows an example of self-assembly

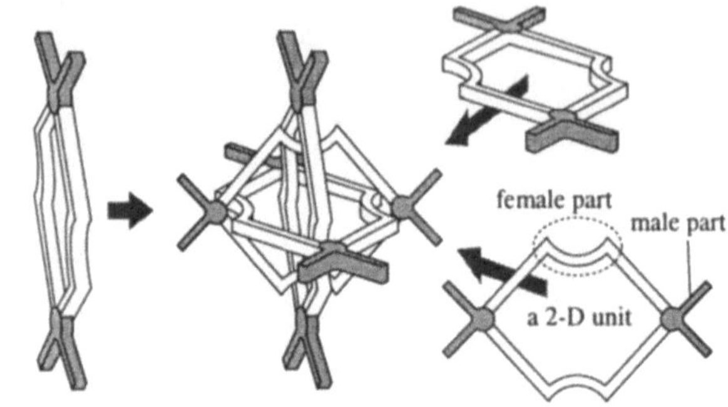


**Fig. 13.** A 3D unit composed of three 2D SMA units.

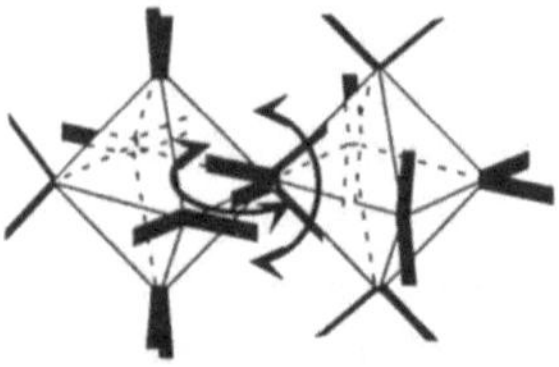

**Fig. 14.** Connected units.

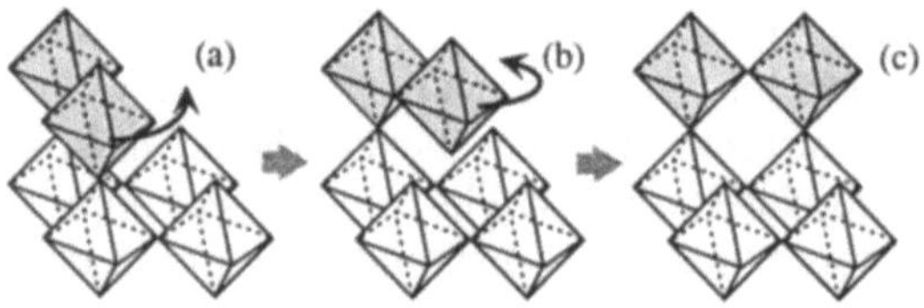


**Fig. 15.** Motion of 3D SMA units.

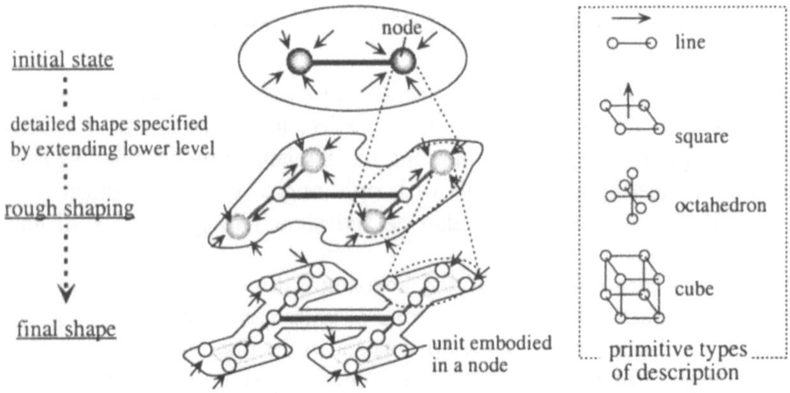


**Fig. 16.** Hierarchical graph structure for self-assembly.

process of the 24-unit planar shape from 27-unit cube, where shaded units have already reached a position in the target shape.

Self-repair is also possible when some of the units are lost or damaged. Upon the detection of fault, surrounding units send messages to spare units in order that they move and repair the faulty part. If the damaged part includes multiple units in different levels of the target description, first the status of the units involved in the self-repair is reset back to the relatively highest level in the target description. Then self-repair is performed from the higher level

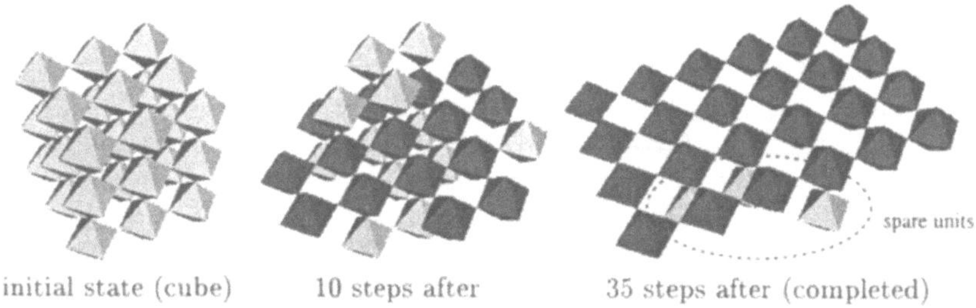


**Fig. 17.** Self-assembly of many-unit structure.

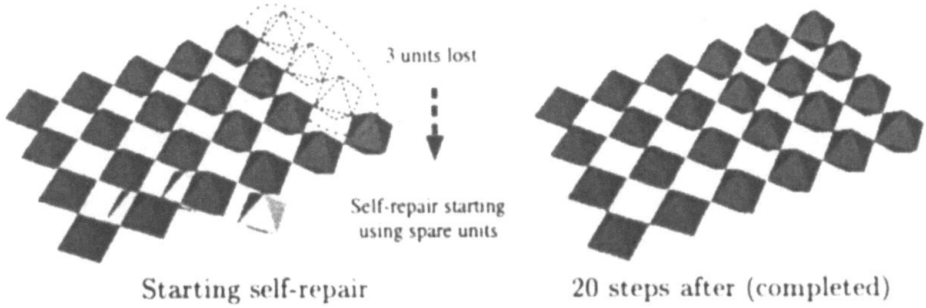


**Fig. 18.** Self-repair of many-unit structure.

down to lower levels. Figure 18 shows a simulation result of self-repair after three units have been removed.

The algorithm can be applied to small-size space structure such as solar panels or antennas. Launched in a compact folded form, they can expand the structure for the mission, and repair themselves if some part is damaged.

## 7 Conclusions

This paper presented micro self-reconfigurable robotic systems using an SMA actuator mechanism. We realized simple 2D mechanical units by using an actuator mechanism composed of SMA torsion coil springs, which can provide sufficient torque and motion range. The fundamental self-reconfiguration function was verified through self-assembly experiments using the first prototype units. Further miniaturization was also investigated by developing the second prototype that demonstrated the basic motion successfully. We have also discussed a 3D reconfigurable system. It was shown that the formerly developed distributed self-reconfiguration algorithm can be applied to the 3D SMA units. Future work includes equipment of controller to the second micro-size unit model and hardware implementation of 3D SMA units.

## References


1. S. Murata, et al. (1998) "A 3-D self-reconfigurable structure," *Proc. IEEE Int. Conf. on Robotics and Automation*, 432–439.
2. H. Kurokawa, et al. (1998) "A 3-D self-reconfigurable structure and experiments," *IEEE/RSJ Int. Conf. on Intelligent Robots and Systems*, 860–865.
3. G. Hamlin and A. Sanderson. (1998) "Tetrobot: A modular approach to reconfigurable parallel robotics," Kluwer Academic Publishers, Boston.
4. K. Kotay, et al. (1998) "The self-reconfiguring robotic molecule," *Proc. IEEE Int. Conf. on Robotics and Automation*, 424–431.
5. C. McGray and D. Rus. (1998) "Self-reconfigurable molecule robots as 3D metamorphic robots," *Proc. IEEE/RSJ Int. Conf. on Intelligent Robots and Systems*, 837–842.
6. P. Will, et al. (1999) "Robot modularity for self-reconfiguration," *Proc. SPIE, Sensor Fusion and Decentralized Control in Robotic Systems II*, 236–245.
7. A. Casal and M. Yim. (1999) "Self-reconfiguration planning for a class of modular robots," *Proc. SPIE, Sensor Fusion and Decentralized Control in Robotic Systems II*, 246–257.
8. C. Ünsal, et al. (1999) "I(CDS)-cubes: a modular self-reconfigurable bipartite robotic system," *Proc. SPIE, Sensor Fusion and Decentralized Control in Robotic Systems II*, 258–269.
9. S. Murata, et al. (1994) "Self-assembling machine," *Proc. IEEE Int. Conf. on Robotics and Automation*, 441–448.
10. G. Chirikjian, et al. (1996) "Evaluating efficiency of self-reconfiguration in a class of modular robots," *J. of Robotic Systems*, **12**-5, 317–338.
11. K. Tomita, et al. (1999) "Self-assembly and self-repair method for distributed mechanical system," *IEEE Trans. on Robotics and Automation*, **15**-**6**, 1035–1045.
12. K. Hosokawa, et al. (1994) "Dynamics of self-assembling systems: Analogy with chemical kinetics," *Artificial Life*, **1**-4, 413–427.
13. P. Green, et al. (1995) "Demonstration of three-dimensional microstructure self-assembly," *J. of Microelectoromechanical Systems*, **4**-4, 170–176.
14. K. Ikuta. (1990) "Micro/miniature shape memory alloy actuator," *Proc. of IEEE Int. Conf. on Robotics and Automation*, 2156–2161.
15. G. Lim, et al. (1996) "Future of active catheters," *Sensors and Actuators*, **A56**, 113–121.
16. Y. Nakamura, et al. (1997) "A three-dimensional shape memory alloy loop actuator," *Proc. IEEE Int. Workshop on Micro Electro Mechanical Systems*, 262–266.
17. E. Yoshida, et al. (1999) "Miniaturized Self-reconfigurable System using Shape Memory Alloy,", *Proc. IEEE/RSJ Int. Conf. on Intelligent Robots and Systems*, 1579–1585.
18. E. Yoshida, et al. (1999) "A distributed method for reconfiguration of 3-D homogeneous structure," *Advanced Robotics*, **13**-4, 363–380.

# Autonomous and Self-Sufficient CONRO Modules for Reconfigurable Robots

Andres Castano, Ramesh Chokkalingam, and Peter Will

University of Southern California
Information Sciences Institute
4676 Admiralty way
Marina del Rey, CA 90292
e-mails: andres@isi.edu, rchokkal@isi.edu, will@isi.edu

**Abstract.** The goal of the Conro project is to build deployable modular robots that can reconfigure into different shapes such as snakes or hexapods. In this paper we present an overview of the first generation of the Conro modules. The modules are designed to work in groups, as part of a larger modular robot. Still, Each module is, itself, a complete autonomous and self-contained robot with its own processor, power supply, communication system, sensors and actuators. We conclude the paper by describing some of the robots that could be built using these modules.

**keywords:** module, reconfigurable, autonomous, self-sufficient.

## 1 Introduction

A reconfigurable robot is a robot that can change its own shape. In contrast to modular robots designed to add versatility to a specific type of robot (e.g., [1]-[3] ), a reconfigurable robot can change its shape at will. It could turn into a snake to reach into narrow places, into a hexapod to carry a load or it may split into many smaller robots to perform a task in parallel. Reconfigurable robots are classified as homogeneous or heterogeneous depending on whether their modules are identical or not. In a homogeneous robot, the position of the module in the robot defines its function, e.g., the module could play the role of head, leg or spine depending on its location in the robot. In a heterogeneous robot, the function of the module defines its position in the robot. For example, a leg module must be used as a leg and a power module must be located in a part of the robot where it is safe from collisions.

Reconfigurable robots can also be categorized according to whether or not their modules are organized in a lattice (either in the plane or 3D space). Lattice-based robots are usually homogeneous and need to reconfigure in order to move, i.e., as they reconfigure, their center of mass translates accordingly (e.g., [4]-[10], ). In contrast, reconfigurable robots that are not based on lattices tend to be heterogeneous and can either translate while reconfiguring or reconfigure and then select a gait that takes advantage of the particular new configuration (e.g., [11]-[14] ).

In this paper we discuss the modules of the Conro robots, which are being designed to be homogeneous reconfigurable robots not based on a lattice. These robots are being designed to be self-contained and to support inter-robot metamorphing, i.e., reconfiguration that involves more than one robot and leads to merging of small robots into a larger one or splitting of a large robot into smaller ones. The basic blocks of these robots, the Conro modules, are themselves self-contained small robots. Thus, a Conro robot is a distributed system of independent modules and its actions require their coordinated movement. Although the individual modules are already fully functional, the software for coordinating the reconfiguration and locomotion of these robots is a work in progress and thus, it is not discussed in the paper. The results of our preliminary work on module design and algorithms for reconfiguration as it applies to the Conro robots can be found in [15].

This paper is organized as follows. In Sect. 2 we summarize the philosophy behind the design of the module. In Sects. 3 and 4 we describe the design of the module from the mechanical and electrical points of view, respectively. In Sect. 5 we describe the resulting module and give some examples of possible Conro robot configurations. In Sect. 6 we present our conclusions.

## 2 Philosophy of Design

The goal of the Conro project is to build deployable reconfigurable robots with inter-robot metamorphic capabilities. The capabilities of these robots are determined by the characteristics and functionality of their modules. A description of the philosophy of design of our modules can be found in [15]. The summary in this section is presented for completeness of the paper.

The specifications of deployability and inter-robot metamorphic capabilities translate into constraints on the levels of self-sufficiency, autonomy and homogeneity of the module, module size and communication capabilities. A deployable robot must be self-sufficient, i.e., capable of untethered operation. In the trivial case, an inter-robot metamorphic split operation may create a robot formed by a single module. Therefore, to guarantee that any robot such created is self-sufficient, each module must be self-sufficient. Likewise, a module must be autonomous with respect to the use of its own resources, e.g., it has exclusive access to its sensors and actuators.

The level of homogeneity of a module determines its capabilities and the functions that it can fulfill. Each module must have a processor, power, sensors, actuators and communication systems to satisfy the self-sufficiency and autonomy constraints. Other components not needed to satisfy these constraints (e.g., cameras, antennas) can be carried by the robot as a load or are piggy-backed on a given module, driven by a generic port. This trade-off between the necessary and desired components of a module reduces its design, manufacturing, testing and programming costs. All the components must fit into a package that is as small as possible to reduce the effect of inertia of the limbs and increase the relative torque-to-robot weight ratio of the actuators.

Finally, we address the communication needs of the module. During inter-robot metamorphing, two robots need to communicate remotely to agree on the merging operation and need a mechanism to guide themselves toward each other. At the local level, each module needs to communicate with its adjacent modules. An infrared (IR) system can satisfy all these requirements; it could be used for remote and local communication and can double as the directional guiding mechanism for both inter-robot and intra-robot dockings.

# 3 Mechanical Design

We now describe our work to build a module that satisfies the goals described in Section 2. Our module achieves self-sufficiency, autonomy, a high level of homogeneity, miniature size and IR-based intra-robot communication and docking. However, in the present state, the module does not support long-range inter-robot communication. Also, the module is not completely homogeneous since, although all the modules have the same mechanical and electrical components, half of the modules have their motors and CPUs on their left side while the other half have them on their right side. This asymmetry was introduced to improve the stability of the snake configuration.

Our implementation of the Conro module has three segments connected in a chain: a passive connector, a body and an active connector, as shown in Fig. 1. At the intersection of the body and the two connectors there are joints that give the module yaw and pitch degrees of freedom. The weight of the module is 115 g (including batteries) and its length is 104 mm excluding the length of the pins protruding from the passive connector. We now describe the parts of the module.

## 3.1 The Module Body

The body is the central part of the module to which the active and passive connectors attach. It is composed of a delrin frame, two servo motors and a printed-circuit board (PCB). The PCB has a hole in its center to allow its accommodation on the frame. When it is in place, the PCB is screwed in position to the frame. The servos fit into cavities of the frame and are held in place by friction.

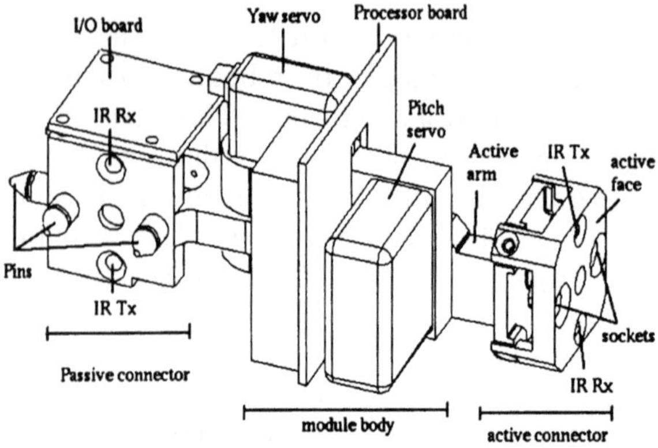


**Fig. 1.** Parts of the Conro module

Commercial-off-the-shelf servos for radio-controlled devices (RC servos) are connected directly to the processor board using 3-pin connectors. Two of the pins provide power to the servo and one carries a pulse width modulated (PWM) signal that defines the position of the axis. The pitch and yaw servos have a torque of 3.7 and 1.7 kg-cm, respectively, and weigh 21 and 9 g, respectively. The output shafts of the servos are connected directly to the active and passive connectors in a direct drive fashion. The servos are back-drivable and thus, under load, they must be powered continuously to maintain a given attitude. The processor board is a two-layer PCB with surface-mounted components that distributes the control signals and power to the rest of the module and serves as holding place for a small 3 v battery.

### 3.2 The Passive Connector

The connectors are the parts of the module that allows it to attach to other modules. The passive connector has no moving components. Its frame is a cube of delrin with a side of 2.54 cm. Three of the vertical faces of the cube have two protruding aluminum pins that fit into the sockets of the active connectors of other modules. The cylindrical pins have a lateral groove to allow the active connector to anchor to them. The particular positions of these pins and sockets permit only connections of modules that lie in the same plane, i.e., modules that are tilted 90 degrees with respect to each other cannot be connected. On each of these faces there is an IR pair used by the module for communication and docking. The fourth vertical face of the cube has a tongue that fits on a fork of the body and allows the module to pivot the connector about the yaw axis.

The frame of the passive connector is hollow and holds the wiring of the IR devices of the faces and the main battery of the module, a 6 v lithium battery of the type K28L (9 g, 2.5 cm height, 1.3 cm diameter). The roof of the cube is a two-layer PCB that is screwed directly onto the cube. This PCB has the input-and-output electronics that drive the IR receivers (RX) and transmitters (RX) of the faces of the connector. The PCB also doubles as the positive contact for the battery. A 14-pin connector is used to transfer the power of the main battery to the processor board and receive the control signals. Finally, the connector has a brass latch at the bottom of the cube that keeps the battery in place and serves as its negative contact. The latch can swing about one of its extremes allowing the removal of the battery. The weight of the passive connector, including the battery, is 31 g.

### 3.3 The Active Connector

The active connector is the part of the module that engages and disengages the pins of the passive connectors of other modules. It weighs 15 g and is composed of an arm and a face frames, both machined in delrin. The body of the module is connected to the active connector by the arm. The active connector can rotate about a pitch axis located at the intersection of the arm and the body. The face of the active connector has the same dimensions as those of the faces of the passive connector. It also has an IR pair but the locations of the transmitter and receiver are the reverse of those of the faces of the passive connector to allow communication between connected modules.

The process of connecting two modules involves the active and passive connectors of the modules. Figure 2a shows a passive connector approaching an active connector in a docking trajectory. The active face has two sockets to

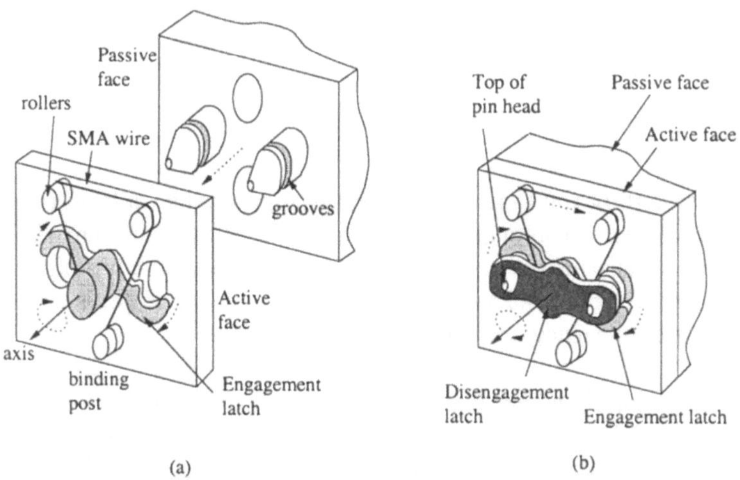


**Fig. 2.** Stages of the docking procedure. **(a)** Engagement. **(b)** Disengagement

receive the pins of the passive face. As the pins slide inside the sockets, their dome-shaped heads force an engagement latch to rotate in a direction normal to the trajectory of the pins. When the pins are fully inserted, the engagement latch is forced by a spring action into their lateral grooves (the spring is not shown in the figure). This docking process is completely mechanical.

The process of disconnecting two modules is initiated by the active face. As shown in Fig. 2a, the engagement latch can be rotated using a shape-memory alloy (SMA) wire. The wire is attached between a fixed binding post and a cylinder attached to the latch. We use two rollers to establish the path of the SMA wire and to extend its working length. When the SMA is contracted, it pulls the cylinder against a spring, forcing the latch to retract and freeing the pins. However, freeing the pins is not the same as disconnecting the modules. The SMA can be activated only for a fraction of a second because it consumes a large amount of power. Thus, it is possible that the modules fail to move away from each other before the SMA is de-energized, re-engaging the pins.

The disconnection process must guarantee that the modules are be free when it is finished. Figure 2b shows the faces of two connected modules. We have added a view of the disengagement latch, a plate with two holes that, during engagement, allows the heads of the pins to go through. When the modules are connected, the engagement latch is pressed against the pins, into their grooves. To disconnect the modules, we contract the SMA wire as described before. As both latches rotate together, first the engagement latch frees the pins and then the disengagement latch pushes the dome-shaped heads of the pins out of the sockets. The latch pushes the pins about 0.125 mm. This displacement is enough to guarantee that the latch will not be able to re-engage the pins when the SMA relaxes. After this process, the modules are disconnected and can be moved away from each other at any moment.

## 4 Electrical Design

The electrical system of the Conro module must support the control of the sensors and actuators, a communication system and a power system. The objectives of the design of the system are to minimize the number of discrete components, their overall weight and their power consumption while preserving the self-sufficiency and autonomy of the module.

A functional diagram of the electric system of the Conro module is shown in Fig. 3. Each module has a processor that gives it control over its sensors and actuators. The processor is defined by the use of one of three different single-chip micro-controllers: a stamp II based on a PIC16C57 processor, a stamp II-SX based on a SCENIX SX28AC/SS processor or a Basic-X based on a ATMEL AT90S8535. The use of a zero-insertion force socket allows for the manual removal of the processor for replacement or programming. The

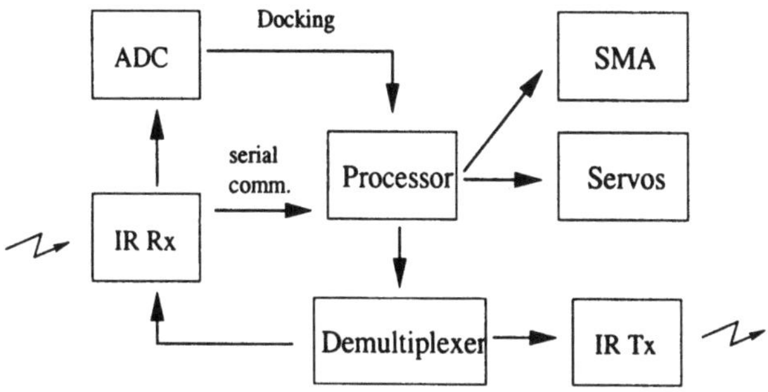


**Fig. 3.** Functional block diagram of a module

three processors are pin-compatible but differ in speed, memory capacity, programming capabilities and power consumption. Thus, the micro-controller of the module can be selected to suit a particular application.

The processor has exclusive access to the actuators of the module. The SMA wire is activated with a fixed current during a programmable period of time. The servos require a PWM signal generated in software because none of these micro-controllers has a suitable PWM circuit. The number of digital outputs was increased using a demultiplexer that is used to control the IR devices. The micro-controllers can set any of their port pins to behave as serial receivers or transmitters. The receiver circuit can direct the input signals to either a low-impedance input pin of the micro-controller or to a high-impedance input pin of an 8-bit analog-to-digital converter (ADC), depending on whether the IR receiver is being used for serial communication or as an IR sensor. This latter state is used during the docking of two modules, where a module uses its IR transmitter as a beacon and the other module uses its IR receiver as an analog directional sensor. The two modules that are docking can belong to the same robot or to two different robots. The combined use of these IR pairs provides the feedback necessary for the modules (or the robots) to approach to each other and dock. However, the working range of the docking sensors is too small (in the order of cms) to serve the general purposes of inter-robot communication.

The Conro module uses two lithium batteries: a 6 v K28L battery and a 3 v K58L cell, each one with a capacity of 160 mA-h. The batteries set up a 9 v high-voltage low-current node to power the micro-controller and a 6 v low-voltage high-current node to power all other components. The use of the two batteries prevents large voltage drops at the micro-controller that would appear when devices like the SMA or the servos are used. The batteries were selected for their voltage, size, weight, capacity, drain and the flatness of their discharge curves; lithium batteries are a good compromise between these features. Rechargable batteries, although desirable for a robotics project, have an energy density that is very inferior to that of the lithium chemistry.

# 5 The Conro Module

We have built twenty modules that follow the description discussed in this paper. Figure 4a shows a close view of a module. Because each module is

**Fig. 4.** The Conro module. (**a**) Close-up view. (**b**) Executing motions while untethered

self-contained and autonomous, it is a robot by its own right and thus, it is possible to program it to execute motions and to react to stimuli. Figure 4b shows the module, completely untethered, running a program that rotates its connectors in sequence. Using the PIC-based micro-controller, the program can run for 35–40 mins continuously. The module consumes 50–70 mA-h at 6 v when it is not moving. The use of the motors adds 70–80 mA-h to this total. Due to the high cost of running experiments using non-rechargable batteries, the modules are tested using external power supplies.

Although the modules can run by themselves, they were designed to work in groups, connected to each other forming large robots. We now give a brief overview of the present state of the robots and the work in progress. At this moment, the robots are configured manually but our goal is to achieve automatic robot reconfiguration eventually. The priority of the modules of a robot is to communicate efficiently with their adjacent modules. The modules do not share a clock signal and thus, robot actions that require the synchronized motions of different modules rely on the quality of the communication. The programming of the communication network is complex because, due to the lack of interrupt mechanisms in our micro-controllers, the module has to poll the ports in a round-robin fashion. At this moment, the IR communication between adjacent modules is a 9600 Bd inverted serial connection with flow control. We are in the process of modifying the modules to support 68HC12 processors that support interrupt-driven communication.

The control of a set of modules can be performed using a distributed control or a centralized control based on a master-slave hierarchy. The robots shown in Fig. 5 are controlled using a master-slave approach where the master is a remote host with a large computational capability. The snake Conro robot

**Fig. 5.** Conro robots executing motions. (**a**) Snake. (**b**) Hexapod

shown in Fig. 5a is formed by eight modules connected in series. This particular head-to-tail configuration simplifies the communication mechanism because each module needs to communicate to, at most, two other modules. Figure 5a shows the snake while executing sinusoidal motions which are the basic steps of the traveling wave gait (e.g., see [16]).

Figure 5b shows a Conro hexapod after standing up by its own means. It is composed of nine modules: six modules play the role of legs and three modules form the spine. In this case, the distribution of the communication load is uneven because the modules that form the spine have to handle up to four communication ports while the modules that form the legs need to handle a single port.

## 6 Conclusion

We have introduced the Conro module, the basic block of a reconfigurable robot that is designed for deployability and inter-robot reconfiguration. It is, to our knowledge, the first module for reconfigurable robots that is self-sufficient and autonomous. The module shows a large potential as the building block for deployable reconfigurable robots.

The Conro modules are homogeneous, self-contained, autonomous, miniature and use an IR system for communication that doubles as a tracking system. We have described the implementation of the design in the form of actual modules that can be used to build large robots. These robots are already being used as a research platform for both distributed and centralized control systems. At the moment, we are implementing the gaits for snake and hexapod robots and the automatic reconfiguration algorithms that were tested on robot prototypes and are discussed in our preliminary work [15].

## 7 Acknowledgments

We thank Robert Kovac and Behrokh Khoshnevis for their contributions to the development of the module. This research is being sponsored by DARPA/MTO under contract number DAAN02-98-C-4032.

## References

1. Will P., Grossman D. (1975) An experimental system for computer controlled mechanical assembly. IEEE Trans Computers, **(C-24)9**:879-888
2. Fukuda T., Kawauchi Y. (1990) Cellular robotic system (CEBOT) as one of the realization of self-organizing intelligent universal manipulator. In Proc. IEEE Intl. Conf. Robot. Autom., 662-667
3. Schonlau W. J. (1999) MMS: A Modular robotic system and model based control architecture. In Proc. SPIE, vol. 3839, 289-296
4. Murata S., Kurokawa H., Kokaji S. (1994) Self-assembling machine. In Proc. IEEE Intl. Conf. Robot. Autom., 441-448
5. Pamecha A., Chiang C. J., Stein D., Chirikjian G. (1996) Design and implementation of metamorphic robots. In Proc. ASME Design Engineering Tech. Conf. and Comput. in Engineering Conf., aug, 1-10
6. Chirikjian G., Pamecha A., Ebert-Uphoff I. (1996) Evaluating efficiency of self-reconfiguration in a class of modular robots. J. Robotic Systems **(13)5**:317-338
7. Murata S., Kurokawa H., Yoshida E., Tomita K., Kokaji S. (1998) A 3-D self-reconfigurable structure. In Proc. IEEE Intl. Conf. Robot. Autom., 432-439
8. Kotay K., Rus D., Vona M., McGray C. (1998) The self-reconfiguring robotic molecule. In Proc. IEEE Intl. Conf. Robot. Autom., 424-431
9. Yim M., Lamping J., Mao E., Chase J. G. (1997) Rhombic dodecahedron shape for self-assembling robots. Xerox PARC, SPL TechReport P9710777
10. Ünsal C., Kiliççöte H., Khosla P. (1999) I(CES)-cubes: a modular self-reconfigurable bipartite robotic system. In Proc. SPIE, vol. 3839, 258-269
11. Fujita M., Kitano H., Kageyama K. (1998) Reconfigurable physical agents. In Proc. Intl. Conf. Autonomous Agents, 54-61
12. Farritor S., Dubowsky S., Rutman N., Cole J. (1996) A systems-level modular design approach to field robotics. In Proc. IEEE Intl. Conf. Robot. Autom., 2890-2895
13. Hamlin G., Sanderson A. C. (1995) Tetrobot: A modular system for hyper-redundant parallel robotics. In Proc. IEEE Intl. Conf. Robot. Autom., 154-159
14. Yim M. (1994) New locomotion gaits. In Proc. IEEE Intl. Conf. Robot. Autom., 2508-2514
15. Castano A., Shen W-M., Will P. (2000) CONRO: Towards deployable robots with inter-robot metamorphic capabilities. Accepted for publication in Autonomous Robots Journal **(8)3**
16. Poi G., Scarabeo C., Allota B. (1998) Traveling wave locomotion hyper-redundant mobile robot. In Proc. IEEE Intl. Conf. Robot. Autom., 418-423

# Motion Planning for a Modular Self-Reconfiguring Robotic System

CEM ÜNSAL, HAN KILIVÇÇÖTE, MARK E. PATTON, and PRADEEP K. KHOSLA

Institute for Complex Engineered Systems, Carnegie Mellon University, Pittsburgh, PA, 15213-3890, USA (*unsal@ices.cmu.edu)

**Abstract.** In this paper, we address the issue of motion planning for a bipartite class of modular self-reconfiguring robotic system (*I-Cubes*) that is a collection of active elements providing reconfiguration (3-DOF manipulators called *links*) and passive elements acting as connectors (*cubes*). The links, capable of attaching/detaching themselves from/to cubes, can position and orient the cubes.
Current solutions for motion planning and reconfiguration of I-Cubes include multi-layered planners that divide a given problem into tractable sub-problems to be evaluated using heuristic methods. The system implementation, its representation in simulation, search algorithms resulting from this representation, and simulation examples are presented.

**Keywords:** self-reconfiguration, metamorphic robots, path planning

## 1. Introduction

Statically stable gaits for mobile robots include wheels, threads, and similar mechanisms. These methods, although providing fast locomotion on level ground, generally do not provide capabilities such as climbing stairs, or moving over obstacles. A robot that can move with relative ease on flat terrain and an ability to climb over large obstacles is yet to be designed. Recent progress in component and sensing technologies move research thrusts on small mobile robots toward smaller scales, and multi-sensor/multi-actuation capabilities further. Applications for groups of small mobile robots working in unstructured environments are slowly emerging.

A sufficient number of robotic modules combined as a single entity would be capable of self-reconfiguring themselves into defined shapes. A large group that can change its shape according to the locomotion, manipulation or sensing task at hand will then be capable of transforming into a snake-like robot to travel inside a air duct or tunnel, a legged robot to move on unstructured terrain, a climbing robot that can climb walls or move over large obstacles, a flexible manipulator for space applications, or an extending structure to form a bridge. Designing a modular system with identical elements has several advantages over large robotic systems. The homogeneity can provide faster production at a lower cost. A modular self-reconfiguring system would be capable of removing malfunctioning elements from the group and reconfigure its elements.

Previous research on modular robotic systems includes manipulators that can be designed according to task specification [1,2], and cellular systems as self-organizing manipulators [3]. These and similar ideas on modularity have been extended to modular structures that can self-reconfigure. Existing 2-D self-reconfiguring systems include *Inchworm* [4], and self-organizing robots [5] moving in vertical planes, *Fractum* [6], and metamorphing hexagonal modules [7] moving in horizontal plane. Recent 3-D systems are *Polypod* and *Polybot* that combine different gaits [8, 9], robotic *molecule* [10] and self-reconfigurable structure [11] that are both capable of changing shape using neighboring elements as pivot points.

In the following sections, we discuss the 3-D reconfiguration for a bipartite class of modular self-reconfiguring robotic system. Our aim is to combine different gaits and task-oriented modules with self-reconfiguration capabilities.

## 2. I-Cubes: A 3-D Modular Self-Reconfigurable System

For self-reconfiguration, a modular system must have several essential properties, such as geometric, physical, and mechanical compatibility among individual modules. Furthermore, several design issues need to be considered for the system to be autonomous. *I-Cubes* are a bipartite modular system, *i.e.*, a collection of independently controlled mechatronic manipulators (*links*) and passive connection elements (*cubes*). Links are capable of connecting to and disconnecting from the faces of the cubes; they can therefore move from one cube to another, or move a cube (Figures 1 & 2). We envision that all links and cubes are capable of permitting power and information flow to attached modules. The system has the following properties:

- Modules can be independently controlled.
- All modules have the same characteristics, and are compatible.
- The 3-D structure fits a cubic lattice to guarantee interlocking of neighboring modules, i.e., distance between cubes in idle position is constant.
- Links have sufficient degrees of freedom to complete motions in 3-D.

Since the actuation for self-reconfiguration is provided by the links, cubes can be used to provide computation, sensing and power resources. If the modules are designed to exchange power and information, the cubes may be equipped with batteries, microprocessors and sensors to be the decision-making elements while the links become the 'muscles' of the system. It is also possible to remove some of the attachment points on the cubes to replace them with wheeled or threaded systems for faster locomotion. Specifically, we envision small robots that can reposition themselves to form a group that is capable of self-reconfiguring in order to move over obstacles that a single robot cannot overtake. Similar scenarios that combine different gaits with shape reconfiguration include stair climbing and traversing pipes.

As seen in Figure 1, the size definitions for the links and the cubes are dependent. If the length of a cube edge is $L$, then the links should have four sections with indicated lengths. The three rotational degrees of freedom for the links are provided by the joint J2 at the middle, and the joints located at end segments (J1 and J3). J1 and J3 are both capable of providing continuous 360-degree rotations, while J2 can only rotate 270 degrees. The design parameters given above and the attachment capabilities enable links to (a) move from one cube face to another, (b) move one cube while attached to another, and (c) move from one cube to another (Figure 2). A cube consists of six faceplates with attachment points for link connectors. Cubes do not contribute to the self-reconfiguring motions with the exception of the motion to lock the link connector in place. A cube attached to a link can be (i) rotated, (ii) translated in vertical or horizontal plane, or (iii) act as a pivot point for a moving link. The role of the cube depends on the position and motion of the active link as well as the connections formed by all modules.

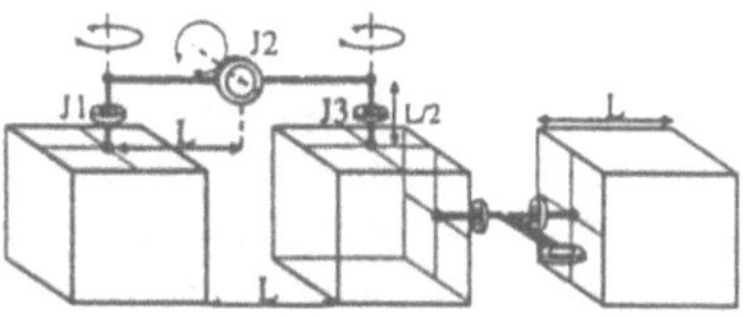


Fig. 1. Geometric constraints and joint definitions for the links

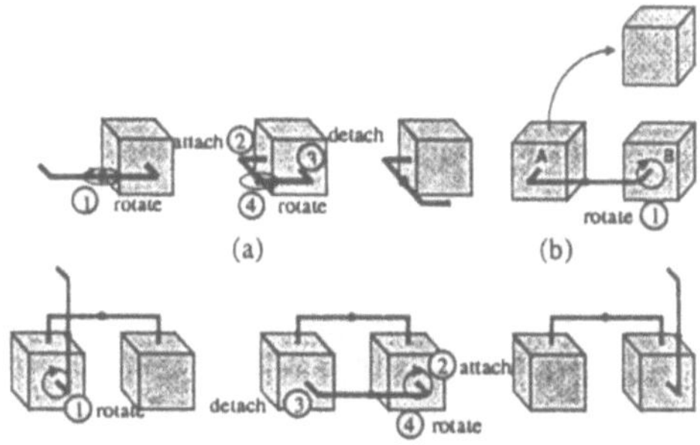


(c)

Fig. 2. Examples of link motions

## 3. Implementation and Experiments

For the prototypes shown below, the module bodies are created using a fused deposition modeling (FDM) machine. The link and cube bodies, generated in *ProEngineer*©, are assembled with off the shelf components. A link equipped with all mechanical and electrical components weights approximately 370gr. The length $L$ is 8cm. A cross-shaped attachment mechanism designed to connect the links and the cubes is actuated using an SMA wire (*Nitinol*) and spring mechanism. One faceplate for a cube weights approximately 20 gr. The depth of a faceplate is 1cm. Therefore, a volume of 6×6×6 $cm^3$ is available for computational, sensing elements and power source. The actuators on the links and the cubes are controlled by microprocessors connected to a graphic user interface (GUI). Two 6V 650mAh rechargeable batteries are used to power all servos on the links, the attachment mechanisms, and the microprocessors. The details of the mechatronic design are given in [12].

The prototypes are used to test the feasibility of the system. Tests show that a link can transfer from a horizontal plane to vertical plane, and vice versa, using three available degrees of freedom. Similarly, a link can move from one cube face to another as well as from one cube to another (Figure 3a-b) The links are also capable of moving and exchanging cubes. As seen in Figure 3c-d, one link (left) moves the cube into position and orients it for another link to attach to and move the cube to its next destination.

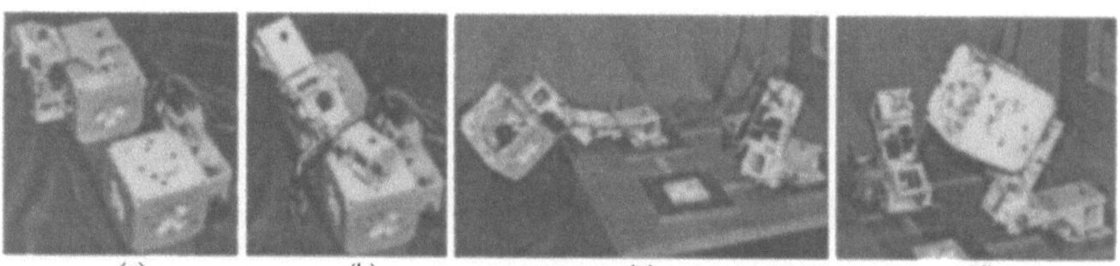

(a) (b) (c) (d)

Fig. 3. (a-b) A link transferring from one cube to another; (c-d) two links exchanging a cube.

## 4. Three-dimensional Reconfiguration

By combining several link actions (attachment/detachment and joint rotations), it is possible to find solutions for a group of cubes and links moving and/or self-reconfiguring from one position/shape to another. Figures 4 and 5 show two groups of four cubes and four links (4C4L). Figure 4 illustrates a scenario where a 4C4L group is moving to higher ground. During reconfiguration, the connections between modules kept such that the group forms a single connected graph at any time. We assume that a link can only carry a cube and another link attached to this cube. This manual solution takes into account the orientation of the cubes, and keeps the same cube faces on the ground. In Figure 5, the group on the ground is moving to the right. The sequence of actions depicted here is one of many possible solutions.

Another example that combines a self-reconfiguration and faster locomotion is shown in Figure 6. The leftmost robot in the first image has a camera, but cannot see what is on/behind the obstacle. These wheeled robots can move into position to form a group and self-reconfigure to lift the camera-equipped robot. Required number of faces with attachment points is two for the camera-equipped robot. It is also possible for a larger group to self-reconfigure into a tower to lift a surveillance robot to see beyond relatively larger obstacles [13]. This scenario illustrates an important characteristic of the system. A heterogeneous group of small robots combines individual robot capabilities with self-reconfiguration to complete a task that is not possible for individual robots.

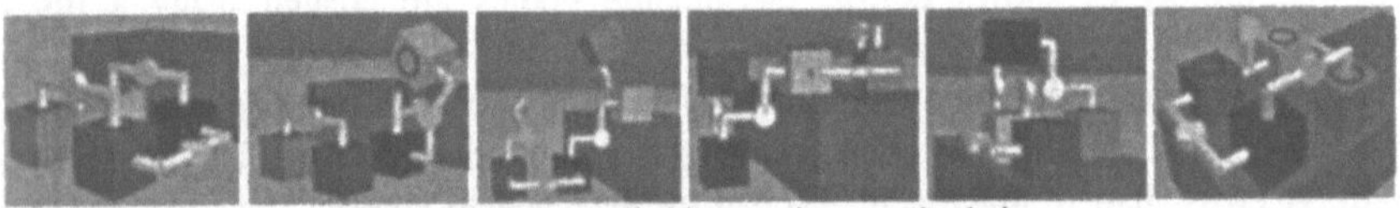

**Fig. 4.** 4L4C group climbing a stair: manual solution.

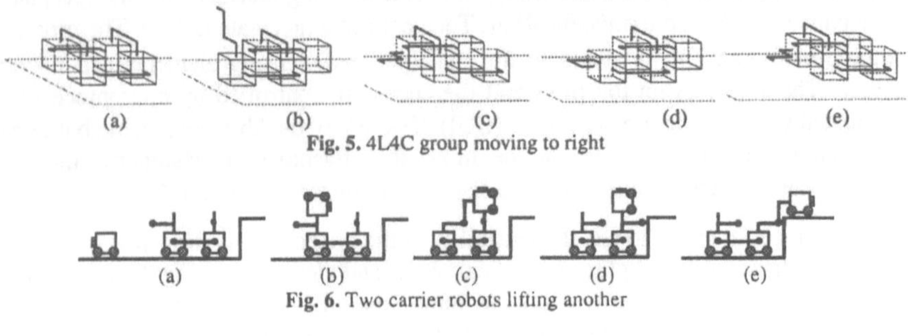

(a) (b) (c) (d) (e)

**Fig. 5.** 4L4C group moving to right



(a) (b) (c) (d) (e)

**Fig. 6.** Two carrier robots lifting another

### 4.1 System Representation for Planning

In this section, we introduce a grid representation for groups of $m$ cubes and $n$ links. For a given $m$-cube $n$-link combination, there are multiple states that the group can take depending on the relative positions and shape of the modules. The number of possible states grows exponentially with the number of cubes and/or links. A group of $m$ cubes and $n$ links can be represented with a three-dimensional cell grid of size $d \times d \times d$ where $d = (2m\text{-}1) + 4$ (Fig. 7). Each cell may be occupied only by a single cube or link. Each cube can occupy only one cell, while the links occupy two or three cells depending on their shapes (See Fig. 7). We discretize the link positions at 90-degree joint angles.

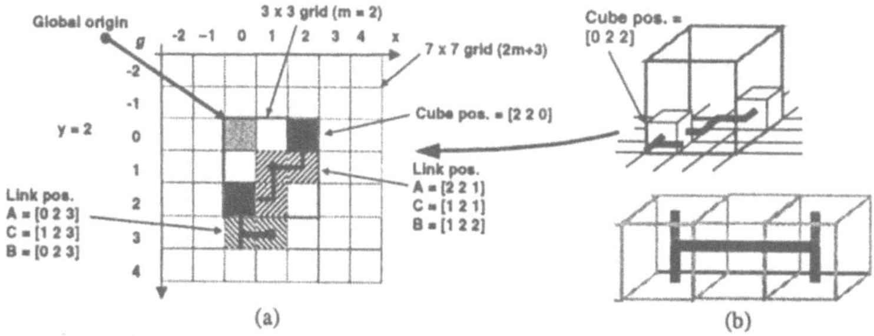


**Fig. 7.** (a) Grid definition, and (b) 2 of 4 possible shapes: link 'looking' up or down at the same position

The cells in the 3-D grid are represented by their local coordinates where the origin is defined as the cell closer to a global origin. Global positions of the links and cubes can be evaluated by combining the local coordinate system and the global position of the grid. We define the following based on the link and cell definitions: (i) A *state S* is a quadruple $\{C, L, f, d\}$ where $C$ and $L$ are sets of vectors indicating cube positions and link positions and shapes, $f$ is a binary relation between $C$ and $L$ indicating the connections, and $d$ defines the size of the grid. (ii) A *node N* is a couple $\{S, g\}$ where $S$ is a state and $g$ is its global position.

A given problem with initial and final configurations can be represented as a pair $\{N_o, N_f\}$ and have multiple solutions based on the sequence of link actions. For a given state $S$, there are 24 actions for each link. These actions define a sequence of link detachments and rotations. We assume that all links automatically attach to a cube at the end of an action if there is such a cube.

All possible actions $a_{ij}$ ($i$=1,...,24, $j$=1,...,$n$) are relations from one state to another, and are associated with translations $t_{ij}$ indicating possible change in global coordinates of the node $N$ associated with the state $S$. For example, the action in Figure 2b is "*detach others at A, fix B, B +*" where all the links except the active link connected to the moving cube are detached, the end $B$ is assumed to be fixed, and the joint at end $B$ is rotated clockwise. These actions $a_{ij}$ are actually a group of actuation commands for attachment mechanisms on the cubes, and for servos on the links. For a link to be able to move, the desired location and the path of the link (and the cube that may be attached) must be clear: the cells associated with new location, and path of the link (and cube), and some of the cells next to the initial and final cube positions should not be occupied by links, cubes or obstacles.

An action $a_{ij}$ acting on a state will create a new state if a link or cube changes position. The associated translation $t_{ij}$ may have non-zero elements if there is a cube relocation that changes the definition of the cube closer to the origin. A *state transition diagram* is a graphical representation of all the states for an *nCmL* combination with the pairs ($a_{ij}$, $t_{ij}$) defined between states. Figure 8a shows part of the transition diagram for *1L2C* case, where each edge indicates one or more action-translation pairs. Total number of possible states for a 1L2C group is 36 if the link is not allowed to detach from the cubes. If link can detach from the cubes, the number is 1404. For an attached 2L2C group, total number of states is 4518.

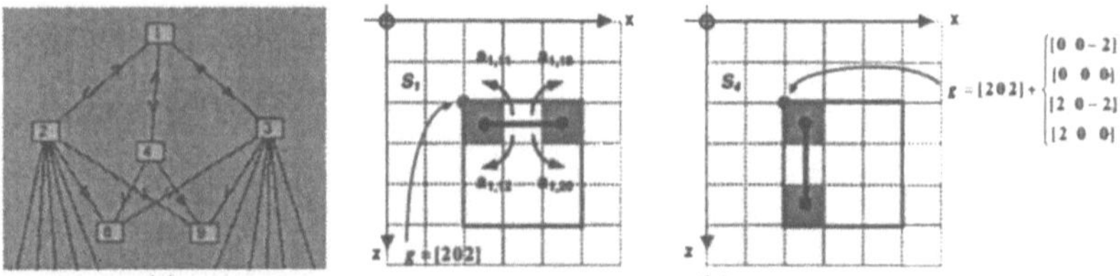


(a) (b)

**Fig. 8.** (a) Transitions from state #1 and (b) action/translation definitions from state #1 to state #4

## 4.2 Complexity

The number of possible states increases exponentially with the number of cubes and links. Note that for a given position of cubes, there may be several states represented by the same connection matrix *f*, but have different link positions and shapes. Each connection between states on the transition diagram represents one or more actions and their associated state-to-state translation in 3-D. For example, as seen in Figure 8b, there are four possible actions from state #1 to state #4, each associated with a different translation value.

While solving a problem for a large group of cubes and link, one has to generate the search graph of nodes $N_i$ evaluating all possible states located at all possible positions. This new graph of nodes is significantly larger that the state transition diagram. Finding a solution with minimal number of actions corresponds to a shortest path problem on this graph of nodes. This problem can be solved using known methods such as A* search where the cost of actions is associated with 90-degree link motions and the cost estimate for reaching the goal can be chosen as the total or average distance of all cubes and/or links to their goal positions. The computation of this search graph is computationally exhaustive due to exponential growth in graph size based on the number of cubes and links.

Finding an upper bound on the number of link actions for a given cube motion proves to be difficult. However, following the discussion of [14], we can state that the number of cube motions for any given problem with a single cube relocation cannot be greater than the perimeter, *i.e.*, neighboring cells of the group of cubes. This perimeter for groups of cubes in 3-D is found to be $10n+8$. Therefore, the total number of cube motions for a single cube relocation is at most $O(n)$. For each cube move, the total number of links that detach/attach or rotate is *m* (*i.e.*, the number of links) at maximum. This is assuming all the links need to moved; the actual number is usually much less than *m*. For each link, the maximum number of 'jumps' from one cube to another is limited to $n$-1. While attached to a single cube, a link can reach any position with a maximum face transition of five (Again, the actual number is less than this). Therefore, a loose upper bound on the number of link motions for a single cube relocation is $O(m.n^2)$. The total number of cube motions depends on the initial and final configurations. For the problems solved by the planner, we see that the number of link motions is much less than the possible orders of magnitudes given above for relatively small groups. Note that the discussion here does not consider interleaved link actions, which is the case most of the time.

## 5. Algorithms

Finding a sequence of actions for a given problem involving $m$ cubes and $n$ links where $m, n > 2$, is fairly difficult due to the nature of the problem as indicated above. Evaluating all possible states/nodes for large group of modules is not possible. However, we might be able find solutions by carefully evaluating states/nodes. For example, the states and nodes necessary to solve a problem can be generated only when necessary as indicated in Section 6. It is also possible to divide the problem into *sub-problems* that can be solved by using heuristics methods. If we are to generate a sequence of actions for a given Problem $\{N_o, N_f\}$, this solution can be divided into subsequences so that the difference between the states corresponding to the beginning and end of each subsequence is a single cube relocation. If there is a solution from $N_o$ to $N_f$, then it can be divided into subsequences that are solutions to less complex and 'local' problems. Each of these sub-problems can be solved using standard heuristic methods. For a given single cube motion, simple heuristics can find a solution for the necessary link motions using the evaluation functions mentioned above. To generate the sub-problems, i.e., to find the individual cube motions to be considered, we can use the methods described in [13] or take the approach given in Section 5.1.

### 5.1 Two-Level Planner

The planner for cube and link motions attacks the problem in two steps. At the lower level, the sub-problems involving limited cube motions are solved, finding a sequence of link actions leading to the solution. At the higher level, individual cube motions are evaluated based on the goal position and shape. The results presented here are generated by two similar approaches. The first approach (we call *SolverIC*) evaluates the desired position for each cube based on occupied goal positions and cube distances to unoccupied positions. These positions are sorted and updated based on the feasibility of the required motions (e.g., cubes must clear all obstacles and other cubes, and keep a maximum distance of 4 using Manhattan metric as well as maximum distance of 2 in each direction to the closest neighbor), possible conflicts between two desired positions, and the requirements for a cube to be in a specific position as pivot. After the negotiation phase, the cubes that are permitted move to their new positions and the cycle is repeated. One or more cubes can move at a time; maximum number of cubes allowed to move is a user-defined variable.

The second approach (*SolverFS*) uses a slightly different approach in evaluating the desired cube motions. All possible cube motions are again filtered using the motion requirements, and resulting nodes are sorted using an evaluation function that defines the distance to the goal node. The cube motion that is evaluated to be the best action is carried out if the pivot cube requirements are met. This algorithm identifies a single cube to be moved at a given step. Both algorithms provided very fast solutions for the problems we have tested. The second approach usually returns better solutions for cases similar to one given in Figure 5, because it evaluates resulting nodes with respect to their distance to goal position. However, the first approach (SolverIC) can handle larger groups of cubes evaluating simultaneous cube motions. Figure 9 shows initial and final conditions for a 27-cube problem and the snapshots of a solution for the given problem. Total number of steps is 22. Note that their success rate is based on the number and diversity of the allowed cube motions.

**Fig. 9.** 27 cubes dissolving: snapshots of a solution with simultaneous cube motions

### 5.2 Pruning the Search Space

The motion planner, while evaluating states (and nodes) resulting from specific actions uses several rules to prune some of these nodes from the search space. Although the new states are added to the transition diagram, if the action is found to be not permissible for the present node (i.e., present state at the given global position), the resulting node is not added to the search list. Possible reasons for a new node to be rejected are:

- Link/cube motion does not clear obstacles
- Cube is swept on the ground or slid over obstacles
- There are constraints on the total number of supported cubes
- Cubes move beyond the user-defined ranges
- The overturning moment resulting for link/cube motion causes instability

All of these rules except the last are easy to check. For the stability analysis, the calculations are slightly more complex, and the order of complexity of the algorithms defined below are $O(n^2)$. The current stability analysis is limited to overturning, and does not consider sliding stability. The surface on which the robot rests is assumed to be flat, level, and a hard stable surface capable of supporting the robot without deflection or distortion. The analysis does take into account the cubes placed on obstacles at different elevations, but does not account for sloped surfaces. The analysis is independent of the plane of analysis; it can be performed for any type of force, similar to gravity, static in nature, irrespective of its direction.

The stability analysis is simplified by the geometry of the robot. All cube faces are either perpendicular or parallel to the faces of adjacent cubes to which they are linked. The link geometry in terms of arm length is fixed, and movements can only result in motions that end in cubes being translated a fixed distance (on a plane perpendicular to one of the default axes) or rotated by 90 degrees at its center. The location of center of gravity on the plane normal to the direction of gravity is compared to the footprint of the robot, which will be stationary during a projected move. If the location of the center of gravity of the entire robot including the moving portion falls within the base of the stationary portion during the course of the movement, the move is found to be safe and will not cause the robot to overturn. If the location of the center of gravity moves to outside the base area, the motion will cause the robot to turn over.

The effective base area of the robot is the area of the stationary portion of the robot that is defined by the boundary formed by the outermost points. These points are simply the corners of the cubes that lie on the ground, or obstacles, and are stationary for the particular move for which this analysis is done. The base area calculation involves finding the 'convex hull' of the base points that form the outer boundary of the cubes that do not move. As a first step, the base cubes are identified. From the base cubes, a collection of the points forming the convex hull is sorted out. Note that the changes in the convex hull at each time step are due to at most one cube relocation. The results of the algorithm are compared to several cases (such as the one in Figure 10d) evaluated in Working Model 3D.

## 6. Simulation Results

The planners described above are used to generate the solutions given in this section. Figure 5 shows a 4C4L group that moves to the right. The difference between initial and final positions is 4 units. The total number of 90-degree link actions to move the group to its desired position is 22. This solution is generated using a high-level SolverFS. Only the final positions of the cubes are defined. For sub-problems, only the cube positions are considered even if final link positions are given. The planner created four sub-problems to be solved by the low-level solver. The maximum number of states and nodes evaluated for the sub-problems are 453 and 108 respectively. Average Manhattan distance over all cubes is used to evaluate the node distance to goal. The snapshots in Figure 5 belong to this solution.

We have also tried to solve this problem using a single A* solver with node-to-node transition cost of 1 and a factor of 2 multiplied with the average cube distance to the goal positions (using Manhattan metric). The total number of actions is again 22, with 2169 nodes and 1542 states generated. Only 581 nodes are evaluated to reach the solution. It is interesting to note that these and the solution given by 3-level solver defined in [13] are found to be the same.

Figure 10 shows snapshots of a solution to stair climbing problem of Figure 4 obtained using a high-level SolverFS. Only the final positions of the cubes are defined in the problem. Total number of link motions to reach the goal is 56. Four sub-problems (initial states shown in Fig. 10a, b, c, and e) are generated. Most difficult sub-problem is the third. Total number of nodes generated for this section is 8501 of which 3244 are evaluated. There were 7555 states associated with this sub-problem. The snapshots show the beginning node of each sub-problem and the final position. A solution for this problem could not be found using the algorithm described in [13] due to the lack of additional rules for higher level solvers.

Similarly, this new approach is capable of finding exact solutions to other problems given in [13] as well as the scenario described in Figure 6. The total number of actions in the computer-generated solution for this case is 10.

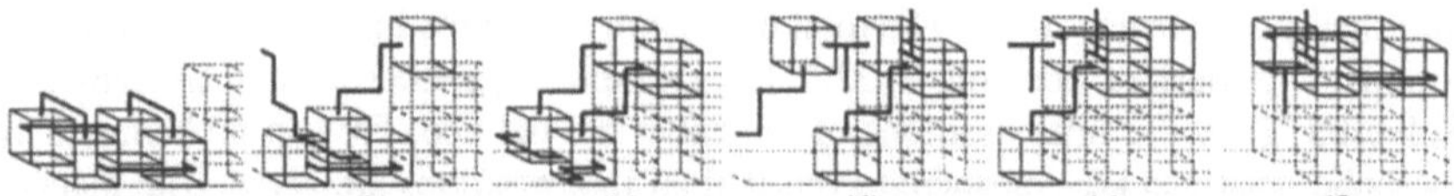

(a) (b) (c) (d) (e) (f)

**Fig. 10.** A solution using FS-A* pair: (a) Initial, (f) final, and (b,c,e) beginning nodes for sub-problems

## 7. Concluding Remarks

We have introduced our results of a divide-and-conquer approach to motion planning for a class of modular self-reconfiguring robotic system. Our attempts to find fast solutions take advantage of defining tractable sub-problems for a given pair of initial and final conditions. Separating the problem into smaller, less complex cases has its advantages; however, issues such as backtracking, algorithmic complexity for low-level planning are yet to be answered. Note that the algorithms described here are capable of finding solutions for initial and final conditions with no overlapping link or cube positions.

In many cases, it is advantageous to combine two or more sub-problems involving single cube motion. In addition, we have found that the high-level planner sometimes returns sub-problems that lead to unnecessary link motions when combined. It is possible to design decision mechanisms to combine sub-problems, reconsider cube motions, or refine the sequence of link motions evaluated by low-level planner. However, the design of these mechanisms is not straightforward. Furthermore, the planner still suffers from the lack of backtracking capabilities for cube motions. We are currently working on incorporating semantic representation of the group into low-level search for speeding up the search process. Our plans also include extending our motion planners using other pruning methods and finding alternatives to heuristic search carried out at the lower level.

## References

[1] Chen I.-M., J. W. Burdick. (1993) Enumerating Non-Isomorphic Assembly Configurations of a Modular Reconfigurable Robotic System, IEEE Intl. Conf. on Intel. Rob. & Syst., 1985-92.

[2] Paredis, C.J.J., H. B. Brown, and P. K. Khosla. (1997) A Rapidly Deployable Manipulator System. Robotics and Autonomous Systems, 21, 289-304.

[3] Fukuda, T., and Y. Kawauchi. (1990) Cellular Robotic System as One of the Realization of Self-Organizing Intelligent Universal Manipulator, IEEE Conf. on Rob. & Auto., 662-667.

[4] Kotay, K., and D. Rus. (2000) Inchworm Robot: A Multi-functional System, Autonomous Robots, 8, 53-69.

[5] Hosokawa, K., *et al.* (1998) Mechanisms for self-organizing robots which reconfigure in a vertical plane. Distributed Autonomous Robotic Systems, 3. Springer-Verlag, NY. 111-118.

[6] Yoshida, E., *et al.* (1998) Experiments of Self-Repairing Modular Machine. Dist. Autonomous Robotic Systems 3. Springer-Verlag, NY. 119-128.

[7] Pamecha, A., *et al.* (1996). Design and Implementation of Metamorphic Robots, Proc. ASME Design Engineering Technical Conf. & Computers in Engineering. Conf.

[8] Yim, M. (1994). New Locomotion Gaits, Proc. IEEE Intl. Conf. on Rob. & Auto.

[9] Yim, M., D. Duff and K. Roufas. (2000) PolyBot: a Modular Reconfigurable Robot, Proc. IEEE Intl. Conf. On Robotics & Automation, 514-520.

[10] Kotay, K., *et al.* (1998) The Self-reconfiguring Molecule: Design and Control Algorithms. Algorithmic Foundations of Robotics. A.K. Peters.

[11] Murata, S., *et al.* (1998) A 3-D Self-Reconfigurable Structure, IEEE Intl. Conf. on Robotics & Automation, 432-439.

[12] Ünsal, C., and P. Khosla (2000) Mechatronic Design of a Modular Self-Reconfiguring Robotic System, Proc. IEEE Intl. Conf. on Robotics & Automation, 1742-1747.

[13] Ünsal, C., *et al.* (2000). A Modular Self-reconfigurable Bipartite Robotic System: Implementation and Motion Planning, Tech. Rep., ICES, CMU, Pittsburgh, PA.

[14] Chirikjian G.S., and A. Pamecha. (1995) Bounds for Self-reconfiguration of Metamorphic Robots. JHU Technical Report, RMS-9-95-1.

# Chapter 6

# Localization

# Distributed Multi-Robot Localization

Stergios I. Roumeliotis and George A. Bekey

Robotics Research Laboratories
University of Southern California
Los Angeles, CA 90089-0781
stergios|bekey@robotics.usc.edu

**Abstract.** This paper presents a new approach to the cooperative localization problem, namely *distributed multi-robot localization*. A group of M robots is viewed as a single system composed of robots that carry, in general, different sensors and have different positioning capabilities. A single Kalman filter is formulated to estimate the position and orientation of all the members of the group. This centralized schema is capable of fusing information provided by the sensors distributed on the individual robots while accommodating independencies and interdependencies among the collected data. In order to allow for distributed processing, the equations of the centralized Kalman filter are treated so that this filter can be decomposed into M modified Kalman filters each running on a separate robot. The distributed localization algorithm is applied to a group of 3 robots and the improvement in localization accuracy is presented.

## 1 Introduction

In order for a mobile robot to autonomously navigate, it must be able to localize itself [6]; i.e. to know its position and orientation (pose). Localization has always been a problem for both indoor and outdoor mobile robots. Different types of sensors [7] and techniques have been employed to attack this problem (e.g. [5], [4], [14], [19]). The basic idea behind most of the current localization systems is to combine measurements from proprioceptive sensors that monitor the motion of the vehicle with information collected by exteroceptive sensors that provide a representation of the environment and its signals.

Many robotic applications require that robots work in collaboration in order to perform a certain task [8], [15]. Most existing localization approaches refer to the case of a single robot. Even when a group of, say $M$, robots is considered, the group localization problem is usually resolved by independently solving $M$ pose estimation problems. Each robot estimates its position based on its individual experience (proprioceptive and exteroceptive sensor measurements). Knowledge from the different entities of the team is not combined and each member must rely on its own resources (sensing and processing capabilities). This is a relatively simple approach since it avoids dealing with the complicated problem of fusing information from a large number of independent and interdependent sources. On the other hand, a more coordinated schema for localization has a number of advantages that can compensate for the added complexity.

First let us consider the case of a homogeneous group of robots. As we mentioned earlier, robotic sensing modalities suffer from uncertainty and noise. When a number of robots equipped with the same sensors detect a particular feature of the environment, such as a door, or measure a characteristic property of the area, such as the local vector of the earth's magnetic field, a number of *independent* measurements originating from the different members of the group is collected. Properly combining all this information will result in a single estimate of increased accuracy and reduced uncertainty. A better estimate of the position and orientation of a landmark can drastically improve the outcome of the localization process and thus this group of robots can benefit from this collaboration schema.

The advantages stemming from the exchange of information among the members of a group are more crucial in the case of heterogeneous robotic colonies. When a team of robots is composed of different platforms carrying different proprioceptive and exteroceptive sensors and thus having different capabilities for self-localization, the quality of the localization estimates will vary significantly across the individual members. For example, a robot equipped with a laser scanner and expensive INS/GPS modules will outperform another member that must rely on wheel encoders and cheap sonars for its localization needs. Communication and flow of information among the members of the group constitutes a form of sensor sharing and can improve the overall positioning accuracy.

## 2 Previous Approaches

An example of a system that is designed for cooperative localization is presented in [12]. The authors acknowledge that dead-reckoning is not reliable for long traverses due to the error accumulation and introduce the concept of "portable landmarks". A group of robots is divided into two teams in order to perform cooperative positioning. At each time instant, one team is in motion while the other remains stationary and acts as a landmark. In the next phase the roles of the teams are reversed and this process continues until both teams reach the target. This method can work in unknown environments and the conducted experiments suggest accuracy of 0.4% for the position estimate and 1 degree for the orientation [11]. Improvements over this system and optimum motion strategies are discussed in [10]. A similar realization is presented in [16], [17]. The authors deal with the problem of exploration of an unknown environment using two mobile robots. In order to reduce the odometric error, one robot is equipped with a camera tracking system that allows it to determine its relative position and orientation with respect to a second robot carrying a helix target pattern and acting as a portable landmark. Both previous approaches have the following limitations:

(a) Only one robot (or team) is allowed to move at a certain time instant, and (b) The two robots (or teams) must maintain visual contact at all times.

A different implementation of a collaborative multi-robot localization schema is presented in [9]. The authors have extended the Monte Carlo localization algorithm to the case of two robots when a map of the area is available to both robots. When these robots detect each other, the combination of their belief functions facilitates their global localization task. The main limitation of this approach is that it can be applied only within known indoor environments. In addition, since information interdependencies are being ignored every time the two robots meet, this method can lead to overoptimistic position estimates.

Although practices like those previously mentioned can be supported within the proposed ***distributed multi-robot localization*** framework (Section 5), the key difference is that it provides a solution to the most general case where all the robots in the group can move simultaneously while continuous visual contact or a map of the area are not required. In order to treat the group localization problem, we begin from the reasonable assumptions that the robots within the group can communicate with each other (at least 1-to-1 communication) and carry two types of sensors: 1. Proprioceptive sensors that record the self motion of each robot and allow for position tracking, 2. Exteroceptive sensors that monitor the environment for (a) (static) features and identities of the surroundings of the robot to be used in the localization process, and (b) other robots (treated as dynamic features). The goal is to integrate measurements collected by different robots and achieve localization across all the robotic platforms constituting the group.

The key idea for performing ***distributed multi-robot localization*** is that the group of robots must be viewed as one entity, the "group organism", with multiple "limbs" (the individual robots in the group) and multiple virtual "joints" visualized as connecting each robot with every other member of the team. The virtual "joints" provide 3 degrees of freedom $(\Delta x, \Delta y, \Delta \phi)$ and thus allow the "limbs" to move in every direction within a plane without any limitations. Considering this perspective, the "group organism" has access to a large number of sensors such as encoders, gyroscopes, cameras etc. In addition, it "spreads" itself across a large area and thus it can collect far more rich and diverse exteroceptive information. When one robot detects another member of the team and measures its relative pose, it is equivalent to the "group organism's" joints measuring the relative displacement of these two "limbs". When two robots communicate for information exchange, this can be seen as the "group organism" allowing information to travel back and forth from its "limbs". This information can be fused by a centralized processing unit and provide improved localization results for all the robots in the group. At this point it can be said that a realization of a two-member "group organism" would resemble the multiple degree of freedom robot with compliant linkage shown to improve localization implemented by J. Borenstein [1], [2],

[3].

The main drawback of addressing the cooperative localization problem as an information combination problem within a single entity ("group organism") is that it requires centralized processing and communication. The solution would be to attempt to decentralize the sensor fusion within the group. The ***distributed multi-robot localization*** approach uses the previous analogy as its starting point and treats the processing and communication needs of the group in a distributed fashion. This is intuitively desired; since the sensing modalities of the group are distributed, so should be the processing modules. At this point it is worth mentioning that a decentralized form of the Kalman filter was first presented in [21] and later revisited in its inverse (Information filter) formulation in [13] for sequential processing of incoming sensor measurements. These forms of the Kalman filter are particularly useful when dealing with asynchronous measurements originating from a variety of sensing modalities (an application of this can be found in [18]). As it will be obvious in the following sections, our formulation differs from the aforementioned ones on its starting point. It is based on the unique characteristic of the multi-robot localization problem that the state propagation equations of the centralized system are decoupled while state coupling occurs only when relative pose measurements become available. Our focus is distributed state estimation rather than sequential sensor processing. Nevertheless, the latter can be easily incorporated in the resulting distributed localization schema.

In order to deal with the cross-correlation terms (localization interdependencies) that can alter the localization result [20], the data processed during each ***distributed multi-robot localization*** session must be propagated among all the robots in the group. While this can happen instantly in groups of 2 robots, in the following sections we will show how this problem can be treated by reformulating the ***distributed multi-robot localization*** approach so it can be applied in groups of 3 or more robots.

## 3 Problem Statement

We state the following assumptions:

1. A group of $M$ independent robots move in an $N-$ *dimensional* space. The motion of each robot is described by its own linear or non-linear equations of motion,
2. Each robot carries proprioceptive and exteroceptive sensing devices in order to propagate and update its own position estimate. The measurement equations can differ from robot to robot depending on the sensors used,
3. Each robot carries exteroceptive sensors that allow it to detect and identify other robots moving in its vicinity and measure their respective displacement (relative position and orientation),

4. All the robots are equipped with communication devices that allow exchange of information within the group.

As we mentioned before, our starting point is to consider this group of robots as a single centralized system composed of each and every individual robot moving in the area and capable of sensing and communicating with the rest of the group. In this centralized approach, the motion of the group is described in an $N \times M$-dimensional space and it can be estimated by applying Kalman filtering techniques. The goal now is to treat the Kalman filter equations of the centralized system so as to distribute the estimation process among $M$ Kalman filters, each of them operating on a different robot. Here we will derive the equations for a group of $M = 3$ robots. The same steps describe the derivation for larger groups. The trajectory of each of the 3 robots is described by the following equations:

$$\boldsymbol{x}_i(t_{k+1}^-) = \Phi_i(t_{k+1}, t_k)\boldsymbol{x}_i(t_k^+) + B_i(t_k)\boldsymbol{u}_i(t_k) + G_i(t_k)\boldsymbol{n}_i(t_k), \quad i = 1..3 \tag{1}$$

where $\Phi_i(t_{k+1}, t_k)$ is the system propagation matrix describing the motion of vehicle $i$, $B_i(t_k)$ is the control input matrix, $\boldsymbol{u}_i(t_k)$ is the measured control input, $G_i(t_k)$ is the system noise matrix, $\boldsymbol{n}_i(t_k)$ is the system noise associated with each robot and $Q_{di}(t_k)$ is the corresponding system noise covariance matrix.

# 4 Distributed Localization after the first Update

In this section we present the propagation and update cycles of the Kalman filter estimator for the centralized system *after the first update.*[1] Since there have been introduced cross-correlation elements in the covariance matrix of the state estimate, this matrix would now have to be written as:

$$P(t_{k+1}^-) = \begin{bmatrix} P_{11}(t_{k+1}^-) & P_{12}(t_{k+1}^-) & P_{13}(t_{k+1}^-) \\ P_{21}(t_{k+1}^-) & P_{22}(t_{k+1}^-) & P_{23}(t_{k+1}^-) \\ P_{31}(t_{k+1}^-) & P_{32}(t_{k+1}^-) & P_{33}(t_{k+1}^-) \end{bmatrix} \tag{2}$$

## 4.1 Propagation

Since each of the 3 robots moves independent of the others, the state (pose) propagation is provided by Equations (1). The same is not true for the covariance of the state estimate. In [20], we derived the equations for the propagation of the initial, fully decoupled system. Here we will examine how the Kalman filter propagation equations are modified in order to include the cross-correlation terms introduced after a few updates of the system. Starting from:

$$P(t_{k+1}^-) = \Phi(t_{k+1}, t_k)P(t_k^+)\Phi^T(t_{k+1}, t_k) \; + \; Q_d(t_{k+1}) \tag{3}$$

[1] Due to space limitations the propagation and update equations of the Kalman filter before and up to the first update are omitted from this presentation. The interested reader is referred to [20] for a detailed derivation.

and substituting from Equation (2) we have:

$$P(t_{k+1}^-) = \begin{bmatrix} \Phi_1 P_{11}(t_k^+)\Phi_1^T + Q_{d1} & \Phi_1 P_{12}(t_k^+)\Phi_2^T & \Phi_1 P_{13}(t_k^+)\Phi_3^T \\ \Phi_2 P_{21}(t_k^+)\Phi_1^T & \Phi_2 P_{22}(t_k^+)\Phi_2^T + Q_{d2} & \Phi_2 P_{23}(t_k^+)\Phi_3^T \\ \Phi_3 P_{31}(t_k^+)\Phi_1^T & \Phi_3 P_{32}(t_k^+)\Phi_2^T & \Phi_3 P_{33}(t_k^+)\Phi_3^T + Q_{d3} \end{bmatrix} \quad (4)$$

Equation (4) is repeated at each step of the propagation and it can be distributed among the robots after appropriately splitting the cross-correlation terms. For example, the cross-correlation equations for robot 2 are:

$$\sqrt{P_{21}(t_{k+1}^-)} = \Phi_2\sqrt{P_{21}(t_k^+)}\ ,\ \sqrt{P_{23}(t_{k+1}^-)} = \Phi_2\sqrt{P_{23}(t_k^+)} \quad (5)$$

After a few steps, if we want to calculate the (full) cross-correlation terms of the centralized system, we will have to multiply their respective components. For example:

$$P_{32}(t_{k+1}^-) = \sqrt{P_{32}(t_{k+1}^-)}\ \sqrt{P_{23}(t_{k+1}^-)}^T = \Phi_3\sqrt{P_{32}(t_k^+)}(\Phi_2\sqrt{P_{23}(t_k^+)})^T =$$

$$\Phi_3\sqrt{P_{32}(t_k^+)}(\sqrt{P_{23}(t_k^+)})^T\Phi_2^T = \Phi_3\sqrt{P_{32}(t_k^+)}\sqrt{P_{32}(t_k^+)}\Phi_2^T = \Phi_3 P_{32}(t_k^+)\Phi_2^T \quad (6)$$

This result is very important since the propagation Equations (1) and (5) to (6) allow for a **fully distributed estimation algorithm during the propagation cycle**. The computation gain is very large if we consider that most of the time the robots propagated their pose and covariance estimates based on their own perception while updates are usually rare and they take place only when two robots meet.

### 4.2 Update

If now we assume that robots 2 and 3 are exchanging relative position and orientation information, the residual covariance matrix:

$$S(t_{k+1}) = H_{23}(t_{k+1})P(t_{k+1}^-)H_{23}^T(t_{k+1}) + R_{23}(t_{k+1}) \quad (7)$$

is updated based on Equation (2) as:

$$S(t_{k+1}) = \begin{bmatrix} 0 & I & -I \end{bmatrix} \begin{bmatrix} P_{11}(t_{k+1}^-) & P_{12}(t_{k+1}^-) & P_{13}(t_{k+1}^-) \\ P_{21}(t_{k+1}^-) & P_{22}(t_{k+1}^-) & P_{23}(t_{k+1}^-) \\ P_{31}(t_{k+1}^-) & P_{32}(t_{k+1}^-) & P_{33}(t_{k+1}^-) \end{bmatrix} \begin{bmatrix} 0 \\ I \\ -I \end{bmatrix} + R_{23}(t_{k+1}) =$$

$$= P_{22}(t_{k+1}^-) - P_{32}(t_{k+1}^-) - P_{23}(t_{k+1}^-) + P_{33}(t_{k+1}^-) + R_{23}(t_{k+1}) \quad (8)$$

where $R_{23}(t_{k+1})$ is the measurement noise covariance matrix associated with the relative position and orientation measurement between robots 2 and 3. In order to calculate matrix $S(t_{k+1})$, only the covariances of the two meeting robots are needed along with their cross-correlation terms. All these terms can be exchanged when the two robots detect each other, and then used to calculate the residual covariance matrix $S$. The dimension of $S$ is $N \times N$, the same as if we were updating the pose estimate of one robot instead of three. (In the latter case the dimension of matrix $S$ would be $(N \times 3) \times (N \times 3)$). As we will see in Equation (9), this reduces the computations required for calculating the Kalman gain and later for updating the covariance of the pose estimate. The Kalman gain for this update is given by $K(t_{k+1}) =$

$$P(t_{k+1}^-)H_{23}^T(t_{k+1})S^{-1}(t_{k+1}) = \begin{bmatrix} P_{11}(t_{k+1}^-) & P_{12}(t_{k+1}^-) & P_{13}(t_{k+1}^-) \\ P_{21}(t_{k+1}^-) & P_{22}(t_{k+1}^-) & P_{23}(t_{k+1}^-) \\ P_{31}(t_{k+1}^-) & P_{32}(t_{k+1}^-) & P_{33}(t_{k+1}^-) \end{bmatrix} \begin{bmatrix} 0 \\ I \\ -I \end{bmatrix} S^{-1}(t_{k+1})$$

$$= \begin{bmatrix} (P_{12}(t_{k+1}^-) - P_{13}(t_{k+1}^-))\, S^{-1}(t_{k+1}) \\ (P_{22}(t_{k+1}^-) - P_{23}(t_{k+1}^-))\, S^{-1}(t_{k+1}) \\ -(P_{33}(t_{k+1}^-) - P_{32}(t_{k+1}^-))\, S^{-1}(t_{k+1}) \end{bmatrix} = \begin{bmatrix} K_1(t_{k+1}) \\ K_2(t_{k+1}) \\ K_3(t_{k+1}) \end{bmatrix} \tag{9}$$

The correction coefficients (matrix elements $K_i(t_{k+1}), i = 2, 3$, of the Kalman gain matrix) in the previous equation are smaller compared to the corresponding correction coefficients calculated during the first update [20]. Here the correction coefficients are reduced by the cross-correlation terms $P_{23}(t_{k+1}^-)$ and $P_{32}(t_{k+1}^-)$ respectively. This can be explained by examining what is the information contained in these cross-correlation matrices. As it is described in [20], the cross-correlation terms represent the information common to the two meeting robots acquired during a previous direct (robot 2 met robot 3) or indirect (robot 1 met robot 2 and then robot 2 met robot 3) exchange of information. The more knowledge these two robots (2 and 3) already share, the less gain can have from this update session as this is expressed by the values of the matrix elements of the Kalman filter (coefficients $K_i(t_{k+1}), i = 2, 3$) that will be used for update of the pose estimate $\widehat{\mathbf{x}}(t_{k+1}^+)$. In addition to this, by observing that $K_1(t_{k+1}) = (P_{12}(t_{k+1}^-) - P_{13}(t_{k+1}^-))\, S^{-1}(t_{k+1})$ we should infer that robot 1 will be affected by this update to the extent that the information shared between robots 1 and 2 differs from the information shared between robots 1 and 3.

Finally, as it is shown in [20], the centralized system covariance matrix calculation can be divided into $3(3 + 1)/2 = 6$, $N \times N$ matrix calculations and distributed among the robots of the group.[2]

# 5 Observability Study

## 5.1 Case 1: At least one of the robots has absolute positioning capabilities

In this case the main difference is in matrix H. If we assume that robot 1 for example has absolute positioning capabilities then the measurement matrix $H$ and the observability matrix $M_{DTI}$ would be:

$$H = \begin{bmatrix} I & 0 & 0 \\ I & -I & 0 \\ 0 & I & -I \\ -I & 0 & I \end{bmatrix}, \quad M_{DTI} = \left[\begin{array}{cccc|cccc|cccc} I & I & 0 & -I & I & I & 0 & -I & I & I & 0 & -I \\ 0 & -I & I & 0 & 0 & -I & I & 0 & 0 & -I & I & 0 \\ 0 & 0 & -I & I & 0 & 0 & -I & I & 0 & 0 & -I & I \end{array}\right] \tag{10}$$

[2] In general $M(M+1)/2$ matrix equations distributed among M robots, thus $(M+1)/2$ matrix calculations per robot.

The rank of the $M_{DTI}$ matrix is 9 and thus the system is observable when at least one of the robots has access to absolute positioning information (e.g. by using GPS or a map of the environment).

### 5.2 Case 2: At least one of the robots remain stationary

If at any time instant at least one of the robots in the group remains stationary, the uncertainty about its position will be constant and thus it has a direct measurement of its position which is the same as before. This case therefore falls into the previous category and the system is considered observable. Examples of this case are the applications found in [12], [11], [10], [16], [17].

## 6 Simulation Results

The proposed *distributed multi-robot localization* method was implemented and tested in simulation for the case of 3 mobile robots. The most significant result is the reduction of the uncertainty regarding the position and orientation estimates of each individual member of the group.

The 3 robots start from 3 different locations and they move within the same area. Every time a meeting occurs, the two robots involved measure their relative position and orientation. Information about the cross-correlation terms is exchanged among the members of the group and the distributed modified Kalman filters update the pose estimates for each of the robots. In order to focus on the effect of the *distributed multi-robot localization* algorithm, no absolute localization information was available to any of the robots. Therefore the covariance of the position estimate for each of them is bound to increase while the position estimates will drift away from their real values.

At time t=320 robot 1 meets robot 2 and they exchange relative localization information. At time t=720 robot 1 meets robot 3 and they also perform *distributed multi-robot localization.* As it can be seen in Fig. 1, after each exchange of information, the covariance, representing the uncertainty of the position and orientation estimates, of robots 1 and 2 (t=320) and 1 and 3 (t=720) is significantly reduced. Robot 1 that met with other robots of the group twice, has significantly lower covariance value at the end of the test (t=1000).

## References

1. J. Borenstein. Control and kinematic design of multi-degree-of freedom mobile robots with compliant linkage. *IEEE Transactions on Robotics and Automation*, 11(1):21–35, Feb. 1995.
2. J. Borenstein. Internal correction of dead-reckoning errors with a dual-drive compliant linkage mobile robot. *Journal of Robotic Systems*, 12(4):257–273, April 1995.

3. J. Borenstein. Experimental results from internal odometry error correction with the omnimate mobile robot. *IEEE Transactions on Robotics and Automation*, 14(6):963–969, Dec. 1998.
4. J. Borenstein and L. Feng. Gyrodometry: A new method for combining data from gyros and odometry in mobile robots. In *Proceedings of the 1996 IEEE International Conference on Robotics and Automation*, pages 423–428, 1996.
5. J. Borenstein and L. Feng. Measurement and correction of systematic odometry errors in mobile robots. *IEEE Transactions on Robotics and Automation*, 12(6):869–880, Dec. 1996.
6. I. J. Cox. Blanche-an experiment in guidance and navigation of an autonomous robot vehicle. *IEEE Transactions on Robotics and Automation*, 7(2):193–204, April 1991.
7. H. R. Everett. *Sensors for Mobile Robots*. AK Peters, 1995.
8. M.S. Fontan and M.J. Mataric. Territorial multi-robot task division. *IEEE Transactions on Robotics and Automation*, 14(5):815–822, Oct. 1998.
9. D. Fox, W. Burgard, H. Kruppa, and S. Thrun. Collaborative multi-robot localization. In *In Proc. of the 23rd Annual German Conference on Artificial Intelligence (KI)*, Bonn, Germany, 1999.
10. R. Kurazume and S. Hirose. Study on cooperative positioning system: optimum moving strategies for cps-iii. In *Proceedings of the 1998 IEEE International Conference in Robotics and Automation*, volume 4, pages 2896–2903, Leuven, Belgium, 16-20 May 1998.
11. R. Kurazume, S. Hirose, S. Nagata, and N. Sashida. Study on cooperative positioning system (basic principle and measurement experiment). In *Proceedings of the 1996 IEEE International Conference in Robotics and Automation*, volume 2, pages 1421–1426, Minneapolis, MN, April 22-28 1996.
12. R. Kurazume, S. Nagata, and S. Hirose. Cooperative positioning with multiple robots. In *Proceedings of the 1994 IEEE International Conference in Robotics and Automation*, volume 2, pages 1250–1257, Los Alamitos, CA, 8-13 May 1994.
13. E.M. Nebot M. Bozorg and H.F. Durrant-Whyte. A decentralised navigation architecture. In *In Proceedings of the IEEE International Conference on Robotics and Automation*, volume 4, pages 3413–3418, Leuven, Belgium, 16-20 May 1998.
14. C.F. Olson and L.H. Matthies. Maximum likelihood rover localization by matching range maps. In *Proceedings of the 1998 IEEE International Conference on Robotics and Automation*, pages 272–277, Leuven, Belgium, 16-20 May 1998.
15. L.E. Parker. Alliance: An architecture for fault tolerant multirobot cooperation. *IEEE Transactions on Robotics and Automation*, 14(2):220–240, April 1998.
16. I.M. Rekleitis, G. Dudek, and E.E. Milios. Multi-robot exploration of an unknown environment, efficiently reducing the odometry error. In M.E. Pollack, editor, *Proceedings of the Fifteenth International Joint Conference on Artificial Intelligence (IJCAI-97)*, volume 2, pages 1340–1345, Nagoya, Japan, 23-29 Aug. 1997.
17. I.M. Rekleitis, G. Dudek, and E.E. Milios. On multiagent exploration. In *Visual Interface*, pages 455–461, Vancouver, Canada, June 1998.
18. S. I. Roumeliotis, G. S. Sukhatme, and G. A. Bekey. Sensor fault detection and identification in a mobile robot. In *Proceedings of the 1998 IEEE/RSJ International Conference on Intelligent Robots and Systems*, volume 3, pages 1383–1388, Victoria, BC, Canada, 13-17 Oct. 1998.

19. S.I. Roumeliotis and G.A. Bekey. Bayesian estimation and kalman filtering: A unified framework for mobile robot localization. In *Proceedings of the 2000 IEEE International Conference on Robotics and Automation*, pages 2985–2992, San Fransisco, CA, April 24-28 2000.
20. Stergios I. Roumeliotis. *Robust Mobile Robot Localization: From single-robot uncertainties to multi-robot interdependencies.* PhD thesis, University of Southern California, Los Angeles, California, May 2000.
21. H. W. Sorenson. *Advances in Control Systems*, volume 3, chapter Kalman Filtering Techniques. Academic Press, 1966.

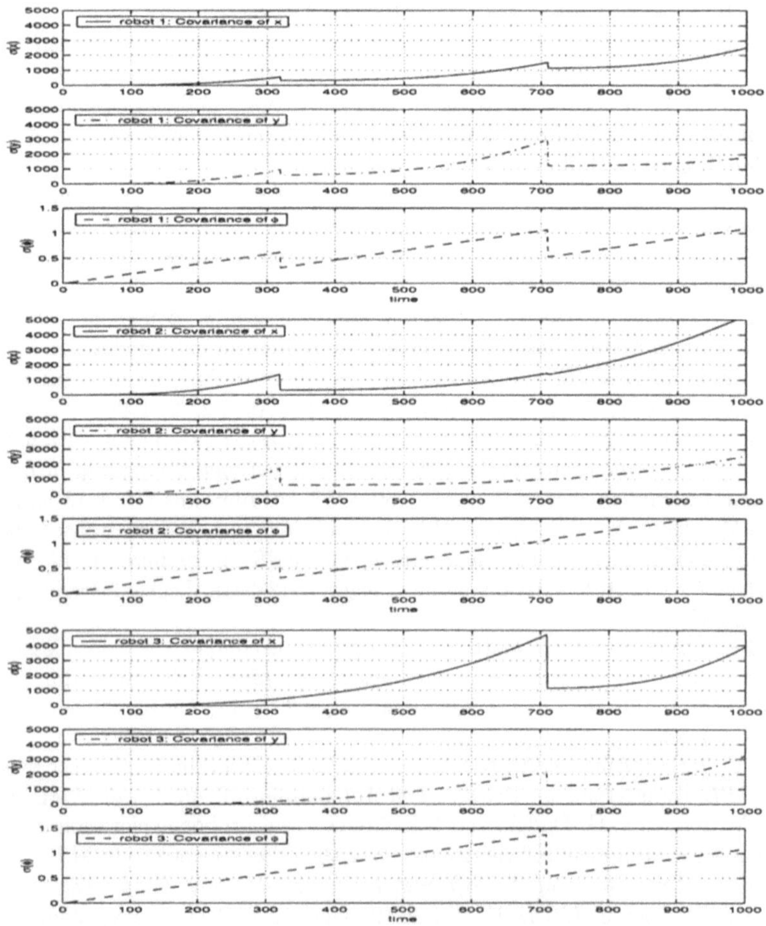


**Fig. 1.** Distributed multi-robot localization results: The covariance of the $x$ (plots 1, 4, 7), $y$ (plots 2, 5, 8), and $\phi$ (plots 3, 6, 9) estimates for each of the three robots in the group.

# Distance Measurment in Multi-Robot Systems based on Time Shared Scheduling

Yoshikazu ARAI[1], Hajime ASAMA[2], Hayato KAETSU[2], and Isao ENDO[2]

[1] Iwate Prefectural University, Sugo 152-52, Takizawa, Iwate 020-0173, Japan
[2] The Institute of Physical and Chemical Research (RIKEN), Hirosawa 2-1, Wako, Saitama 351-0198, Japan

arai@soft.iwate-pu.ac.jp, {asama, kaetsu, endo}@cel.riken.go.jp

**Abstract.** In this paper, we discuss a method of distance measurement in multi-robot systems. Distance information for surrounding environment of a robot is very useful to plan robot's paths. Ultrasonic sensors and some kind of active sensors are used for the purpose actually. However it is difficult for the sensors to measure distances accurately in an environment in which multiple robots exist because emission from the sensors interfere each other. In this paper, we introduce a time shared scheduling for ultrasonic sensors to emit ultrasonic waves without interferences in such environment. In this method, a sensor occupies a space momentarily and is able to measure distance without interferences of emission. After the concept of the LOCISS we have developed to achieve local communication between robots, time shared scheduling is described. It is shown that a robot can measure distances to surrounding environment in which another robot exits, and map the environment based on proposed method.

**Keywords:** Disatance Measurement, Time Shared Scheduling, Sensor Fusion, Multi-Robot System, Local Communication

## 1 Introduction

To recognize a working environment and plan paths of movement, distances between a robot and surrounding objects are very useful information. Ultrasonic sensors are used widely for this purpose [5,6] However, in Distributed Autonomous Robotic Systems (DARS) [2,3,8], it is difficult for the sensors to measure distances accurately. This problem is caused by interferences of ultrasonic waves which multiple sensors emit at the same time. It is very important to resolve this problem for distance measurement. Kawabata et al. have proposed a method which extracts a target from interfered ultrasonic waves [7]. In this method, a special signal processing is required to calculate a cross correlation between interfered waves and a wave which a robot emitted. By scheduling timing, in which sensors of multiple robots emit ultrasonic waves, based on time sharing scheme, the sensors can measure distances without any special signal processing because an emitted wave occupies a space momentarily. In this case, a synchronization of timing between multiple sensors is very important and robots have to communicate

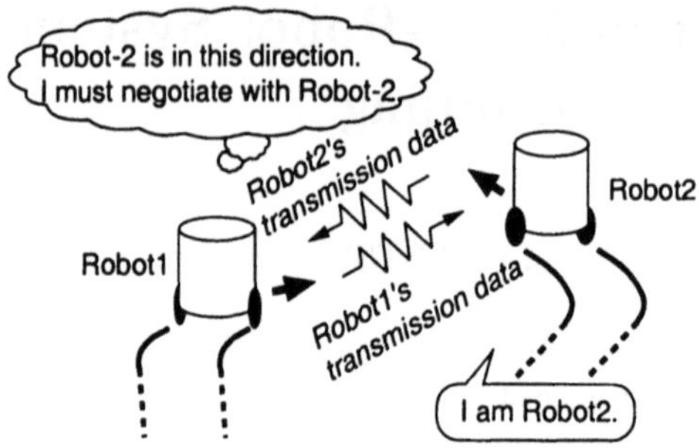


**Fig. 1.** Concept of discriminating surrounding robot

each other to carry out the synchronization. As a method to exchange information between robots, global communications have been used generally [4,10]. These kinds of methods are not suitable for above use which requires frequent communication in a short cycle because overheads of communication are large. In addition, since robots which should synchronize each other are close, it is reasonable to communicate locally only between them. We have developed the local sensing device called "LOCISS (LOcally Communicable Infrared Sensory System)" [9] and applied this device for a collision avoidance problem in multi-robot environment [1]. By the LOCISS, it is possible to achieve multiple local communication between robots at the same time in a working environment.

In this paper, we propose a method for distance measurement in multi-robot systems based on time shared scheduling by synchronizing timing for emitting ultrasonic waves with local communication between robots. By applying this method, robots can measure accurate distance to surrounding objects and recognize the working environment even in multi-robot systems without complex signal processings. The concept of the LOCISS and time shared scheduling to synchronize timing of emissions. Finally, a validity of proposed method is shown through an experiment for a distance measurement and a mapping of the environment using actual robots.

## 2 Structure of Sensor Unit

The LOCISS is a sensing system which utilizes infrared light as media and achieves a local communication between robots by adjusting intensity of emissions. By receiving a signal of the LOCISS, a robot detects surrounding objects. When an object is detected by the LOCISS, it is important to discriminate whether the object is other robot or an obstacle to know how the robot should schedule a timing of emission for distance measurements as described later. For this purpose, an unique ID number for each robot (robot ID) is transmitted as a code which shows a source of signal as shown in Fig. 1. When the robot receives other robot's ID number, it can recognize

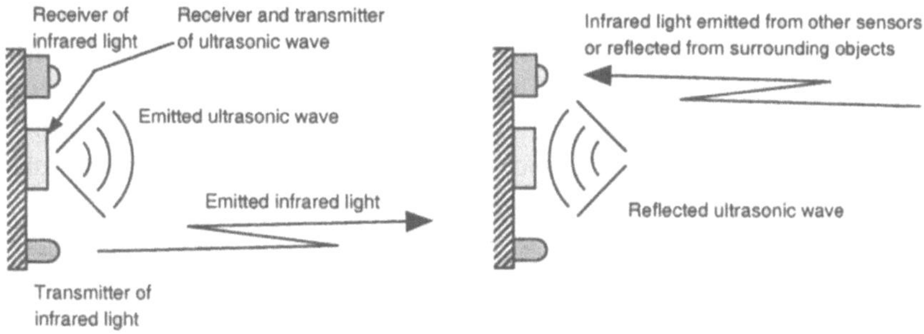


**Fig. 2.** Structure of LOCISS and ultrasonic sensor

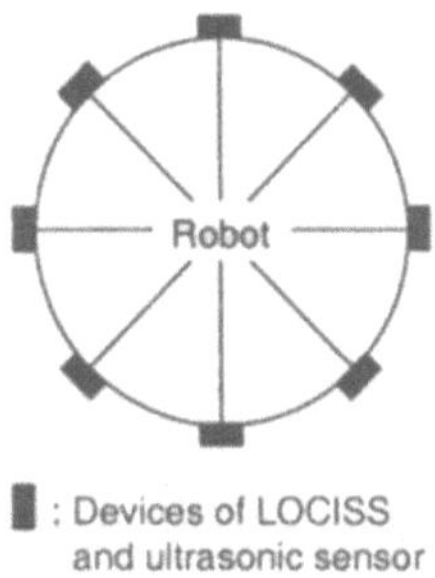


(a) Configration of sensing devices

(b) Sensor unit mounted on robot

**Fig. 3.** Sensor unit constructed with eight pairs of sensing devices

and discriminate the other robot besides it. When the robot receives its own ID number, it can recognize an obstacle because the signal is reflected by the obstacle.

A transmitter and a receiver of sensors are configured as shown in Fig. 2. The devices of an infrared and an ultrasonic sensors are synchronized and repeat transmitting (Fig. 2(a)) and receiving (Fig. 2(b)) in a same sampling cycle. A sensor unit has eight pairs of such devices radially, as shown in Fig. 3(a). Each pairs of infrared devices in the LOCISS has a directivity of 45 degrees to communicate with robots in every direction by eight pairs of devices. A communicable area of the device is limited to about 1 [m] according to characteristics for movement of a robot. The sensor unit mounted on the omni-directional mobile robot is shown in Fig. 3(b).

## 3 Strategy of Distance Measurement

To recognize working environment, a distance measurement by ultrasonic sensors is very useful. However it is difficult for the sensors to measure accurate distance in multi-robot systems due to interferences of ultrasonic waves emitted from themselves. One of the simplest method to resolve this problem is to control timing of emissions exclusively as only an ultrasonic wave emitted from a sensor occupies momentarily. Namely, by scheduling timing for emitting based on time sharing scheme, a distance measurement is accomplished simply without special signal processings even in such environment.

### 3.1 Switching of scheduling methods

Although the scheduling of timing for emitting based on time sharing scheme (time shared scheduling) expects that accurate distance measurement is accomplished, it also brings reduction of samplings. Therefore application of this scheduling should be minimized and the scheduling of timing for continuous emitting (continuous scheduling) should be usually applied. The necessity of the time shared scheduling depends on an existence of surrounding robot. Therefore a robot detects the surrounding robot by using the LOCISS and switches above two scheduling policies. The robot emits ultrasonic waves based on the timing planed on the scheduling.

### 3.2 Time Shared Scheduling

While a robot detects no surrounding robot, the robot emits ultrasonic waves continuously based on the continuous scheduling on one-sided matter. Therefore, to accomplish the time shared scheduling, it is assumed that both robots detect each other. However there is a case, that one dose not detect the other though the other detects one, due to positional gaps between the robots. In this case, the other must stop emitting and wait for one to detect the other because one has no timing to stop emitting. After both robots detect each other, they alternate emitting and waiting frequently. Based on above consideration, following states and behaviors which a robot should assume and execute in the state are designed to accomplish the time shared scheduling.

- $S_0$ : No surrounding robot is detected $\rightarrow$ emitting
- $S_1$ : The robot detects surrounding robot, but is not detected $\rightarrow$ waiting
- $S_2$ : The time shared scheduling is accomplished and the robot should emit in this turn $\rightarrow$ emitting
- $S_3$ : The time shared scheduling is accomplished and the robot should wait in this turn $\rightarrow$ waiting

In addition, the factors which lead changes between above states are also designed as follows.

- $d_0$ : The robot detects the surrounding robot
- $d_1$ : The robot is detected by the surrounding robot
- $d_2$ : robot ID of the robot $<$ robot ID of the surrounding robot

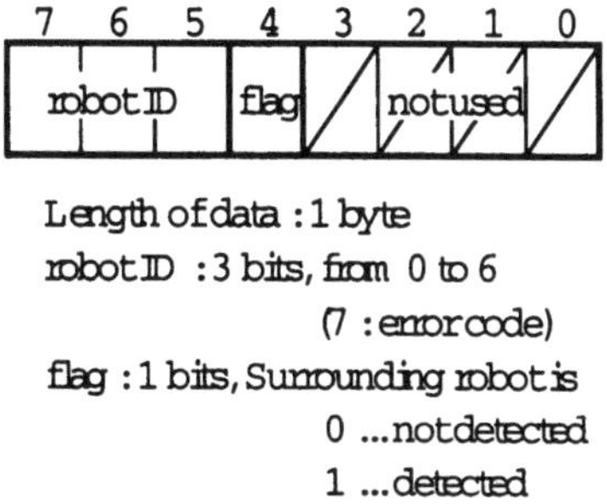


**Fig. 4.** Transition data of LOCISS

where $d_2$ denotes a factor decides which robot emits firstly.

Figure 4 shows the transmission data of the LOCISS to achieve a detection of above factors. The length of data is 8 bits. The robot ID and the detection flag are assigned into upper 3 bits and next 1 bit, respectively. Although the robot ID is designed as 3 bits long to discriminate seven robots, this is able to be expanded easily if necessary. The code $111_{(2)}$ of the robot ID is used as error code. The detection flag is set to ON when the robot receives a signal from surrounding robot, or to OFF when the robot receives no signals. The flag shows a state for detection of surrounding robot.

By catching above factors as inputs to occur transitions of states, the time shared scheduling model is represented as a kind of automaton. The transition diagram of this automaton is shown in Fig. 5. The initial state of the diagram is state $S_0$. $\overline{d_n}(n = 0, 1, 2)$ means negatives of factors $d_n(n = 0, 1, 2)$, respectively. The inputs have higher priority in order of $d_0$, $d_1$, $d_2$ and they are evaluated until the transition of state is decided based on the diagram in the order. The details of each state are as follows.

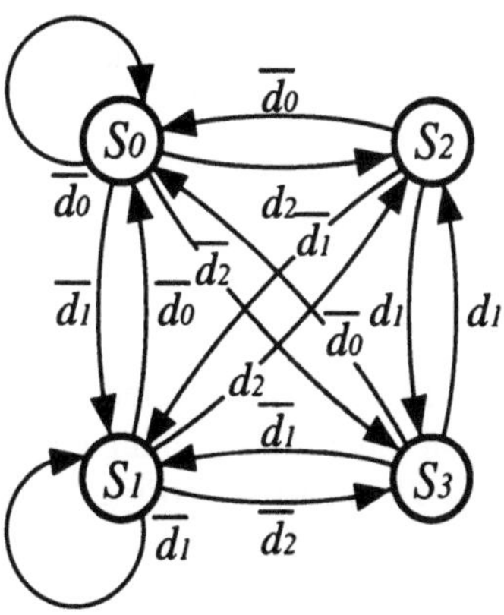

**Fig. 5.** Transition diagram for time shared scheduling

In the state $S_0$, there is no surrounding robot and the robot emits an ultrasonic wave continuously. This state is repeated while no surrounding robot is detected by the LOCISS, that is, the input $\overline{d_0}$ is true. When surrounding robot is detected, the detection flag which is included in detected signal is referred. If the flag is OFF, it means that the robot is not detected by surrounding robot, that is, the input $\overline{d_1}$ is true, and the state is changed into $S_1$ because the time shared scheduling is not accomplished yet. When the flag is ON, it means both robots detect each other and the robot IDs of both robots are referred to decide which robot emits an ultrasonic wave firstly. If the robot's one is smaller than surrounding robot's one, that is, the input $d_2$ is true, the state is changed into $S_2$ and the robot emits firstly. If the surrounding robot's one is smaller, that is, the input $\overline{d_2}$ is true, the state is changed into $S_3$ and the robot waits for the surrounding robot to emit.

In the state $S_1$, the robot detects surrounding robot, but the robot is not detected by the surrounding robot. In this case, the surrounding robot emits an ultrasonic wave continuously and the robot waits without emitting to avoid interferences of ultrasonic waves. This state is repeated while the robot is not detected by the surrounding robot, that is, the input $\overline{d_1}$ is true. When it is confirmed that the robot detected according to the detection flag, the state is changed into $S_2$ or $S_3$ based on robot IDs of both robots. On the other hand, if the signal from the surrounding robot stops, that is, the input $\overline{d_0}$ is true, the state is changed into $S_0$ and the robot resumes continuous emission.

In the state $S_2$, the time shared scheduling is accomplished and the robot emits ultrasonic waves. The state is changed into $S_3$ to alternate emit while the detection flag is ON, that is, the input $d_1$ is true. If the flag is OFF, that is, the input $\overline{d_1}$ is true, the state is changed into $S_1$ to wait. In the case that the signal from the surrounding robot stops, that is, the input $\overline{d_0}$ is true, the state is changed into $S_0$ to emit continuously, no matter what the surrounding robot exists.

In the state $S_3$, the time shared scheduling is also accomplished and the robot waits for the surrounding robot to emit. The state is changed into $S_2$ to alternate emit while the detection flag is ON, that is, the input $d_1$ is true. In other case, the state is changed like in the state $S_2$.

## 4 Experimental results

To show the efficiency of the proposed method, a distance measurement experiment using real robots is conducted. In this experiment, a sampling cycle of infrared and ultrasonic sensing is designed to 69 [msec] according to a performance of sensor unit's controller. Therefore the robots can measure distances approximately 14 times a second and 7 times a second based on the continuous scheduling and the time shared scheduling, respectively.

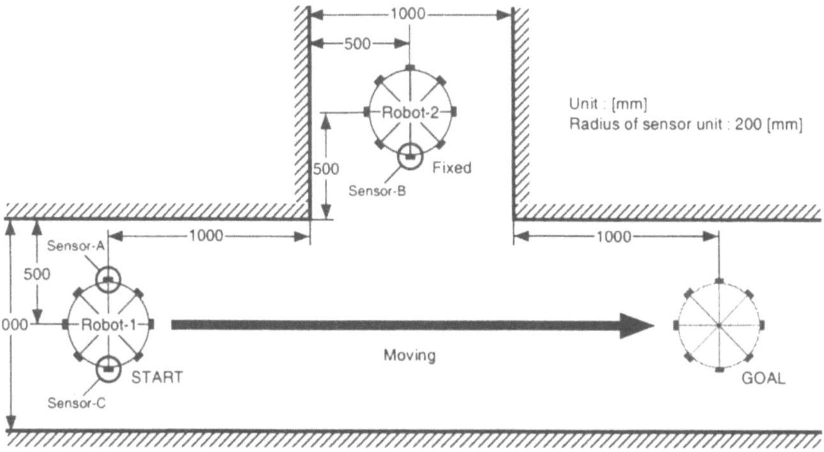


**Fig. 6.** Environment of experiment

The environment of the experiment is shown in Fig. 6. In this environment, there are two robots in a passage which is 1000 [mm] in width. The robots measure distances based on proposed method in eight directions in above-mentioned sampling cycle. The robot-1 moves at 100 [mm/sec] toward its goal from start position along the center of the passage. On its way to move, the robot-1 encounters the robot-2 which is on branch. An interference of ultrasonic waves occurs here and the time shared scheduling between the robots is applied then.

Figure 7 shows the result of the experiment. In this figure, the sensor-A and the sensor-B represent sensors of the robot-1 and the robot-2, which face each other when the robots encounter, as shown in Fig. 6.

The transition of measured distances and negotiation states for the sensor-A of the robot-1 for time series are shown in Fig. 7(a) and (b), respectively. In the figure for measured distances, a distance of 0 [mm] means the sensor can't measure distances within a range of 1000 [mm] or don't measure based on time shared scheduling. In terms from 0 to 10 [sec] and from 20 to 30 [sec], the sensor-A of the robot-1 measure distances to a wall, which is in front of the sensor at interval of 300 [mm], without interfering with ultrasonic waves from other robots. On the other hand, in a term from 10 to 20 [sec], the sensor is influenced by the sensor-B of the robot-2. However it is confirmed that the sensor-A measures distances to the robot-2 which is in front of the sensor at interval of approximately 600 [mm]. At the same time, the state $S_2$ and $S_3$ alternate frequently.

Figure 7(c) and (d) show the transition for the sensor-B of the robot-2. In a term from 10 to 20 [sec], the sensor-B measures distances to the robot-1 moving across in front of the robot-2. Moreover the state $S_2$ and $S_3$ also alternate frequently. Then it is confirmed the state $S_2$ and $S_3$ in Fig. 7(b)

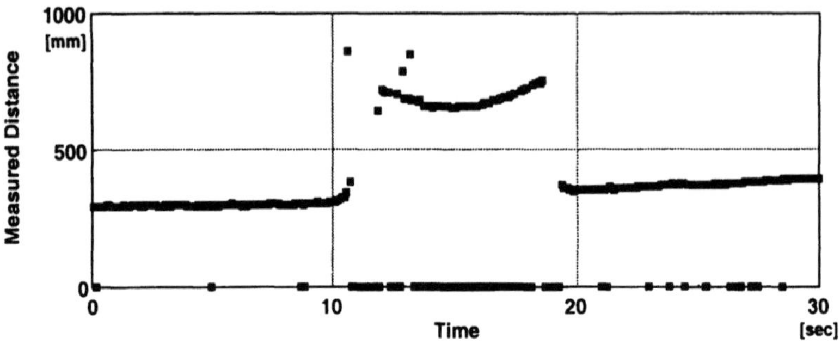


(a) Distance measured by sensor-A of robot-1

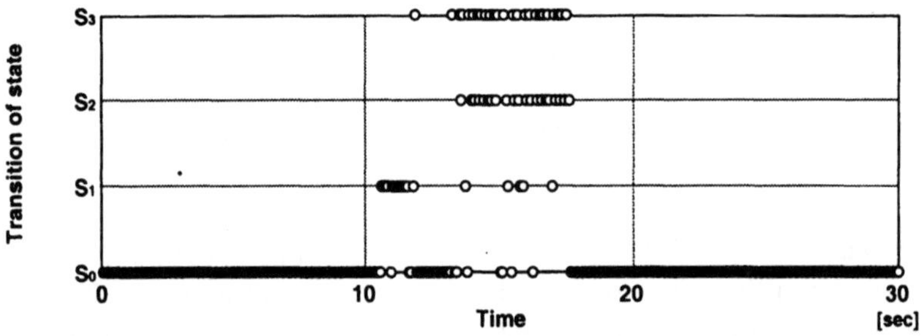


(b) Situation transition of sensor-A of robot-1

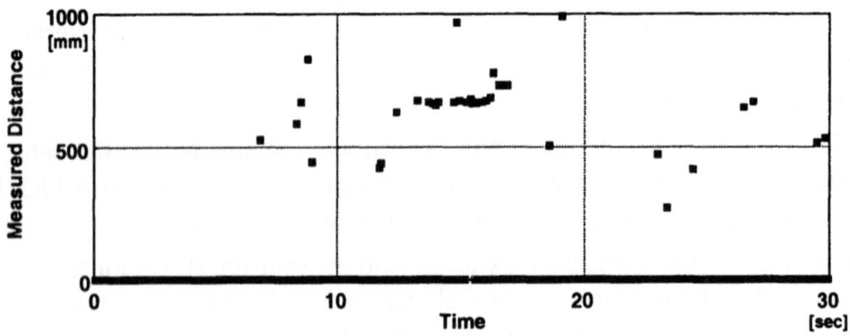


(c) Distance measured by sensor-B of robot-2

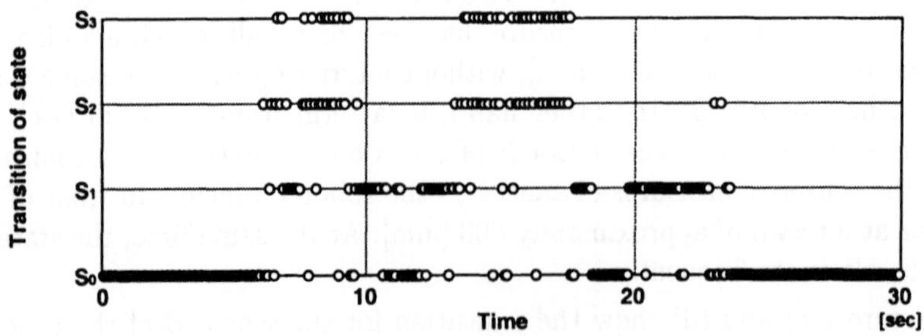


(d) Situation transition of sensor-B of robot-2

**Fig. 7.** Result of distance measurement

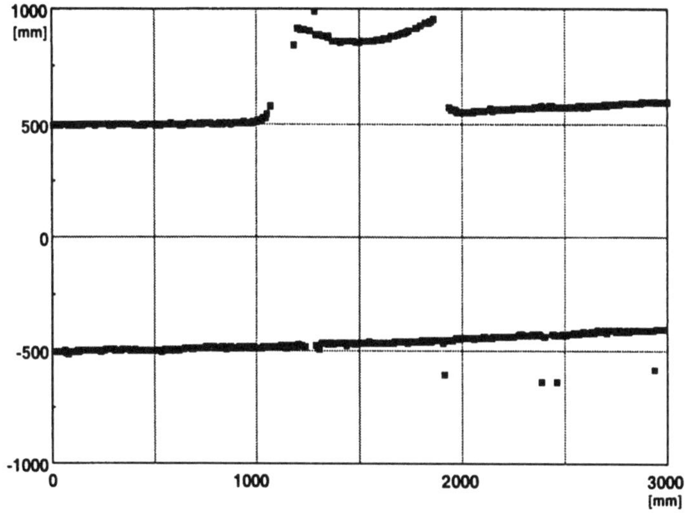


**Fig. 8.** Result of environment mapping based on distances measured by sensor-A and sensor-C of robot-1

accord with ones in Fig. 7(d) by exchanging the states. These means the distance measurement based on time shared scheduling is carried out. The transitions of states in terms from 6 to 10 [sec] and from 20 to 24 [sec] mean the time shared scheduling with next sensors of the sensor-A is accomplished.

Figure 8 shows an environment map drawn based on distance information measured by the sensor-A and the sensor-C of the robot-1. The sensor-C represents the sensor of the robot-1 which faces right wall for moving direction of the robot as shown in Fig. 6. In this figure, walls on measured distances are plotted when the start point of the robot-1 is defined as an origin of two axes, that is, one is along the direction of robot-1's movement and the other crosses one. It is confirmed that the experimental environment as shown in Fig. 6 is reproduced accurately. Although the passage slants, this is caused that the robot slides from center of the passage when it moves toward its goal.

Consequently, the distance measurement and the mapping of the environment based on the time shared scheduling in the environment in which two robots exist are accomplished. It is shown that proposed method is efficient in multi-robot systems.

# 5 Conclusion

We proposed a new distance measurement method based on time shared scheduling in an environment in which multiple robots exist. By using local communication to detect surrounding robot and negotiate with the detected robot, the time shared scheduling is accomplished. The scheduling can let an ultrasonic wave occupy a space momentarily and allow sensors to measure accurate distances even in multi-robot systems. By applying proposed method, it was shown that robots can measure distances to each other in the situation in which ultrasonic sensors are face to face. Although it is assumed that measured distances are enough accurate in this experiment, measured errors are corrected by introducing some signal processing methods, if necessary. For future works, methods on distributed environment mapping by multiple robots and collision avoidance among multiple robots based on distance information should be discussed to accomplish flexible function by multi-robot systems.

## References


1. Arai, Y., T. Fujii, H. Asama, H. Kaetsu and I. Endo (1998). Robust collision avoidance in multi-robot systems - implementation onto real robots -. In: *Distributed Autonomous Robotic Systems 3*. Springer Berlin. pp. 23–33.
2. Asama, H., Fukuda, T., Arai, T. and Endo, I., Eds.) (1994). *Distributed Autonomous Robotic Systems*. Springer Tokyo.
3. Asama, H., Fukuda, T., Arai, T. and Endo, I., Eds.) (1996). *Distributed Autonomous Robotic Systems 2*. Springer Tokyo.
4. Asama, H., K. Ozaki, A. Matsumoto, Y. Ishida and I. Endo (1991). Collision avoidance among multiple mobile robots based on rules and communication. In: *Proceedings of IEEE/RSJ International Workshop on Intelligent Robots and Systems*. pp. 1215–1219.
5. Borenstein, J. and Y. Koren (1988). Obstacle avoidance with ultrasonic sensors. *IEEE Journal of Robotics and Automation* **4**(2), 213–218.
6. Buchberger, M., K.-W. Jorg and E. von Puttkamer (1993). Laserradar and sonar based world modeling and motion control for fast obstacle avoidance of the autonomous mobile robot mobot-iv. In: *Proceedings of the 1993 IEEE International Conference on Robotics and Automation*. pp. 534–540.
7. Kawabata, K., N. Nishioka, P. C. Lin, H. Nakamura and H. Kobayashi (1996). Distance measurement method under multiple ultra sonic sensors environment. In: *Proceedings of the 22nd Annual International Conference of the IEEE Industrial Electronics Society*. pp. 812–816.
8. Lueth, T., Dillmann, R., Dario, P. and Wörn, H., Eds.) (1998). *Distributed Autonomous Robotic Systems 3*. Springer Berlin.
9. Suzuki, S., H. Asama, A. Uegaki, S. Kotosaka, T. Fujita, A. Matsumoto, H. Kaetsu and I. Endo (1995). An infra-red sensory system with local communication for cooperative multiple mobile robots. In: *Proceedings of the 1995 IEEE/RSJ International Conference on Intelligent Robots and Systems*. pp. 220–225.
10. Yuta, S. and S. Premvuti (1992). Coordination of autonomous and centralized decision making to achieve cooperative behaviors between multiple mobile robots. In: *Proceedings of IEEE/RSJ International Workshop on Intelligent Robots and Systems*. pp. 1566–1574.

# Relative Position Localizing System for Multiple Autonomous Mobile Robots in Distributed Robotic System
# ---Algorithm to Detect Overlapped SLPM Patterns ---

Suparerk Premvuti, Warin Wattanapornprom,
Mattana O-Sot and Jittipan Wattanawejwijit

Sirindhorn International Institute of Technology, Thammasat University,
Rangsit Campus, Patumtani 12121, THAILAND suparerk@siit.tu.ac.th

## Abstract

Each mobile robot in the distributed robotic system using the proposed localizing method is equipped with a rotating laser beacon which project vertical slit laser beam. Below the beacon, photo diodes are installed in circular array to detect laser beam from any direction around the robot. Measurement of distance is based on detecting linear velocity of a vertical slit laser beam from a rotating beacon that projected on the photodiode array mounted on a robot. Direction is also detected by locating area on the sensor array where the beam is projected and finding the center of that area.

Output from photodiode array are acquired at a constant sampling rate and stored in a matrix named 'Swept Laser Projection Matrix' (SLPM). However, due to the omni-directional sensing capability, this sensor allows multiple laser beams to be swept on a sensor array. The paper proposes an algorithm to extract overlapped SLPM patterns. The algorithm is designed to be able to handle errors and noises while using as simple algorithm as possible so that real-time processing can be achieved.

**Key words:** localization, positioning sensor, cooperation of multiple robots

# 1 Introduction

## 1.1 Principle and configuration of the sensor system

Coordination strategies and algorithms for a system with multiple autonomous mobile robots often require a robot to know the relative position of other robots in its vicinity [1][2][3]. Many experimental systems employ centralized localization mechanisms, which are invalid for a fully distributed robotic system. An ideal mechanism for relative localization is a fully autonomous passive sensing system [4][5], such as stereo vision, which involves great degree of sophistication and high cost. The authors had proposed an alternative autonomous relative localization strategy [6][7].

A mobile robot is equipped with a rotating beacon which sweeps a vertical slit laser beam at a known constant angular speed. The distance from the robot to a point in its vicinity can be calculated based on linear velocity that the beacon swept through the point measured along the tangent line.

Rotating speed of the beacon is constant and known. Photodiodes are installed on the circular ring around the robot forming the circular sensor array, or the sensor belt, for detecting linear speed of a vertical slit laser beam that swept through. The distance can be calculated from the known rotating speed of the beacon and linear speed measured by the sensor array. An immediate advantage of the proposed method is that the positioning system can be constructed without requiring a great deal of processing power and memory as normally needed by a positioning system using passive or active vision method.

All robots in the multiple autonomous mobile robot system have to be equipped with a set of rotating laser beacon and the circular sensor array of photodiodes. Each robot operating in the system always emits a vertical slit laser beam rotating at a fixed angular velocity. The laser beam periodically projects on the sensor array of neighboring mobile robots as shown in Fig. 1.

## 1.2 "Swept Laser Projection Matrix" (SLPM) and "SLPM patterns"

The sensor array around the robot body is sampled repetitively to form a Swept Laser Projection Matrix (SLPM) as shown in Fig. 2. If the sensor array is in the range of laser beacon from a robot, detected signals of photodiodes in the sensor array will form a nearly straight line pattern which is named as "SLPM pattern". The distance of a neighboring robot is determined by the "slope" of its SLPM pattern. The direction is indicated by location of the photodiode at the center of the SLPM pattern.

The simple principle and straightforward system arrangement make the proposed system possible to be implemented upon the current technology with reasonable cost. The system satisfies accuracy and speed requirements for most coordination tasks designed for autonomous mobile robots operating under distributed robotic systems.

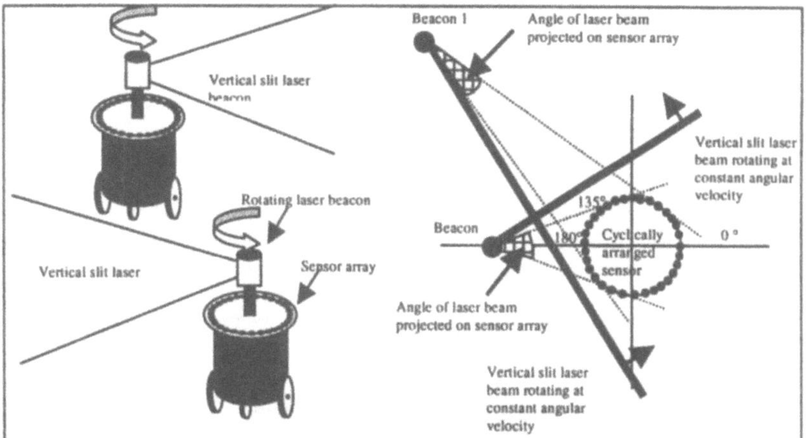


(a) Autonomous robots equipped with the proposed localizing system

(b) Rotating laser beams swept on the sensor array of a robot

**Fig. 1.** Proposed relative localization strategy

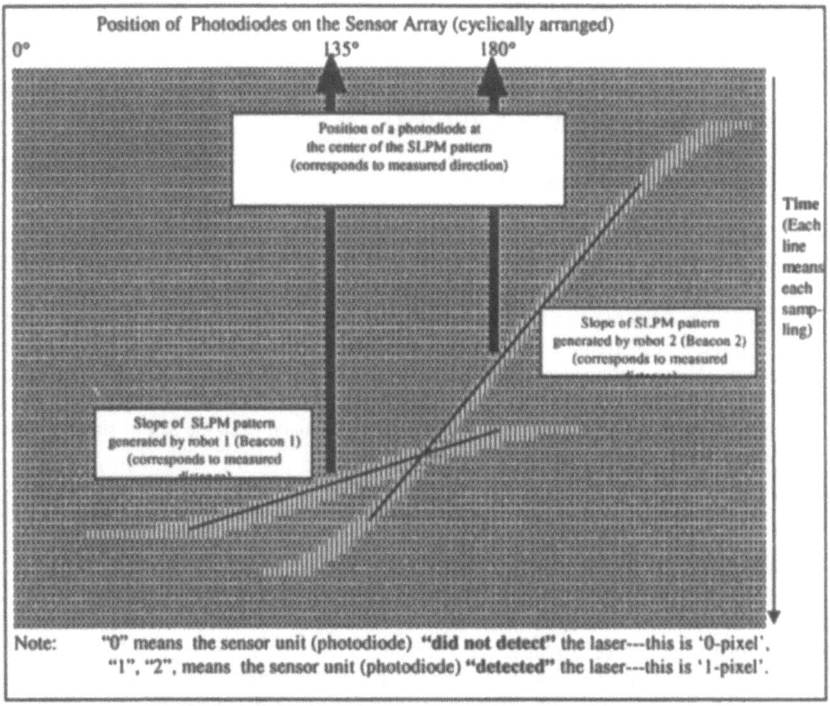


**Fig. 2.** Swept Laser Projection Matrix (SLPM) and two overlapped SLPM patterns

### 1.3 The reason why multiple SLPM pattern detection is important

While the sensor system is operating, all robots keep emitting rotating laser slit beams. For a robot to detect positions of other robots, it must keep taking samplings from all sensor units periodically to obtain Swept Laser Projection Matrix. Then, the sensor system must extract SLPM patterns to calculate distance and direction of the source of rotating laser beams.

Since each beacon is rotating independently, there are possibilities that two laser beams or more may project to the same sensor array at the same time. This could result in overlapped SLPM patterns in Swept Laser Projection Matrix as shown in Fig. 2.

A simple algorithm to detect the begin and the end of an SLPM pattern cannot be used to detect overlapped SLPM patterns and the sensor will give erroneous reading.

### 1.4 Simplicity and real-time processing is also important

The design of algorithm aims at simplicity since the sensor must keep taking samplings and process the data in real-time. After the interfacing circuit latches the outputs of all photodiodes, the data is transferred to the memory. The program will read output of each photodiode and compute the begin and the end position of each SLPM pattern. After the begin and the end position of each SLPM patterns are identified, distance and direction of the source laser beam can be calculated.

## 2 The proposed algorithm to extract multiple line-patterns in Slit Laser Projection Matrix (SLPM)

### 2.1 Characteristics of SLPM pattern

The gradient of SLPM pattern indicates linear speed of the laser that swept through a sensor array. Due to the circular shape of sensor array, linear speed of the laser beam along the circumference of the sensor array varies slightly. At the beginning of SLPM pattern where the laser beam starts to project on the sensor array, linear speed is higher than at the center of SLPM pattern. This results in the gradient of SLPM pattern at the beginning is lower than at the center. This situation also happens near the end of SLPM pattern where the laser beam is about to depart the sensor array. Due to the above reason, the SLPM pattern looks like an alphabet 'S' with gentle curves. The SLPM patterns obtained from the SLPM simulator show this effect in Fig. 2.

### 2.2 Line tracking based on the slope, predicted pixel and actual pixel

In Swept Laser Projection Matrix, a photodiode that is not hit by the laser beam is called "0-pixel". On the other hand, a "1-pixel" indicates a photodiode that detected the laser beam.

Each horizontal line of Swept Laser Projection Matrix is scanned for a "1-pixel". The program will try to fit the "1-pixel" to any existing SLPM patterns. If this failed, the program will consider the "1-pixel" as the starting point of a new SLPM pattern.

To fit a "1-pixel" to an existing SLPM pattern, the program must determine a "predicted 1-pixel" then compare the "actual 1-pixel" to the "predicted 1-pixel". The "temporary last 1-

pixel" of the previous horizontal line and the "first 1-pixel" are used in calculation to find the "current slope" that leads to the "predicted 1-pixel".

If the position error between the "actual 1-pixel" and "predicted 1-pixel" is within a pre-defined limit, the "actual 1-pixel" will be used in calculation to find the "temporary last 1-pixel" of the current horizontal line. The "temporary last 1-pixel" is calculated with a "bias" factor to control the influence from the "actual 1-pixel" and the "predicted 1-pixel". The pixel with higher influence will be closer to the "temporary last 1-pixel".

The use of bias factor allows tracking along a nearly straight line with gentle curves. The influence of "predicted 1-pixel" makes it possible to terminate the tracking when the slope of the line changes rapidly. This indicates overlapped SLPM patterns.

Based on the above method, an SLPM pattern will continue to expand when a new "1-pixel" matches its "predicted 1-pixel". Each SLPM pattern is recorded with the following data,
(1) Position of "first 1-pixel"
(2) Position of "temporary last 1-pixel".

When there in no "actual 1-pixel" that fits the SLPM pattern, the "temporary last 1-pixel" will be considered as the "last 1-pixel". Then, the "first 1-pixel" and the "last 1-pixel" are used in an equation to find the slope and the midpoint of that SLPM pattern.

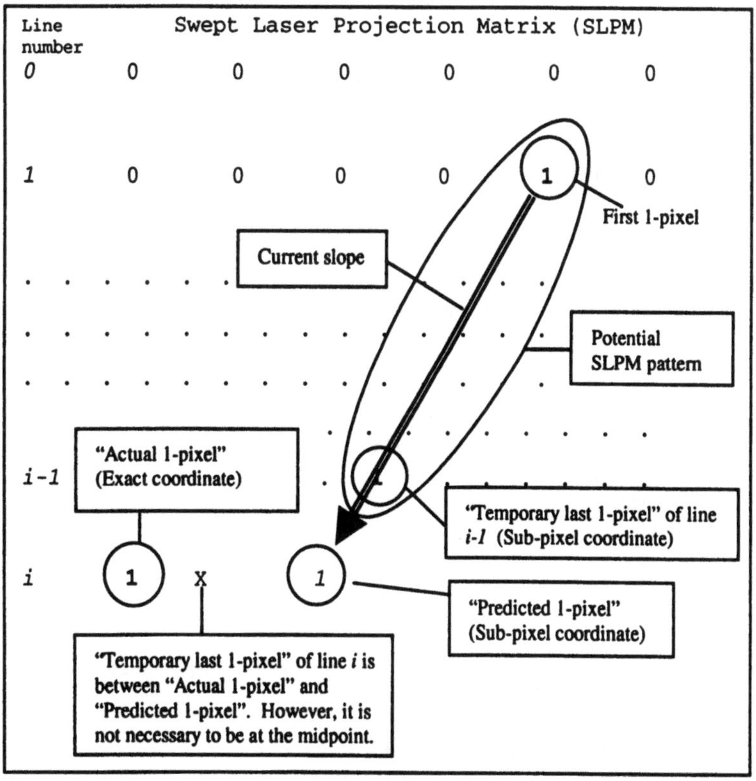


**Fig. 3.** How to process an "actual 1-pixel"

### 2.3 How to handle a "1-pixel" which does not belong to any existing SLPM pattern

If the position error between the "actual 1-pixel" and "predicted 1-pixel" is more than a pre-defined limit, the "actual 1-pixel" will be considered as the beginning pixel of a new SLPM pattern. Due to the characteristic of SLPM pattern, limit of position error is designed so that it will be adjusted automatically. An SLPM pattern with higher slope will have smaller error limit and vice versa.

### 2.4 Sub-pixel utilization

"Actual 1-pixel" uses exact coordinate while "predicted 1-pixel" uses sub-pixel coordinate. The use of sub-pixel increases the resolution of the slope.

### 2.5 Influence from "actual 1-pixel" and "predicted 1-pixel" on the "temporary last 1-pixel" can be adjusted by a "bias" factor

The algorithm is designed to calculate the "temporary last 1-pixel" so that this pixel is between the "actual 1-pixel" and the "predicted 1-pixel". To cope with slightly changing of slope of an SLPM pattern, the influence of "actual 1-pixel" and "predicted 1-pixel" on the "temporary last 1-pixel" can be adjusted with a "bias" factor.

### 2.6 Capability of handling error and noisy input

The algorithm allows a number of missing "1-pixel" along an SLPM pattern. Noises are also filtered through discarding of potential SLPM patterns with too short length.

## 3 Experiment Results

### 3.1 Experiment with overlapped SLPM patterns from the SLPM simulator

To verify the proposed algorithm, the developed program to detect SLPM patterns processes sets of data obtained from the SLPM simulator. Parameters of the simulator are selected so that the SLPM patterns are overlapped. The results from the SLPM patterns detection program are displayed in the format that indicate the SLPM patterns (A,B,C,.....) of each 1-pixel. For a 1-pixel which is member of more than one SLPM pattern, the 1-pixel is shown as an integer. The integer value will indicate lines which the 1-pixel belongs to. Each line has its own weight which can be added and displayed in the result.
Weight of each line is as follows,

| Line | Weight |
|---|---|
| A | 1 |
| B | 2 |
| C | 3 |
| D | 4 |
| .... | ..... |

For example, a 1-pixel belongs to both line A and line C will be displayed as an integer 4. (Weight 1 of A is added to weight 3 of C.)

For the sake of clarity, "0-pixels" are not displayed.

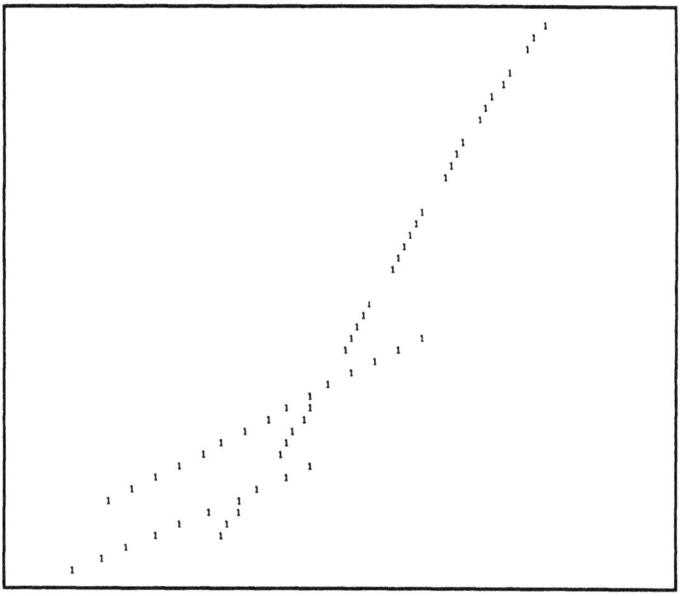

**Fig. 4.** Overlapped SLPM patterns obtained from the SLPM simulator

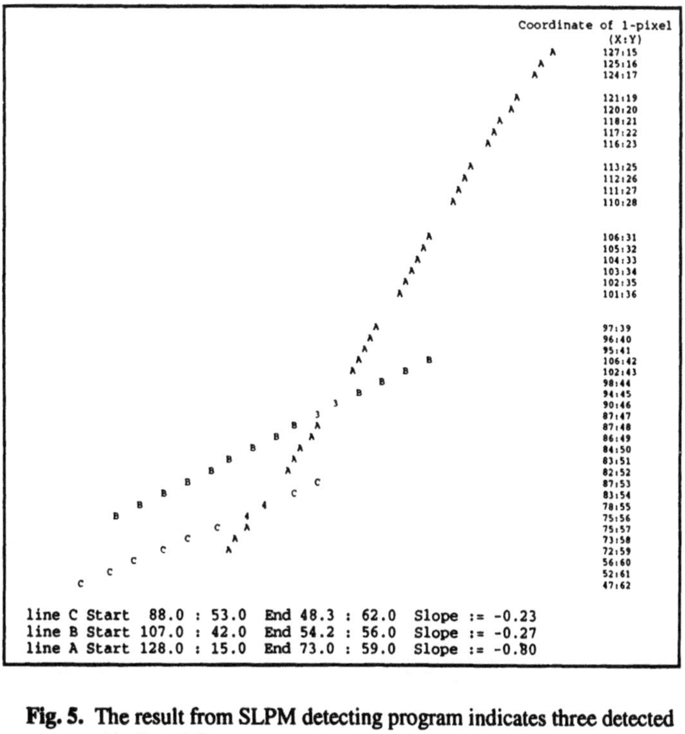


**Fig. 5.** The result from SLPM detecting program indicates three detected patterns (A, B and C)

### 3.2 Experiment with overlapped SLPM patterns synthesized from actual sensor data

Based on the prototype sensor which is being developed for PICO-BOT (Photo 1), the sensor is capable of detecting distance and position of a single robot within its vicinity [7]. To verify the proposed overlapped SLPM detection algorithm, actually detected SLPM patterns are stored in files. Then, two SLPM pattern files are merged together to obtain a file of overlapped SLPM patterns. This off-line overlapped SLPM patterns file is input into the developed program.

**Photo 1.** Mobile robot "PICO-BOT" with sensor array and rotating laser beacon installed on top

**Table 1.** Parameters of the prototype

| Radius of circular sensor array (meter) | Number of photodiodes (unit) | Beacon frequency frequency (Hertz) | Sampling period (milisecond) |
|---|---|---|---|
| 0.13 | 90 | 0.08 | 13.5 |

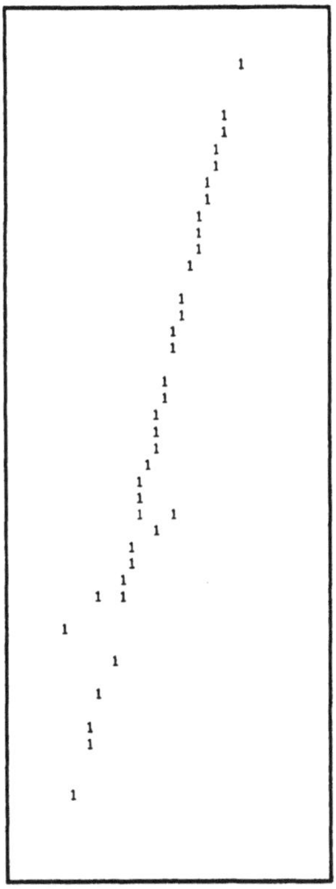

Fig. 6. Overlapped SLPM patterns obtained from the prototype

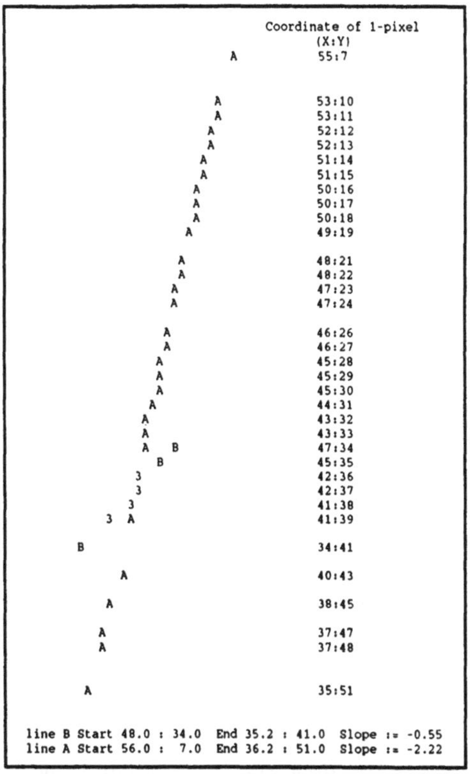


Fig. 7. The result from SLPM detecting program indicates two detected patterns (A and B)

## 4 Discussion

The proposed sensor is based on patterns of slit laser beam that swept through circular array of laser detector. For single SLPM pattern, the detection can be done with simple algorithm of detecting the begin and end point of the pattern. However, to detect multiple robots, ability to extract individual line from overlapped SLPM patterns is crucial.

As seen from the experiment results, the effectiveness of the proposed strategy and algorithm are verified. The algorithm is designed based on the characteristics of the sensor that keep taking samplings from the sensor array and process each sampling in real-time.

The proposed active sensor for localizing relative position of other robots in multiple mobile robot system has the following advantages.

- Omni-direction sensing capability
- No complicated lens system is required
- Less moving parts (Beacon is the only moving part.)
- Simple algorithm
- Low cost

# 5 Conclusions

The proposed sensing method can locate relative position of other robots in the same vicinity by utilizing an integrated sensor system which consists of rotating beacons and cyclically arranged photodiode sensor arrays. Measurement of distance utilizes linear velocity of a vertical slit laser beam, generated from a beacon rotating at a constant speed, when observed at a distance from the beacon. The circular sensor array of photodiodes will function as an omni-directional sensor for detecting speed of the vertical slit laser beam that swept through it.

Sensor array is sampled at a constant rate. Plotting of outputs from each sampling versus time gives a matrix called Swept Laser Projection Matrix or SLPM. Detection of a complete sweeping from a beacon will form a pattern on the matrix which is called SLPM pattern. An SLPM pattern is a collection of '1-pixels' (Photo diodes with valid output.) in almost straight line on the background of '0-pixels' (Photo diodes without output.). Detection of a discrete SLPM pattern can be done with a simple algorithm to detect the begin and the end of the line. However, overlapped SLPM patterns cannot be detected easily.

The paper proposed an algorithm to extract discrete line from overlapped SLPM patterns. The algorithm of line tracking is based on the slope, predicted pixel and actual pixel. Usage of a bias factor during calculation of temporary last 1-pixel allows tracking of an SLPM pattern with gentle curves.

# References


1. G. Beni and J. Wang. (1991) Theoretical Problems for the Realization of Distributed Robotic Systems. Proceedings of the IEEE International Conference on Robotics and Automation, Sacramento, CA, USA, 1914-1920.

2. J. Wang. (1993) DRS Primitives based on Distributed Mutual Exclusion. Proceedings of IROS-93, Yokohama, Japan, 1085-1090.

3. J. Wang and S. Premvuti. (1994) Resource Sharing Based on a Medium Access Protocol (CSMA/CD-W). Proceedings of IROS-94, Munich, Germany, 784-791.

4. V. Graefe. (1989) Dynamic vision systems for autonomous mobile robots. Proceedings of the IEEE International Conference on Robotics and Automation, 12-23.

5. J. Miura and Y. Shirai. (1994) Modeling Obstacles and Free Spaces for a Mobile Robot using Stereo Vision with Uncertainty. Proceedings of the IEEE International Conference on Robotics and Automation, San Diego, CA, USA, 3368-3373.

6. S. Premvuti and J. Wang. (1996) Relative position localizing system for multiple autonomous mobile robots in distributed robotic system --System design and simulation--. Robotics and Autonomous Systems, Vol. 18, Elsevier, 319-326.

7. S. Premvuti, C. Sittisombat, K. Aunvichit and T. Jenjiratthitikarn. (1999) Relative Position Localizing System for Multiple Autonomous Mobile Robots in Distributed Robotic System --- System Design and Prototype Development ---. Proceedings of 9$^{th}$ International Conference on Advanced Robotics (ICAR'99), Tokyo, Japan, 277-282.

# Development of a Multiple Mobile Robot System Controlled by a Human
# - Realization of Object Command Level -

Akio Nakamura[1], Tamio Arai[1],
Takenori Hiroki[2], and Jun Ota[1]

[1] Department of Precision Machinery Engineering, The University of Tokyo, 7-3-1 Hongo, Bunkyo-ku, Tokyo 113-8656, JAPAN
*E-mail address: nakamura@prince.pe.u-tokyo.ac.jp*
[2] FA Division, Hitachi Construction Machinery Co., Ltd.
650 Kandatsu-machi, Tsuchiura-shi, Ibaraki 300-0013, JAPAN

**Abstract.** Until now, we have been developing a system for multiple mobile robot operation by a single human user. In order that robots may achieve tasks in unknown environments, we must solve two problems; one is an environment recognition problem and the other is a self-positioning problem. Concerning the former issue, the two robots that equip a monocular camera on each acquire a 3-dimensional position of a point in an environment with a stereo vision method. As for the latter issue, we set landmarks on the robots by which the robots can measure their relative positions. In each phase mentioned above, the human user gives commands to the robots appropriately. Solving the above problems, we can make a virtual environment in the computer. The virtual environment displays remote site information to the user. We can thus implement object-command level at which the user operates a simple model of an object in the virtual environment, and robots manipulate the real object as the user says. We verified effectiveness of the proposed method by experiments.

**Keywords.** Human Robot Interaction, Object Control, Multiple Mobile Robots, Cooperation, Environment Recognition

## 1. Introduction

In the near future, a robot with moving facility is expected to play important roles in many fields of a society. For example, a mobile robot undertakes dangerous tasks instead of humans in disadvantageous environments, such as nuclear power plants, space, deep sea, and so on. It may also help people at home. In that situation, if there are multiple robots in a robot system, some merits will be expected compared to use of only one intelligent robot; the systems become more robust against some failures and can achieve various kinds of tasks. Furthermore, if the human factors (flexible judgement, ability of solving complicated problems, and so on) can be introduced into the system, we make the system more efficiently.

Previous works concerning human-machine systems mostly focused on cooperation between a human operator and a single robot or a manipulator, and there were not many methods that targeted multiple robots. Some research have been done in teleoperation field of multiple robots [5-12]. Though these systems work well, they seem to be limited regarding the environments for which they are suitable. In addition, commands for the robots are not designed based on clear reasons.

So far we also do research for the purpose of building a system that enables a human user to control multiple robots appropriately. Until now, the user has commanded not each robot but a group of robots in our system. We design commands for the robots systematically [1-4] and develop a GUI (Graphical User Interface) to input the commands easily. We also introduce a simple motion prediction function into our system [3].

However, even if the user interface works well, a burden on the human user is still big. The user wants to control not motions of the robots, but an object path or motion. If the user controls the object via robots with commands for the robots, he/she may slightly feel troublesome. In this paper, we implement a higher command level; the user controls motions of the object and then the robots achieve that task autonomously. We call this level *object-command level.*

Some teleoperation systems have already realized something like the object-command level [5,6]. In many of their systems it was assumed that the information of the work area had been already known. However, a system applicable to an unknown environment is more desirable in the real world.

In order to implement the object-command level in an unknown environment, we must solve some problems detailed below; the robot must know its own pose (pose: position and orientation). And it has to know environment information, that is, objects' poses, obstacles' poses, wall position and so on. The term "unknown" used in this paper means that the environment where robots work is not structured suitable for the robots in advance. For example, the environmental information or landmarks for robot self-positioning are not acquired in advance. If the system has a ceiling camera, a bird's eye view of the work area tells the robots that information. However, an assumption of the ceiling camera existence limits the environment. Accordingly, the robots need to adapt to the unknown environment for wide use. These problems are categorized into two aspects;

1. Environment recognition
2. Self-positioning

Concerning the former issue, two robots that equip a monocular camera on each acquire a 3-dimensional position of a point in the environment with a stereo vision method. Measurement by the stereo vision method with cameras mounted on robots is slightly more inaccurate than that with fixed cameras because of robot motion errors. We consider, however, mobility of robots is important. To complement measurement errors, the robots behave reactively with their sensors near the object.

According to the latter issue, we set landmarks on the robots so that the robots measure their poses each other. It is very difficult that a system solves the problems of self-positioning automatically. If the user can give the system some appropriate instructions, the problems become easier.

Our last target is to implement a system with which a single user controls multiple mobile robots. To improve ease of commanding the robots is important. Object-command level is one of solutions. In this paper, we propose solving methods for that command level. There have been individual solutions for the problems until now [13-16]; integration as one system for multiple mobile robots is very important. We build a real system with the object-command level. A task is object transportation with pushing. Experiments demonstrate that our system, embedding our proposed methods, works appropriately.

## 2. Realization of the object-command level

As mentioned above, in order to implement the object-command level, environment recognition and self-positioning are necessary. Until now, we set a CCD camera over a work area. Through that camera we get poses of robots or objects and monitor tasks [1-3]. No information can be obtained without the camera in an unknown environment. However, in that situation, obtained information is limited to 2-dimension. It is ideal that robots acquire information about the environment with their sensors and achieve tasks in a general environment where we do not equip with special devices or do not tell the robots information in advance.

### 2.1 Environment recognition

Environment information includes 3-dimensional poses, shapes of obstacles or objects. It is very difficult to obtain that information automatically in an unknown environment. Many researchers deal with this topic and some methods are proposed, such as a stereo vision method, a laser-range finder modeling, a ultra-sonic sensing, and so on. The stereo vision method is one of the useful methods; however, it is difficult to detect the corresponding features of two images (right and left images). Furthermore, if an occlusion exists, the problem gets more complicated. There are some studies in which the corresponding problem is solved automatically; however, many of them are available in the limited conditions, for example, a simple shape object or constant light.

In this paper, a human user directly indicates a point on an object (or obstacle) with a laser pointer and two robots in a robot group measure that point with the stereo vision method. Merits of our method are detailed below:

- Corresponding problem can be avoided.
- Acquisition of useless information can be omitted and image processing becomes easier and faster. This is superior to automatic environment recognition methods.
- The user can directly indicate one point in a 3-dimensional space. This means the user can transmit his/her abstract decision such as "go there!"

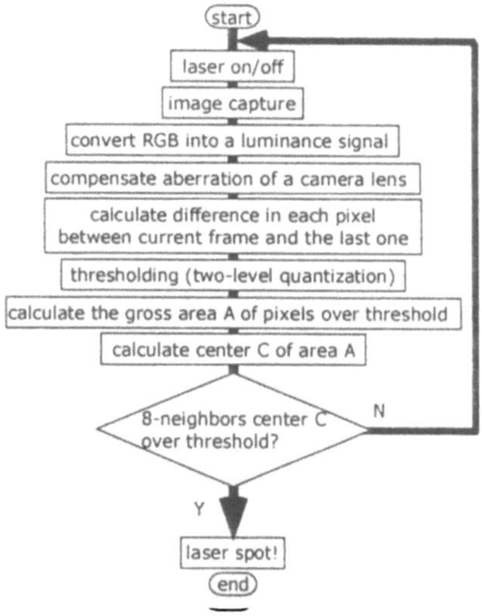


Fig. 2. Laser spot detection flow chart

We call two robots that measure a point, *worker robots*. And we mount a laser pointer on another robot, we call it *manager robot*. The manager robot is an origin of a work area. Each robot has a CCD camera. In the future, all robots will have laser pointers and change their roles (worker, manager) flexibly according to user's instructions. Schematic view of our proposed system is illustrated in Fig. 1.

Flow of the laser spot detection is illustrated in Fig. 2. The laser pointer flashes on and off so that the spot can be detected in bright ambient light. After the two robots detect the laser spot, the 3-dimensional position of the spot is calculated by the stereo vision method.

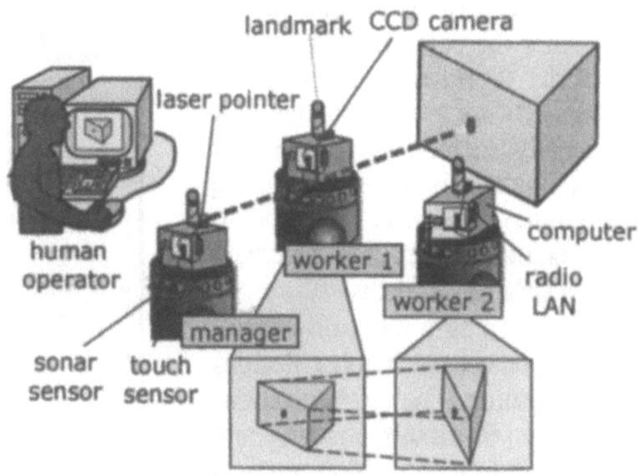


Fig. 1. Proposed system

Compared to a case of using fixed cameras, our method that a camera is mounted on each robot may not be so accurate because of robot motion errors. It is ideal that robots can move anywhere, even if they are strangers there. We target to develop a system with which we only throw robot groups in an environment the robots can achieve tasks with human help.

## 2.2 Self-positioning in a robot group

In order to obtain a 3-dimensional pose of a point in an environment with our proposed method detailed in section 2.1, it is necessary that poses of two robots (worker robots) are known. There are some methods to measure pose of a robot in 3-dimensional space; for example, dead-reckoning, CPS [14], or landmark methods.

As for the dead-reckoning, error accumulation can not be neglected. CPS is a useful method; however, it needs a laser range finder. And it is suitable for a long distance movement of robots. It is slightly excessive for tasks with short distance movement dealt in this paper. A more simple reasonable method is necessary. About the landmark method, we must set landmarks in the environment and know the absolute positions of them in advance. If we do so, the environment information is not any longer "unknown." Then, we mount a landmark on each robot in this research. The manager robot observes landmarks of the worker robots from backward and calculates relative poses of them. The manager robot is a "base" during the task execution. And in a remote site, it is also an "agent" of the human user in the local site. If the relative pose from the manager robot is known, the task can be achieved. In case that the manager robot moves, the worker robots stop and observe a landmark on the manager robot. This part in our method is similar to the CPS method. These processes reduce errors and enable the robots to decide their positions as precisely as possible. Concerning the landmark, we adopt a "signboard" system developed by Arai, et al. [15] However its available range of measurement is not so wide. Then we are now designing a new landmark referring to refs. [16].

## 2.3 Operational steps

The relationship between the user operation and the robot motion is shown in Fig. 3. The user gives commands to the robots, watching a virtual environment displayed on a monitor. The virtual environment is detailed in chapter 3. First, the robot pose is reflected in the virtual environment based on dead-reckoning data or observation of the landmark by a manager robot. The user watches the virtual environment and images through CCD cameras mounted on the robots. The user operates the manager robot and points at any point on an object with laser. Then he/she operates worker robots so that they can detect laser spot (this phase will be automated). After some laser spots are observed, the system can build an object shape model. Once the model is built, the user only operates it and the robots transport the real object automatically.

## 2.4 Movement of robot group

After the user indicates motion of an object (translation and rotation), the robots achieve a transportation task automatically. As for a path planning of the robot, linear paths and spin turns are adopted in the translation. And when it curves (rotation), the path is generated as a set of short linear paths.

Concerning a robot disposition around the object, each robot decides its target position. The target positions are equally distributed along and slightly apart from an object surface. The distance between the target position and the object surface is equal to the radius of the body of robot. After the robots touch the object, the robots make a parallel movement to the object motion indicated by the user and transport the object.

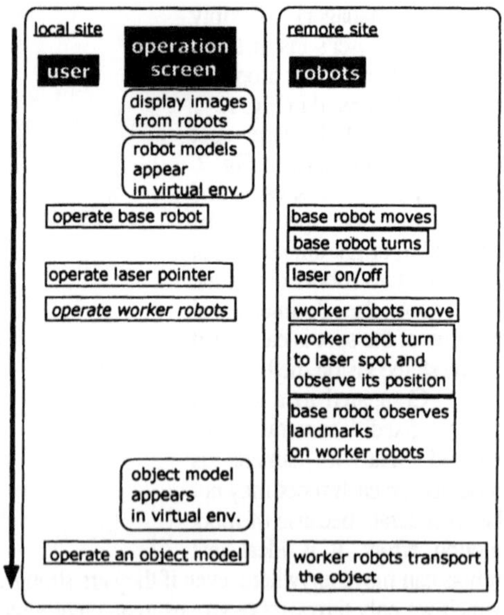


Fig. 3. User operation and robot motion

## 3. System architecture

Figure 4 shows our system architecture. We use a workstation as a host computer that manages a human-robot interface program and communicates with computers on robots via wireless LAN. A PC (IBM AT compatible machine, OS: Linux) is mounted on each robot and it controls the robot, processes images, and so on.

The robot used in our system is Nomad Scout (Nomadic Technologies, Inc., USA), which has a differential drive system. The Scout has ultrasonic and tactile sensing modules. And it has a monocular CCD camera and a laser pointer (Now, the laser pointer is on the manager robot only).

An image taken through the CCD camera is processed by the PC and it is simultaneously radioed to the host computer. The human user can watch all the images through the cameras on the robots (four images are displayed in one window) and a virtual environment. Figure 5 (a) shows an operation screen. Figure 5 (b) shows the window of 4 images, one of which the user can select to enlarge and look at. However, the image through the camera on the robot is a side view and containing local information. Though the user must guess a global view, it is difficult to grasp the environmental situation globally only by side views. Then, to give the user global information, the system makes a virtual environment with the information acquired with the stereo vision method mentioned in section 2.1, and shows it to the user. The user can change viewpoints of the virtual environment and he/she can also select a bird's eye view. Object motion is guessed by the motion of the robots contacting the object and reflected to the virtual environment.

(a) Operation screen (b) Images through the cameras on the robots

Fig. 5. Screen snapshots

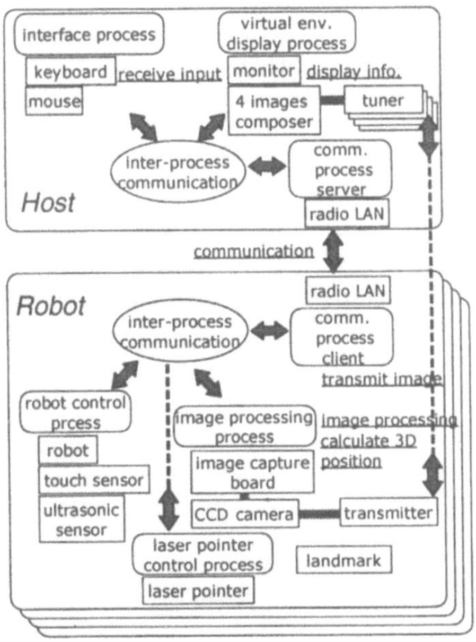


Fig. 4. System architecture

Figure 6 shows the human user in a local site and the robots in a remote site. The manager robot locates on the left in the image and other two robots are worker robots. The manager robot points at one point in a space with laser under the user supervision. The user selects necessary points to build the virtual environment according to a task, thereby it is unnecessary to recognize the whole environment information.

(a) local site (b) remote site

Fig. 6. View of the system

## 4. Experiments

### 4.1 3-dimensional position measurement

#### 4.1.1 Simulation

As mentioned above, in order to build the virtual environment, the robots need to observe poses of some points in a 3-dimensional space. However, the robot motion always has motion errors. Consequently, those errors will influence calculation accuracy when a laser spot position is calculated. And the robots recognize the laser spot in an image taken through a CCD on the robot with quantization errors. In order to make the accurate virtual environment with those errors, we decide an appropriate disposition of the robots to observe the laser spot by a simulation.

Figure 7 shows parameters that should be designed in the simulation. We assume that an object shape to be modeled is a plane perpendicular to a horizontal one. The user thus points at only two points on the object to make an object shape model in this case (Our method is also available to a general shape object modeling). We change each parameter variously. We give each robot pose errors (maximum position error: $\Delta r =$ 20[mm], orientation error: $\Delta\theta = \pm 5$[degree], see Fig. 7 (b)) and set the quantization error ±1[pixel]. These error values are decided by a preliminary robot experiment, which are not explained in this paper. There are twelve error positions for each position and two kinds of error orientation for each orientation. That is, there are twenty-four kinds of error poses for each robot. And consequently, there are 576 error poses for each disposition of two robots.

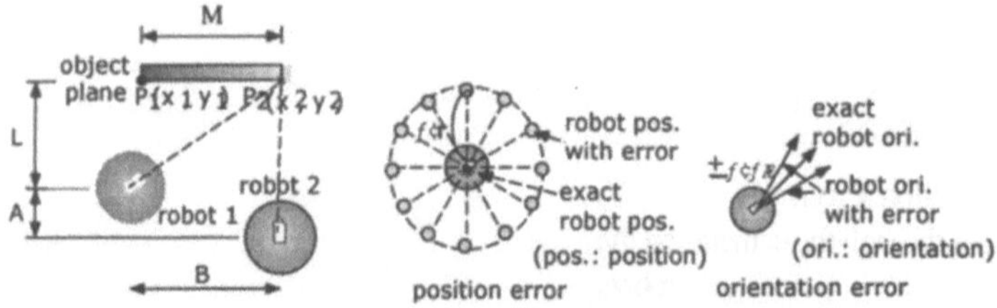


(a) disposition parameters (b) error parameters

Fig. 7. Simulation parameters

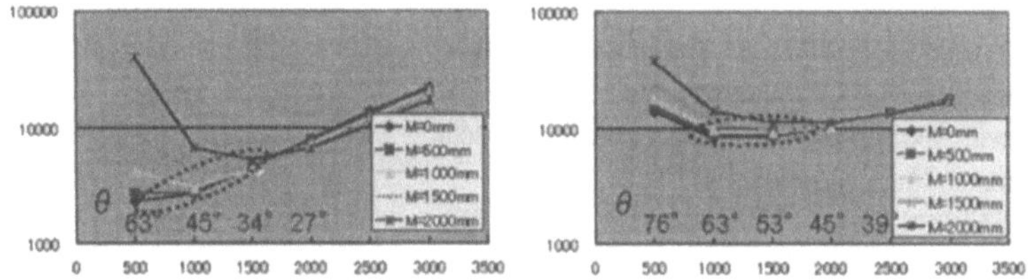

(a)L=500[mm], A=0[mm] (b)L=1000[mm], A=0[mm]

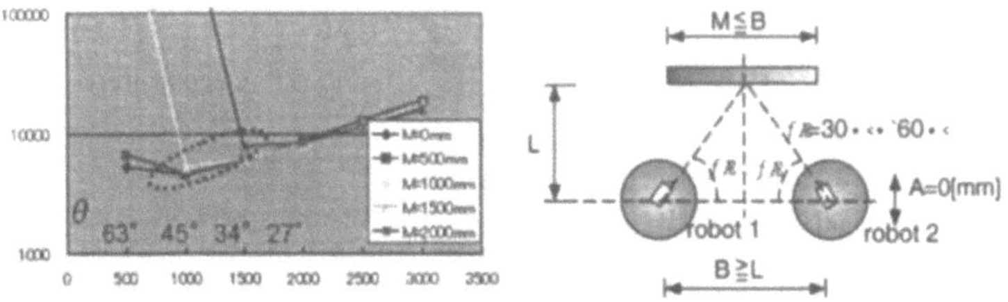

(c)L=500[mm], A=500[mm] (d) robot disposition

Fig. 8. Simulation results

We calculate an error ellipse from variance and covariance, and adopt it as an evaluation index of the robot disposition. The disposition that has the smallest area of the error ellipse is the most appropriate and robust one for observing the laser spot.

Figure 8 shows examples of simulation results. Y-axis represents the error ellipse value and X-axis represents a distance between two robots B. Analyzing Fig. 8 (a)-(c), we can find the appropriate disposition condition as follows (see Fig. 8 (d)):

- The segment combined two robots is parallel to the object surface to be observed (the robots should form a line parallel to the object surface).
- B (the distance between two robots) should be longer than L (the distance between robots and the surface).
- $\theta$ in Fig. 8 (d) should be within 30-60 [degree].
- M (a measurement distance (area) on the object surface) should be smaller than B.

These conditions agree with other simulation results under different error conditions. By making use of these results, the robots will move in and observe the environment with steps as follows:

1. The robots make a preliminary observation at initial random positions.
2. Based on that observation result, the robots move to appropriate positions agreeing with simulation results and observe the object.

These steps are not yet implemented. At the present stage the human user move the robots to the appropriate disposition.

### 4.1.2 **Experiment**

We verify accuracy of the proposed method through a real robot experiment and compare the results with the simulation results. The experiment parameters in Fig. 9 are shown as follows:

M= 500[mm]. L= 500[mm]. B= 1000[mm]. A= 0[mm]. r= 1000[mm].

These parameters satisfy the conditions of the appropriate robot disposition decided by the simulation mentioned in section 4.1.1.

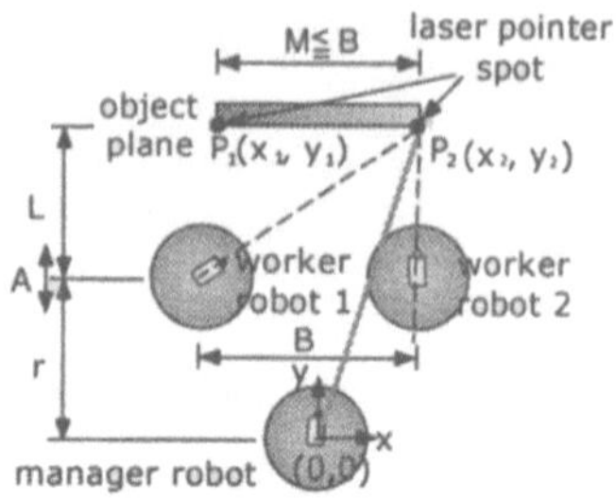


Fig. 9. Disposition of robots in the experiment

We point a laser pointer at features •$P_1$•$P_2$• on the object surface. M (the distance between $P_1$ and $P_2$) is longer than B (the distance between robots). The poses of worker robots are observed by the manager robot; the worker robot has a landmark which was designed by Arai, et al. [15]. The worker robots observe each point feature 20 times. We calculate average and 3• along X-axis or Y-axis illustrated in Fig. 9. Table 1 shows the experimental results. The right column shows simulation results under the same condition.

You can see that the average error along X-axis is about 40-50 [mm] and that along Y-axis is about 50-80 [mm]. They are so small compared to the distance 1500 [mm](L+r). We think these errors depend on many kinds of errors, such as the setting error of CCD camera, the error of focal length of a camera lens, the setting error of the landmark on the worker robot, and so on. However, these errors are not so critical and do not seriously influence achievement of transportation task. If the robots approach the object, they can move to avoid collision and make compensation for these errors owing to the model errors by using the equipped touch sensors and ultrasonic proximity sensors.

Table 1. The result of experiment

| | $P_1$ ($x_1$•$y_1$) | $P_2$ ($x_2$•$y_2$) | Simulation |
|---|---|---|---|
| Theoretical [mm] | ••250•1500• | •250•1500• | •±250•1500• |
| Average [mm] | ••296.8•1587.0• | •289.9•1551.9• | •±250.2•1499.0• |
| 3•[mm] | •33.3•58.9• | •28.6•42.8• | •45.6•63.6• |

## 4.2 **Object transporting demonstration**

We perform a simple task by the robots. Figure 6 shows a general view of the whole system taken while the robots performing the demonstration. First, the robots get the surface pose of the object that the robots transport. Figure 10 shows a virtual environment and the image of the work area during the demonstration. The white plane displayed at the center of the image is the object shape model. The robots approach the object (Fig. 10 (c)). The user operates the object model to translate forward. The robots achieve transporting task automatically (Fig. 10 (d)).

(a) Virtual environment

(b) Start

(c) Approach

(d) Finish

Fig. 10. Virtual environment and a View of the work area

In this demonstration, the shape model of the object is a plane. It seems to be very simple; it is sufficient for this transportation task. In case that the robots perform more complicated tasks, the user indicates many points on the object and the system makes a more precise model in the virtual environment.

## 5. Conclusions

In this paper, we proposed a method to realize the object-command level for multiple mobile robot operation in an unknown environment. And we built a basic actual system. Compared to the case that the user commands the robots directly, he/she only operates motion of the object. Thus, a burden on the user reduces.

In order to implement the object-command level, 1) environment recognition problem and 2) self-positioning problem were solved. Concerning the former issue, the stereo vision method with human help was used. We proposed a relative position calculation method using landmarks on the robots for the latter problem. And we integrated the proposed methods to one system.

From now on we improve a user interface (GUI) and make the system easier to use. For example, we plan to use a special force-feedback joystick that can act as a controlling device to move an object. We are trying to expand our system that do not only involve the pushing objects but also achieve more various tasks, such as assembling parts by robot groups.

## References


1. Nakamura, A., *et al.* (1998) Multiple Mobile Robot Operation by Human, *Proc. of the IEEE Int. Conf. on Robotics and Automation*, 2852-2857.
2. Beltrán-Escavy, J., *et al.* (1998) Human-Robot GUI System with Group Control for Multiple Mobile Robot Systems •Graphic Interface and Comparison with Previous Joystick-based System•, *Int. J. of the Japan Society for Precision Engineering*, **32**, 4, 291-296.
3. Beltrán-Escavy, J., *et al.* (1999) Human-Robot GUI System with Group Control for Multiple Mobile Robot Systems •System Monitoring•, *Int. J. of the Japan Society for Precision Engineering*, **33**, 2, 149-155.
4. Nakamura, A., *et al.* (1999) Development of a system for multiple mobile robots and a human, *Proc. of the JSME Robotics and Mechatronics Conference*,•CD-ROM••1P1-30-036, **in Japanese**.
5. Hirai, S. and Sato, T. (1988) Object Model for Telerobot, *Proc. IEEE Int. Workshop on Intell. Robots and Systems (IROS '88)*, 701-706.
6. Kosuge, K., *et al.* (1990) Control of Single-Master Multi-Slave Manipulator System Using VIM, *Proc. of the 1990 IEEE Int. Conf. on Robotics and Automation*, 1172-1177.
7. Koide, S., *et al.* (1993) Semi-autonomous Teleoperation Control System with Layered Architecture •An Application to Space Robots•, *Proc. IEEE/RSJ Int. Conf. on Intell. Robots and Systems (IROS '93)*, **3**, 1995-2001.
8. Tsumaki, Y., *et al.* (1996) Virtual Reality Based Teleoperation which Tolerates Geometrical Modeling Errors, *Proc. IEEE/RSJ Int. Conf. on Intell. Robots and Systems (IROS '96)*, **3**, 1023-1030.
9. Hasegawa, K., *et al.* (1996) Novel Teleoperation System for Multi-Micro Robots based on Intention Understanding through Operator's Intuitive Behaviors, In *J. of the Robotic Society of Japan*, 14, 4, 567-573, **in Japanese**.
10. Arkin, R. C. and Ali, K. S. (1994) Integration of Reactive and Telerobotic Control in Multi-agent Robotic Systems, *From Animals to Animats 3: Proc. the Third Int. Conf. Simulation of Adaptive Behavior*, 473-478.
11. Adams, J. A., *et al.* (1995) Cooperative Material Handling by Human and Robotic Agents: Module Development and System Synthesis, *Proc. IEEE/RSJ Int. Conf. on Intell. Robots and Systems (IROS '95)*, 1, 200-205.
12. Suzuki, T., *et al.* (1996) Cooperation between a Human Operator and Multiple Robots for Maintenance Tasks at a Distance, *Distributed Autonomous Robotic Systems 2*, Springer-Verlag, 50-59.
13. Hasegawa, T. and Givert, D. (1988) An Interacitve Environment Modeling system with Geometric Reasoning, In *J. of the Robotic Society of Japan*, **6**, 6, 471-479, **in Japanese**.
14. Kurazume, R. and Hirose, S. (1998) Study on Cooperative Positioning System •Optimum Moving Strategies for CPS-III•, *Proc. of Int. Conf. on Robotics and Automation*, 4, 2896-2903.
15. Arai, T., *et al.* (1993) Real-Time Measuring System of Relative Position on Mobile Robot System, *Proc. Int. Symp. Ind. Robots*, 931-938.
16. Bison, P., *et al.* (1998) A "Barber Pole" Beacon for Mobile Robot Cooperation, In *Distributed Autonomous Robotic Systems 3*, Springer-Verlag, 193-202.

# Chapter 7

# Exploration, Mapping, and Model Acquisition

# Autonomous Mobile Robots and Distributed Exploratory Missions*

Dean F. Hougen, Michael D. Erickson, Paul E. Rybski, Sascha A. Stoeter, Maria Gini, and Nikolaos Papanikolopoulos

Center for Distributed Robotics, Dept. of Computer Science and Engineering, University of Minnesota, Minneapolis, MN 55455, USA
hougen—merickso—rybski—stoeter—gini—npapas@cs.umn.edu

**Abstract.** This paper presents our design philosophy for robotic teams involved in distributed exploratory missions, such as military reconnaissance, scientific exploration, and security patrols. It outlines the fundamental requirements of these mission types, finds commonalities among them, analyzes the roles robots can play, and notes research areas requiring greater attention.

**Key Words:** Robotic exploration, Heterogeneous robot teams

## 1 Introduction

We define *exploratory missions* as those missions in which the primary goal is to gather information about some structure, environment, region, individual, or group. Some manipulation of the environment to aid in information gathering may occur (such as drilling a core sample) but modification of the environment is not in itself a goal of an exploratory mission. Similarly, exploration may be facilitated by the movement of resources (such as positioning a recharging station in the vicinity of the exploration site) but movement of resources is, likewise, not in itself a goal of an exploratory mission. This definition distinguishes exploratory missions from such missions as manufacturing, construction, transportation, and shipping.

While all exploratory missions cover a significant area relative to the size of the robot – requiring robot mobility to complete – in some missions the environment will be inherently distributed. This may mean that the mission requires sensing at many places simultaneously or that other requirements (such as time or fuel available) do not permit a single robot to effectively cover the points of interest. We term such missions *distributed exploratory missions*. By definition, then, distributed exploratory missions require teams of agents distributed throughout the environment. In most cases these will be teams of robots, although robots could be teamed with non-robotic agents in some cases as well.

* This material is based upon work supported by the Defense Advanced Research Projects Agency, Microsystems Technology Office (Distributed Robotics), ARPA Order No. G155, Program Code No. 8H20, issued by DARPA/CMD under Contract #MDA972-98-C-0008.

**Fig. 1.** A scout robot shown next to a quarter for scale

**Fig. 2.** A team for distributed exploratory missions - a ranger (center) with four scouts (foreground)

We have designed and built a team of robots suitable for a range of distributed exploratory missions. This team is based on a group of small sensor robots we call "scouts" (see Fig. 1) which are detailed elsewhere [1]. The scouts are teamed with larger utility platforms known as "rangers" (see Fig. 2) which provide greater range and computational power to the team [2]. This team can autonomously engage in distributed exploratory missions such as covert surveillance [3].

Others have also built robot teams for exploratory missions, such as aiding in security patrols [4], or to study particular tasks that might be useful in exploratory missions, such as map making [5] or "bounding overwatch" (a team travel method that keeps one group on the lookout while the other group moves) [6].

In the present paper, we take a broader view of distributed exploratory missions to identify fundamental requirements of these mission types, find commonalities among them, analyze the necessary and desirable abilities and characteristics of robots intended for such missions, and note research areas requiring greater attention.

## 2 Types of distributed exploratory missions

While we do not seek to exhaustively categorize all possible distributed exploratory missions, a selection of common mission types will be useful in our analysis of the shared and distinct elements of various distributed exploratory missions.

**Evaluation of accidental releases** These are missions to quickly gather information about accidental releases of hazardous materials, such as toxic chemicals or nuclear waste. We envision an exploratory mission as most likely in the event that the release takes place during transport, since releases at production or consumption locations are likely to be monitored by previously emplaced sensors.

**Scientific exploration** These are missions in which the primary purpose is to collect information of scientific interest. In general, these missions will take place in hard to reach or dangerous environments which preclude human exploration. These include lunar, planetary, volcanic, polar, and deep sea environments.

**Traditional tactical military reconnaissance** These missions are for gathering information on terrain and enemy positions. Traditional tactical military reconnaissance excludes urban zones as these have been historically avoided.

**Urban reconnaissance** Urban reconnaissance serves the same purpose as traditional military reconnaissance but provides additional challenges. The close quarters of urban settings put troops at great risk and urban building materials can hamper radio communications. Also, the immediacy of the enemy means that reconnaissance may need to be done under fire. Finally, urban reconnaissance may be useful for law enforcement groups, such as Special Weapons And Tactics (SWAT) teams, as well as the military.

**Surveying** Surveying is preparatory to map making. The maps created are highly accurate with regard to the distances and angles recorded during the surveying missions and, therefore, differ from most of the maps made in other mobile robotics missions.

**Security patrol** Security patrols are surveillance missions in which the objective is to keep an area (such as a home, office, factory, or military encampment) safe. Unlike the use of emplaced security devices (such as burglar alarms or smoke detectors), security patrols involve moving throughout the area to be protected.

**Survivor detection and location** Disasters, such as earthquakes, explosions, or chemical spills can leave survivors in places that are difficult or dangerous to traverse. Finding survivors quickly and determining their locations to bring help is paramount in saving lives.

### 2.1 Mission facets

From the mission type descriptions above, we can see distinct facets of the various missions. These include:

- the necessity for covert operation,
- an expectation of opposition,
- a requirement for special deployment methods,

- the possibility that the environment is hazardous to people,
- the difficulty that the environment is not controlled, and
- the fact that the environment is not known in advance.

These facets are presented in such a way that having one or more of them is likely to make the mission more difficult to complete.

The necessity for covert operation means that the mission may be compromised if one or all of the robots is observed by individual(s) not part of the mission group. For example, a security patrol may fail to detect an intruder if the intruder sees the patrol coming and is able to hide away until the patrol has passed.

Opposition means that there are agents actively trying to thwart the mission, perhaps by impeding travel, blocking communications, or even damaging or destroying the robots. These enemies may, however, only seek to evade detection or otherwise provide inaccurate information, as more aggressive actions may be too informative for the robots (e.g., by revealing enemy positions). Evasion will often be the favored mode of opposition in security missions.

While all robots need to be deployed into their mission areas, for some missions it will be enough to have the robots carried into position and deployed by standard means (e.g., transportation by trucks, vans, or other standard vehicles; unloaded by human operators) but for other missions special deployment will be necessary. Special deployment includes being launched from rockets, dropped from aircraft, and released underwater. Special deployment can also refer to the deployment of only some members of a robotic team, such as when a larger robot deploys sensor robots.

If an environment is hazardous to people, then the robots will either not be able to work closely with people or the people must be specially equipped to enter the environment. In either case, the need for robot autonomy is increased. (While the expectation of opposition could be thought of as providing one case of a hazardous environment, we will restrict our definition of a hazardous environment to exclude active opposition and keep these notions separate.)

If the environment is controlled, it can be modified or adapted to the robots. A security patrol in a factory, for example, can be supported by providing landmarks or wireless communication nodes where needed. If the environment is not under the control of the mission team, such modifications are not possible.

Finally, if the environment is largely known, plans can be made ahead of time and/or with a global view in mind. Otherwise, the robot must discover the large features of its environment as it explores.

## 2.2 Mission types vs. mission facets

The mission types discussed above (Sect. 2) share many similarities yet have many differences as well. Their similarities and differences with respect to

**Table 1.** Mission Types vs. Mission Facets

| | Covert Operation | Opposition Expected | Deployment Issues | Hazardous Environment | Uncontrolled Environment | Unknown Environment |
|---|---|---|---|---|---|---|
| Accidental Releases | N | N | Y | Y | Y | ? |
| Scientific Exploration | N | N | Y | Y | Y | Y |
| Trad. Reconnaissance | ? | Y | Y | ? | Y | N |
| Urban Reconnaissance | ? | Y | Y | ? | Y | ? |
| Surveying | N | N | N | ? | Y | N |
| Security Patrol | ? | Y | N | ? | N | N |
| Survivor Detection | N | N | Y | Y | Y | ? |

the mission facets of Sect. 2.1 are summarized in Table 1. In the table, 'Y' indicates the mission has this facet, 'N' that it does not, and '?' that it may vary and is, therefore, discussed below. Because these facets add difficulty to a mission, more 'Y' entries in the row for a mission type indicate more challenges that need to be overcome.

Evaluation of accidental releases has only one '?' symbol in the table. These missions may vary as to whether a map is available, in which case the environment is known. While it is unlikely that a map designed specifically for robotic use would be available for such a mission, maps intended for human use are likely to be available. For this and other robotic exploratory missions, it would be worthwhile to develop appropriate technology to allow robots to use maps made for people.

Traditional military reconnaissance may be carried out covertly although it is not necessary to mission success in all cases. It may also take place in areas that have been contaminated (for example, by chemical or biological weapons), if robots are used rather than human troops.

Like traditional military reconnaissance, urban reconnaissance may be done covertly in some cases. However, in cases where urban reconnaissance is done under fire, covert operations are not possible. Also as with traditional military reconnaissance, urban reconnaissance may take place in areas contaminated by hazardous materials when chemical or biological weapons are in use. Unlike traditional military reconnaissance, however, urban reconnaissance is likely to take place in areas where the basic layout of the physical features is unknown.

Surveying is not normally done in hazardous environments. However, this may change if robots are the only ones at risk during such missions.

Like military reconnaissance, security patrols may be carried out covertly. However, they may also be done openly to discourage illegal activity. Robotic security patrols may occur in places that pose intrinsic hazards to human health or that may become hazardous in some circumstances (e.g., when toxic fumes are released by burning materials).

As with evaluation of accidental releases, a map designed for robotic use in the detection and location of disaster survivors is unlikely to be available, although maps intended for human use are likely to be available. However, in disaster situations previous knowledge about the area may not prove useful after the disaster strikes since earthquakes, floods, etc. may radically alter the physical layout of the environment.

## 3 Robot abilities and characteristics

Because distributed exploratory missions require multiple agents, we will generally consider teams of robots rather than single robots. However, it is important to consider robot abilities and characteristics individually as well as collectively. We can then match these individual and team abilities and characteristics with distributed exploratory missions types.

Robot abilities or characteristics needed or useful for distributed exploratory missions include:

- movement,
- sensing,
- reporting,
- information evaluation,
- localization,
- navigation,
- map following,
- map making,
- heterogeneity of the robots,
- the ability to team with non-robotic agents.
- special deployment methods,
- group movement methods,
- planning,
- coordination,
- the ability to communicate throughout an area.

To complete any distributed exploratory mission, the robots must be able to move, sense, and report back. Naturally, the movement and sensing modalities must be appropriate for the particular mission(s) undertaken. Reporting of collected information may be piecemeal to ensure that information is received as soon as possible and to prevent total mission failure should disaster occur. Alternately, collected information may be stored for later analysis in cases where transmission is impossible or undesirable (such as covert missions where radio silence must be maintained).

The information collected may be needed by the robots to complete the mission itself, in which case it needs to be evaluated by the robots in real time – that is, *perceived* – or may be useful for later evaluation by people or more powerful computers. In the missions we envision, at least some of the information will need to be used for perception (for example, to visually guide a robot to a location) while other information will be used elsewhere and/or for less immediate purposes.

Localization is important to all exploratory missions but the type of localization can vary. The robot may need to know its coordinates relative to the world or structures in it, may need to know its coordinates relative to one or more other robots in the group, may not need to know its own coordinates as long as others do, or even may not need to know its own coordinates at all as long as others can find it (e.g., by activating a homing beacon once a survivor has been found in a rescue mission).

Navigation, interpreted broadly to mean intentional movement from place to place, is likewise important to all exploratory missions. This may include random walks executed for area coverage but in most cases will be more directed. Navigation may be based on information including landmarks, sensor gradients, and maps.

Map following, then, can be seen as a special case of navigation. Maps may be topographical or may be more abstract and relational. Not all exploratory missions will involve map following.

Some exploratory missions may involve map making. Whether map making is a part of a given mission will depend on such factors as whether a map is already available, whether a map is necessary to complete the mission itself, and whether constructing a map is a goal of the mission. In some cases, such as locating earthquake survivors in rubble, the environment may be too complex or confused to allow for mapping.

Heterogeneity is a characteristic of teams or groups. The creation of heterogeneous robot teams involves the production of robots that are specialists, either in terms of their hardware or their roles within the team, rather than generalists.

Similarly, specialization may be brought in by taking heterogeneity one level higher and teaming the robots with non-robotics agents, such as people or dogs.

As mentioned in Sect. 2.1, some exploratory missions may require special deployment methods. Likewise, special group movement methods, such as platooning, may increase team effectiveness. Planning and coordination may be needed in many missions, although in some cases totally reactive behaviors may be sufficient. Finally, the ability of robots to act as communication relays can keep a team intact over larger areas or where the environment severely limits radio range.

**Table 2.** Mission Types vs. Robot Characteristics.

| | Map Making | Map Following | Heterogeneity of Robots | Teaming with Non-Robots | Group Movement | Planning | Coordination | Communication Relays |
|---|---|---|---|---|---|---|---|---|
| Accidental Releases | ? | Y | Y | N | Y | Y | Y | Y |
| Scientific Exploration | Y | Y | Y | N | Y | Y | Y | Y |
| Trad. Reconnaissance | N | Y | Y | ? | Y | Y | Y | Y |
| Urban Reconnaissance | ? | ? | Y | Y | Y | ? | Y | Y |
| Surveying | Y | Y | ? | ? | ? | Y | Y | Y |
| Security Patrol | N | Y | Y | ? | Y | Y | Y | ? |
| Survivor Detection | ? | ? | Y | Y | Y | Y | Y | Y |

### 3.1 Mission types vs. robot characteristics

As mentioned in Sect. 3, all missions require the robots to be able to move, sense, report, evaluate at least some of the information sensed, localize, and navigate. Table 2 summarizes whether other abilities or characteristics are important for each mission type. In this table, 'Y' indicates the mission requires robots or teams to have this ability or characteristic, 'N' that it does not, and '?' that it may vary and is, therefore, discussed below.

While making a map is generally more difficult than not making one, map following may actually be easier than other forms of navigation. Producing specialized robots may be easier than producing robots that can perform all required tasks, but heterogeneity generally means more planning and coordination is needed to get the proper robot to the proper place. Teaming with non-robots means that some specializations or abilities do not need to be embodied in robot agents but adds difficulties in terms of interactions and shared information. Group movements can simplify some logistics and increase efficiency but may require hardware or software adaptations to allow it. A need for planning and coordination adds difficulty, as does the requirement that some or all of the robots be able to act as communication relays.

With evaluation of accidental releases, map making is not an intrinsic part of the mission but may be required if a map is not available to follow. It may be noted that this mission type would then require both map making and map following. This is not as contradictory as it might first appear as it is possible to follow partial maps to frontiers of unexplored space and incrementally build larger maps [7]. The build-and-follow strategy would be present in scientific exploration and possibly urban reconnaissance missions as well. For traditional military reconnaissance, the basic layout of the environment is known and overall maps do not need to be made. However, reconnaissance allows changes to be noted and details to be added to the existing map. This is also true of most surveying missions.

Traditional military reconnaissance could be done by robots alone but could also be done by mixed human/robot teams. In urban reconnaissance, on the other hand, people will always be present, often including bystanders. Also unlike traditional reconnaissance missions, urban reconnaissance will often proceed without a map.

Surveying by definition involves map making and following. Surveying also requires planning and coordination but the survey team composition and movement types can vary.

As mentioned previously, in disaster situations the state of the environment will determine if it is possible to make or even follow an available map.

## 4 Overall similarities

Two mission facets are common to many of the mission types we have discussed. These are:

- a need for special deployment methods (true in five of seven mission types), and
- the fact that the environment is not controlled and therefore not subject to prior adaptations on behalf of the robots (true in six mission types).

Additionally, a hazardous environment is a likelihood in three mission types and a possibility in the rest.

Similarly, some abilities and characteristics of robots or robotic teams were found to be common in the mission types we have examined. These are:

- map following (needed in five mission types and possibly needed in two others),
- heterogeneity of the robots in a team (definitely valuable in six mission types and potentially valuable in the seventh),
- specialized group movements (definitely useful in six mission types and potentially so in the seventh),
- planning (needed in six missions types and possibly needed in the seventh),
- coordination (needed in all seven missions types), and
- communication relay ability (needed in six missions types and possibly needed in the seventh).

These commonalities – together with the universals of movement, sensing, reporting, perception, localization, and navigation – should not only be seen as areas in which individual research should be pursued but should be taken together when designing robotic teams intended for exploratory missions.

## 5 Conclusions

We have given an overview of exploratory missions, examining inherent facets of the missions and the abilities and characteristics needed by robots and robotic teams in order to accomplish these missions. We have found many similarities in both mission facets and robotic requirements across the mission types we have considered, and have drawn attention to them.

While we have considered only those missions in which exploration and transmission of gathered information are goals, exploration may be a part of a larger mission, such as search and retrieval [8]. In these cases a distributed, heterogeneous robotic team may also prove quite useful. By distributing numerous, remote, mobile, sensor robots throughout the area to be searched, the objects to be retrieved can be located quickly. If speed is a high priority, then it will not do to have a few, large retrieval robots spending their time searching when they could be kept busy transporting items, if only they knew where the items were. The relatively small cost of scouting robots is more than repaid by the greater coverage area of all their sensors put together.

## References

1. D. F. Hougen, S. Benjaafar, J. C. Bonney, J. R. Budenske, M. Dvorak, M. Gini, D. G. Krantz, P. Y. Li, F. Malver, B. Nelson, N. Papanikolopoulos, P. E. Rybski, S. A. Stoeter, R. Voyles, and K. B. Yesin. A miniature robotic system for reconnaissance and surveillance. In *Proc. of the IEEE Int'l Conf. on Robotics and Automation*, pages 501–507, San Francisco, CA, April 2000.
2. D. F. Hougen, J. C. Bonney, J. R. Budenske, M. Dvorak, M. Gini, Donald G. Krantz, F. Malver, B. Nelson, N. Papanikolopoulos, P. E. Rybski, S. A. Stoeter, R. Voyles, and K. B. Yesin. Reconfigurable robots for distributed robotics. In *Gov't Microcircuit Applications Conf.*, pages 72–75, Anaheim, CA, March 2000.
3. P. E. Rybski, S. A. Stoeter, M. D. Erickson, M. Gini, D. F. Hougen, and N. Papanikolopoulos. A team of robotic agents for surveillance. In *Proc. of the Int'l Conf. on Autonomous Agents*, pages 9–16, Barcelona, Spain, June 2000.
4. H. R. Everett and D. W. Gage. From laboratory to warehouse: Security robots meet the real world. *Int'l Journal of Robotics Research*, 18(7):760–768, July 1999.
5. W. Burgard, D. Fox, M. Moors, R. Simmons, and S. Thrun. Collaborative multi-robot exploration. In *Proc. of the IEEE Int'l Conf. on Robotics and Automation*, pages 476–481, San Francisco, CA, April 2000.
6. L. Parker. Multi-robot team design for real-world applications. In H. Asama, T. Fukuda, T. Arai, and I. Endo, editors, *Distributed Autonomous Robotic Systems 2*, pages 91–102. Springer-Verlag, Tokyo, 1996.
7. B. Yamauchi. A frontier-based approach for autonomous exploration. In *Proc. IEEE Int'l Symposium on Computational Intelligence in Robotics and Automation*, pages 146–151, Monterey, CA, July 1997.
8. P. E. Rybski, A. Larson, M. Lindahl, and M. Gini. Performance evaluation of multiple robots in a search and retrieval task. In *Workshop on Artificial Intelligence and Manufacturing*, pages 153–160, Albuquerque, NM, August 1998.

# Terrain Model Acquisition By Mobile Robot Teams and $n$-Connectivity

Nageswara S. V. Rao

Center for Engineering Science Advanced Research, Computer Science and Mathematics Division, Oak Ridge National Laboratory, Oak Ridge, TN 37831-6355; email:raons@ornl.gov.

**Abstract.** The connectivity of the configuration space has been a valuable concept in the motion planning for single robots in both known and unknown terrains. We show here that $n$-connectivity plays a similar role for mobile robot teams in providing algorithms for terrain model acquisition in unknown terrains. A bound on the connectivity degree of the free-space, reflected in that of a navigation course, provides us an estimate of the size of a robot team that is effective for the terrain. We consider an unknown planar polygonal terrain. The robots are point-sized and equipped with visual sensors which acquire all visible parts of the terrain by scan operations executed from their locations. The performance is measured by the number of sensor (scan) operations which are assumed to be the most time-consuming of all the robot operations. We show that the Voronoi diagrams and trapezoidal decomposition methods yield solutions for efficient terrain model acquisition by a 2- and 3-robot team using visual sensors.

## 1 Introduction

The terrain model acquisition problem deals with robots autonomously acquiring a complete model of an unknown terrain by systematically visiting portions of it. There are several motivations for this problem. Once the terrain model is completely acquired, the navigation algorithms of known terrains can be employed for path planning with two potential advantages. First, the sensors may be switched off (at least in theory) in future navigation, saving the time and resources involved in the acquisition and processing of sensor data. Second, navigation paths with the shortest distance between the start and goal positions may be computed using the terrain model. In applications involving mobile robots in indoor environments for repetitive operations, typically a human operator is in charge of model building, which is tedious and time-consuming. Robots capable of autonomous terrain model acquisition (even in only small parts of the terrain) can assist in the model building process.

The terrain model acquisition problem (TMAP) for polyhedral or polygonal terrains has been solved using a single robot in [14,13,18,12] for the case of a discrete vision sensor. TMAP for a robot equipped with a continuous vision sensor has been solved in [10,12]. TMAP for *team* of robots has been recently addressed by a number of investigators. Ishioka *et al.* [7] described a cooperative map generation by heterogeneous autonomous mobile robots (also see Dudek *et al.* [3]). A cooperative recognition system for the environment using multiple robots has been developed by Ishiwata *et al.* [8]. Rao et al [17,16] addressed this problem in the formulation of convergent algorithms using visibility graphs. In particular, if all obstacles are convex, the sensing time was shown to be essentially reduced by a factor of $1/n$ for $n = 2, 3, 4$. The disadvantage of the visibility graph method is that robots must be capable of navigating very close to the obstacles. The problem of incidentally building the map of a convex-polygonal environment by a team of mobile robots, while they perform other tasks, has been addressed by Inoue *et al.* [6] using the visibility graphs.

In this paper, we show that the navigational structures that keep the robots away from the obstacle boundaries can also be utilized by an $n$-robot team for $n = 2, 3$. There are other problems in unknown terrains studied for mobile robot teams. The study by Harinarayan and Lumelsky [5] indicates that the simultaneous navigation of two robots cannot be solved if no "cooperation" is present between them. Note that our overall objective is different from theirs in two ways, namely: (a) we are interested in terrain model acquisition, and (b) we wish to explore the cooperation mechanisms so that the objective can be achieved more effectively by a team of robots instead of a single robot.

There are obvious limiting cases in deploying mobile robot teams for the terrain model acquisition. In "very bad " cases, e. g. the entire team of robots is initially located at one end of a "long narrow polygonal corridor", there may not be any advantage in employing a team. Robots are forced to form a single cluster, thereby obviating any advantages of a team. However, if the terrain has "branches", a team is likely to acquire the terrain faster than one robot. In this paper, we discuss general results that indicate the conditions under which a robot team is more effective from an algorithmic perspective. We show that for 2- and 3-robot teams the navigation structures based on the Voronoi diagram and trapezoidal decomposition methods reduce the sensing time, while keeping the robots wary from obstacle boundaries. More generally, we show that the $n$-connectivity of the navigation course, which reflects the connectivity structure of the configuration space, enables an $n$-robot team to reduce the sensor time by a factor of $1/n$ compared to a single robot. We show that Voronoi diagram and dual graph based on trapezoidal decomposition are 2-connected if the obstacles are all convex. For 3-robot team, these navigational courses offer good solutions although they do not reduce the sensor time by $1/3$. For general polygonal terrains, the reductions may not be as dramatic, but a reduction in sensor time is achieved depending on the connectivity structure of the configuration space.

Preliminaries are described in Section 2. Algorithmic methods for terrain model acquisition by a single robot are discussed in Section 3. The concepts of $n$-connectivity and its implications on robot team exploration are discussed in Section 4. The cases of convex and non-convex polygonal obstacles are discussed in Section 5 and 6, respectively.

## 2 Preliminaries

We consider a finite two-dimensional terrain cluttered by a finite and non-intersecting set of polygonal obstacles. An obstacle vertex is *convex* if the angle included inside the obstacle by the edges that meet at this vertex is less than 180 degrees; otherwise the obstacle vertex is *concave*. Two points $p$ and $q$ in plane are *visible* to each other if the line segment joining $p$ and $q$, denoted by $\overline{pq}$, lies entirely outside the interior of all obstacle polygons.

The robot, denoted by $R$, is point-sized and equipped with a vision sensor. A *discrete vision sensor* is characterized by a *scan* operation: a scan operation performed from a position (point) $p$ returns the *visibility polygon* of $p$ that consists of all points in the terrain visible to $p$ (Fig. 1). We assume that the most time consuming-part of the robot operation is the scan operation. In vision-based robots, each scan may take several minutes including the time required to acquire and process the sensory data. The total *sensing time* is given by the number of scan operations performed by the robot(s) in sequence. The robots communicate with each other via a wireless network.

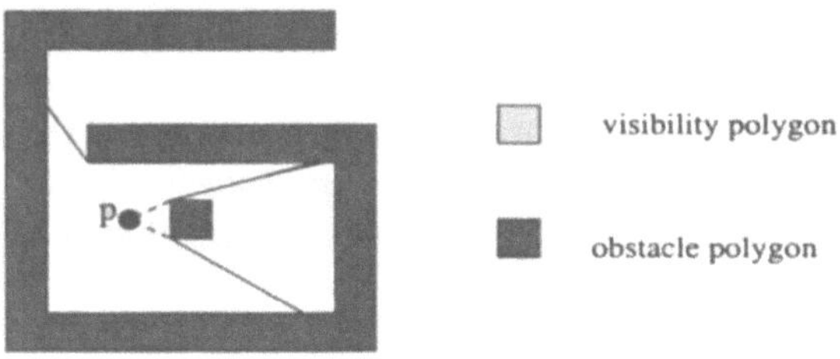


**Fig. 1.** Visibility polygon from location $p$.

## 3 Navigational Courses For TMAP

An algorithmic paradigm to solve TMAP has been proposed by Rao [11]. A finite graph called the *navigation course* $\xi$, is used as an underlying structure. Initially navigation course is not known, but it is incrementally constructed from the sensor operations by the robot systematically visiting all vertices of the navigation course.

In order that a graph exploration algorithm terminates, the navigation course must contain a finite number of edges and vertices, i.e., must satisfy *finiteness* property. It must satisfy the *terrain-visibility* property which requires that every point in the free-space is visible from some vertex of the navigation course. It must also satisfy the *connectivity*, more precisely

1-connectivity, property which requires that every pair of vertices be connected by a graph path on the navigation course. We require that adjacency list of a $\xi$-vertex can be constructed from the information of a single scan; this property is called the *local-constructibility*. For a navigation course $\xi$ that satisfies the properties of finiteness, connectivity, terrain-visibility and local-constructibility, any graph search algorithm (e.g. depth-first search) can be employed to solve the navigation and the terrain model acquisition problems.

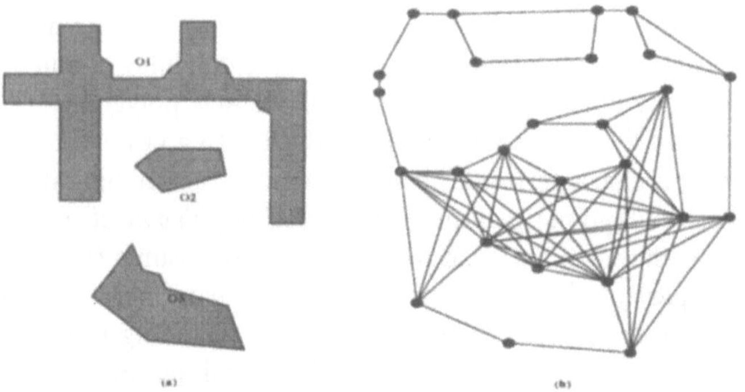

**Fig. 2.** Restricted visibility graph.

We now discuss three examples of the navigation course. The ***restricted visibility graph***, $RVG = (V, E)$ is defined as follows [13]: (a) $V$ is the set of all convex obstacle vertices, (b) an edge $(v_1, v_2)$, for $v_1, v_2 \in V$ represents the fact that the line joining $v_1$ and $v_2$ either corresponds to an obstacle edge or does not intersect any obstacle polygon. See Fig. 2 for an example of $RVG$.

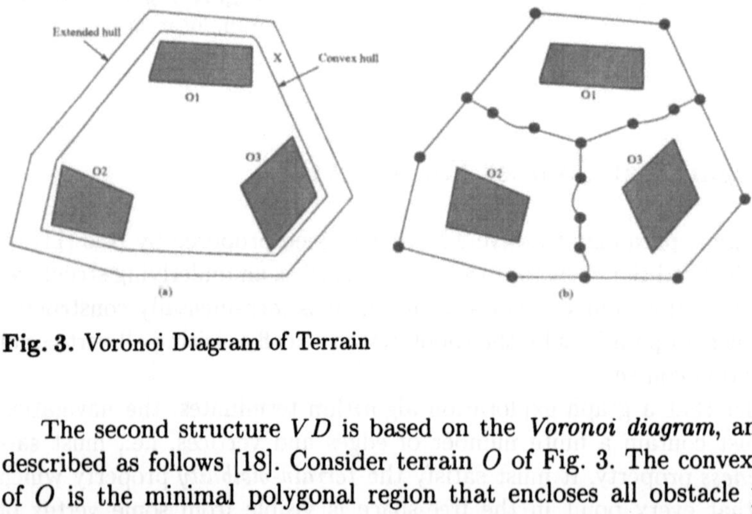


**Fig. 3.** Voronoi Diagram of Terrain

The second structure $VD$ is based on the *Voronoi diagram*, and can be described as follows [18]. Consider terrain $O$ of Fig. 3. The convex hull $CH$ of $O$ is the minimal polygonal region that encloses all obstacle polygons. The extended hull $EH$ is the convex polygonal region obtained by pushing out the edges of $CH$ by a distance $x$. The Voronoi diagram of the terrain is the locus of points that are closest to at least two points on the obstacle boundary. The Voronoi diagram consists of straight line segments and second order curve segments. The $VD$ is obtained by taking the union of the Voronoi diagram contained in $EH$, and the boundary of $EH$ as in Fig. 3(b).

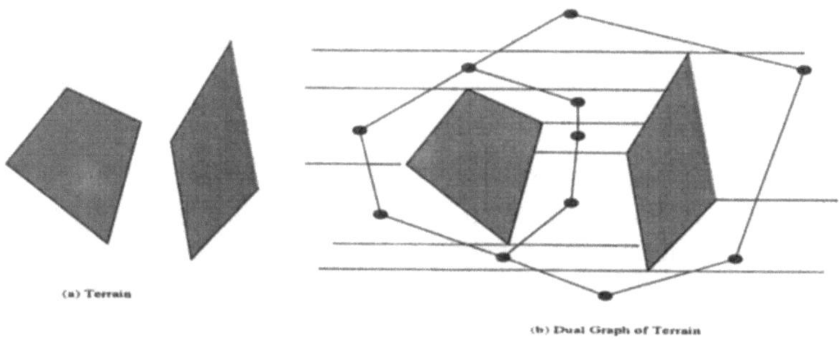


**Fig. 4.** Dual Graph of Terrain

There are many other ways of generating navigational courses, based on dual graphs corresponding to decompositions such as trapezoidal decomposition, triangulation, etc., of free space [12]. Consider the terrain of Fig. 4. We decompose free-space into trapezoids by sweeping a horizontal line. When this line reaches a vertex, we extend (at most two and at least one) line segments from this vertex into free-space until obstacle boundary is reached or to infinity. The free-space is then decomposed into trapezoids by these line segments. Then the *dual graph* DG is obtained by denoting each trapezoid by a node and joining two nodes by an edge if and only if the corresponding trapezoids share a boundary edge.

The terrain model acquisition problem was first formulated and solved for two and three dimensional terrains by Rao et al [14] using the visibility graph. When the robot is circular in shape, Rao and Iyengar [13] proposed a navigation course based on the visibility graph. For a robot of polygonal shape with an ability to translate in any direction, Foux et al. [1993] proposed a navigation algorithm. See [15] for a survey (also see [2]). The terrain model acquisition method based on the extended visibility graph method was proposed for a team of two, three, and four robots in [16].

## 4 n-Connectivity and TMAP

For single robots, 1-connectivity in sufficient to solve TMAP. More generally, connectivity of the configuration space plays a vital role in path planning for single robots [9]. However, 1-connectivity is not sufficient to ensure the effectiveness of a team, especially in explorations. Consider the scenario depicted in Figure 5, where node $v$ is an *cut point* [4] of the navigation course, whose removal results in a disconnected navigation course. The robots, represented by unfilled circles, are located on one side of $v$ and the vertices to be explored (shown in dark circles) are located on the other side. The vertex $v$ must be explored first before getting to the other unexplored vertices, and hence acts as a bottleneck.

A graph $G = (V, E)$, for $|V| \geq n + 2$ is *n-connected* if it cannot be disconnected by removing $n - 1$ vertices [1]. For example, a cycle is 2-connected since removal of any vertex still leaves the rest of the cycle connected. Note that if the graph has a cut point, it is not 2-connected. The following are important consequences of $n$-connectivity:

(a) The degree of every vertex is at least $n$, if not by removing at most $n - 1$ neighbors of this vertex, it can be disconnected from the rest of the graph.
(b) If the graph has at least $2n$ nodes, then for every two disjoint sets $V_1$ and $V_2$, there exist $n$ disjoint paths joining these two sets of points [4].

The first consequence shows that RVG, VD and DG are not 3-connected. RVG is shown to be 2-connected [17], and we will show that VD and DG are 2-connected in the next section.

The second consequence is important in robot team exploration as follows. Consider two robots $R_1$ and $R_2$ located at the vertices of a 2-connected navigation course. There are two vertices $u$ and $v$ to be explored such that they are adjacent to explored part of the graph. Consider that $R_1$ has chosen to visit $u$ next. Then $R_2$ can get to $v$ without going through $u$. In other words, $R_2$ can work around $R_1$, when the latter makes the decisions on its own about which vertex to explore next. Assuming that travel time is small, the robots can perform scan operations in parallel, in which case the overall exploration time is halved.

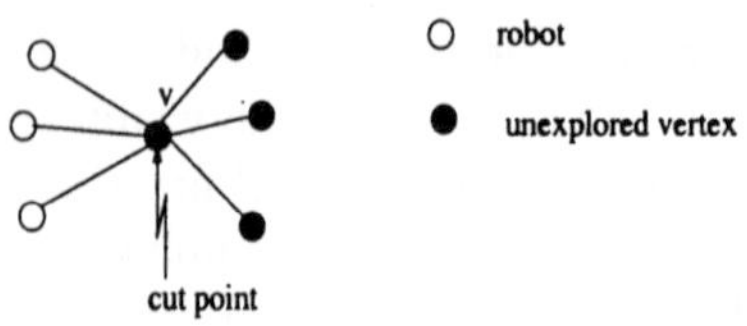


**Fig. 5.** Illustration of limitations of 1-connectivity.

The overall algorithm for a team of $n$ robots is based on the robots executing a graph search algorithm in a cooperative manner using an $n$-connected navigation course. At any step, each robot has the same version of an incomplete navigation course. For the team of robots $R_1, R_2, \ldots, R_n$, let $R_1$ have the highest priority, $R_2$ have second highest priority, and so on. Each robot performs a scan operation and obtains the resultant visibility polygon. Each robot computes its own adjacency list and communicates it to the other robots. $R_1$ sends to $R_2$ its next destination $d_1$ which is one of the nodes adjacent to its present location. Then $R_2$ marks $d_1$ as visited and computes its next destination $d_2$. $R_2$ communicates $d_1$ and $d_2$ to $R_3$, and $d_2$ to $R_1$. This process is repeated until $R_n$ computes its destination. Then the robots move to their chosen destinations and repeat the algorithm. Since the navigation course is $n$-connected, all the $n$ robots can get to their respective chosen vertices without being blocked by an unexplored vertex. Informally speaking, $R_2$ works around $R_1$, $R_3$ works around $R_2$ and $R_1$, and so on. This algorithm terminates when all known vertices have been visited; the connectivity property together with the terrain-visibility property ensure that the terrain model is completely acquired.

## 5 Convex Polygonal Terrains

In this section, we consider terrains composed of *convex* polygonal obstacles, and the navigational courses $RVG$, $VD$ and $DG$.

### 5.1 Two-Robot Team

In the algorithm of previous section, $R_1$ is guaranteed to find a destination at each step, due to the connectivity of the navigation course, RVG, VD or DG. We now establish that $R_2$ can always find its destination, by showing the 2-connectivity.

RVG is shown to be 2-connected in [16]. For $VD$ and $DG$, 2-connectivity can be shown as follows. The proof method is similar in both cases, and we illustrate the case of VD. For each convex obstacle $O_1$, one can trace the cycle of VD that immediately surrounds it. Consider an adjacent obstacle $O_2$ such that it's cycle shares at least an edge with the that of $O_1$; then there are three cycles corresponding to each of $O_i$'s and the cycle formed by the union of the two cyles minus the common edges. By extending this argument, it follows that any two vertices of VD belong to some cycle, possibly formed from cycles of different obstacles. Thus, there are two vertex disjoint paths between every pair of vertices, which shows that VD is 2-connected (Theorem 5.10 [4]).

Note that 2-connectivity implies that the navigation course cannot be disconnected by removing a single vertex. Indeed, let the next destination chosen by $R_1$ at any step be denoted by $v$. By connectivity, if there is an unvisited node (other than $v$), then there is at least one unvisited node adjacent to the paths traced by $R_1$ or $R_2$. If not, all the unvisited nodes can only be reached via $v$, which makes $v$ a cut point; this in turn contradicts the 2-connectedness. Thus by the time $R_1$ performs $\lceil N/2 \rceil$ scan operations, $R_2$ would have performed scan operations from the remaining nodes, where $N$ is the number of nodes of the navigation course.

### 5.2 Three-Robot Teams

The Extended Visibility Graph (EVG) is the RVG augmented as follows. The extended hull is obtained by expanding the convex hull of the obstacles by a fixed non-zero amount. Then the vertices of the RVG on the convex hull are connected to the corresponding vertices on the extended hull as shown in Fig. 6. The EVG of a terrain cluttered by a finite number of convex polygonal obstacles is shown to be 4-connected and, in general is not 5-connected [16].

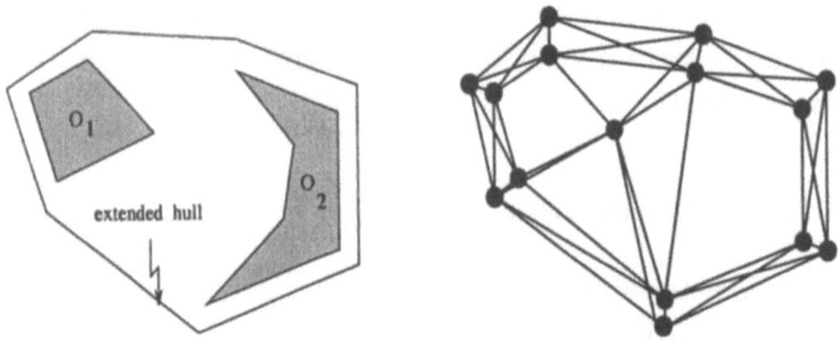


(a) terrain (b) EVG

**Fig. 6.** Definition of the EVG.

Although, VD and DG are only 2-connected, a 3-robot team can still be effective as shown in Figure 7 for DG. Here $R_1$ and $R_2$ are forced to stay together at vertices $u$ and $v$, and after that the three robots explore in parallel. In this case $N = 11$, and the sensing time is 4, even though there are only two vertices with degree 3. A bound on the sensing time can be obtained as follows. Let $N_2$ denote the vertices with degree 2. And $N - N_2$ is the number of vertices with degree at least 3. Nodes with degree 2 can force two robots to stay together, and the other nodes do not. Hence the total sensing time is upper-bounded by

$$\frac{N - N_2}{3} + N_2/2 = N/3 + N_2/6.$$

Note that for 3-robot team the best possible sensing time is $N/3$. If $N_2 = 0$, i.e. the graph is 3-connected, the bound is $N/3$. If $N_2 = N$, i.e. the graph is 2-connected, the bound is $N/2$, which is the best possible. In a general case, the effect of the additional term $N_2/6$, which adds to the sensing time, is quite moderate.

## 6 General Polygonal Terrains

For a general polygonal terrain, the navigational course consists of trees which are only 1-connected. The performance of the algorithm for general terrains depends on the number of concave regions and their depths. To tackle this situation, a hierarchical decomposition of the navigation course into $n$-connected and $(n-1)$-or-less-connected components was proposed in [17]; although their main focus is on visibility graph, their decomposition is valid for any navigation course. The performance for the $n$ $(= 2, 3)$ robot team is expressed in

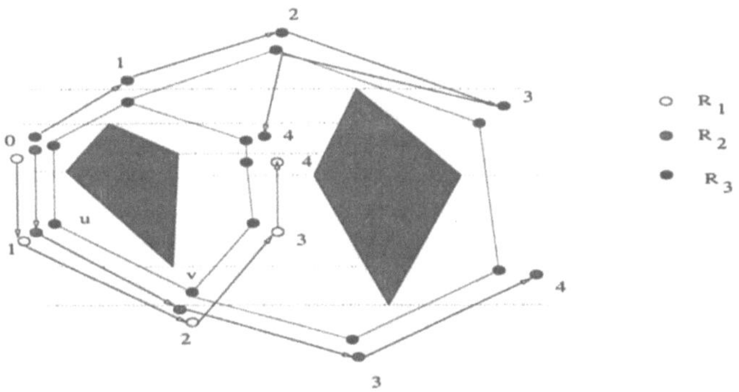


**Fig. 7.** 3-robot team using DG.

terms of the sizes of $n$-connected components, and the sizes and diameters of $(n-1)$-or-less connected components. This analysis highlights the critical properties such as 2- and 3-connectivity, depth of hierarchy, etc. that support or impede the parallel acquisition of the terrain model.

# 7 Conclusions

The connectivity of the configuration space has been a valuable concept in the terrain model acquisition by single robots. We showed here that the $n$-connectivity plays a similar role for robot teams. The methods based on visibility graphs, Voronoi diagrams and trapezoidal decompositions are shown to yield solutions for efficient terrain model acquisition by a 2- and 3-robot team using visual sensors. This paper is only a first step towards establishing the algorithmic and analytical aspects of mobile robot teams for terrain model acquisition. More generally, it would be of future interest to investigate the role of $n$-connectivity for other applications such as cooperative navigation and search. It would be interesting to see if $n$-connectivity holds for general configurations for non-point robots. Here, only the sensor time is considered here as a measure of performance and the estimates are conservative. We believe that alternative characterizations and better performance estimates are possible. The recently studied class of competitive algorithms for the TMAP by guarantee that the distance traversed by a single robot is bounded by a factor times the minimum possible value achieved if the terrain model is available. Improving the performance of this type of algorithms by employing a team of robots will be of future interest.

**Acknowledgements**

This research is sponsored by the Engineering Research Program of the Office of Basic Energy Sciences, U.S. Department of Energy, under Contract No. DE-AC05-00OR22725 with UT-Battelle, LLC.

## References


1. G. Chartrand and L. Lesniak. *Graphs and Digraphs.* Wadsworth and Brooks/Cole Advaced Books and Software, 1986.
2. H. Choset and J. Burdick. Sensor based motion planning: The hierarchical generalized Voronoi diagram. In *Workshop on Algorithmic Foundations of Robotics.* 1996.
3. G. Dudek, M. Jenkins, E. Milios, and D. Wilkes. Robust positioning with a multi-agent robotic system. In *1993 IJCAI Workshop Series: Dynamically Interacting Robots*, pages 118–123, 1993. Working Notes.
4. F. Harary. *Graph Theory.* Addison-Wesley Pub., 1969.
5. K. R. Harinarayan and V. J. Lumelsky. Sensor-based motion planning for multiple mobile robots in an uncertain environment. In *Proceedings of IEEE/RSJ/GI Int. Conf. on Intelligent Robots and Systems*, pages 1485–1492, 1994.
6. K. Inoue, J. Ota, T. Hirano, D. Kurabayashi, and T. Arai. Iterative transportation by cooperative mobile robots in unknown environments. In *Distributed Autonomous Robotic Systems 3*, pages 2–12. Springer-Verlag, 1998.
7. K. Ishioka, K. Hiraki, and Y. Anzai. Cooperative map generation by heterogeneous autonomous mobile robots. In *1993 IJCAI Workshop Series: Dynamically Interacting Robots*, pages 57–67, 1993. Working Notes.
8. Y. Ishiwata, M. Inaba, and H. Inoue. Cooperative recognition of environments by multiple robots. In *Proc. of JSME Annual Conf. on Robotics and Mechatronics*, pages 79–84, 1992.
9. J. C. Latombe. *Robot Motion Planning.* Kluwer Academic Pub., Boston, 1991.
10. V. Lumelsky, S. Mukhopadhyay, and K. Sun. Dynamic path planning in sensor-based terrain acquisition. *IEEE Transactions on Robotics and Automation*, 6(4):462–472, 1990.
11. N. S. V. Rao. Algorithmic framework for learned robot navigation in unknown terrains. *IEEE Computer*, 22:37–43, 1989.
12. N. S. V. Rao. Robot navigation in unknown generalized polygonal terrains using vision sensors. *IEEE Transactions on Systems, Man and Cybernetics*, 25(6):947–962, 1995.
13. N. S. V. Rao and S. S. Iyengar. Autonomous robot navigation in unknown terrains: Visibility graph based methods. *IEEE Transactions on Systems, Man and Cybernetics*, 20(6):1443–1449, 1990.
14. N. S. V. Rao, S. S. Iyengar, B. J. Oomen, and R. L. Kashyap. On terrain model acquisition by a point robot and polyhedral obstacles. *IEEE Journal of Robotics and Automation*, 4:450–455, 1988.
15. N. S. V. Rao, S. Kareti, W. Shi, and S. S. Iyengar. Robot navigation in unknown terrains: An introductory survey of non-heuristic algorithms. Technical Report ORNL/TM-12410, Oak Ridge National Laboratory, Oak Ridge, TN, 1993.
16. N. S. V. Rao, N. Manickam, and V. Protopopescu. Connectivity of visibility graphs and terrain model acquisition by robot teams. *Congressus Numerantium*, 120:129–137, 1996.
17. N. S. V. Rao, N. Manickam, and V. Protopopescu. Cooperative terrain model acquisition by a team of two or three point-robots. In *Proc. Int. Conf. on Robotics and Automation*, pages 1427–1433, 1996.
18. N. S. V. Rao, N. Stoltzfus, and S. S. Iyengar. A 'retraction' method for learned navigation in unknown terrains. *IEEE Transactions on Robotics and Automation*, 7(5):699–707, 1991.

# Graph-Based Exploration using Multiple Robots

Ioannis M. Rekleitis[1], Gregory Dudek[1], and Evangelos E. Milios[2]

[1] Centre for Intelligent Machines, McGill University,
3480 University St., Montreal, Québec, CANADA H3A 2A7
[2] Faculty of Computer Science, Dalhousie University,
6050 University Avenue Halifax, Nova Scotia, Canada B3H 1W5

**Abstract.** We present an approach to multi-robot exploration of large environments. Our method is designed to be robust in the face of arbitrarily large odometry errors or objects with poor reflectance characteristics. The algorithm achieves its robustness by using a team of cooperating agents. The critical aspect of our method is the use of a vision system that sweeps areas of free space and generates a graph-based description of the environment. This graph is used to guide the exploration process and can also be used for subsequent tasks such as place recognition or path planning. As a result of the guidance provided by the dual graph of the triangulated environment, our system can guarantee complete exploration without any overlaps.

We present an algorithmic solution, simulation results, as well as a cost analysis and experimental data. In this approach a pair of robots observe one another's behavior, thus greatly reducing odometry errors. We assume the robots can both directly sense nearby obstacles and see one another (if their view is not obstructed). We have implemented both these capabilities with actual robots in our lab.

**Keywords:** Multi-robot, exploration, mapping, cooperation, sensor fusion.

## 1 Introduction

We present an approach to the exploration and mapping of large environments that functions despite arbitrarily large odometry errors or objects with poor reflectance characteristics. We also present experimental results that validate this approach and discuss specific implementation issues for a suitable sensor.

The exploration is done by two robots which jointly assist one another. We utilize a vision sensor (**Robot Tracker**) with a dual purpose: localization and mapping. At any time, one of the robots is stationary while the other robot is moving. The stationary robot acts as an artificial landmark in order for the moving robot to recover its pose with respect to it. Therefore, a detectable landmark is provided without any modification of the environment. We call this approach *Cooperative Localization*. Moreover, when one robot is moving maintaining uninterrupted the line of visual contact with the stationary robot, it effectively maps the area covered by the line of visual contact. As the two robots move through the environment, they map areas of free space using the fact that they constantly are able to "*see*" each other. Our approach is sufficiently robust to be able to cope with environments that may have uneven or slippery terrains, or whose surface reflectance properties are not well suited to conventional sensors.

Observe that conventional approaches to robotic mapping and navigation are typically applied to test environments of rather limited size. Further, the sensing techniques used to both explore the environment and position the robot often make rather optimistic assumptions about the environment: diffuse visual reflectors, substantial reflectivity, etc. In practice, some surfaces may either be specular (mirror-like) reflectors or be hard to detect due to low reflectance, and some parts of the environment may have frictional properties that make large-scale odometry difficult.

We deal with these diverse issues in two ways, both based on a polygonal approximation to the environment and the detection of convex (reflex) vertices. The presence of reflex vertices is critical since it is these reflex vertices that determine the occlusion of regions of the environment with respect to one another. We use a pair of robots observing one another to build a map and circumvent problems of object visibility. The exploration process is based on the dual graph of the triangulation using an environment decomposition attached to reflex vertices.

This paper builds on previous work in which we have examined algorithms for multi-robot exploration in a theoretical context [9]. Now we consider how to realistically implement a *low-cost* multi-robot position tracker and evaluate its performance, and we examine the empirical performance of multi-robot localization and exploration in a simulation context.

In the next subsection, we will briefly discuss background research. In Section 2 we discuss multi-robot localization and exploration including, in Subsection 2.1, an example of a low-cost visual tracker that we have used to implement the algorithm described in the paper. Section 4 contains an overview of the exploration algorithm. In Section 4 we present experimental results from simulations and laboratory experiments.

### 1.1 Background

There are two major approaches to localization while mapping for a mobile robot. The first approach to localization is to use landmarks in the environment in order to localize frequently and thus reduce the odometry error [3]. A common technique is to select a collection of landmarks in known positions and inform the robot beforehand [5]. Another approach is to let the robot select its own landmarks [11] according to a set of criteria that optimize its ability to localize, and then use these landmarks to correct its position. A variation in this theme is to perform a matching of the sensor data collected at the current location to an existing model of the environment. In addition to vision, sonar and laser range finder data have been matched to geometrical models [11,6,8], and images have been matched to higher order configuration space models [1] in order to extract the position of the robot.

A related dichotomy is between natural and artificial landmarks; rather than exploit landmarks already occurring in the environment, whether learned or manually selected, it has sometimes been found preferable to physically modify the environment to ensure that good landmarks are present. Artificial landmarks that have been used include visually detectable ones, sonar beacons, and radio beacons. The key advantage of artificial landmarks is not only that they assure the presence of the landmarks, but that the landmarks are assured to be detectable and unambiguous.

For navigation, it is often desirable to combine information from multiple sensors rather than using a single modality. In particular, the results of odometry can be combined with external sensing. A common mechanism for this is the Extended Kalman filter [1].

Several authors have examined the issue of exploring space with one or more robots [7]. In general, multi-robot exploration techniques have tended to focus on models with limited coordination or communication between the robots [2]. In contrast, we consider a tight coupling between the exploring robots in the interest of greater accuraey of more efficient behavior. Related work deals with exploring spaces large enough that the robots cannot see one another across the environment [9]. In this work, we consider the case where the robots do not lose visual contact as long as their view of one another is not occluded.

## 2 Cooperative Localization

Localization is achieved in our approach by tight collaboration of the robots. Since sensing the other robot is used in order to correct position estimation errors, the main source of error in the cooperative localization of the robots is the inaccuracy of the "robot tracker" sensor. This sensor is used to update/correct the position of the moving robot relative to the position of the stationary robot. At any given time one robot is stationary and the other robot is moving (later they exchange roles). After the robot moves it tracks the stationary robot and updates its position estimate according to the data from the robot tracker. Note that information from both sensing and odometry could be readily combined in practice using either extended Kalman filtering or optimal multiscan alignment [6].

There are three potential sources of information for the localization of the moving robot. First, the odometry measurements provide a base estimate of the moving robot's position $\hat{X}_{odom}(t)$ (with high uncertainty $\sigma_o$). Second, the different objects in the environment, when sensed from different positions, provide updates in the robot position [8,11]. Finally, the robot tracker

[1] In this paper, while we acknowledge the desirability of such sensor fusion, we will not consider it in detail for reasons of brevity.

$\hat{X}_{track}(t)$ provides measurements relative to the position of the stationary robot $\hat{X}_{stat}(t)$. In practice, over large scale environments, the estimates of the position of different objects drift over time and cannot provide safe position updates. On the other hand, the estimate of the robot tracker is influenced by the uncertainty in the position of the stationary robot $\sigma_s$ plus the error of the tracker $\hat{X}_{track}(t)$. The accumulation of uncertainty on the position of the stationary robot depends only on the number of role exchanges the two robots had. Consequently, over large open spaces where the odometry error grows unbounded the moving robot could always reference back to a stationary landmark (played by the second robot).

$$\hat{X}(t) = \frac{\sigma_s(\hat{X}_{track}(t) + \hat{X}_{stat}(t))}{\sigma_s + \sigma_o} + \frac{\sigma_o \hat{X}_{odom}(t)}{\sigma_s + \sigma_o} \tag{1}$$

## 2.1 Tracker implementation

There are many sensors that could be used for the robot tracker. Our current implementation is based on visual observation of a geometric target on the robot [4]. (Alternative possible implementations use retroreflectors or laser light striping – our actual robot is equipped with such technologies.) Each robot is equipped with a camera that allows it to observe its partner. The robots are both marked with a special pattern for pose estimation.

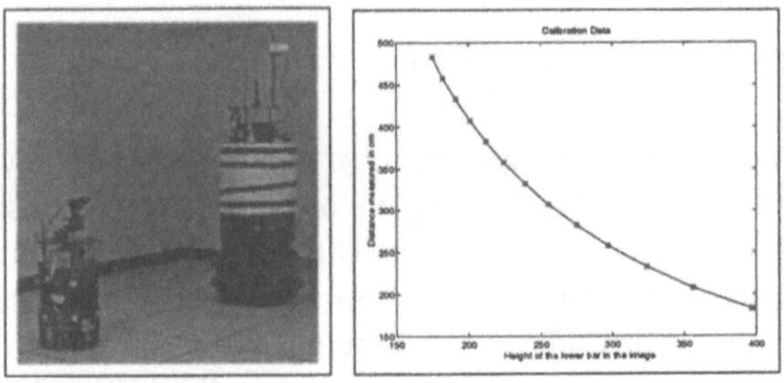


(a) (b)

**Fig. 1.** (a)The visual robot tracker system (camera mounted on one robot, helix target pattern mounted on the second robot. (b)Calibration data for the distance estimation relating observed image position to actual distances

By mounting the observing camera above (or below) the striped pattern, the distance from one robot to the other can be inferred from the height of the stripe pattern in the image, due to perspective projection (scaling of the pattern could also be used). The difference in heights between the observing camera and the target can be manipulated to provide a tradeoff between range of operation and accuracy (see Fig. 1b). One advantage of the helical target for orientation estimation is that it functions correctly even at very large distances (although with reduced accuracy, of course).

# 3 Outline of the exploration algorithm

In [9] we presented an algorithm for mapping the interior of an environment. The size of the area should be small enough to be covered by the range of the robot tracker sensor. Two mobile robots equipped with two different types of sensors are used in close cooperation to completely map the free space. Both robots use a traditional range finder in order to detect obstacles that are very close to them and, subsequently, to follow the object perimeter during the exploration. In addition, each robot has a robot tracker sensor that provides the pose of the other robot. This establishes a **line of visual contact**. If the view from one robot to the other is interrupted, we can determine the presence of an obstacle somewhere along this line.

For environments of limited size, we will show how this line of visual contact can be exploited to precisely explore the environment and determine its geometry. The essential operation in our strategy is triangulation: that is, the decomposition of the environment into triangular regions. The triangulation is accomplished by exploring one triangle of free space each time. One robot is stationary and acts as a landmark while the other robot moves mapping a triangle of free space. The stationary robot does not accrue odometry error and it can determine the position of the moving robot (while sweeping out free space with the line of visual contact).

## 3.1 Triangulation

The moving robot follows a wall of the environment while the stationary robot is positioned at a corner. If the two robots maintain visual contact during the exploration of a wall then the line of visual contact sweeps through a triangle of free space. These triangles of free space are the basic building blocks of the map. Any environment can be described using a polygonal approximation, a triangulation of a polygon is a well known spatial decomposition of polygons providing full coverage of the interior of the polygon. As can be seen in Fig. 2b, the moving robot explores a triangle at a time until it covers the complete triangulation.

In order to complete the exploration without any overlaps the dual graph of the triangulation is used. Every triangle corresponds to a vertex in the dual graph. If two triangles share a common internal diagonal, then an edge connects the corresponding vertices. Fig. 2a presents a polygonal world with an obstacle and its triangulation with the dual graph embedded in it. During the exploration, the two robots traverse the dual graph. It is worth noting that the reflex vertices of the environment produce decision points due to the break of the line of visual contact. Therefore, when the moving robot reaches

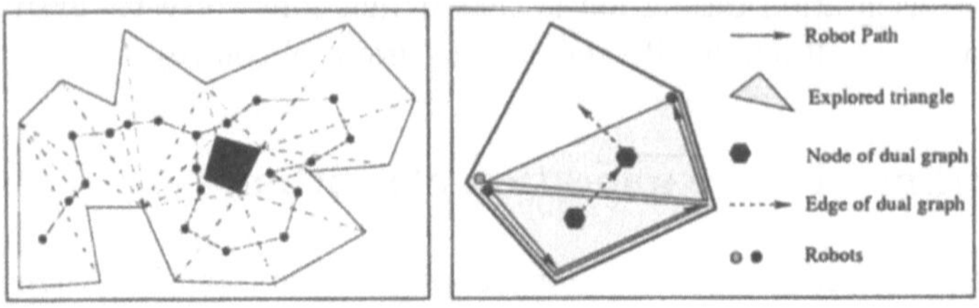


(a) (b)

**Fig. 2.** (a) A polygonal environment, the triangulation and the embedded graph. (b)Exploration of two triangles in a simple environment

a reflex vertex that would interrupt the line of visual contact, the two robots exchange roles and the stationary robot starts moving. If the line of visual contact is interrupted by a reflex vertex at the middle of exploring an edge, then an internal triangle is constructed and the dual graph bifurcates. The moving robot proceeds to the occluding vertex, marking one triangle as a gate to an unexplored area, and maps the third triangle adjacent to the internal triangle (see Fig. 2a, the internal triangle attached to the left corner of the obstacle).

### 3.2 Illustrative Example

Fig. 3 illustrates the operation of the algorithm (simulation results). Figures 3(a-o) present snapshots of the exploration in three columns. The first column presents the world as perceived by *Robot 0*. The second column presents the world as perceived by *Robot 1*. Finally the last column presents the resulting map up to that point showing the triangulation and the dual graph. The two robots exchange roles when the line of visual contact breaks. In the first row an early phase of the exploration is presented. The two robots have exchanged roles once, *Robot 0* is stationary (Fig. 3a), while *Robot 1* explores the third triangle (Fig. 3b). Consequently, in the second row *Robot 1* is stationary after it reaches a reflex vertex that interrupts the line of visual contact. *Robot 0* is mapping the sixth triangle. The third row demonstrates the case of an occluding obstacle. *Robot 0*, after losing visual contact with the stationary *Robot 1*, moves towards it mapping the occluding vertex. The resulting map contains an internal triangle and a bifurcation is introduced to the underlying graph (see Fig. 3i). The guiding graph contains two possible paths, *Robot 0* explores the right path keeping the line of visual contact unbroken with the stationary *Robot 1*. The fourth row presents the exploration at a later stage. *Robot 0* is stationary at a reflex vertex while *Robot 1* explores the other side of the environment. The fifth row illustrates the final stages of the exploration where *Robot 1* explores the final parts of the environment using *Robot 0* as a reference.

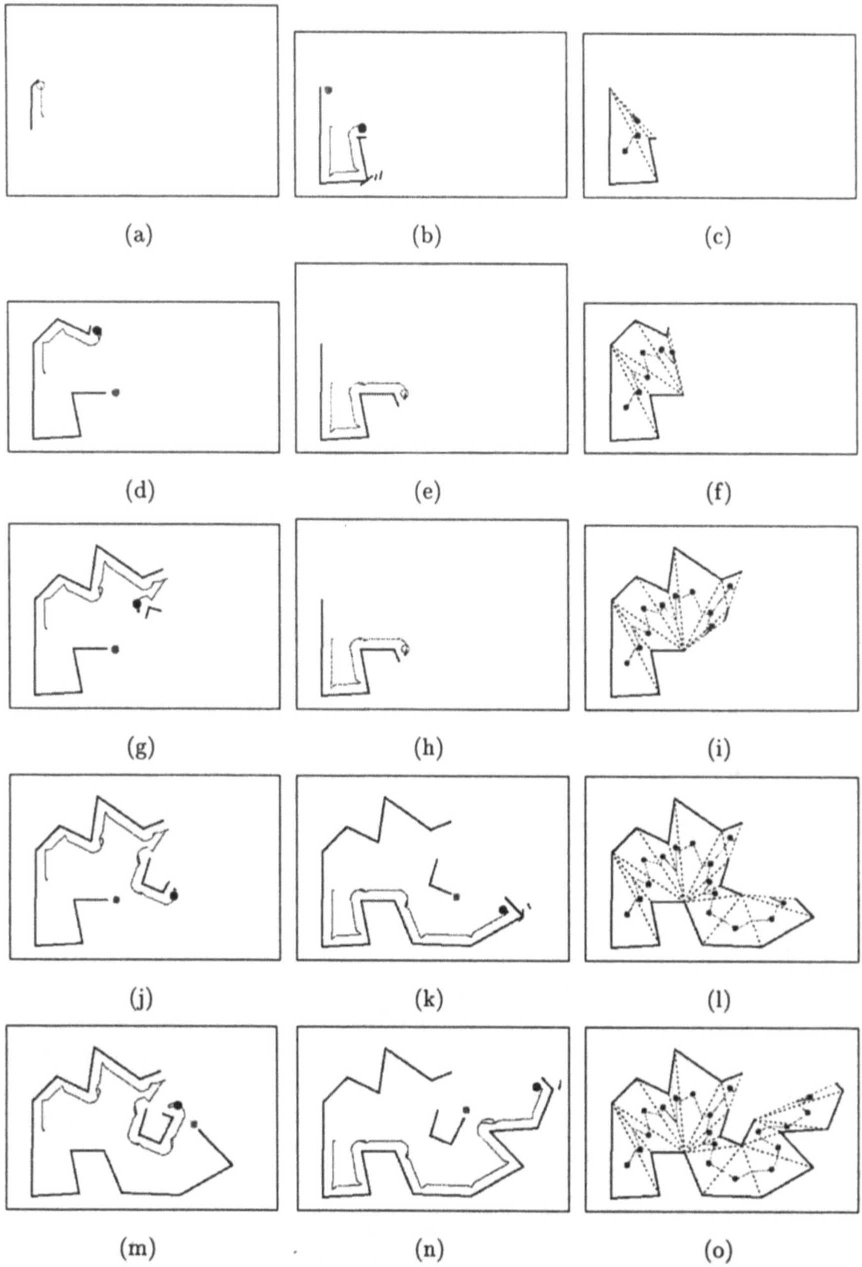

(a) (b) (c) (d) (e) (f) (g) (h) (i) (j) (k) (l) (m) (n) (o)

**Fig. 3. Exploring an unknown environment with one obstacle:** The first column illustrates the trajectory of **Robot 0**. (subfigures a,d,g,j,m). The second column illustrates the trajectory of **Robot 1** (b,e,h,k,n). Finally the third column presents the **map** up to that point (c,f,i,l,o)

# 4 Experimental results

Several sets of experiments have been conducted in order to validate our approach. Experiments in a simulated environment (using the RoboDaemon package, see Fig. 4) provided verification in a variety of model worlds. In addition, laboratory experiments with the real robots helped us estimate realistic values for the uncertainty of the sensors and the odometry.

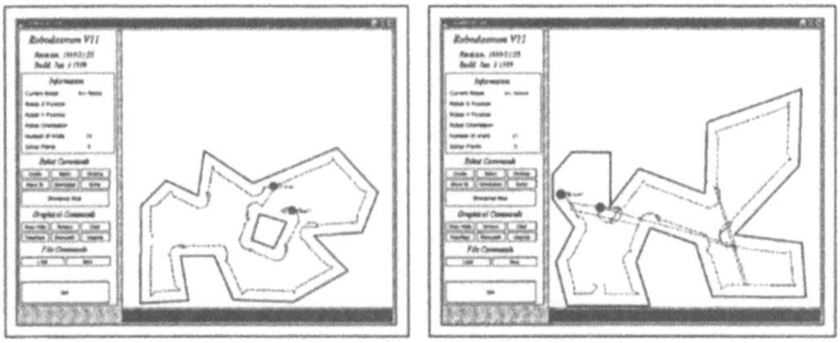

(a) (b)

**Fig. 4.** The paths of the two robots after the completion of the exploration

*Simulation:* Extensive experiments have been conducted using the robotic simulation package RoboDaemon. The simulations allowed us to examine the impact of different values of the different parameters such as odometry error, robot-tracker uncertainty and the complexity of the explored environments. Typical values for the standard deviation of the odometry errors are: 3cm/m in translation 3°/360° in rotation and $2°/m$ drift during translation. Fig. 4a presents a typical environment used in the simulations and the path the two robots followed. An obstacle is placed in the middle of the environment and the two robots construct the map by moving around it. As seen earlier (see Fig. 3(a-o)) the two robots successfully mapped this model world. Fig. 4b presents a complex environment with two occluding vertices that create bifurcations in the graph. In small worlds and/or cluttered environments, multiple observations of the same object could be used in order to correct the positioning of the moving robot.

*Physical Validation:* In order to demonstrate the effectiveness of the proposed approach, several preliminary exploration tests were carried out in our laboratory in workspaces of roughly 16 $m^2$. This comparatively small testbed allowed us to control various factors such as inhomogeneities in the terrain as a function of trajectory and obtain ground truth data. Using this testbed we compared the time, accuracy, and robustness of different exploration strategies. In several trials the role of the Nomad robot is played by a tripod mounted camera at the same height as the Nomad. This allowed us to more reliably and repeatably verify ground truth. In preliminary experiments using this arrangement, we have verified that our approach to collaborative exploration can improve pose estimation and map generation even over fairly short trajectories (of as little as 2m).

## 5 Discussion

The triangulation of the environment provides a complete decomposition of the free space; no overlapping occurs and no area is left uncovered. The dual graph acts as a guide during the exploration, and after the completion of mapping provides a topological map in addition to the metric map constructed.

The dual graph can be used for efficient path planning after the exploration phase. For any start and finish point, the corresponding triangles could be marked and different strategies could be applied to estimate a path. When the objective is to find the shortest path, a visibility graph could be calculated directly from the triangulation. Moreover, the embedded graph could contain additional information regarding the centroid of each triangle and the estimated odometry errors along this portion of the environment as recorded during the exploration. Thus, the robot path could be used to subsequently compute optimal paths in terms of either robustness or efficiency.

Our exploration/localization approach is especially useful when the internal motion estimates are completely unreliable, as is the case with many legged robots. In that case cooperative localization could provide a robust pose estimation thus allowing the robots to explore or navigate in an unknown environment despite that almost complete absence of accurate odometry.

## 6 Conclusions

In this paper, we have described an approach to exploring and navigating in *large scale spaces* where positioning and obstacle detection might be difficult using traditional methods. In fact, such difficulties are likely to arise in many real-world environments.

Our approach is based on exploiting a line-of-sight constraint between two robots to achieve exploration with reduced odometric error. This approach can also cope with obstacles with hard-to-sense reflectance characteristics. A key requirement is that the environment be small enough so that the robots can see one another from any two points on its boundary that are not occluded from one another (i.e. they are never unable to see one another simply because they are too far away).

We are currently planning large-scale experiments of this strategy in a real physical environment. One issue in this context is that it is difficult to obtain accurate ground-truth to validate the performance of our approach over a large terrain. A standard practice is to simply observe the "clean-ness" of the resulting map and use this as a performance metric [11]. However we expect that the triangulation-based mapping we perform will yield results whose accuracy may be too great in polyhedral environment for such qualitative evaluation methods to be satisfactory.

In prior work, we have considered alternative strategies for environments where the distance is too large to permit reliable operation of the tracker across the workspace [9]. An open issue is how to automatically detect such situations *efficiently* during exploration and switch strategies, or switch back-and-forth between strategies based on local properties of the environment.

We are also considering combining this approach with more traditional localization methods (such as landmarks [10]) where they can be used effectively. Doing this efficiently appears feasible but has not been achieved yet.

## References


1. Ivan A. Bachelder and Allen M. Waxman. A view-based neurocomputational system for relational map-making and navigation in visual environments. *Robotics & Autonomous Systems*, 16:267–289, 1995.
2. Tucker Balch and Ronald C. Arkin. Communication in reactive multiagent robotic systems. *Autonomous Robots*, 1(1):27–52, 1994.
3. J. Borenstein, H R. Everett, L. Feng, and D. Wehe. Mobile robot positioning: sensors and techniques. *Journal of Robotic Systems*, 14(4):231–249, Apr 1997.
4. Gregory Dudek, Michael Jenkin, Evangelos Milios, and David Wilkes. Experiments in sensing and communication for robot convoy navigation. In *Proc.Int. Conf. on Intelligent Robots & Systems (IROS)*, v. 2, p. 268–273, 1995.
5. C. Lin and R. Tummala. Mobile robot navigation using artificial landmarks. *Journal of Robotic Systems*, 14(2):93–106, 1997.
6. F. Lu and E. Milios. Globally consistent range scan alignment for environment mapping. *Autonomous Robots*, 4:333–349, 1997.
7. V. Lumelsky, S. Mukhopadhyay, and K. Sun. Dynamic path planning in sensor-based terrain acquisition. *IEEE Trans. on Robotics & Automation*, 6(4):462–472, 1990.
8. Paul MacKenzie and Gregory Dudek. Precise positioning using model-based maps. In *Proc. of the Int. Conf. on Robotics & Automation*, 1994.
9. Ioannis Rekleitis, Gregory Dudek, and Evangelos Milios. Multi-robot exploration of an unknown environment, efficiently reducing the odometry error. In *Int. Joint Conference in Artificial Intelligence*, v. 2, p. 1340–1345, 1997.
10. Robert Sim and Gregory Dudek. Learning visual landmarks for pose estimation. In *Proc. of Int. Conf. on Robotics & Automation (ICRA)*, May 1999.
11. S. Thrun, D. Fox, and W. Burgard. A probabilistic approach to concurrent mapping and localization for mobile robots. *Machine Learning & Autonomous Robots*, 31,5:29–53,253–271, 1998.

# Landmark-based Matching Algorithm for Cooperative Mapping by Autonomous Robots

Göksel Dedeoglu and Gaurav S. Sukhatme

Robotics Research Laboratory
Computer Science Department
University of Southern California
Los Angeles, CA 90089-0781

dedeoglu , gaurav @ robotics.usc.edu

**Abstract.** This paper describes a landmark-based algorithm for map matching, and demonstrates its use in efficiently combining topological maps of indoor environments built by autonomous mobile robots. Results of the first implementation are presented, in which two robots with different mechanics and sensory capabilities independently explore their environments with no a priori maps. Each robot is initially unaware of the other's relative position and orientation. Using the match algorithm, they eventually merge their maps into a topologically correct single map in real time, based only on the feature sets they have discovered independently.

## 1 Introduction

As autonomous mobile robots are increasingly expected to share dynamic, un-instrumented, indoor spaces with humans and other robots, purposeful navigation and mapping capabilities arise as key requirements [2]. Broadly speaking, research efforts have led to two distinct mapping techniques: Metric and topological maps. Metric approaches [5] are typically implemented as occupancy grids, first introduced by [10] and [7], and later used by [13,14]. On the other hand, topological maps view the world as a collection of interconnected landmarks [8,9,4,1], often augmented with distance information and defined using probability distributions [3,12,11].

Our research addresses the development of algorithms for distributed, anytime, topological map building that can efficiently scale to groups of heterogeneous robots. This paper builds upon our existing mapping strategy [6], and demonstrates a simple, yet robust starting point for the task of cooperative mapping with multiple robots. An algorithm for combining two topological maps is evaluated, and preliminary results with two robots are shown.

Section 2 highlights the topological map components that underly the algorithm, which, in turn, is described in Section 3. A number of map combination experiments are reported in Section 4, providing evidence for the feasibility of the approach.

## 2 Mapping System

The mapping and navigation algorithms used in this work have been described in detail in [6]. The two mobile robots used in this study are Pioneer 1-AT (Vespucci) and Pioneer 2-DX (Milka) platforms with different kinematics and sensory equipment. Vespucci uses a gyroscope to enhance its position tracking, and uses a Pan-Tilt-Zoom camera connected to the FastTrack vision system to detect visual color blobs. In contrast to Vespucci which has only seven sonar transducers, Milka has sixteen, and carries a SICK laser range finder as well.

The Building blocks of our topological mapping scheme are corners, T-junctions, end-of-corridors and closed doors, and links interconnecting them, as shown in Figure 1. Based on the perception of these features, each robot is capable of generating and updating a topological map augmented with metric information [6], and refer back to it for enhancing its estimated position.

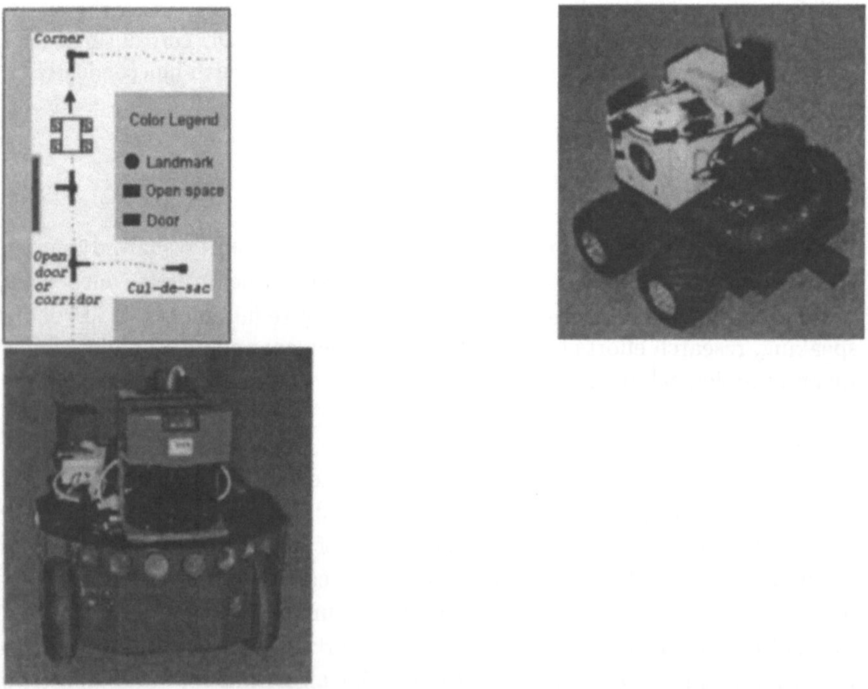


**Fig. 1.** On the left, topological landmarks and their attributes are illustrated on a sample map. Robots Vespucci (middle) and Milka (right) are shown.

Both Vespucci and Milka use their sonar arrays to detect corridors, openings that can be either open doors or hallways, junctions, and end-of-corridors.

While Vespucci seeks for color blobs for detecting closed doors, Milka looks for signatures in laser scans for the same purpose.

As reported in [6], over a set of 300 sonar-based feature detection cases, the rate for correct detection was found to be 81%, along with 20% of false alarms. For closed door detection, the color-blob detector had a success of 92% correct detection with 3% false alarm rate over 180 cases. Laser scanner-based door detector performed at a correct detection rate of 85% over 130 cases. The associated signal processing and basic feature detection is described in [6] as is the algorithm for single robot mapping and navigation.

### 2.1 Data Structures

Our map is a collection of landmark elements that are the *nodes* of an augmented graph. A *node* carries the following attributes:

```
struct node {
 id
 type              // corner, junction, door
 x,y               // Approximate coordinates
 struct link[4]    // 4 possible directions
 visit_counter
 detection_counter
}
```

A variation of landmark types can be obtained based on the *link* structure's *type* attribute. A corner has two consecutive neighboring links as *blocked*, whereas the others will be *open*. An end-of-corridor has only one link that is *open*, which connects it to other nodes.

```
struct link {
 connected_to_id
 type              // open space, blocked, door
 heading           // in local reference frame
 compass
 distance
 travel_counter
}
```

## 3 Map-Matching Algorithm

### 3.1 Heuristics

The goal of the map matching algorithm is to find the transformation $T_{12}$ that relates a robot's ($R_1$) coordinate system with that of another ($R_2$), thus enabling the merging of the two maps. The basic intuition is that, if the robots have visited at least partially overlapping spaces and registered enough of the same features in their maps, it should be possible to find and measure the correlation between their maps. Using this idea, the search space of our algorithm is the set of all landmark pairs ($L_{R1}$,$L_{R2}$), since each landmark discovered by $R_1$ could potentially correspond to any other reported by $R_2$. Furthermore, since the landmarks carry heading and position information, each pair defines a unique match, called a *candidate transformation.* Our algorithm uses two important heuristics to reduce this search space.

**H1** Only landmarks of the same *type* attribute are paired up (*e.g.*, doors with doors, corners with corners)

**H2** Only landmarks that describe spatial features (*i.e.* T-junctions, corners and end of corridors) are considered as candidates.

The reasoning behind **H1** is that, over the period of time in which the robots build and maintain their maps, a relatively small number of features change their attribute (open vs. closed door) or disappear totally. In other words, the heuristic assumes that the dynamics of the environment are slow relative to the speed of exploration. In addition, **H2** cuts the search space down further by relying on those features which are not only less prone to false alarms, but also represent an inherently sparser set in typical indoor environments (*e.g.*, merely a handful of corners and T-junctions as opposed to dozens of doors on a given floor).

### 3.2 Map Matching

The preprocessing step applies the heuristics discussed above and produces a reduced search space, populated by the so-called *selected* pairs of landmarks. At this stage, these pairs are further tested for possible mismatches in attributes that reflect absolute measurements. When available, compass readings are used to match approximately (+/- 30 degrees), or the candidate is eliminated. In case the landmarks were registered while the robots traveled in opposite directions, a hypothesized compass value is used for comparison.

Finally, for all remaining candidate transformations, translation and rotation components are computed by requiring both the coordinates and the *arm-type* attributes of the two landmarks ($L_{R1}$,$L_{R2}$) to coincide. The latter is then evaluated by applying it to all elements of $R_2$'s map, and counting the number of overlapping and matching landmarks. Having tried all transformations, the one that yields the highest number of match-counts is selected, and the resulting combined map is presented to the user as the *best match.* In the next section, we analyze 26 such match trials and evaluate the matching algorithm.

## 4 Experimental Evaluation

### 4.1 USC Salvatori Computer Center

The map combination results reported in this section are based on data obtained on the second floor of the USC Salvatori Computer Center, where a

| Combined Data Sets | Direction of Travel | Total Number of Candidates | Number of Selected Candidates | Correctness of the Match (%) | Eliminated False Alarm (%) |
|---|---|---|---|---|---|
| $D_1$ - $D_2$ | same | 15 | 4 | 100 | - |
| $D_{15}$ - $D_{16}$ | opposite | 15 | 4 | 100 | - |
| $D_3$ - $D_4$ | same | 30 | 8 | 100 | - |
| $D_{16}$ - $D_{17}$ | same | 33 | 6 | 100 | 100 |
| $D_3$ - $D_5$ | same | 40 | 11 | 100 | 100 |
| $D_4$ - $D_5$ | same | 48 | 8 | 100 | 100 |
| $D_1$ - $D_6$ | opposite | 51 | 4 | 100 | - |
| $D_1$ - $D_7$ | opposite | 54 | 5 | 100 | - |
| $D_{15}$ - $D_{16}$ | opposite | 55 | 11 | 100 | 100 |
| $D_1$ - $D_8$ | same | 60 | 5 | 100 | 100 |
| $\mathbf{D_1}$ - $\mathbf{D_9}$ | opposite | 60 | 6 | 100 | 100 |
| $D_{10}$ - $D_{11}$ | same | 65 | 9 | 50 | 50 |
| $D_{11}$ - $D_{12}$ | opposite | 70 | 7 | 100 | - |
| $D_{12}$ - $D_{13}$ | same | 70 | 10 | 100 | 100 |
| $D_2$ - $D_6$ | opposite | 85 | 12 | 75 | - |
| $D_{18}$ - $D_{19}$ | opposite | 88 | 6 | 86 | 80 |
| $D_2$ - $D_7$ | opposite | 90 | 15 | 100 | - |
| $D_{12}$ - $D_{14}$ | opposite | 91 | 9 | 100 | 100 |
| $D_2$ - $D_8$ | same | 100 | 13 | 60 | 50 |
| $D_2$ - $D_9$ | opposite | 100 | 16 | 100 | 100 |
| $D_{11}$ - $D_{13}$ | opposite | 100 | 11 | 100 | 100 |
| $D_{12}$ - $D_{13}$ | opposite | 130 | 12 | 100 | - |
| $D_6$ - $D_7$ | same | 306 | 20 | 92 | 0 |
| $D_6$ - $D_9$ | same | 340 | 21 | 100 | 33 |
| $\mathbf{D_7}$ - $\mathbf{D_9}$ | same | 360 | 25 | 92 | 66 |
| $D_8$ - $D_9$ | opposite | 400 | 22 | 70 | 33 |

**Table 1.** 26 experiments of map combination.

complete loop corresponds to approximately 80 meters of traveled distance. As summarized in Table 4.1, a total of 26 map combination experiments were computed offline, combining various pairs from a total set of 19 available maps, $D_1$ through $D_{19}$. Some of these combinations involved map pairs built by the same robot at different runs as well, which does not restrict the algorithm's generality.

The third and fourth columns of Table 4.1 indicate the original size of the search space, and its reduced value after applying heuristics **H1** and **H2**. For comparison, these two numbers, and their relatively stable ratio are plotted in Figure 2.

The column indicating the *Correctness of the Match* is the ratio of correct matches to all reported matches. Finally, the last column gives a measure of the false alarms that were successfully eliminated by combining the maps

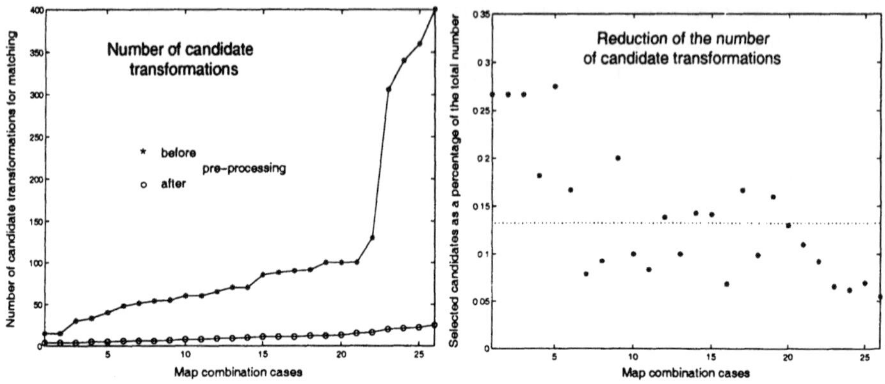


**Fig. 2.** For all map combination cases summarized in Table 4.1, the heuristics reduce the number of transformation candidates drastically (left), also plotted as a percentage of the initial set (right).

built by different robots (or, by the same robot at different times). Since the probability of both robots experiencing the same false positive at the same location is lower than it is in the single robot case, multi-robot mapping is inherently well-suited to eradicating false positives (*e.g.*, the feature detector can mistakenly signal the presence of a door due to noisy sensory input).

The first example of map combination is depicted on Figures 3 through 4. At a first glance, the maps $D_9$ and $D_7$ may seem to have most landmarks of the building in common. Yet, a closer look reveals that in $D_9$, the robot explored a cul-de-sac in the lower right corner of the map, which does not exist in $D_7$. Furthermore, their sets of detected doors do not perfectly overlap. Despite all the discrepancies, the matching process was able to collect enough evidence for the transformation correctly proposed in Figure 4.

A more interesting example of map matching involves $D_1$, which is an unfinished map as shown in Figure 5 (left). In $D_1$, the robot has only had a chance to explore the northern corridor of the same area as in $D_9$, but traveling in the opposite direction. As can be seen in Figure 5, $D_1$ and $D_9$ are still matched successfully. From the robot's point of view, this allows time-critical localization decisions to be made with respect to other robots or a global map.

## 4.2 Preliminary on-line multi-robot trials

Our matching algorithm was also tested on-line with robots building maps in parallel and independently from each other. The data included here is a representative case of many multi-robot mapping experiments performed during the DARPA Tactical Mobile Robotics experiments in July 1999, which took place in an empty hospital building in Fort Sam Houston, San Antonio. Figure 6 shows full-screen snapshots of the maps built by Vespucci and Milka. In this experiment, Vespucci first explored the West wing of the hospital

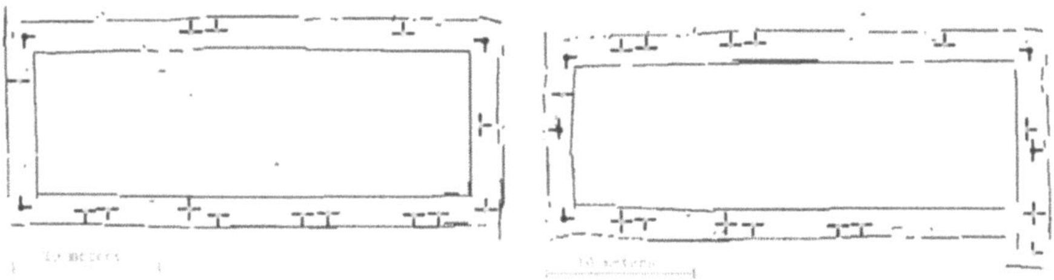

**Fig. 3.** Map $D_9$ (left) of the Salvatori Center. The robot traveled in a counter-clockwise direction, registering doors and junctions as they are detected. In an independent mapping experiment, the robot detected a slightly different set of landmarks as shown in map $D_7$ (right).

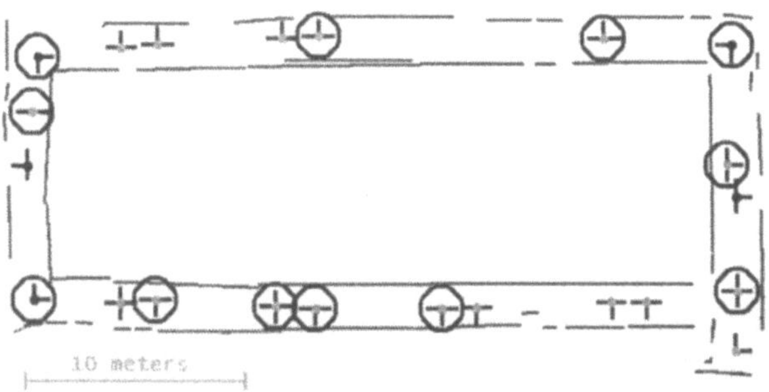


**Fig. 4.** Maps $D_7$ and $D_9$ of the same environment are matched against each other and their combination is correctly generated. Circles indicate features that were registered in both maps.

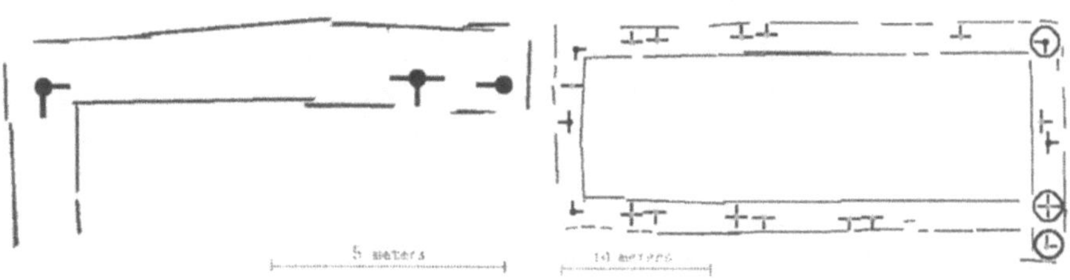


**Fig. 5.** A snapshot of the map built by Milka ($D_1$, left), which has so far only explored the northern corridor of the Salvatori Center, navigating in a clockwise direction. Salvatori maps $D_1$ and $D_9$ are correctly combined (right).

building, and then made a tour around the central elevator block. On the other hand, Milka started its journey at the East wing, also followed by a tour around the elevator block. Figure 7 shows how their maps were combined. Based on the evidence that both robots collected while turning around the central block, they were cooperatively able to produce a complete map of the scene.

Finally, a short remark about the metric accuracy of the maps is in order. In Figure 8, the combined map of Figure 7 was overlaid on the floor plan of the hospital. Over a stretch of more than 50 meters of corridor, all mapped wall segments were within 1 meter of their corresponding position according to the blue prints. This rough measure indicates a metric accuracy of about 2% of our mapping scheme.

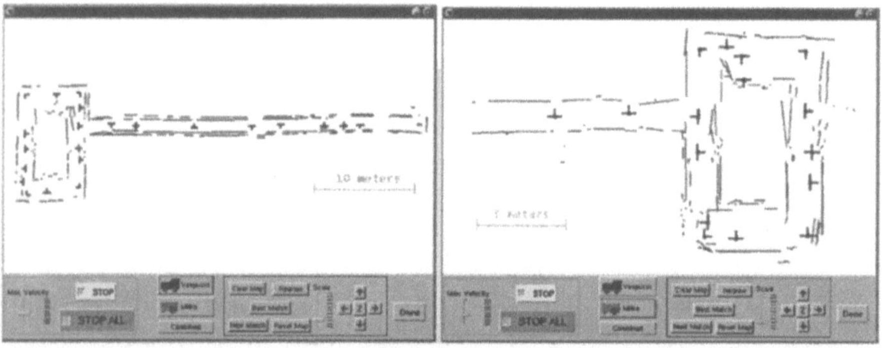

**Fig. 6.** Map built by Vespucci (left), starting from the West wing of the building. The second map was built by Milka (right), which started from the East wing of the building.

## 5 Conclusions

We have proposed and evaluated a map matching algorithm that allows mobile robots to combine their augmented topological maps with those of others. Extensive offline tests of the algorithm suggest that the approach will efficiently scale to the multi-robot case, and also provide a reasonably cheap framework for a group of robots to localize themselves with respect to each other. Preliminary results from the first online, two-robot implementation are also presented.

The extension of this algorithm to the multi-robot case is currently underway. Future work on the mapping scheme will include the use of probability distributions for feature sets. For the multi-robot case, a generic version of the algorithm will be made part of the existing mapping system so that multiple robots can dynamically combine their maps to produce a unified representation of their common surroundings.

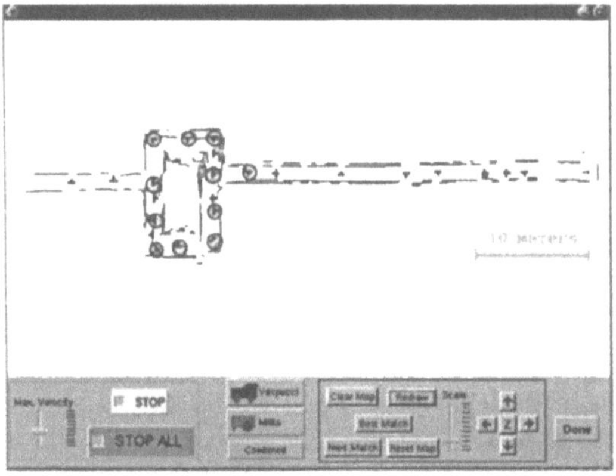


**Fig. 7.** Combined maps of Vespucci and Milka provides the complete picture that was not available to individual robots.

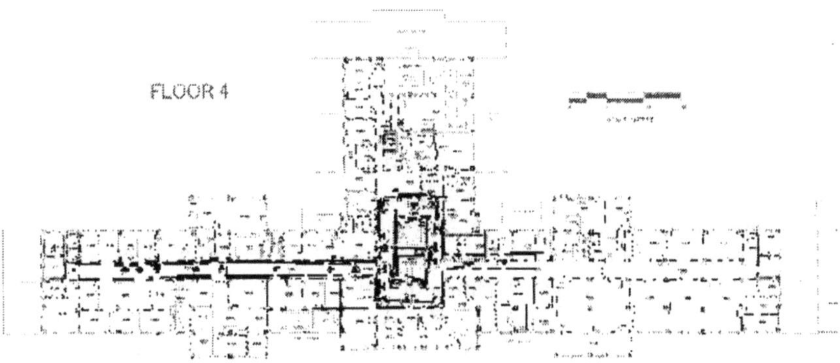


**Fig. 8.** Combined maps laid over the 4th floor plan of the hospital building.

## Acknowledgments

The authors would like to thank Kale Harbick for hardware and operating system support, and Richard Vaughan for helping with demonstrations. This work is supported by contracts F04701-97-C-0021 and DAAE07-98-C-L028 from DARPA under the TMR program, and by grant DABT63-99-1-0015 from DARPA under the MARS program.

# References

1. G. Beccari, S. Caselli, and F. Zanichelli. Qualitative spatial representations from task-oriented perception and exploratory behaviors. *Robotics and Autonomous Systems*, 25:147–157, 1998.
2. J. Borenstein, B. Everett, and L. Feng. *Navigating Mobile Robots: Systems and Techniques.* A.'K. Peters Ltd., Wellesley, MA, 1996.

3. Anthony R. Cassandra, Leslie Pack Kaelbling, and James A. Kurien. Acting under uncertainty: Discrete bayesian models for mobile robot navigation. In *Proceedings of IEEE/RSJ International Conference on Intelligent Robots and Systems*, 1996.
4. Howie Choset. *Sensor Based Motion Planning: The Hierarchical Generalized Voronoi Graph.* PhD thesis, California Institute of Technology, 1996.
5. Ingemar J. Cox. Blanche - an experiment in guidance and navigation of an autonomous robot vehicle. *IEEE Transactions on Robotics and Automation*, 7(2):193–204, April 1991.
6. Göksel Dedeoglu, Maja J. Mataric, and Gaurav S. Sukhatme. Incremental, online topological map building with a mobile robot. In *Proceedings of the Photonics East Conference*, Boston, MA, 1999. SPIE.
7. Alberto Elfes. *Occupancy grids: A Probabilistic framework for robot perception and navigation.* PhD thesis, Carnegie Mellon University, 1989.
8. Benjamin Kuipers and Yung-Tai Byun. A robot exploration and mapping strategy based on a semantic hierarchy of spatial representations. *Robotics and Autonomous Systems*, 8:47–63, 1991.
9. Maja J. Mataric. Integration of representation into goal-driven behavior-based robots. *IEEE Transactions on Robotics and Automation*, 8(3):304–312, 1992.
10. Hans P. Moravec and Alberto Elfes. High resolution maps from wide angle sonar. In *Proceedings of the IEEE International Conference on Robotics and Automation*, pages 116–121, St. Louis, Missouri, 1985.
11. Hagit Shatkay and Leslie Pack Kaelbling. Learning topological maps with weak local odometric information. In *Proceedings of the 15th International Joint Conference on Artificial Intelligence (IJCAI-97)*, pages 920–929, San Francisco, August 23–29 1997. Morgan Kaufmann Publishers.
12. Reid Simmons and Sven Koenig. Probabilistic robot navigation in partially observable environments. In *Proceedings of the International Joint Conference on Artificial Intelligence*, pages 1080–1087, 1995.
13. Sebastian Thrun, W. Burgard, and Dieter Fox. A probabilistic approach to concurrent mapping and localization for mobile robots. *Machine Learning and Autonomous Robots (joint issue)*, 31/5:29–53 / 253–271, 1998.
14. B. Yamauchi, A. C. Schultz, and W. Adams. Mobile robot exploration and map-building with continuous localization. In *Proceedings of the IEEE International Conference on Robotics and Automation (ICRA-98)*, pages 3715–3720, Piscataway, May 16–20 1998. IEEE Computer Society.

# Comparing Distributed Exploration Strategies with Simulated and Real Autonomous Robots

Adam T. Hayes, Alcherio Martinoli, and Rodney M. Goodman

Microsystems Lab 136-93, California Institute of Technology, Pasadena CA 91125
athayes,alcherio,rogo@micro.caltech.edu
http://www.micro.caltech.edu/micro/CORO/index.html

**Keywords:** Collective Autonomous Robotics, Probabilistic Modeling, Embodied Simulation, Distributed Exploration

**Abstract.** This paper presents an experiment in collective robotics in which a group of autonomous robots searches for an infrared target beacon placed in a corner of the exploration area. This task is a more experimentally tractable version of the plume tracing problem, in which robots search for the source of an odor plume. Two different exploration strategies (collaborative and non-collaborative) are implemented and compared on the basis of several team performance metrics. The collaborative strategy uses a simple, binary signaling schema among robots. The experiment is implemented at three different levels: in a physical setup composed of groups of 1 to 8 Moorebot robots, in Webots, a 3D sensor-based, kinematic simulator, and with probabilistic simulations. Results show that the collaborative approach drastically improves the search across several metrics. Furthermore, the probabilistic model qualitatively and quantitatively reproduces the enhanced team performance via collaboration. Additional investigations using the probabilistic model indicate that the optimal number of robots is a function of the ratio between target and exploration areas.

## 1 Introduction

The application of swarm intelligence principles to autonomous collective robotics aims to develop robust task solving by minimizing the complexity of the individual units [1]. The central idea of the swarm intelligence approach is to distribute control over a group of numerous minimalist robots rather than gathering and redistributing information with the help of a central unit. The main advantages of this approach are two: first, scalability from a few to thousands of units, second, increased system robustness, not only through unit redundancy but also through the unit minimalistic design. Several examples of collective robotics tasks solved with swarm intelligence principles can be found in the literature: aggregation [2,3] and segregation [4], exploration [5], stick pulling [6], and collective transportation [7,8].

One way to increase the performance of a robot swarm is collaboration. In particular, if collaboration is obtained with simple explicit communication schemes such as binary signaling, the team performance can be enhanced without losing autonomy or significantly increasing the complexity at the individual level. The experiments presented in this paper deal with the implementation of distributed exploration strategies for a group of autonomous robots, and compare the effectiveness of the collaborative approach vs. the non-collaborative one. The beacon finding task is meant to address some of the issues present in the more complex task of plume tracing, without having to deal with the experimental complications of creating repeatable chemical plume stimuli.

The aim of this case study is three-fold. Firstly, we want to show that collaboration can drastically improve the team performance. Secondly, we want to show that Webots [9], a 3D sensor-based, kinematic simulator originally developed for Khepera robots [10], can be successfully used for simulating experiments with other, kinematically similar robots. Thirdly, the presented experiments are used as a testbed for the development of a new characterization and prediction tool in the form of a probabilistic model, capable of wider application across tasks and environments than previous models [3,5].

## 2 Non-Collaborative Collective Exploration

### 2.1 Experimental Setup with Real Robots

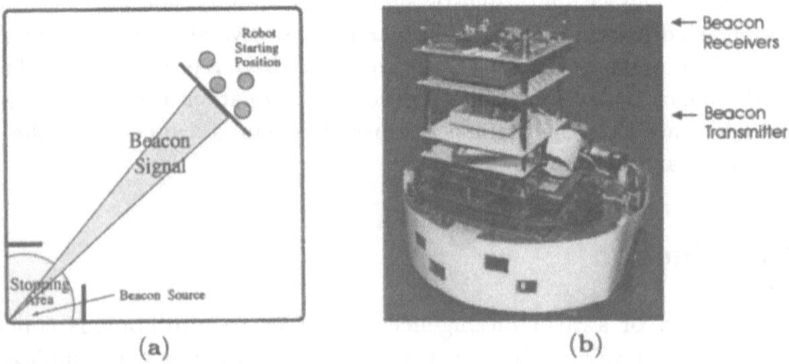


**Fig. 1.** **(a)** The arena layout. **(b)** A Moorebot

The layout of the test arena is shown in Fig. 1a. The size is 6.7 by 6.7 m, and the robots are 24 cm in diameter. The robots begin each trial in the corner of the arena opposite the stopping area. They are started simultaneously, at random positions and orientations within the starting area, and the time required for each to enter the stopping area (as defined by a separate beacon signal) is recorded.

We use Moorebots, as shown in Fig. 1b. In addition to the standard configuration, as described in [11], each robot is equipped with an infrared transmitter and three binary infrared receivers (for determining teammate communication, the target beacon, and the stopping area). All explicit inter-robot communication is performed via the single binary channel.

When exploring the arena, the robots move in straight lines, avoiding walls and teammates with random turns between 0 and 270 degrees. If a signal is detected, the robots move directly toward the source beacon, although obstacle avoidance takes precedence when necessary. In the non-collaborative case, the only beacon signals present are the target and the stopping beacons.

## 2.2 Sensor-Based Simulations

To more systematically investigate the exploration strategies, we also implemented this experiment in Webots, a 3D kinematic simulator originally designed to simulate Khepera robots. Because of the morphological similarities between the two platforms, it is possible to scale environment, robot size, and speed to reflect the relationships of the real robot experiments. In addition to size, the main morphological differences between the Khepera and Moorebot platforms are the position and number of proximity sensors. Moorebot sensor outputs are generated by combining and thresholding the outputs of the Khepera's four front and central built-in proximity sensors, as shown in Fig. 2a. With the creation of a Moorebot-to-Khepera translation library, control algorithms can be transferred from the Moorebots to the simulator with very few changes.

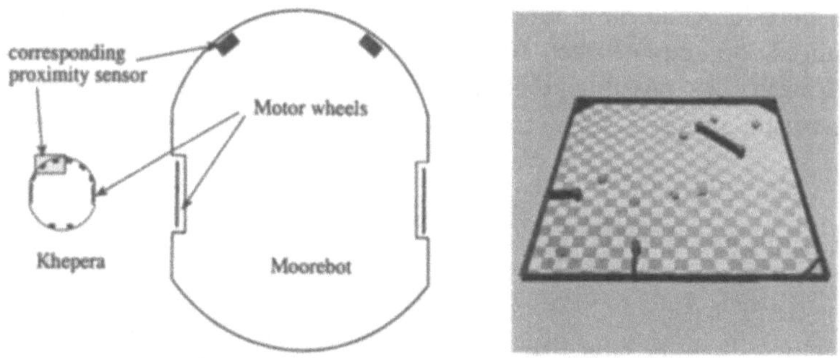


(a) (b)

**Fig. 2.** (**a**) Khepera and Moorebot sensor layouts. (**b**) Implementation of the experiment in Webots

The simulator computes trajectories and sensory input of the robots in an arena corresponding to the physical set-up (Fig. 2b). The simulation is sufficiently faithful for the controllers to be transferred from the simulation to real robots and achieve very similar performance (see Section 4). Working in simulation not only increases the experimental reliability but also saves time. For this experiment with 8 robots, the mean acceleration ratio between Webots and real time is about 14 on a Pentium III 500.

### 2.3 The Probabilistic Model

In an effort to gain an understanding of the essential aspects of the system, as well as to further decrease simulation time, we have developed a probabilistic model which can generate a robot arrival time distribution for a given set of experimental parameters (e.g. arena size, robot size, robot speed, avoidance time, communication range, and number of robots). In order for this model to be useful, it has to produce data similar to that observed in the embodied experiments, predict results of future experiments, and contain no free parameters. All parameters in this model are determined from measurements taken from the arena and experiments with a single robot.

The basic idea behind the model is similar to that of [3,5], where the experiment is treated as a series of stochastic events with probabilities based on geometrical relationships. However, because this experiment depends heavily on the spatial location of the individual robots (their distribution cannot be assumed to be uniform across the arena, as was the case in [3,5]), a different method must be used. To capture spatial information, the arena is divided into four distinct states, as shown in Fig. 3a. At any given point in time each robot occupies one state. As the arena is partitioned with greater granularity, more information is included in the simulator, but more computation is required to complete each simulation. Since this model should be as simple and as fast as possible, the minimum number of states for accurate data generation are used. As a result, it is about 15,000 times faster than Webots.

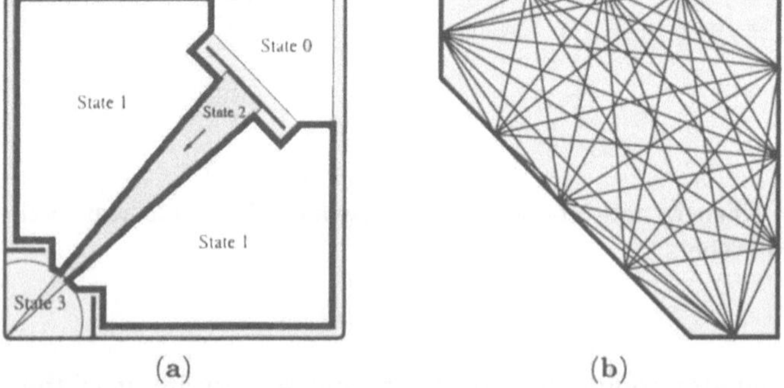


**Fig. 3.** **(a)** Arena partitioning in the non-collaborative experiment. The dark bands represent the distribution of locations occupied by any robot entering State 1. **(b)** Example set of chords in State 0

The transition probabilities between states in which the robots move randomly (States 0 and 1) are calculated from the perimeters of each state, which are offset from the physical perimeters of the arena to account for the spatial extent of the robots. Each time a robot encounters the outer edge of a state, it can transition into another state or bounce off a wall back into its current state (with an avoidance time penalty corresponding to the average avoidance time of a real robot). Collisions are also modeled as an avoidance time penalty, applied stochastically based on state size, effective robot area, and number of robots currently in a given state.

For states in which the robots move in a particular direction (State 2, when a robot sees a beacon signal), the expected time before transitioning into the next state was taken to be the average distance to the next state divided by the robot speed. This assumes that robots never lose an acquired signal, which does not entirely capture the behavior of the real robots, but is a close approximation in this case. In principle, data could be collected from a real robot and this probability of signal loss incorporated into the model.

A difficulty with the probabilistic model is the generation of the perimeter encounter probability. Based on the average robot speed and the average distance across the state (average chord length - see Fig. 3b), the expected time to perimeter encounter can be generated. Treating the system as a simple Markov process would result in a geometric time-to-perimeter-encounter distribution (with a single parameter $p$), as at each time step the probability of encountering the perimeter would be a constant value, as given by:

$$p = v/m \tag{1}$$

where $v$ is the robot speed and $m$ is the average chord length. This distribution does not accurately reflect the spatial realities of the embodied systems, as many times are unreasonably short, and a few are unreasonably long (for instance, if a robot moves at 1 m/s, and an area has a maximum chord length of 10 m, it cannot go for more than 10 seconds without encountering something). Despite the fact that the expected value of the distribution is accurate, overall robot arrival times have bigger variances than those observed in the embodied experiments.

To account for this defect, a hybrid distribution is used– basically a geometric distribution in which the parameter $p$ is a function of the amount of time a robot has been in a given state. If the maximum chord length of a state is $l$, robots must not be able to wait longer than $l/v = k$ seconds between encounters, so $p(k)$ should be set to 1. This suggests an equation of the form:

$$p(x) = x/k \tag{2}$$

In order that the expected value of the distribution can be set independently of this cutoff time, two parameters are needed:

$$p(x) = \frac{x^\alpha}{A} \tag{3}$$

The form of the resulting probability density function is quite similar to that of the geometric distribution:

$$P(x) = p(x) * \prod_{i=1}^{x-1} (1 - p(i)) \tag{4}$$

For the experiments in this paper, the values of $\alpha$ and $A$ are set numerically based on geometrical considerations, as further analysis of this distribution is ongoing. See Fig. 4a to compare the transition distributions from State 0 to State 1 generated by Webots, this hybrid process, and a simple Markov process.

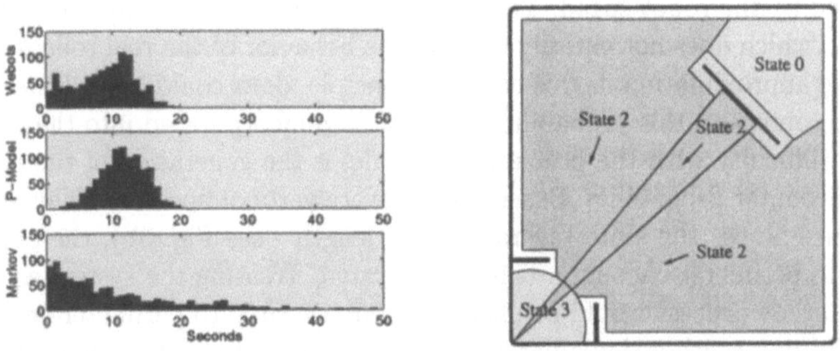


(a) (b)

**Fig. 4.** (**a**) Histogram of the number of encounters as a function of time-to-perimeter for State 0, N = 1000. (**b**) Arena partitioning during collaboration

Note that the Webots data in Fig. 4a includes repeated encounters with the same wall, due to the nature of the collision sensors and avoidance behavior, and this increases the number of small waiting times. Such encounters are not included in the probabilistic model because they are a result of spatial correlation between successive encounters which are not captured within the probabilistic framework. Even though the hybrid distribution generates perimeter encounter data which are closer to the target distribution, it cannot be truly accurate because the model does not incorporate exact robot trajectories.

The lack of complete spatial information also influences the state transition probabilities in a way that is difficult to capture in the model. For example, the location of a robot that has just transitioned into State 1 is represented by a probability distribution at the perimeter of the state, as shown by the dark band in Fig. 3a. Thus it is possible for a robot in the model to transition from State 0 to State 3 more quickly than if it had to follow a real trajectory (as is necessary in the embodied simulator and the real world). To be truly accurate, the state transition probability (which in the model is based on static perimeter ratios) must change with the entrance point and amount of time spent within the state. Calculating the nature of these changes would be complex, and incorporating them into the model would restrict simulation speed. Instead, we model the position dependent effects as a constant time penalty added to the simulated trials (roughly counteracting the fact that robots in the model can change states too fast). Since the behavior of one robot captures the spatial dependencies of the task, we take the time penalty to be the difference between the mean arrival time for one real robot and one uncompensated simulated robot (in this case, 31.5 seconds).

# 3 Collaborative Collective Exploration

## 3.1 Experimental Setup with Real Robots

For the collaborative experiments, when a robot sees either the target beacon or the stop beacon, it turns on a beacon of its own, thus attracting surrounding robots. The range of the robot beacons generally extends across the arena, but due to slight hardware differences (e.g. LED orientation), the communication range is not uniform across all robot pairs. Because these heterogeneities are difficult to quantify, they are not included in either of the simulations. Other than the robot beacon activity, the collaborative and non-collaborative trials are exactly the same.

## 3.2 Sensor-Based Simulations

Beacon signaling was implemented via the external supervisor module of Webots. The supervisor module is based on an External Authoring Interface API similar to the one implemented in VRML97 for enabling Java applets to communicate with VRML 97 scenes within a Web browser. This module can read and edit the world loaded in the simulator, including some customized robot features such as virtual signaling beacons.

For the sake of simplicity, the range of the robot beacons was assumed to be sufficient for communication across the entire arena.

## 3.3 The Probabilistic Model

Collaboration is modeled by reorganizing the distribution of states whenever a robot beacon becomes active (see Fig. 4b). State 1 does not appear because it is assumed that as soon as a robot leaves State 0, it sees a beacon and begins directed motion toward the final state. State 0 slightly changes shape because beacon signals become visible in some areas, so those regions join State 2. Note this reorganization does not wholly capture the behavior of the embodied robots, but it provides a close approximation because it captures the principle features of the collaborative system.

# 4 Results

## 4.1 Exploration with collaborative and non-collaborative strategies

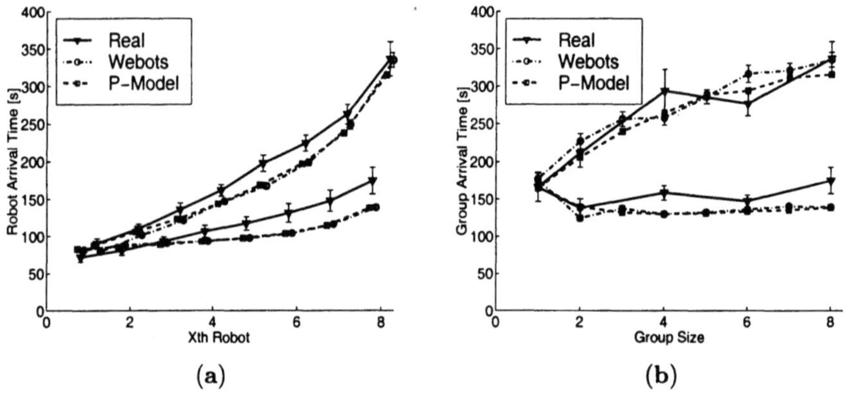


**Fig. 5.** Number of trials per experiment: real=15, Webots=100, p-model=500. (**a**) Arrival times for a group size of 8 robots. The lower set of curves is the collaborative data. (**b**) Group arrival times for groups of 1 to 8 robots.

Figure 5a shows arrival times for both non-collaborative and collaborative trials generated by all three experimental levels for a group size of 8 robots. Figure 5b shows the group arrival times for group sizes from 1 to 8 robots. Figure 6a gives the average power consumption per robot for both types of trials. All error bars represent standard error. Note the good qualitative and quantitative agreement among the different experimental levels.

## 4.2 Increasing the ratio between exploration and goal area

The benefit of using an increasing number of collaborating robots is not clear from the above experiments because performance saturates at a small group size. As shown in Fig. 5b, the 3 point running average of group arrival time in the collaborative case shows less than 1% improvement at group sizes larger than 3 (Webots) or 4 (p-model) robots. Further simulations were performed both in Webots and with the probabilistic model in a larger environment, 20.1 m per side (9 times larger area), but with the start, stop, and target beacon areas kept the same. Figure 6b shows that even in a different environment the probabilistic model shows good correspondence with Webots. Also, in this larger arena, performance (using the same definition as above) saturates at increased group sizes: 7 (Webots) or 9 (p-model). Based on the interaction between increasing search speed and increasing inter-robot interference, we

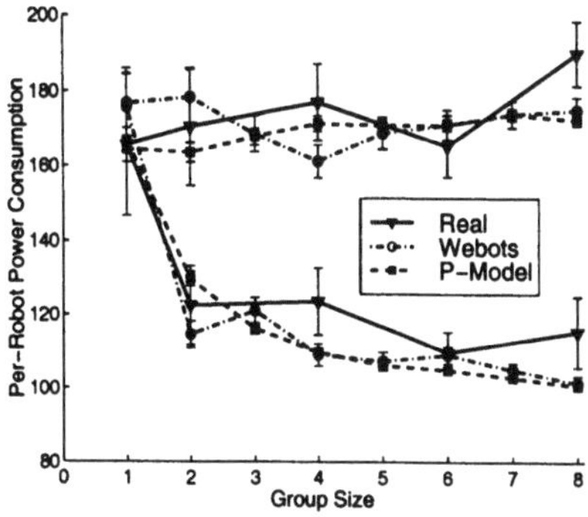


(a)

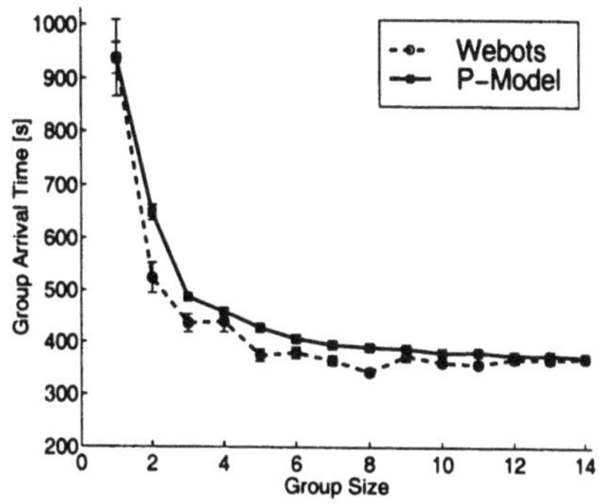


(b)

**Fig. 6.** (**a**) Per-robot power consumption for groups of 1 to 8 non-collaborating (upper curve) and collaborating (lower curve) robots (same trials as Fig. 5). (**b**) Group arrival times for groups of 1 to 14 robots searching a larger arena. Number of trials per experiment: Webots=100, p-model=500.

conjecture that the optimum group size is dependent upon the relationship among the target area, exploration area, and the communication range.

# 5 Discussion and Conclusion

In this paper, we investigated collaborative and non-collaborative exploration strategies for a group of autonomous robots at three different experimental levels: real robots, sensor-based simulations, and probabilistic simulations. The data shows that simple collaboration among robots drastically improves the team performance by reducing the arrival time of the group as well as the group power consumption. Collaborative group performance saturates as the team size grows, and the smaller the ratio between target area and exploration area or the signaling range, the greater the optimal team size.

Furthermore, we have shown that the results obtained with a zero-free parameter probabilistic model are qualitatively and quantitatively in good agreement with those obtained from more sophisticated sensor-based simulations and real robots.

Considering the tremendous acceleration ratio between the probabilistic model and the other implementations, we believe that it represents an extremely useful tool for capturing the role of the crucial experimental parameters in distributed exploration experiments. This could be particularly useful for investigating team performances with thousands of units before actually building them or developing a dedicated sensor-based simulator for the given task. It is unknown how the model will perform on more complex environments and tasks (such as plume tracing), and this is a subject of ongoing research.

## 6 Acknowledgments

We would like to thank Owen Holland, Sanza Kazadi, and Andrew Lundsten for their work on the Moorebot platform. This work is supported in part by the Center for Neuromorphic Systems Engineering as part of the National Science Foundation Engineering Research Center Program under grant EEC-9402726. Other support in part by DARPA under grant DAAK60-97-K-9503, and the Office of Naval Research under grant N00014-98-1-0821 in the Chemical Plume Tracing Program. We also acknowledge support from the DARPA program in Distributed Robotics. Adam Hayes is supported by a National Science Foundation Graduate Research Fellowship.

## References

1. E. Bonabeau, M. Dorigo, and G. Theraulaz. *Swarm Intelligence: From Natural to Artificial Systems.* Oxford University Press, New York, US, 1999.
2. R. Beckers, O.E. Holland, and J.L. Deneubourg. From local actions to global tasks: Stigmergy and collective robotics. In R. Brooks and P. Maes, editors, *Proc. of the Fourth Workshop on Artificial Life*, pages 181–189, Boston, MA, 1994. MIT Press.
3. A. Martinoli, A. J. Ijspeert, and F. Mondada. Understanding collective aggregation mechanisms: From probabilistic modelling to experiments with real robots. *Robotic and Autonomous Systems*, 29:51–63, 1999.
4. O.E. Holland and C. Melhuish. Stigmergy, self-organization, and sorting in collective robotics. *Artificial Life*, 5:173–202, 1999.
5. A. Billard, A. J. Ijspeert, and A. Martinoli. A multi-robot system for adaptive exploration of a fast changing environment: Probabilistic modelling and experimental study. *Connection Science*, 11:359–379, 1999.
6. A. Martinoli and F. Mondada. Collective and cooperative group behaviours: Biologically inspired experiments in robotics. In O. Khatib and J. K. Salisbury, editors, *Proc. of the Fourth International Symposium on Experimental Robotics ISER-95*, pages 3–10, Stanford, U.S.A., June 1995. Springer Verlag.
7. M. J. B. Krieger and J.-B. Billeter. The call of duty: Self-organised task allocation in a population of up to twelve mobile robots. *Robotics and Autonomous Systems*, 30:65–84, 2000.
8. C. R. Kube and E. Bonabeau. Cooperative transport by ants and robots. *Robotics and Autonomous Systems*, 30:85–101, 2000.
9. O. Michel. Webots: Symbiosis between virtual and real mobile robots. In *Proceedings of the First International Conference on Virtual Worlds, VW'98*, pages 254–263, Paris, France, July 1998. Springer Verlag.
10. F. Mondada, E. Franzi, and P. Ienne. Mobile robot miniaturization: A tool for investigation in control algorithms. In T. Yoshikawa and F. Miyazaki, editors, *Proc. of the Third International Symposium on Experimental Robotics ISER-93*, pages 501–513, Kyoto, Japan, 1993. Springer Verlag.
11. A.F.T. Winfield and O.E. Holland. The application of wireless local area network technology to the control of mobile robots. *Microprocessors and Microsystems*, 23:597–607, 2000.

# Chapter 8

# Distributed Sensing

# Distributed Sensing and Data Collection Via Broken Ad Hoc Wireless Connected Networks of Mobile Robots

Alan FT Winfield

Intelligent Autonomous Systems (engineering) laboratory, University of the West of England Bristol, Coldharbour Lane, Frenchay, Bristol BS16 1QY, UK. Email: Alan.Winfield@uwe.ac.uk

**Abstract.** This paper reports on ongoing work to develop *ad hoc* wireless networking for application in distributed mobile robotics. The paper discusses a mission scenario in which a number of mobile robots are required to autonomously disperse into a physically bounded region, take sensor readings at predetermined time intervals, and then communicate the sense data back to a single collection point. There is assumed to be no overall command and control structure. The robots are assumed to be equipped with low-power short-range wireless network interfaces, which only allow direct communication with near neighbours, and overall an *ad hoc* (multi-hop) wireless networking scheme is employed. Furthermore, it is assumed that there are insufficient robots to allow full wireless connectivity through the region so that, at any instant, the robots actually appear as a set of disconnected sub-nets (i.e. a broken *ad hoc* network). The paper proposes a mechanism that exploits the mobility of the robots to overcome the lack of overall wireless network connectivity, resulting in a sub-optimal but robust scheme for garnering sense data. The paper presents simulation results that show how the arithmetic mean data collection delay varies with the number of robots deployed. The paper concludes that in mission scenarios that do not require data collection in real-time and where occasional data loss (erasure) is acceptable, then the proposed data collection mechanism is feasible, robust and economical in terms of the number of robots required.

Key words. Distributed sensing; *ad hoc* wireless networking.

## 1 Introduction

Consider a group of mobile robots that are required to autonomously disperse throughout a region, perform distributed sensing, monitoring or surveillance, and pass the sense data to a single collection point. Assume that the robots are equipped with low-power wireless transceivers whose range is too short to allow direct communication with the data collection point, but sufficient to allow robots to communicate with near neighbours. Under these circumstances the only practical approach to distributed command, control and sensing is to employ an *ad hoc* wireless networking scheme.

The networking literature defines an *ad hoc*, or *multi-hop mobile* wireless network, as a network with a set of geographically dispersed nodes in which each node is able to forward packets for other nodes that cannot communicate directly. Thus each node is able to act as a router, as well as a data source, or sink. A characteristic of *ad hoc* mobile wireless networks is that it is virtually impossible for each node to know the entire network topology at any given time, which means that optimal strategies for message routing in wide area networks cannot be used. Broadcast protocols that do not require knowledge of the entire network topology have been described [1], [2] and recently Basagni *et al* [3] introduced a mobility-transparent deterministic broadcast mechanism for such networks.

Since the advent of high performance wireless local area network technology at relatively low cost, its use for wireless control of mobile robots has become a practical proposition. Wireless local area networked miniature mobile robots, using TCP/IP (i.e. Internet Protocols), have proven highly successful as a platform for experiments in collective robotics [4], [5]. The issues of communication with large populations of mobile robots have received attention, for example in Gage [6]. Genovese *et al* [7] considered both swarm-like self-organisation and communication. Arai *et al* [8] proposed an information diffusion mechanism for distributing command data to multiple robots and analysed the overall time-delays incurred. However, *ad hoc* radio networks based on Internet Protocols have not yet, to the author's knowledge, been applied in the field of distributed mobile robotics.

This paper discusses the feasibility of applying *ad hoc* wireless networking to the problem of garnering sensory data from a spatially dispersed group of mobile robots. The paper looks at perhaps the simplest means of spatial deployment of a group of mobile robots into a physically bounded region, namely by random diffusion. Unlike the network communications literature this paper considers the likely scenario in which the *ad hoc* wireless network becomes disconnected and fragments into smaller sub-nets. The paper proposes a mechanism that exploits the mobility of the robots, and hence the highly dynamic network topology, in order to collect the distributed sense data despite an apparently 'broken' *ad hoc* network. Simulation results are presented, which show how the data collection time delay varies with the number of robots deployed and with each robot's communications range.

## 2 A model of robot connectivity for analysis and simulation

For the purpose of the simulation presented in this paper we shall assume an idealised wireless communications model for the mobile robot. Firstly, we shall assume that the wireless antenna fitted to each mobile robot is uniformly omni-directional in the horizontal plane. This is a sound assumption since the horizontal polar diagram of a simple vertical monopole antenna is a circle [9]. Secondly, we assume that the modulation scheme employed by the wireless transceivers experience a 'threshold effect'. In other words, as the distance between transmitter and receiver increases and the received signal strength reduces according to the inverse square law, the receiver's ability to demodulate the received signal will suddenly fail as the threshold is reached. Thus for any given transmitter power level (or receiver sensitivity) each mobile robot will be at the centre of a circle of radius $r$ (i.e. the communications 'range'). Reliable communications between any two robots separated by less than $r$ is thus guaranteed, whereas two robots more than $r$ apart will be unable to establish a communications link. Again this is a reasonable assumption, since mobile robots are likely to employ Frequency Shift Keying (FSK), Phase Shift Keying (PSK) or one of its many variants, all of which experience the threshold effect [10]. A third assumption is that the wireless network will employ some form of time, frequency or code diversity in order that many robots (wireless nodes) may share the same RF spectrum and interoperate without compromising the first two assumptions. Again this is reasonable if, in practice, we employ either Frequency Hopping (FH) or Direct Sequence (DS) variants of spread spectrum modulation [11] together with some strategy for Carrier Sense Multiple Access (CSMA).

These assumptions provide us with an idealised model of wireless network communications for distributed mobile robots that lends itself to straightforward simulation but with reasonable confidence that the results will transfer successfully to physical implementation. Figure 1 illustrates this connectivity model for a set of 5 randomly dispersed mobile robots in a 2D space, numbered 0 to 4. Each mobile robot is shown with a circle that defines its radio range $r$. It is easy to see by inspection that robot 0 can communicate with robot 1. Robot 1 can communicate directly with robots 0, 2 and 3. Robots 0 and 3 can communicate either with 2 hops via robot 1 or with 3 hops via robots 1 and 2. Robot 4 on the other hand is currently out of range of any other robot but, depending on its heading, might move into range with robots 2 or 3.

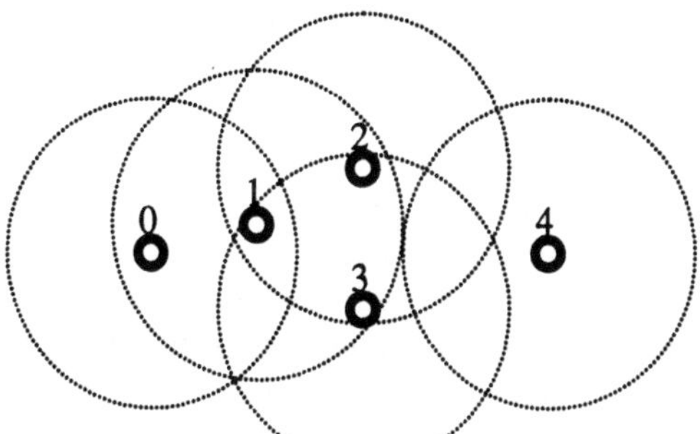


Fig. 1 Wireless range of randomly dispersed mobile robots in 2 dimensions

The networking communications literature conventionally models an *ad hoc* network as an undirected graph, $G=(V,E)$ in which $V=\{p_0,\ldots,p_{n-1}\}$ is the set of wireless nodes and there is an edge $(p_i,p_j)\in E$ if and only if $p_i$ and $p_j$ are in range. In this case $p_i$ and $p_j$ are neighbours. The *distance* $d(p_i,p_j)$ between two nodes $p_i$ and $p_j$, $0\le i,j\le n-1$, is defined as the minimum number of hops between $p_i$ and $p_j$. The maximum distance between any pair of nodes is called the *Diameter D* of the network. Figure 2 shows the same set of five robots in fig. 1 as an undirected graph, or to be precise two undirected graphs, one with four nodes and the other with one. The *Diameter* of the graph which comprises robots 0 to 3, $D = 2$. Additionally we define the *Connectivity C* of the overall network as the number of edges, and it is easy to see that for $n$ robots $0 \le C \le \frac{1}{2}n(n-1)$.

Fig. 2 Figure 1 represented as an undirected graph

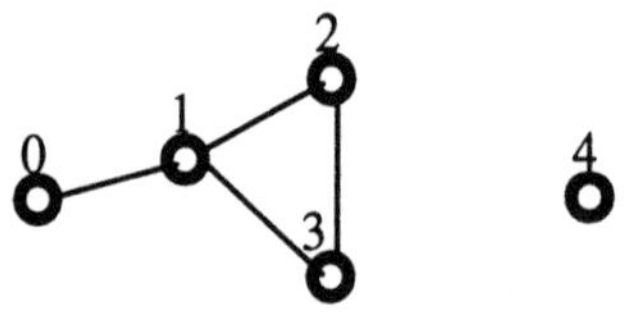


## 3 Spatial organisation of ad hoc networked mobile robots

Consider the following mission scenario for distributed mobile robots. A group of mobile robots is required to perform some sensing or surveillance task that requires that the robots be spatially distributed. Once the robots have been deployed the sensor data from each robot is then moved across the group, via the wireless network, and garnered by a nominated robot (robot 0, say) on the edge of the group. Robot 0 is in turn connected to some base station for forward transmission of the collected sensor data. Consider now the possibility that directly deploying each of these robots into their final spatial distribution might be difficult or dangerous. This might be because they are to be deployed into a hostile environment or simply because there are too many robots.

In order to visualise this scenario consider, for example, that a large number of small submersible mobile robots are required to perform marine pollution monitoring. It would make sense to introduce the robots into the marine environment from the monitoring vessel *en-masse* and then rely on the robots to disperse themselves throughout a specified 3D volume of ocean. A number of small robotic deep-space probes might similarly be deployed from the hold of the *Space Shuttle* and then self-organise into a spatial formation before embarking on their mission. In all of these mission scenarios the ability of the robots to spatially self-organise, relative to each other, without the need for sophisticated means of absolute navigation might be a significant advantage.

Consider also a scenario in which the sensing robots act autonomously. In other words no command, control or time synchronisation messages whatsoever are required, from robot 0, to the rest of the group. In other words all motion control decisions are made locally, on the basis of local information (including, perhaps, information about local wireless connectivity). We shall assume that each robot has an accurate on-board clock, synchronised at the time the robots are deployed, and that data samples are undertaken at predetermined time intervals.

## 4 Spatial distribution within a physically bounded region

Consider now how spatial organisation can be achieved within a physically bounded region. For terrestrial mobile robots operating in, essentially, 2 dimensions the physical bounds might be natural terrain features such as hills or rivers; for laboratory robots the bounds are provided by the walls at the edge of an experimental arena. Submersible pollution monitoring robots operating within an inland lake would similarly be constrained by natural physical barriers. In such a bounded environment the simplest way of achieving spatial self-organisation would be by random diffusion. The robots are simply introduced *en-masse* at one point. Each then chooses a random heading and sets off along that heading. With an appropriate change of heading when a robot approaches a boundary, or when robots collide then, eventually, the robots will diffuse to fill the bounded region.

Fig 3. Initial robot deployment

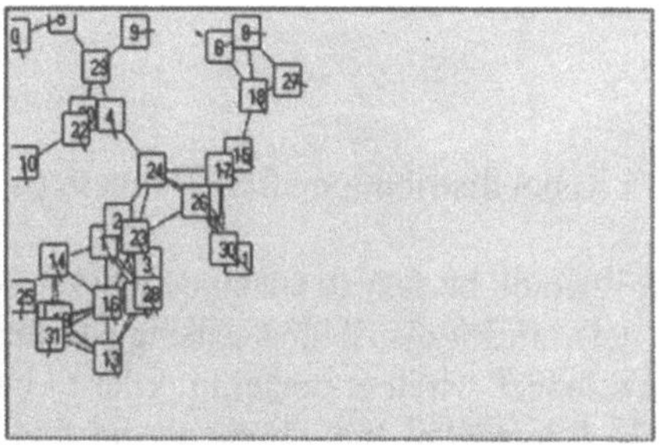

Fig 4. Robot distribution after 40 timesteps

To illustrate random diffusion consider the introduction of 32 robots into one corner of a rectangular arena, as shown in fig. 3. The robots have a wireless range $r = 50$, much smaller than the arena dimensions of 450 x 250. At time = 0 each robot chooses a random heading, from 0 to 359 degrees, and when in collision with the arena wall robots assume a new random heading anywhere within the 180 degree arc facing away from the wall. Initially we have a maximally connected network, $C = 496$, in which each robot is only one network hop from any other. Each robot moves a distance of 5 units along its heading in 1 time step, except robot 0 which remains anchored at its starting position (robot 0 is the fixed data collection point for the whole network). After 40 time steps the robots have formed a distinct wavefront and have filled about one quarter of the arena, as shown in fig. 4. At this point the connectivity has fallen to $C = 76$, but the network is still fully connected; in other words there is still a network path between robot 0 and every other robot. The lines shown joining individual robots in fig. 4 indicate neighbours that are in wireless range (or, edges, of the network graph). The short lines shown pointing from the centres of each robot indicate, compass fashion, that robot's current heading.

Figure 5 shows the situation after 80 time steps. The robots have now almost filled the entire arena, but the level of connectivity has fallen to just 39, less than 8% of the maximal connectivity. Most significant however, is that the *ad hoc* network has now broken into a number of fragments, or sub-nets, with a number of robots out of range of any neighbours at all. There is now no network path back to robot 0. After 120 time steps the robots have diffused throughout the region and any wavefront structure has now disappeared, as shown in fig. 6. The overall level of connectivity is very low at just 23 and there are no less than 7 sub-net fragments. It is also interesting to see how much the sub-net structures have changed between figs. 5 and 6.

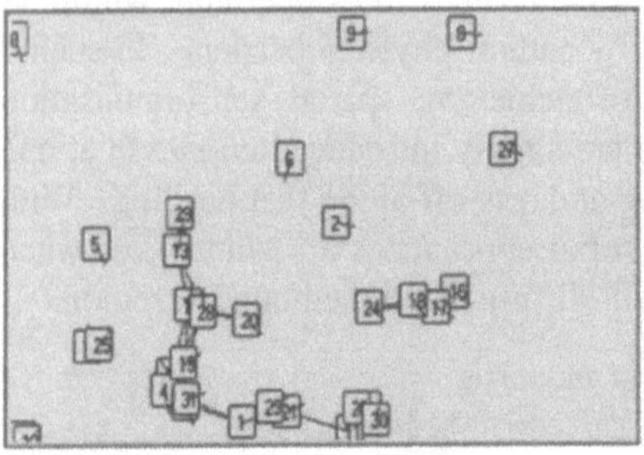

Fig. 5 Robot distribution after 80 timesteps

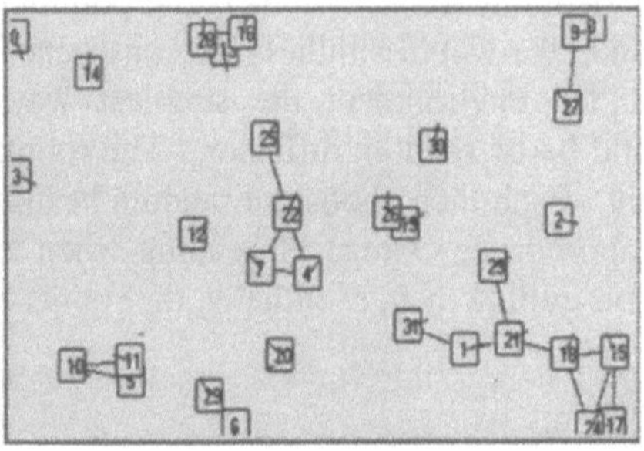

Fig 6. After 120 timesteps

Now it would be easy to conclude that the scenario illustrated in fig. 6 is not viable as a network for distributed sensing and that we need either more robots, or robots with a longer wireless range, in order to build a fully connected network. Such a highly fragmented (i.e. 'broken') *ad hoc* network would, at first sight, appear useless for collecting sensory data. However, this is not a static network but is implemented on mobile robots, whose most important attribute is of course their mobility. Figure 6 is just a freeze-frame of a highly dynamic process in which individual robots, and sub-nets, quickly move in and out of range of each other.

## 5 A mechanism for data collection via a broken ad hoc network

Let us propose a distributed data sensing and collection strategy for sparsely populated ad hoc connected networks of mobile robots, in bounded regions as follows. Assume that sensory data must be collected from every robot and that all data must be garnered at robot 0.

- At a predetermined time every robot simultaneously samples new sensory data (based upon their internal clocks, which are assumed to have been synchronised at time = 0);
- Every robot then immediately broadcasts its sensory data to every one of its neighbours;
- In turn each neighbour broadcasts any received data, together with its own sensory data, to its neighbours where it is stored in a data buffer. In this fashion sensory data is propagated through each of the sub-nets;
- Each time a robot detects that another robot has come within wireless range (and becomes a new neighbour) then all data within its buffer is broadcast to the new neighbour and to its neighbours, and vice-versa;

- Each robot's sensory data is marked with that robot's ID (IP address), and when robot 0 has received sensory data from every robot in the network then the data collection for this sample period is complete.

As a practical proposition this algorithm makes a number of assumptions. Firstly, that data collection is not required in real-time, and that the application can tolerate variable time delays between sensory sampling and completion of data-collection. Secondly, that the bandwidth of the wireless networking is sufficiently high and the volumes of message data sufficiently low, that all message buffers can be propagated between joining sub-nets before robot mobility re-fragments the joined sub-net. This is a reasonable assumption given typical wireless LAN bandwidths of 1-2 Mbytes/s. Finally, the algorithm assumes that each robot has sufficient local storage to be able to buffer messages from a single sample without overflow. This is again a reasonable assumption given current memory technology. The proposed algorithm may result in a robot receiving multiple copies of another robot's sensory data; it will however only need to store one copy in its data buffer, thus placing an upper bound on the size of buffer storage required per sample.

At first sight the use of a 'broadcast' mechanism to propagate sensory data across sub-nets may appear counter-intuitive given that the overall objective is to concentrate distributed sensory data to a single collection node. The rationale behind this approach is that by broadcasting the same sensory data to as many robots as are within sub-net connectivity at any given time, and therefore duplicating the same data widely across the network, we are minimising the time it takes for data to propagate back to the collection node. One way of visualising this approach is to think of the sensory data as 'infecting' the whole of a sub-net. As soon as any member of that sub-net contacts another sub-net, the sensory data will in 'virus fashion' infect the new sub-net, and so on. Thus in the same way that the robots physically diffuse across the bounded region, the sensory data diffuses, in reverse, back to robot 0. Note that we do not need to place a restriction on the sensory data sampling frequency. We do not, for instance, need to wait until data collection for one sensory sample is complete before triggering the next sample. Providing that there is sufficient buffer storage within each robot then we can exploit the highly parallel structure of the distributed robot network to allow multiple samples to be simultaneously propagating back to the collection node.

## 6 Simulation results

To assess the performance of the proposed algorithm the author has constructed a computer simulation in which the number of robots $n$, and their wireless range $r$, can be varied. The key performance indicator is the length of time for data collection to complete, in simulated time steps. Figure 7 shows the data collection time delay for range $r = 100$, with error bars. Figure 8 shows the arithmetic mean data collection delay graphically plotted against the number of robots, for a number of different values of wireless range $r$. All simulations were performed in an arena of 450 x 250 units, and each robot moved 5 units per time step. Each run, for a given number of robots and wireless range was undertaken 100 times, and the arithmetic mean plotted in fig. 7. To ensure that each run gave an independent value for the time to complete data collection, following completion the robots were moved through a further 100 time steps before initiating a subsequent sample and data collection.

As would be expected the data collection delay asymptotically approaches zero as the population of robots in the bounded region increases. (Zero represents the time delay experienced by a fully connected network, i.e. one in which there is, at data sampling time, a multi-hop path from every robot back to robot 0). Also as expected, increasing the radio range *r* reduces the mean data collection delay for a given number of robots. These simulation results demonstrate that, at the cost of some delay time, it is feasible to propose distributed sensing and data collection

**Fig. 7 Data Collection Delay, *r* = 100**

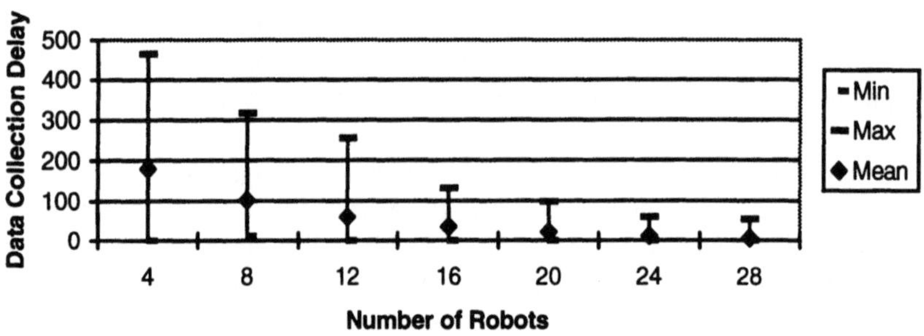


**Fig. 8 Mean Data Collection Delay**

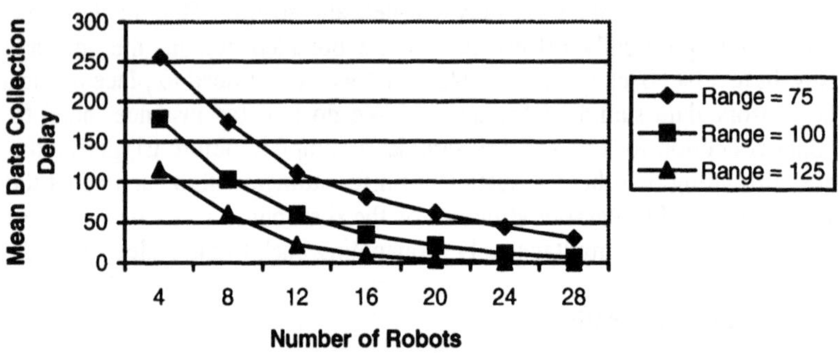


across a 'broken' ad hoc wireless network by exploiting the mobility of the robots. Another way of thinking about the proposed mechanism is that we are trading wireless utilisation and buffer storage (which are high) for robot mobility which is low (compared with a scenario without wireless communications in which each robot would have to physically transport its collected data back to the collection point).

## 7 Discussion and Further Work

Consider now the buffer memory requirement of the proposed data collection mechanism. As already stated, for a single data sample, the buffer memory requirement is bounded by the number of robots. However in a practical distributed sensing scenario samples would be required periodically. This could potentially result in an unbounded buffer requirement because of the unbounded upper limit on the data collection delay time. In practice buffers need to have a fixed size limit and this will result in the possibility of erasures (data loss) because of buffer overflow. The probability of erasure can be minimised by careful design of the buffer size in relation to predicted data collection delays (from fig. 7). The provision of a circular buffer arrangement in which the oldest data is overwritten by the newest would provide a simple strategy for buffer management. It is, however, important to recognise that 'erasure' will very rarely mean the loss of an entire data sample. It is much more likely that the majority of the data from the oldest sample will already have propagated back to the collection node by the time its buffer needs to be overwritten by the latest sample. Thus the proposed data collection mechanism, via a broken *ad hoc* network, is likely to result in - at worst - partial data for occasional samples. In many remote sensing scenarios, for instance, pollution or environmental monitoring, this compromise would be perfectly acceptable.

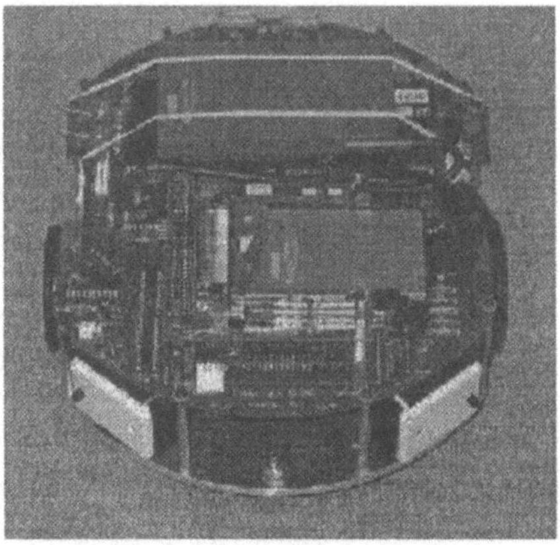

Fig. 9 The LinuxBot

As stated in the introduction, the paper reports on current work on the implementation of *ad hoc* wireless networking on distributed mobile robots. An important aspect of the work of the author's laboratory is the transfer of theoretical concepts to engineering realisation. To this end we are currently implementing an *ad hoc* wireless network, using Internet Protocols, on a fleet of fifteen wheeled mobile robots (the *LinuxBots* [12]). Figure 9 shows a single LinuxBot. Thus, in the laboratory environment, we will be able to compare the real-life performance of the distributed data sensing and collection mechanism discussed in this paper, with simulation and evaluate the effects of noise and interference.

In parallel, we shall extend the strategies for robot deployment and data collection in physically bounded regions, as discussed in this paper, to unbounded regions. Recalling the principle that decisions regarding motion control must be taken only on the basis of local information and without the benefit of received command data, *is it possible to achieve spatial self-organisation in unbounded regions on the basis of network connectivity information alone?*

## References


1. I Chlamtac and S Kutten. (1985) On broadcasting in radio networks – Problem analysis and protocol design, IEEE Trans. Commun., **COM-33**, 1240-1246.
2. I Chlamtac and S Kutten. (1987) Tree-based broadcasting in multihop radio networks, IEEE Trans. Comput., **C-36**, 1209-1223.
3. S Basagni, D Bruschi and I Chlamtac. (1999) A mobility-transparent deterministic broadcast mechanism for ad hoc networks, IEEE/ACM Trans. Networking and Computing, **7/6**, 799-807.
4. R Dawkins, O Holland, A Winfield, P Greenway and S Stephens. (1997) An interacting multi-robot system and smart environment for studying collective behaviours, Proceedings of the 8th International conference on Advanced Robotics, Monterey, CA.
5. AFT Winfield and OE Holland. (2000) The application of wireless local area network technology to the control of mobile robots, Microprocessors and Microsystems, **23/10**, 597-607.
6. DW Gage. (1993) How to communicate with zillions of robots, Proceedings of SPIE Mobile Robots VIII, Vol 2058, 250-257.
7. V Genovese, P Dario, R Magni and L Odetti. (1992) Self organising behaviour and swarm intelligence in a pack of miniature robots in search of pollutants, Proceedings of the 1992 IEEE/RSJ International Conference on Intelligent Robots and Systems, 1575-1582.
8. T Arai, E Yoshida and J Ota. (1993) Information diffusion by local communication of multiple mobile robots, Proceedings of the 1993 IEEE Conference on Systems, Man & Cybernetics, 535-540.
9. FR Connor. (1972) Antennas, Arnold, London.
10. BP Lathi. (1983) Modern Digital and Analog Communication Systems, Holt Rinehart and Winston.
11. P Wong and D Britland. (1995) Mobile Data Communications Systems, Artech House Publishers.
12. http://www.ias.uwe.ac.uk/linuxbot.htm.

# Experiments and Results in Multi-modal, Distributed, Robotic Perception

Andrzej Kasiński and Piotr Skrzypczyński

Technical University of Poznań, Department of Control, Robotics, and Computer Science ; ul.Piotrowo 3A, PL-60-965 Poznań, Poland

**Abstract.** Distributed architecture for cooperative perception processes is described. This is called henceforth a perception net (PN). The case of multiple, autonomous, mobile robots navigating with reference to the multi-form world-model is considered. Information sharing and communication issues are pointed out. Description of the experimental implementation made of two mobile platforms is accompanied with the demonstration of results, namely the maps, which have been obtained by fusing the measurements with the a priori knowledge within the logical framework of the PN. The application have been based on the QNX network with Ethernet links.

## 1 Introduction

Autonomous, mobile robot navigation can be based on the use of the internal world-models (maps) being the abstraction of the robot environment model. For robots operating in industrial environments the world-model can be provided in advance. Nevertheless, the internal model has to be updated on-line to reflect the scene changes which are essential to the feasibility of current tasks. In most cases, single-mode sensors are insufficient to guarantee the required autonomy of mobile robots. Thus a multi-sensor, multi-modal system has been used to provide data on the actual state of the operational environment, enabling recognition of its various features.

In the multi-robot systems the data interchange between robots can help to complete the local information available at particular robot, often incomplete and not sufficient for planning robot operation. Here the goal is to reduce the overall uncertainty of the robot operation by cross-validating partial world-models against the global model of the scene [1], treated as a reference.

Much work has been done so far on the multi-sensor fusion [2], on integration of sensors [3], on the use of different kinds of maps [4], and on the perception and co-ordination in distributed systems. In a multi-sensor system hosted on independent mobile units still a number of specific problems have to be solved, such as : (1) integration of different physical sensors with various principles of operation and diversified abilities to extract features; (2) fusion of multi-sensory data into one coherent world-model; (3) efficient use in the system of all the information available at hand; (4) the *a priori* knowledge embedding within the system. These issues made it clear that new architectures for the perception and world-model maintenance are still required. The new approach described here, integrates the concept of logical sensors with that of the structured semantic network. Moreover a multi-agent paradigm have been applied w.r.t. the communication system and software architecture. The goal of the article is to give an evidence for the viability of the distributed perception solutions and to point out the advantages due to the world-model sharing.

## 2 The concept of a distributed perception network

In the proposed solution we allow for the coexistence of different representations of the world-model, thus enabling a flexible use of sensors with different principles of operation, registration modes, or uncertainty characteristics. To embed the *a priori* known elements of the world-model, the Geometrical Data Base (GDB) has been used [5]. The 2D vector-based and grid-based maps have been applied to account for the dynamic changes of the scene. These maps, being aspects of robot's internal world-models are updated with sensory data and by information exchanged with other robots. In this way, the robots with different perceptual competencies can share the acquired knowledge [1].

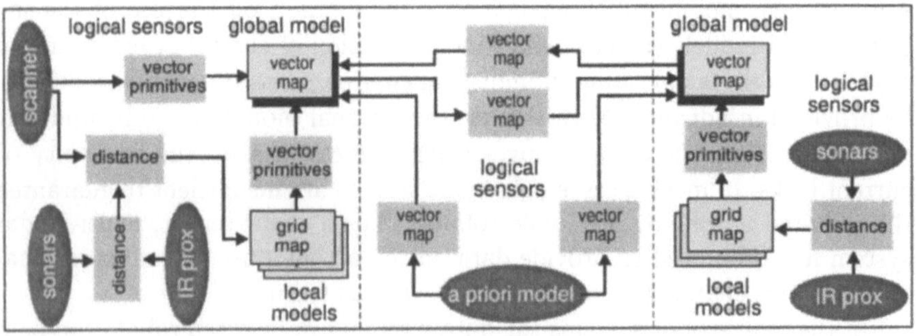


**Fig. 1.** Logical sensors in the structure of the perception network

Unified concept of the *logical sensor* has been applied. Logical sensors have been proposed by Henderson [3] as a generic description of sensors and data processing algorithms for multi-sensor systems. In the system under study, logical sensors are treated as abstract sources of information made of the source (a physical sensor or a data repository, such as a map) and of specific data-processing methods. Logical sensors are used to extract the features out of raw sensory data or from a particular world model, and then to produce the output vectors, which are used at higher processing layers of the system (Fig. 1). In that way, a single physical sensor or the environment model can be used to multiple purposes, depending upon its logical specification.

Unification and a clear structure of the system consisting of many logical sensors and several internal world-models can be achieved by introducing the concept of the *perception network* (PN). Lee *et al.* [6] first have proposed to use this concept in robotics as an architectural framework for robots with multiple sensors. We have extended this idea by associating a world-model building system to support group-navigation of mobile robots. In [6] the perception net has been treated solely as the engine to process on-line the incoming stream of information. We propose an extended PN, being able to accumulate and organize the extracted features. This concept has been worked out to enable consistent management of parallel perception and information-interchange threads. Formally, PN is a directed, acyclic graph with labeled nodes and edges. Its structure reflects relationships among logical sensors and the associated knowledge representation schemes. The PN fuses features at different abstraction levels, implementing the flow of data from sources (i.e. physical sensors), through the intermediate features level, to high-level, abstract models (i.e. maps) already useful for particular robot-tasks. As opposed to [6], we do not insist on the unified representation of the uncertainty in our PN. This is due to the fact, that the operation principles of the on-board sensors applied in our experiments are substantially different, and some of sensors have no sound statistical models (sonars spurious readings). To cope with various uncertainty models associated to the appropriate map classes (parametric or grid-based), we let use of both models: covariance matrices, and certainty factors. Additionally, special operators to perform the mutual conversion between uncertainty models have been proposed. For example : the uncertainty model for the grid-based map is the occupancy probability distribution of grid cells. Therefore, once the vector map has been extracted from the grid-map, we are consequently forced to re-formulate the uncertainty model and to get the covariance matrix for parametric models of elements of the resulting vector map.

## 3 Multi-agent methodology

To manage the information flow in a complex, multi-sensor and multi-robot system, the multi-agent methodology has been adopted [7]. In [1] we have introduced *sensor agents* and *perception agents* as main data-processing elements in the multi-agent perception system. The communication scheme between these agents is similar to the contract net approach [8] with a blackboard as the controller of the virtual communication channels connecting particular agents. Blackboard matches incoming service requests from all agents with services available in the distributed system at a given moment of time.

The agent interested in pursuing a task signals the existence of its task to other agents by putting a proper *request* on the blackboard. Other agents (potential servers) submit to the blackboard their *proposals* of services. The blackboard receives and matches *announcements* containing *requests* from clients to *proposals* from servers. Whenever the matching is successful, the potential server evaluates its ability to perform the specific task [1,7], then it contacts the client. The client evaluates the proposals from particular servers and grants the contract for the execution of the task to the agent with the highest ability to complete this task (e.g. the winner could be an agent which offers the world-model with the lowest uncertainty level). Client and server are thus linked by a contract and they communicate in peer-to-peer mode to establish the transfer of information,for instance a piece of a map. There are some tasks which can not be executed by a single server, because information from several agents is required to complete the task. This is the case when new data has to be embedded into the map. In such a case both data from the external sensors and the actual coordinates of the vehicle are required. In such circumstances the server (sensor agent) may split up the task and grant contracts to some other agents (a self-localization agent). It is then becoming the priority client for such contracts. The described above hierarchical protocol structure is typical for the task-sharing paradigm, widely used in distributed problem solving [8]. The sensor and perception agents are not *a priori* designated as clients or servers. They can assume both roles, depending upon the current circumstances. For example, a perception agent that builds the grid-map acts as client (whenever it requests a new sensory data), or as a server (whenever it provides the data extracted from the map to some other agent to update its map). This communication scheme enhances the operational reliability of the system, as the processing of the sensory data once distributed over the network inhibits the abrupt degradation of performance caused by the failure of a particular agent or physical sensor. Agents which can not process data to yield results in the requested time (e.g. due to the sensor failure) are not considered as potential contractors by a client that follows a particular task. For example, a robot which is using the sonars to build a grid-map can grant a contract to the perception agent of another robot equipped with the laser scanner, whenever in specific configuration of obstacles reliable data can not be provided by the sonars.

# 4 Implementation of the system

## 4.1 Experimental test-bed and the software architecture

The described architecture has been implemented and tested at the Mobile Robotics Lab of the Department. The goal of experiments was to demonstrate the ability of the proposed architecture to support distributed multi-sensor perception and cooperative world-model building. The test-bed consisted of two robots of the *TRC Labmate* class equipped with a low-cost laser scanning range finder, sonars, and IR proximity sensors [9]. Each of vehicles is controlled by its on-board PC computer being a node of the communication network (nodes being linked via wireless Ethernet links). The communication between agents within the system is based on the use of multi-tasking, real time operating system QNX.

Agents are implemented as objects having a *communicator* unit, charged with sending/receiving messages from the other agents and buffering in-going messages. Another unit called *body* consists of a set of data-processing procedures, implementing algorithms specific to the particular agent (preprocessing raw data, extracting features, maintaining the global map). These procedures are invoked by the *event manager* which is a part of the agent's *head* unit, handling the messages emerging in the system. All agents in the system, including the coordinating blackboard and the GDB-based perception agent, have been implemented as processes running on QNX nodes.

### 4.2 The perception network implementation

The perception network has been implemented as a set of feature-extraction and map-building procedures, connected together by the model transformation operators and the rule-based truth maintenance subsystem. The vector-based and grid-based maps are the main nodes of the PN and the data sources for map-based logical sensors. The decision which map is currently used for navigation depends essentially on the type of task. If the high accuracy of path planning and self-localization is required, the exact vector-maps seem to be appropriate. On the other hand, raster-based maps are more useful in sensory data fusion tasks.

Data fusion in the PN is performed with operators which are sub-procedures associated to particular map-builders (for vector- and grid- maps). In the case of 2D vector-maps, the fusion operator matches and merges geometrical primitives extracted by logical sensors from the incoming sensory data, or picked-up from some other representations of the environment by using transformation procedures. In the case of grid-based maps, fusion procedure updates evidence in grid cells by using uncertain data provided by logical sensors. Map-conversion operators enable the information interchange between these robots, which have accumulated already their private environment models in proper forms. These procedures are attached to the particular map-builders and they transform the imported data to the form required at given level of data fusion or enable the check of model consistency at the due level of representation. For example, the procedure extracting geometrical primitives from the grid-based map and putting them onto the vector map is based on the Hough transform as the method for the search of straight lines [5].

Within the PN, the grid-based maps are mainly used as common ground for fusing data from different sensors. These maps tolerate data uncertainty and ambiguity so they can serve also in the case of weakly-structured environments. For occupancy grids we have implemented two different updating procedures: one conventional, implementing a Bayesian integration scheme proposed by Elfes [10], and the second one, based on Dempster-Shafer (D-S) theory, and exploiting the Dempster's rule to combine the belief functions

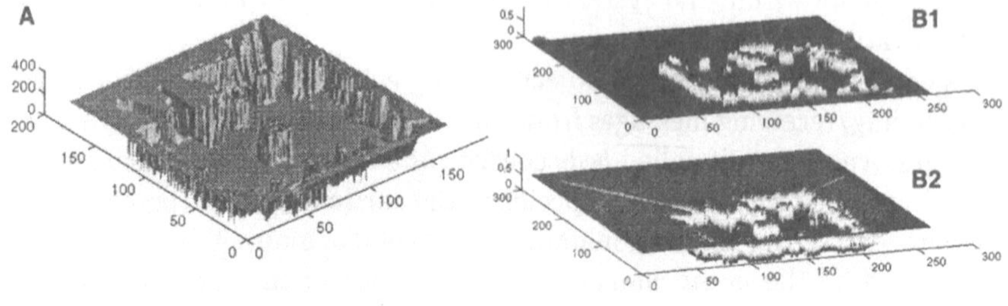


**Fig. 2.** Grid maps of the laboratory built with two different updating schemes

[11]. Figure 2 illustrates the use of these two different evidence-updating schemes. In Fig. 2A there is displayed the probability of occupation according to the Elfes method. In Fig. 2B1 and B2 respectively, the ***support*** and ***plausibility*** of occupation for D-S framework are shown. It can be seen from these figures that the D-S theory-based updating scheme enables better identification of the inconsistent sensory data, by observing the difference between support and plausibility values in the map.

As most elements of the indoor scenes can be roughly represented by 2D polygonal models, the vector-maps composed of geometric primitives are used. The primitives are in turn represented by vectors of parameters accompanied by their covariance matrices (modelling the spatial uncertainty) [9]. The vector map can be updated by the 2D primitives obtained from the raw laser scanner data, or by the line segments extracted from the local grid-map (in the later case also sonars contribute to the primitives). The global vector-map can be fused also with compatible vector-map imported from another robot or the GDB. In all above cases the primitives/maps are provided along with the covariance matrices. The global model of the whole environment explored by the mobile robot is obtained by integrating the line segments and clusters from local maps obtained during the execution of the robot path. Because geometric primitives are considered to be vector random variables disturbed by zero mean Gaussian noise, the fusion is performed with a Kalman filter [5]. This global map building is a part of the perception agent. For the global map, the estimate of the current robot coordinates is required. The perception agent receives this estimate from the robot odometers, or from the self-localization subsystem. The partial maps obtained from the GDB-based agent or another robot are already situated in global coordinates.

Within PN, perception-based models could be incomplete and/or inconsistent. Because of this, the GDB communicates with the internal world-models owned by the particular perception agents to update sensor-based maps and to resolve for the ambiguities. The perception-based model is checked for its internal consistency against the ***a priori*** model. To that goal, the model stored in the GDB is compared with the contents of maps maintained by robots. A set of data abstraction operators, associated with the GDB (being part of the GDB-perception-agent body), can transform the geometrical 3D-model into the form appropriate to be matched and/or fused with current, sensor-based maps. Moreover, the general constraints, originating from the predefined model, reflecting for instance the fact, that walls are immobile, some objects (e.g. desks, bookshelves) can be temporarily dislocated, but keep the shape etc., can be applied. These constraints are introduced into the vector-map builder in a form of production rules to be observed by the data-fusion procedures. The use of such a simple truth maintenance sub-system results with practical elimination of some extracted primitives of the model, that can be raised by spurious measurements but do not satisfy the above constraints of general nature. Only the vector-based maps are updated with the *a priori* knowledge. The grid-map builders have not been provided with such mechanism. This is due to the lack of an explicit representation of objects in these maps.

## 5 The results

Figure 3 shows the process of parametric interpretation of the grid-based model. The experiment has been carried out in a laboratory room, cluttered with obstacles. The room size is 6 × 5 m and it is crowded with desks, chairs, and laboratory equipment (Fig. 3A). Due to a high number of small obstacles (which hardly can be approximated by line segments), it is very hard to build a useful vector model from the raw scanner data (Fig. 3B). The vector-based model has been obtained by transforming the results of measurement accumulated in a occupancy grid (Fig. 3D).

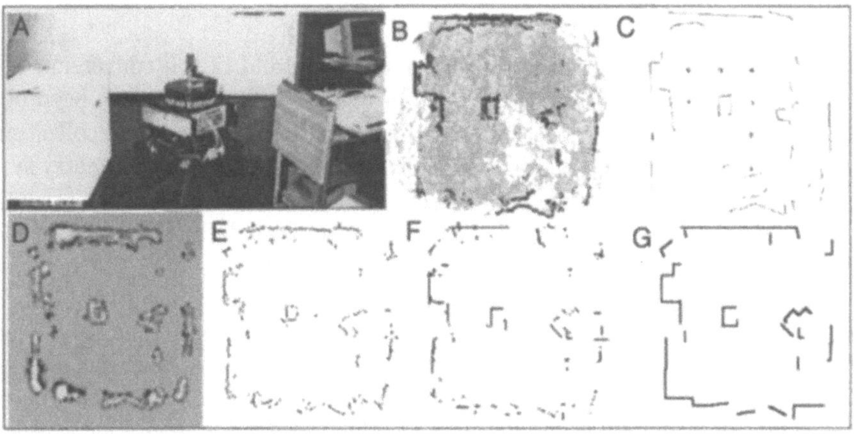


**Fig. 3.** A sequence of grid-map transformation leading to the parametric map

The method is described in [5]. In Fig. 3E the skeleton of the occupancy grid is displayed. In fig. 3F there is the resulting parametric model, and in Fig. 3G, the same map after applying a credibility threshold to line-segments. The final model has even better quality than the map obtained as the result of direct line-based interpretation (Fig. 3C). It is due to the fact, that in this case the geometrical interpretation took place at the much later stage than in the case of the vector-map, and thus much more evidence could be accumulated.

The next example (Fig. 4) illustrates how the world-model can be updated by using the *a priori* knowledge from GDB. In Fig. 4A there is the vector-based map obtained by fusion of two parametric models — the one directly built upon the scanner data and the second one converted from the occupancy grid (cf. Fig. 3). In Fig. 4B there is shown the predefined model of the laboratory room. It contains main structures (walls, a table, several desks) and does not contain many small objects. There is a box in the center of the room (partially visible in Fig. 3A). In Fig. 4C the model obtained by fusing of these two maps is shown. The use of the highly credible *a priori* model resulted with the elimination of the vast majority of spurious primitives.

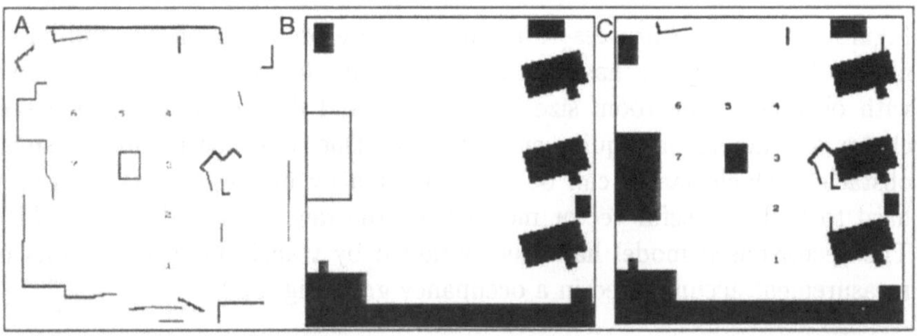


**Fig. 4.** Vector maps obtained from fusing scanner data and GDB information

A further example of the sensory-data fusion and of the conversion between various representations is shown in Figure 5. The maps have been obtained while one of the robots followed a path taking laser scans and collecting sonar data at several predefined positions. In this case only the odometry system has been used to track of the vehicle's position and orientation. Figure 5A shows the grid-based map built by using scanner data. Figure 5B is the parametric model extracted from this map. The grey-lines mark the approximate shape of the real corridor. Figure 5C shows the occupancy grid obtained by fusing data from both scanner and sonars. Some objects that have been not detected by the scanner (dark-grey carton boxes), have been included in this map. The parametric interpretation of this map (Fig. 5D) is much more complete than the vector map obtained solely from the single-mode sensory data.

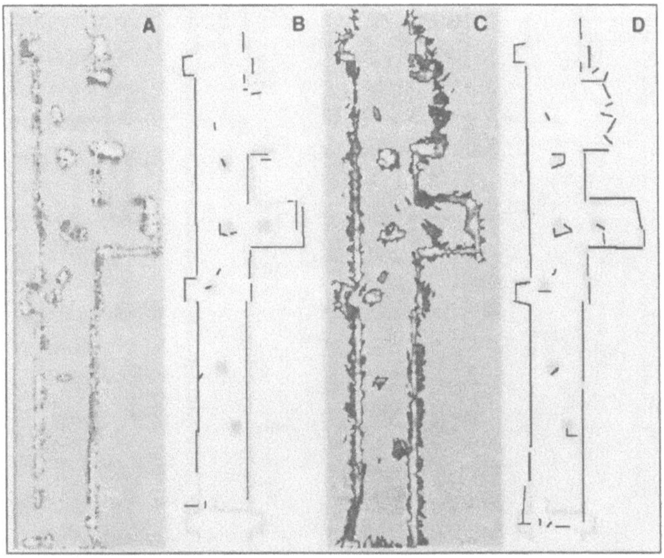


**Fig. 5.** The example of sensory data fusion and conversion

The next experiment illustrates the way in which two robots share their accumulated knowledge on the environment by using PN mechanisms. In this experiment the robot A has been equipped with the laser scanner, while the second one (robot B) carried the belt of 24 sonars. Both robots followed partially overlapping paths in the hallway (robot B) and the laboratory room (robot A) taking measurements at predefined positions. Then, the perception agents of the robots worked grid-based maps. The grid map of robot B (Fig. 6A) has been updated by a part of the occupancy grid transferred from robot A (Fig. 6B). The resulting map is shown in Fig. 6C. Afterwards, this map has been converted to the vector form (Fig. 6D — the uncertainty areas around primitives are drawn with thin lines)

## 6 Conclusions

Distributed perception network has been tested as a concept enabling the consistent management of parallel perception and information interchange between polymorphic world-models. Autonomous agent use this network to share information. The *a priori* known properties of the environment have been introduced into PN as production rules. The improved tolerance to spurious read-outs and sensor failures has been attained, as it can be seen from presented examples of maps. This was mainly due to the elimination of ambiguity made possible by using multiple independent sources of data and alternative information channels in the network.

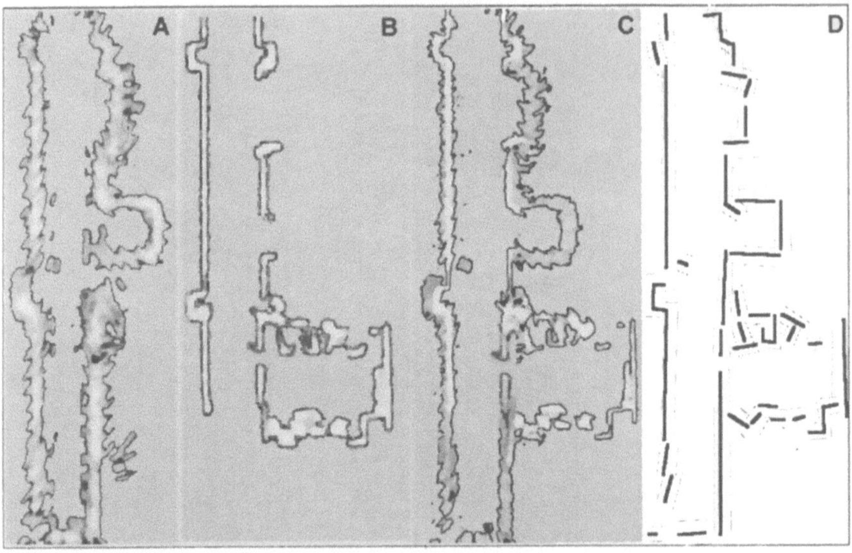


**Fig. 6.** The results of map fusion from two robots with different sensors

## References


1. Kasiński A., Skrzypczyński P. (1998) Cooperative Perception and World-Model Maintenance in Mobile Navigation Tasks, In: Distributed Autonomous Robotic Systems 3, T. Lueth *et al.*, (Eds.), Springer-Verlag, Berlin, 173–182
2. Luo R. C., Kay M. G. (1989) Multisensor Integration and Fusion in Intelligent Systems. IEEE Trans. on SMC, **19(5)**, 61–70
3. Henderson T., Shilcrat E. (1984) Logical Sensor Systems, Journal of Robotic Systems, **2(1)**, 169–193
4. Thrun S., (1998) Learning Metric-Topological Maps for Indoor Mobile Robot Navigation. Artificial Intelligence, **99(1)**, 21–71
5. Skrzypczyński P., Drapikowski P. (1999) Environment Modelling for a Multi-Agent Mobile System, Proc. EUROBOT'99, Zurich, 41–48
6. Lee S., Ro S., Schenker P. (1997) Perception-Net Based Geometric Data Fusion for State Estimation and System Self-Calibration, Proc. IROS'97. Grenoble, 1375a–1375g
7. Laengle T., Lueth T. (1994) Decentralized Control of Distributed Intelligent Robots and Subsystems. AIRTC Symp. on AI in Real-Time Control, Valencia, 447–452
8. Smith R. G., Davis P. (1981) Frameworks for Cooperation in Distributed Problem Solving. IEEE Trans. on SMC, **11(1)**, 61–70
9. Skrzypczyński P. (1997) 2D and 3D World Modelling Using Optical Scanner Data, In: Intelligent Robots: Sensing, Modelling and Planning, R. Bolles *et al.*, (Eds.), World Scientific, 211–228
10. Elfes A. (1987) Sonar-Based Real-World Mapping and Navigation. IEEE J. of Robot. and Autom., **RA3(3)**, 249–265
11. Tirumalai A., Brian B., Jain R. (1995) Evidential Reasoning for Building Environment Maps. IEEE Trans. on SMC, **25(1)**, 10–20

# Cooperative Sentry Vehicles And Differential GPS Leapfrog

John Feddema, Chris Lewis, Robert LaFarge
Intelligent Systems and Robotics Center
Sandia National Laboratories*
Albuquerque, NM 87185-1003

## Abstract

As part of a project for the Defense Advanced Research Projects Agency, Sandia National Laboratories' Intelligent Systems & Robotics Center is developing and testing the feasibility of using a cooperative team of robotic sentry vehicles to guard a perimeter, perform a surround task, and travel extended distances. This paper describes our most recent activities. In particular, this paper highlights the development of a Differential Global Positioning System (DGPS) leapfrog capability that allows two or more vehicles to alternate sending DGPS corrections. Using this leapfrog technique, this paper shows that a group of autonomous vehicles can travel 22.68 kilometers with a root mean square positioning error of only 5 meters.

**Keywords:** Mobile Robots, Cooperative Control, Distributed Autonomous Systems

## Introduction

The field of mobile robotics is quite advanced. The ability to build robotic vehicles that can navigate over long distances either using tele-operation or autonomous control has been demonstrated by a number of researchers, see for instance [1]. In recent years, this field has expanded to consider large numbers or squads of vehicles [2]. The underlying goal of multi-vehicle systems is expanded capability through cooperation. Methods for controlling groups of vehicles range from distributed autonomy [3] to intelligent squad control and general purpose cooperative mission planning [4]. The types of tasks under study range from moving large objects [5] to troop hunting behaviors [6]. Conceptually, large groups of mobile vehicles outfitted with sensors should be able to automatically perform military tasks like formation following, localization of chemical sources, de-mining, perimeter

---

* Sandia is a multiprogram laboratory operated by Sandia Corporation, a Lockheed Martin Company, for the United States Department of Energy under contract DE-AC04-94AL85000. This research is partially funded by the Advanced Technology Office of the Defense Advanced Research Projects Agency under contract e169/09.

control, surveillance, and search and rescue missions [7-10]. Simulation has shown that by sharing concurrent sensory information, the group can better estimate the shape of a chemical plume and therefore localize its source [11]. Similarly, for a search and rescue operation, a moving target is more easily found using an organized team [12-13]. Simulation has also shown that enhanced perimeter control can be achieved by dispersing the group uniformly and by communicating when possible intrusions occur.

As a proof-of-concept, Sandia National Laboratories is developing a squad of semi-autonomous all terrain vehicles for remote cooperative sensing applications (see Figure 1). The system is being used to demonstrate the viability of using a cooperative team of robotic sentry vehicles to investigate alarms from intrusion detection sensors and to surround and monitor an enemy facility.

**Figure 1. 8 RATLER vehicles around the laptop base-station.**

To date, we have developed a robotic perimeter detection system which consists of eight "Roving All Terrain Lunar Explorer Rovers" (RATLER™), a laptop-based base-station, and several Miniature Intrusion Detection Sensors (MIDS). A radio frequency receiver on each of the RATLER vehicles alerts the sentry vehicles of alarms from the hidden MIDS. When an alarm is received, each vehicle decides whether it should investigate the alarm depending on the proximity of itself and the other vehicles to the alarm. As one vehicle attends an alarm, the other vehicles adjust their position around the perimeter to better prepare for another alarm [14].

For the surround task, both potential field and A* search path planners have been added to the base-station and vehicles [15]. At the base-station, the operator specifies goal and exclusion regions on a GIS map. The path planner generates vehicle paths that are previewed by the operator. Once the operator has validated the path, the appropriate information is downloaded to the vehicles. On board the vehicles, the path planner generates the path except that it uses the true location of itself and the nearest neighboring vehicle to avoid collisions.

Most recently, we have added the capability for multiple robots to travel long distances using a differential GPS leapfrog technique. This paper first describes the system and the basic communication modes that have been implemented. This is followed by a detailed description of the differential GPS leapfrog technique.

## System Description

Eight RATLER™ vehicles have been built at Sandia as a test platform for cooperative control and sensing applications. These electric, all wheel drive vehicles consist of two composite bodies joined by a passive central pivot. This flexible structure when combined with an aggressive asymmetric tread on custom carbon composite wheels provides agile off road capabilities. The RATLER vehicles are equipped with a PC104 form factor Intel 80486 processor for control. This computer interfaces to a wide range of sensors and peripherals. Software on the vehicles is currently a single-threaded DOS-based application for simplicity. The vehicles have been programmed to operate either through tele-operation or autonomously. The RATLER vehicles rely heavily on Radio Frequency (RF) signals for communications. Currently, the vehicles are outfitted with differential GPS sensors, and two spread spectrum RF modems. One modem is for inter-vehicle and base-to-vehicle communication and the other is for the differential GPS signal. Video cameras communicate to the base-station via a separate RF video link.

A laptop computer is used as the base-station. A Windows NT application was written to control the vehicles from the base-station. A Graphical User Interface (GUI) displays vehicle status information and allows the operator to monitor the vehicles positions on a Geographic Information System (GIS) map – either aerial photo or topological data, as well as view the live video from a selected vehicle. Mission specific control modes such as tele-operation, formation following, autonomous navigation, and perimeter detection can be initiated and monitored using this GUI interface.

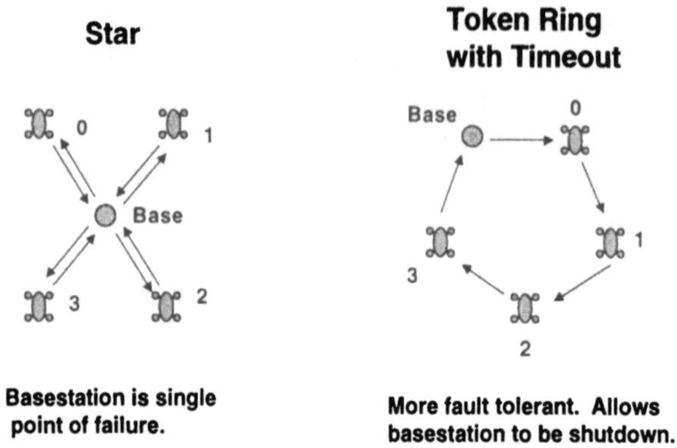


**Figure 2 Two communication options are available. The star network requires the base-station to coordinate all communication. The token ring network provides a decentralized solution.**

## Communication

The RF communication network between the RATLER vehicles and the base-station can be configured as either a star or token ring (see Figure 2). With the star network, all communication is coordinated by the base-station. The base-station sends messages to each vehicle in turn, and each vehicle responses with a broadcast to all the network nodes (the base-station and all the other vehicles). With the token ring network, the base-station and vehicles each take turns broadcasting messages over the network. Each node (either vehicle or base-station) speaks only when it receives the token. The token ring network is more fault tolerant than the star network because there is no single point of failure as there is with the star network. Also, the token ring network allows the vehicles to continue operating even if the base-station is shutdown.

## Differential GPS Leapfrog

In order to understand the differential GPS leapfrog capability, we must first review the basic concepts of the Differential Global Positioning System (DGPS). The standard GPS receiver measures the time of flight (or raw pseudorange $\rho$) of transmitted RF messages from four or more satellites [15]. This measured value is corrupted by noise. In general, the measured pseudorange is equal to the true range D from the user to the satellite plus an unknown offset between the user clock and the satellite clock. Additional time delays are caused by the ionosphere and troposphere, as well as noise multipath, and/or interchannel errors in the user's receiver. The corrected pseudorange is

$$\rho_c = \rho - c\left(\hat{I} + \hat{T} - \hat{B}\right) \tag{1}$$

where c is the speed of light; and $\hat{I}$, $\hat{T}$, and $\hat{B}$ are the estimate time delays caused by the ionosphere, troposphere, and satellite clock. Table 1 shows the estimated effects of these delays.

**Table 1. Typical error in meters per satellite.**

| | Standard GPS | Differential GPS |
|---|---|---|
| Satellite Clocks | 1.5 | 0 |
| Orbit Errors | 2.5 | 0 |
| Ionosphere | 5.0 | 0.4 |
| Troposphere | 0.5 | 0.2 |
| Receiver Noise | 0.3 | 0.3 |
| Multipath | 0.6 | 0.6 |
| SA | 30 | 0 |

Using a least squares solution, the estimated user position $\hat{\vec{r}}_u$ and user clock offset $\hat{b}_u$ is

$$\hat{\vec{x}} = G^{+}\left(A\vec{R} - \vec{\rho}_c\right) \tag{2}$$

where

$$\hat{\vec{x}} = \begin{pmatrix} \hat{\vec{r}}_u \\ -c\hat{b}_u \end{pmatrix}, G = \begin{bmatrix} \hat{\vec{1}}_{s1}^T & 1 \\ \hat{\vec{1}}_{s2}^T & 1 \\ \vdots & \end{bmatrix}, A = \begin{bmatrix} \hat{\vec{1}}_{s1}^T & & 0 \\ & \hat{\vec{1}}_{s2}^T & \\ 0 & & \ddots \end{bmatrix}, R = \begin{pmatrix} \hat{\vec{r}}_{s1} \\ \hat{\vec{r}}_{s2} \\ \vdots \end{pmatrix}$$

The matrices G and A are composed of the estimated directions $\hat{\vec{1}}_{sn}$ of each of the visible satellites. The vector R is the estimated satellite positions that have been received as part of the satellite broadcast. The vector $\vec{\rho}_c$ is the corrected pseudorange to each satellite, arranged as a vector.

In a DGPS system, a reference receiver is put in a known location so that the corrections to the raw pseudorange can be found by:

$$\Delta\hat{\vec{\rho}}_R = A\vec{R} - \vec{\rho} - G\vec{x}_R \tag{3}$$

The reference receiver transmits this correction to all users. It is assumed that if the reference receiver and the user are near each other, then they both have the same pseudorange correction. As shown in Table 2, the horizontal position accuracy improves from 50 meters (without differential corrections) to approximately 0.67 meters (with differential corrections). Even the military's Precision Positioning Service, with a normal positioning accuracy of 15 meters, is improved to less than 1 meter with DGPS corrections.

**Table 2. Typical Position Accuracy.**

| | Standard GPS | Differential GPS |
|---|---|---|
| Horizontal | 50 | 0.67 |
| Vertical | 78 | 2.0 |
| 3-D | 93 | 2.8 |

While performing coordinated robotic vehicle maneuvers such as formation following or perimeter control, it is important to be able to precisely locate each individual vehicle's absolute location. For this reason, DGPS is used on most robotic vehicle platforms for position control. However, the one downfall to DGPS is that it requires that one GPS receiver (typically at the base-station) be surveyed to a know location. Another alternative to a surveyed point is keep a receiver in a single position for a long period of time (as much as 24 hours), and use the average position as the "true" location when determining the differential correction.

Neither of these options is viable for most military operations where there are no surveyed points and there is no time to wait. For these situations, we have developed a DGPS leapfrog technique where one vehicle stops and becomes a base GPS unit while the others get corrections from it. The process is initialized by one GPS receiver (this could be the base-station or a vehicle) having a known location, either surveyed or time averaged. Corrections from this receiver are transmitted to the others. As the other vehicles reach the communication range of the first, another vehicle becomes the base GPS unit using the differentially corrected position from the first vehicle. This frees the first vehicle to again become a free roaming vehicle using differential corrections from the second. This process may be repeated indefinitely; however, eventually the accumulated error will be above an acceptable level. At which point, the newly assigned GPS base vehicle is required to stay in one position and average GPS measurements until the absolute positioning error is back to an acceptable level.

The question arises as to how many leapfrog steps are acceptable? The accuracy of the leapfrog technique may be analyzed using a random walk. Assume that each time two vehicles exchange roles as GPS base, the expected RMS positioning error of the second vehicle relative to the first is $r_{rms}$. After N leapfrog steps, the total expected RMS error is given by

$$R^2 = E\left\{(\Delta x_1 + \ldots + \Delta x_N)^2 + (\Delta y_1 + \ldots + \Delta y_N)^2\right\} \tag{4}$$

where $(\Delta x_1, \Delta y_1)\ldots(\Delta x_N, \Delta y_N)$ are the errors in x and y positions. Assuming each of the N leapfrog steps is independent, this error reduces to

$$R^2 = E\left\{\Delta x_1^2 + \Delta y_1^2\right\} + \ldots + E\left\{\Delta x_N^2 + \Delta y_N^2\right\} = N r_{rms}^2 \,. \tag{5}$$

Solving for the number of leapfrog steps,

$$N = \left(R/r_{rms}\right)^2 \ . \tag{6}$$

Therefore, if the acceptable accumulated error of the robot vehicles is 5 meters, and the RMS error of each step is 0.67 meters, then two or more vehicles can travel 22.86 kilometers in 56 steps, each step being 0.4 kilometers long.

This DGPS leapfrog mode of operation has been installed on our RATLER test bed. Figure 3 shows the base-station's graphical user interface being used to direct two RATLER vehicles along a test path while a third, stationary vehicle monitors the accumulated error in the corrections. To run the test, the operator first draws a path for both vehicles. At intermediate waypoints, the operator alternates selecting each vehicle to be the GPS base-station (denoted by green triangles in the figure). Currently, the operator must be sure to select adjacent GPS base positions that are within range of each vehicle's radio that transmits and receives the GPS corrections. This process could be automated in the future.

To execute the test path, the waypoints and exchange point are downloaded to the robots over the communications link. Then, both vehicles are activated. Initially they both use the command station's GPS receiver for corrections. When the first vehicle reaches a waypoint node designated by the operator as a DGPS node, it stops, and uses its corrected GPS position to re-initialize its GPS receiver to act as a correction station. It also sends a command to the base station to turn off the base station GPS receiver. From there on, the base station is not necessary to the operation. The vehicles execute exchanges by sending commands to each other autonomously. For the test, a third motionless vehicle's position was monitored and recorded throughout the course. The accumulated change in its recorded position was expected to follow a random walk model.

The path including 12 exchanges is shown in Figure3. It was executed 5 times. The position data recorded from the stationary vehicle is shown in Figure 4. The first test run accumulated an error of 2 meters, while each of the following 4 runs only accumulated a 1 meter error.

## Conclusion

This paper described a squad of mobile robotic vehicles being developed at Sandia National Laboratories to demonstrate cooperative robotic sensing capabilities. The squad currently consists of eight RATLER™ vehicles, a laptop base-station, and a battery powered antenna array. In the past, this system has been used to demonstrate autonomous navigation of a cooperative team of robots and their use for surround and perimeter detection missions. This paper highlights the use of a DGPS leapfrog technique that allows a squad of vehicles to travel long distances without maintaining communication to a base-station

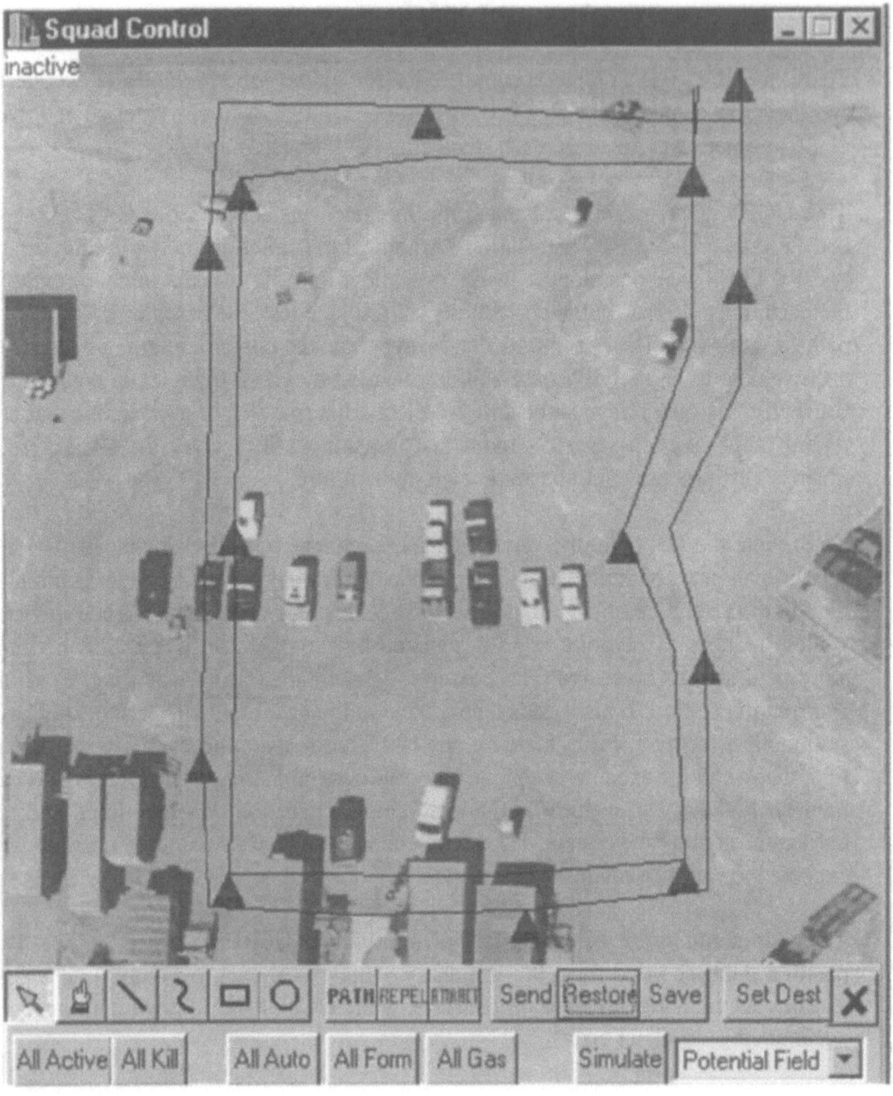


**Figure 3 GUI used to designate paths and exchange points for two vehicles to demonstrate the DGPS leapfrog technique**

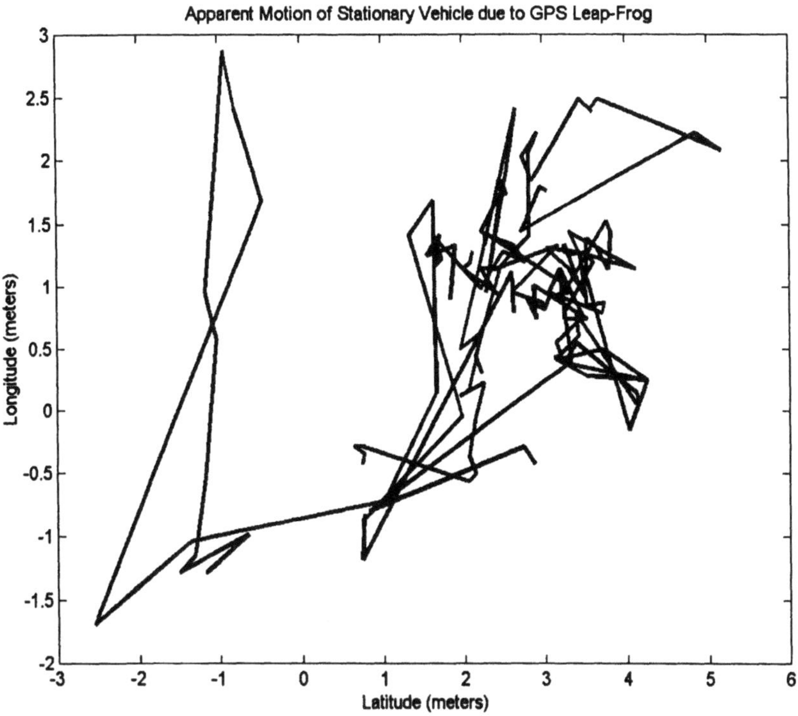


**Figure 4 GPS errors accumulated by a stationary vehicle using corrections from a pair of mobile vehicles leapfrogging through a path with 12 exchanges.**

## References


1. D Bapna, E et .al., "The Atracama Desert Trek: Outcomes," *Proceedings of the 1998 Conference on Robotics & Automation,* Leuven, Belgium, May 1998, 597-604.
2. Y. Uny Cao, Alex S. Fukunaga, and Andrew B. Kahng, "Cooperative Mobile Robotics: Antecedents and Directions," *Proceedings of the 1995 IEEE/RSJ IROS Conference,* 226-234.
3. T. Fukuda, et al , "Evaluation on Flexibility of Swarm Intelligent System, "*Proceedings of the 1998 Conference on Robotics & Automation*, Leuven, Belgium, May 1998, 3210-3215.
4. B. Brummitt and Anthony Stentz, "GRAMMPS: A Generalized Mission Planner for Multiple Mobile Robots In Unstructured Environments," *Proceedings of the 1998 Conference on Robotics & Automatio*n, Leuven, Belgium, May 1998, 1564-1571.
5. K. Kosuge, T. Oosumi, M. Satou, K. Chiba, K. Takeo, "Transporation of a Single Object by Two Decentralized-Controlled Nonholonomic Mobile Ro-

bots," *Proceedings of the 1998 Conference on Robotics & Automation*, Leuven, Belgium, May 1998, 2989-2994.

6. Hiroaki Yamaguchi, "A Cooperative Hunting Behavior by Mobile Robot Troops," *Proceedings of the 1998 Conference on Robotics & Automation,* Leuven, Belgium, May 1998, 3204-3209.
7. F.R. Noreils, "Multi-Robot Coordination for Battlefield Strategies," *Proceedings of the 1992 IEEE.RSJ International Conference on Intelligent Robots and Systems,* Raleigh, NC, July 1992, 1777-1784.
8. Q. Chen, J.Y.S. Luh, "Coordination and Control of a Group of Small Mobile Robots," *Proceedings of the 1994 IEEE.RSJ International Conference on Intelligent Robots and Systems,* 2315-2320.
9. J.P. Desai, J. Ostrowski, V. Kumar, "Controlling Formations of Multiple Mobile Robots," *Proceedings of the 1998 Conference on Robotics & Automation*, Leuven, Belgium, May 1998, 2864-2869.
10. H. Yamaguchi, J.W. Burdick, "Asymptotic Stabilization of Multiple Nonholonomic Mobile Robots Forming Group Formations," *Proceedings of the 1998 Conference on Robotics & Automation*, Leuven, Belgium, May 1998, 3573-3580.
11. J.E. Hurtado, R.D. Robinett, C.R. Dohrmann, S.Y. Goldsmith, "Distributed Sensing and Cooperating Control for Swarms of Robotic Vehicles," *Proc. IASTED Conference Control & Applications*, Honolulu, Hawaii, Aug. 12-14, 1998.
12. J.S. Jennings, G. Whelan, W.F. Evans, "Cooperative Search and Rescue with a Team of Mobile Robotis," *Proc. IEEE International Conference of Advanced Robotics*, Monterey, CA, 1997.
13. S. Goldsmith, J. Feddema, R. Robinett, "Analysis of Decentralized Variable Structure Control for Collective Search by Mobile Robots," SPIE98, *Proc. Sensor Fusion and Decentralized Control in Robotic Systems*, Boston, November 1-6, 1998.
14. C. Lewis, J.T. Feddema, P. Klarer, "Robotic Perimeter Detection System," *Proceedings of SPIE Vol. 3577*, Boston, MA, November 3-5, 1998, pp. 14-21.
15. J. Feddema, C. Lewis, P. Klarer, "Cooperative Robotic Sentry Vehicles," *Proceedings of SPIE Vol. 3839*, Boston, MA, September 19-20, 1999.
16. Bradford W. Parkinson and Per K. Enge, "Differential GPS," *Global Positioning System: Theory and Applications, Volume II*, Bradford W. Parkinson, James J. Spiker Jr., Editors, Progress in Astronautics and Aeronautics, Vol. 164., American Insitute of Aeronautics and Astronautics, Inc., 1996.

# Chapter 9

# Multi-Robot Motion Coordination and Tracking

# Motion coordination in formations of multiple mobile robots using a potential field approach

F.E. Schneider, D. Wildermuth, and H.-L. Wolf

Research Establishment for Applied Science; Research Institute for Communication, Information Processing, and Ergonomics; Department of Ergonomics and Information Systems; Neuenahrer Straße 20, 53343 Wachtberg, Germany
e-mail: {schneid1 | dennis | wolf}@fgan.de

**Abstract** – This paper presents a novel approach for motion coordination in formations of multi-robot-systems (MRS). First a short introduction to the application of MRS navigation is given. For the special sub-problem of moving in formation, a solution based on a potential field approach is presented. Different forces belonging to other robots, obstacles and the aspired shape of formation are combined and used to move each robot to its desired position inside the formation. While moving in formation the group is able to avoid obstacles and approach towards a specified target. First results in simulation show that a group of robots can form up a rectangle or circle even in the presence of obstacles.

**Keywords** – multi-robot, navigation, formation, potential field

## 1 Introduction

The idea of multiple robots working together in order to accomplish a task efficiently has become increasingly popular in the last several years. Among these tasks, many situations can be found in which the sub-problem of generating some kind of formation and then maintaining it during a movement must be solved. Possible application areas for multi-robot-systems generating and moving in a formation (which is also called *marching* sometimes) are

- reconnaissance, surveillance and target acquisition (RSTA),
- ordnance disposal, demining and ABC decontamination,
- security, defence and sniper discovery,
- transport and convoying.

These applications differ in the strictness of the formation constraints. While e.g. a transport mission requires very tight constraints on the robot pattern to share the workload equally among all robots, other applications like reconnaissance and surveillance only need the robots to be spread more or less uniformly over a specified area. Some tasks like convoying or demining build a formation but allow a distortion or partition of the pattern, e.g. while passing obstacles. Under the latter condition four basic subtasks can be identified for a mission that requires moving in a formation:

- assemble at the starting point and/or establish the formation,
- maintain formation while moving,
- avoid obstacles while occasionally splitting/deforming and then re-establishing the formation,
- change the shape of the formation.

Several approaches to the establishing and maintaining subtasks can be found in the literature which mostly do not address the problem of avoiding obstacles. Examples include papers by Wang [1], who uses neighbour tracking and a mathematical model of the leader-follower co-ordination strategy, and Suzuki et al. [2, 3, 4, 5]. The latter works treat the problem of formation generation in terms of distributed algorithms using idealised robots with perfect sensors and no physical dimension. Chen and Luh [6], like some other authors, improve these algorithms with regard to usefulness for physical robots and also propose a method to move a group of mobile robots in strict formation [7]. Yamaguchi and Arai [9] define a potential field spread on the space of relative distances between neighbours, which makes the robots converge into a predefined shape.
Parker [8] discusses the proper balancing between local and global knowledge while moving in a formation. Global information about goals and obstacles should influence the robots local behaviour when avoiding obstacles. Balch and Arkin present a behaviour-based approach to robot formation keeping [10]. Reactive behaviours, implemented as so-called motor schemas, correspond to the different influences, which have an effect on the movements of the robots: move-to-goal, avoid-obstacle, avoid-robot and maintain-formation. Several authors worked on the somewhat related problem of pattern generation in Cellular Robotic Systems[1] [12, 13, 14].

If strict formation constraints are not required and deforming of the formation in the presence of obstacles is allowed, two opposite influences on the resulting paths for each robot can be identified:

- reaching a goal point on the shortest possible path and
- maintaining the formation.

Obviously, these two demands contradict in many environments, e.g. at doorways or in narrow corridors. Without distortion of the formation a detour around such a doorway and thereby a longer path is necessary. In order to model this situation we use a potential field approach as described in [15] and therefore assign each robot a specific position inside the formation. As an extension to the similar method presented in [9] not only a shape-forming force is used to generate the formation,

---

[1] The concept of Cellular Robotic Systems (CRS) and its relation to other robotic research fields is, for example, described in [11].

but also additional types of forces are modeled and combined to achieve more functionality and flexibility. These different virtual forces are calculated dynamically in order to move a robot formation towards a specified goal point. Depending on the distance and relative direction between each robot and its neighbours attractive and / or repulsive forces are generated to establish the formation and keep each robot at its specified position relative to the others. It is even possible to change the goal formation temporarily, e.g. when reaching a small corridor. During movement, other robots and obstacles produce repulsive forces to prevent collisions. Additionally a goal force attracts the whole group towards a target point. From a special point of view, these virtual forces correspond to the mentioned motor schemas presented in [10]. Different weightings of the virtual forces lead to a different behaviour with respect to the foregoing discussion of the resulting paths.
A current requirement for the calculation of the virtual forces is the presence of a common X/Y co-ordinate system, which is communicated among the robots. A common co-ordinate system or at least an exact relative position estimation between the robots is quite simple to achieve in simulation. But as we are going to implement our method on physical robots this precondition is subject to ongoing research. The following section describes the potential field method used to generate and maintain the formation. Chapter 3 presents a short summery of the first simulation results and finally chapter 4 gives some concluding remarks.

## 2 Control of formations

The concept of using a potential field method for controlling a formation is based on the view that not only obstacles and goals can exert forces onto a robot [16, 17, 15]. The basic idea is that each robot in a formation can exert repulsive and attractive forces to other robots. These virtual forces can be used to keep the robots in the desired formation.

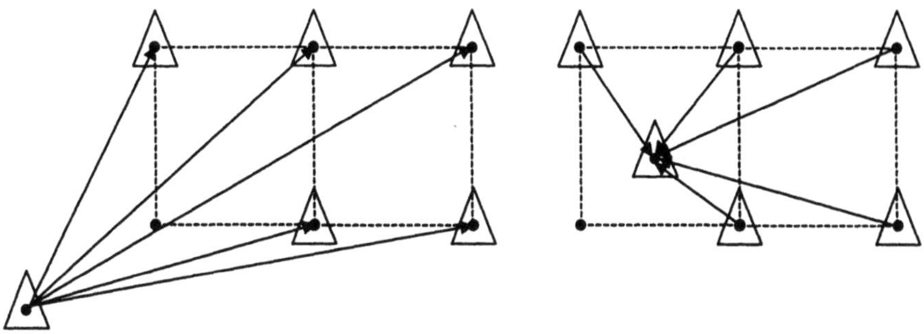

**Fig. 1.** Robot outside (left) and inside (right) the formation

The first example in Figure 1 shows a robot that is outside the formation. The attractive forces of the other robots will "pull" it back into the formation. On the other hand, the attractive force of the robot itself will cause the formation to move towards its position. In the second example, a robot has moved to close to the

formation. The repulsive forces of the other robots will "push" it away. Additionally the repulsive force of the robot itself pushes the formation away.
In both cases, the forces will move the robot and the formation into a constellation where the desired formation is re-established. Two forces have to be considered when the formation is forming up or in motion:

1. the repulsive obstacle forces and
2. the forces that establish and maintain the formation.

The following two sections will briefly discuss these different forces more in detail.

### 2.1 Repulsive obstacle force

The method for generating the repulsive forces is a modified approach similar to the one proposed by Borenstein and Koren [16]. One of the major variations is that no certainty grid is used. Since only local obstacle avoidance[2] is carried out and all robots are equipped with 360° laserscanners, a map is not absolutely necessary.
The repulsive forces that should effect the robot are directly generated from the laserbeams. The magnitude of each of these forces is inversely proportional to the square of the distance measured. By this method, close objects (small distances) exert strong repulsive forces and produce weak forces when they are further away. The repulsive force $O_{ij}$ between robot i and endpoint laserbeam j is defined as

$$O_{ij} = \left( \frac{x_i - (s_j \sin\omega_j)}{s_{ij}^{\;2}}, \frac{y_i - (s_j \cos\omega_j)}{s_{ij}^{\;2}} \right),$$

where

| | |
|---|---|
| $x_i, y_i, \theta_i$ | denotes position and orientation of robot i, |
| $s_j$ | the sensor reading beam j, and |
| $\omega_j$ | the orientation of sensor beam j. |

The resultant repulsive force $O_i$,

$$O_i = \sum_j O_{ij},$$

describes the sum of the individual forces generated by all laserbeams taken at a time *t*.
It is assumed that a robot senses other robots as obstacles. Therefore, special collision avoidance between the robots is not necessary. However, in the case that

[2] Since this paper is focused on the formation problem under consideration of local obstacles higher path planing problems are beyond the scope of this paper.

some of the robots are not senseable (e.g. because they are too small) an additional repulsive "robot" force has to be calculated. Let

$$OR_{ij} = \left( \frac{x_i - x_j}{d_{ij}^{\;2}}, \frac{y_i - y_j}{d_{ij}^{\;2}} \right)$$

be the repulsive force between robot i and j, where $d_{ij}$ denotes the distance between robot i and robot j.

The resultant repulsive force $OR_i$, defined as

$$OR_i = \sum_j OR_{ij} ,$$

is the sum of the individual forces caused by all other robots inside a defined window. Finally, the summation of $O_i$ and $OR_i$ yields the overall repulsive force $RS_i$:

$$RS_i = O_i + OR_i .$$

## 2.2 Shape forming forces

The shape forming forces are virtual forces between the robots of a formation. Each robot applies attractive or repulsive forces onto the other robots to generate and maintain the shape of the formation. The presented approach uses two distinct forces. The force that directs the robot to the desired position inside the formation and the force that turns the robot into the direction of the formation (see Fig. 2).

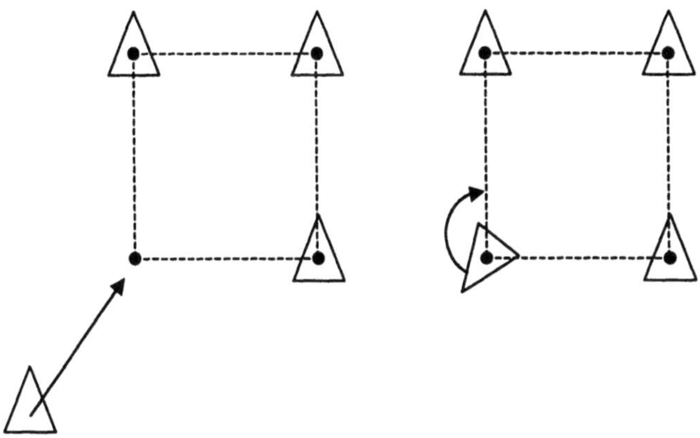

**Fig. 2.** Position force (left) and orientation force (right)

### 2.2.1 Position force

The positioning of the robots should be independent of the orientation of the formation. This can be achieved by specifying the desired distances to the neighbours in the formation, which are angle-independent. In some cases this will lead to a potential filed typical local minimum situation. To avoid this case the position force $P_{ij}$ is "directed" with the desired formation angle between the robots i and j:

$$P_{ij} = \left(\left(d_{ij} \sin\left(\delta_{ij}\right)\right) - \left(z_{ij} \sin\left(\sigma_{ij} + \theta_j\right)\right), \left(d_{ij} \cos\left(\delta_{ij}\right)\right) - \left(z_{ij} \cos\left(\sigma_{ij} + \theta_j\right)\right)\right).$$

The resultant position force $P_i$,

$$P_i = \left(\frac{\sum_j P_{ij}}{j}\right),$$

is the sum of the individual forces caused by all other robots inside a defined window and scaled by the number of those neighbouring robots, whereby

| | |
|---|---|
| $d_{ij}$ | denotes the actual distance between robot i and robot j, |
| $\delta_{ij}$ | denotes the actual angle between robot i and robot j, |
| $\sigma_{ij}$ | the formation angle between robot i and robot j, and |
| $z_{ij}$ | the desired formation distance between robot i and robot j. |

### 2.2.2 Orientation force

After calculating the position force, the robot has to get a new orientation that depends on the position of the rest of the formation and the desired orientation inside the formation. Let

$$\delta_{ij} = \text{atan}\,2\left(\left(y_j - y_i\right), \left(x_j - x_i\right)\right)$$ [3] and

$$\varphi_{ij} = \left(\theta_i + \delta_{ij}\right) - \sigma_{ij},$$

where $\varphi_{ij}$ means the repulsive force between robot i and j and the other variables defined as noted above. Then the overall orientation force $A_i$, defined as

[3] The atan2() function computes the principal value of the arc tangent of y/x, using the signs of both arguments to determine the quadrant of the return value. Upon successful completion, atan2() returns the arc tangent of y/x in the range [-pi, pi] radians. If both arguments are 0.0, then 0.0 is returned.

$$A_i = \left( \frac{\sum_j \varphi_{ij}}{j} \right),$$

is the sum of the individual forces caused by all other robots inside a defined window and scaled by the number of the other robots.

### 2.3 Resultant force

The vectorial sum of all forces, repulsive from the obstacles and shape forming from the formation, produces a resultant force vector S:

$$S = P_i + RS_i .$$

This force can now be used for generating the steering commands for the robot with respect to the orientation force $A_i$.

## 3 Experimental Results

This section shows the implementation of the presented approach in a plain simulation. A number of robots can be positioned at arbitrary starting points. Additionally an arbitrary number of obstacles can be placed around them in the simulated plane. In the current version of the simulation, only the use of round obstacles is implemented. Figure 3 shows the movement of a group of six robots in order to generate a square shaped formation.

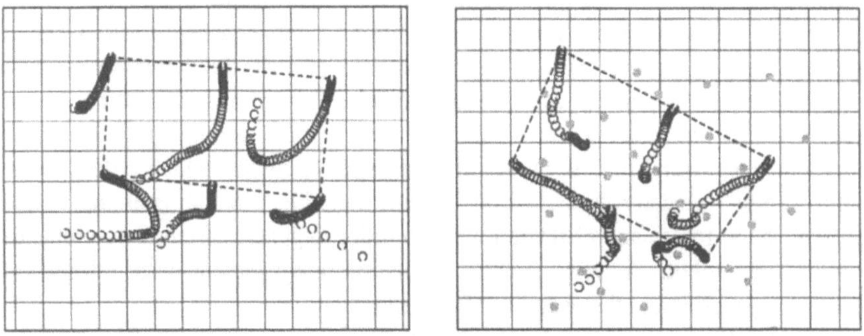

**Fig. 3.** Forming a rectangle shape without (left) and with obstacles (right)

For simplicity, the continuous movement of the robots is partitioned into discrete intervals of length $\Delta t$. Starting at time $t_0$ the resultant force for each robot is calculated as described in chapter 2. Since the robots are modelled as free-floating objects with no constraints on their ability to move always to any given direction, the resultant force is simply translated into a movement for the next time interval. Thus the new positions at time $t_0 + \Delta t$ can be computed and the simulation step can be repeated. With the robots' positions at each time step $t_i$ being printed on the screen, it is easy to imagine the driven path and the changes in velocity of every robot. Any number of robots and arbitrary shapes of formation can be handled within the simulation. Figure 4, for example, shows a group of eight robots moving towards their specified relative positions and thus creating a circle formation, on the left side without and on the right side while avoiding several obstacles.

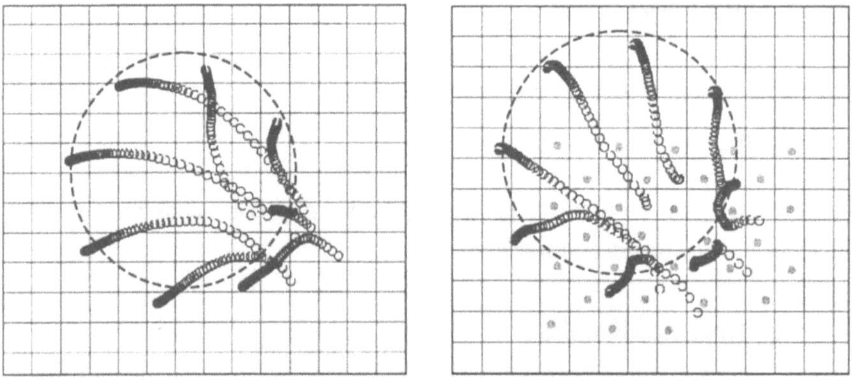

**Fig. 4.** Forming a circle shape without (left) and with obstacles (right)

## 4 Conclusion and Future Work

In this paper, a new method for controlling a formation of multiple mobile robots is presented. Because in our context a formation is not regarded as strict but as capable of deforming and splitting in the presence of obstacles we make use of a virtual force field in order to establish and maintain a particular formation. In our view obstacles and goal points as well as other robots belonging to the same formation exert virtual forces. These different types of forces, which influence the resulting paths for each robot, are modelled and combined, including attractive and repulsive shape forces and obstacle forces, in order to generate steering commands to move the robot. A very simple simulation environment was utilised to demonstrate in principle the usability of this approach.

In a next step the missing goal force mentioned in chapter 1 will be added to make the robot formation move towards any given target point. Additional simulation experiments are necessary to find out suitable weightings of the different forces in order to adjust and steer the movement of the robots. Afterwards the algorithms have to be transferred to our existing multi-robot simulation environment, which models the behaviour of physical robots. Thereby the simple view of a robot being a free-floating object has to be re-considered. Finally, the concept shall be implemented on real robots.

## 5 Acknowledgements

The authors wish to thank Prof. Dr. W. Burgard (University of Freiburg) and Dr. D. Fox (Carnegie Mellon University) for their helpful comments and support.

## 6 References


[1] P. K. C. Wang, *Navigation Strategies for Multiple Autonomous Mobile Robots Moving in Formation*, Journal of Robotic Systems 8(2), 1991, pp. 177-195.

[2] K. Sugihara and I. Suzuki, *Distributed Motion Coordination of Multiple Mobile Robots*, Proc. of 5$^{th}$ IEEE International Symposium on Intelligent Control, Philadelphia, September 1990, pp. 138-143.

[3] I. Suzuki and M. Yamashita, *Formation and Agreement Problems for Anonymous Mobile Robots*, Proc. of 31$^{th}$ Annual Allerton Conference on Communication, Control and Computing, University of Illinois, Urbana, September 1993, pp. 93-102.

[4] I. Suzuki and M. Yamashita, *A Theory of Distributed Anonymous Mobile Robots – Formation and Agreement Problems*, Technical Report TR-94-07-01, Department of Electrical Engineering and Computer Science, University of Wisconsin, Milwaukee, Juli 1994.

[5] H. Ando, I. Suzuki, M. Yamashita, *Formation and Agreement Problems for Synchronous Mobile Robots with Limited Visibility*, Proc. of International Symposium on Intelligent Control, Monterey, August 1995, pp. 453-460.

[6] Q. Chen and J. Y. S. Luh, *Distributed Motion Coordination of Multiple Robots*, Proc. of International Conference on Intelligent Robots and Systems, Federal Armed Forces University Munich, September 1994, pp. 1493-1500.

[7] Q. Chen and J. Y. S. Luh, *Coordination and Control of Small Mobile Robots*, Proc. of 11$^{th}$ IEEE International Conference on Robotics and Automation, San Diego, May 1994, pp. 2315-2320.

[8] L. E. Parker, *Designing Control Laws for Cooperative Agent Teams*, Proc. of 10$^{th}$ IEEE International Conference on Robotics and Automation, Atlanta, May 1993, pp. 582-587.

[9] H. Yamaguchi and T. Arai, *Distributed and Autonomous Control Method for Generating Shape of Multiple Mobile Robot Group*, Proc. of International Conference on Intelligent Robots and Systems, Federal Armed Forces University Munich, September 1994, pp. 800-807.

[10] T. Balch and R. C. Arkin, *Behavior-based Formation Control for Multi-robot Teams*, to be published in IEEE Transactions on Robotics and Automation, 1999/2000.

[11] G. Beni, *The Concept of Cellular Robotic Systems*, Proc. of 3rd IEEE International Symposium on Intelligent Control, Arlington, August 1988, pp. 57-62.

[12] J. Wang and G. Beni, *Pattern Generation in Cellular Robotic Systems*, Proc. of 3rd IEEE International Symposium on Intelligent Control, Arlington, August 1988, pp. 63-69.

[13] V. Genovese, P. Dario, R. Magni, L. Odetti, *Self-organizing Behavior and Swarm Intelligence in a Pack of Mobile Miniature Robots in Serach of Pollutants*, Proc. of IEEE International Conference on Intelligent Robots and Systems, Raleigh, July 1992, pp. 1575-1582.

[14] P. Liang and G. Beni, *Robotic Morphogenesis*, Proc. of 12th IEEE International Conference on Robotics and Automation, Nagoya, May 1995, pp. 2175-2180.

[15] Y. Koren and J. Borenstein, *Potential Field Method and Their Inherent Limitations for Mobile Robot Navigation*, Proc. of 8th IEEE International Conference on Robotics and Automation, Sacramento, April 1991, pp. 1398-1404.

[16] J. Borenstein and Y. Koren, *Real-time Obstacle Avoidance for Fast Mobile Robots*, IEEE Transactions System, Man and Cybernetics, Vol. 19, No. 5, Sept./Oct. 1989, pp. 1179-1187.

[17] J. Borenstein and Y. Koren, *Real-time Obstacle Avoidance for Fast Mobile Robots in Cluttered Environments*, Proc. of 7th IEEE International Conference on Robotics and Automation, Cincinnati, Ohio, May 1990, pp. 572-577.

# Fly spy: lightweight localization and target tracking for cooperating air and ground robots

Richard T. Vaughan, Gaurav S. Sukhatme, Francisco J. Mesa-Martinez, and James F. Montgomery

Robotics Research Laboratories, University of Southern California, Los Angeles, CA 90089-0781

vaughan|gaurav|mesamart|monty@robotics.usc.edu

**Abstract.** Motivated by the requirements of micro air vehicles, we present a simple method for estimating the position, heading and altitude of an aerial robot by tracking the image of a communicating GPS-localized ground robot. The image-to-GPS mapping thus generated can be used to localize other objects on the ground. Results from experiments with real robots are described.

## 1 Introduction

Payload volume, mass and power consumption are critical factors in the performance of aerial robots. Very small "micro" air vehicles (MAVs) are particularly desirable for reasons of cost, portability and, for military applications, stealth. Building on several years work with small robot helicopters, we are interested in minimalist approaches to localization which reduce the number of sensors that such robots carry with them. This paper presents a simple method for estimating the position, heading and altitude of an aerial robot using a single on-board camera. Many applications of aerial robots such as reconnaissance and target tracking already require an on-board camera. We exploit this sensor to simultaneously perform localization. Since ground platforms are more suited to carry larger sensor payloads, and are typically localized, collaboration and data sharing between ground and aerial robots is employed to provide the extra information needed to localize the air vehicle.

## 2 Related work

The work that most closely parallels the system discussed in this paper is the "visual odometry" system at CMU [1] which can visually lock-on to ground objects and sense relative helicopter position in real time. The system tracks image pair templates using custom vision hardware. Coupled with angular rate sensing, the system is used to stabilize small helicopters over reasonable speeds.

Approaches to automate helicopter control include both model-based and model-free techniques. Pure model-based RC helicopter control systems have

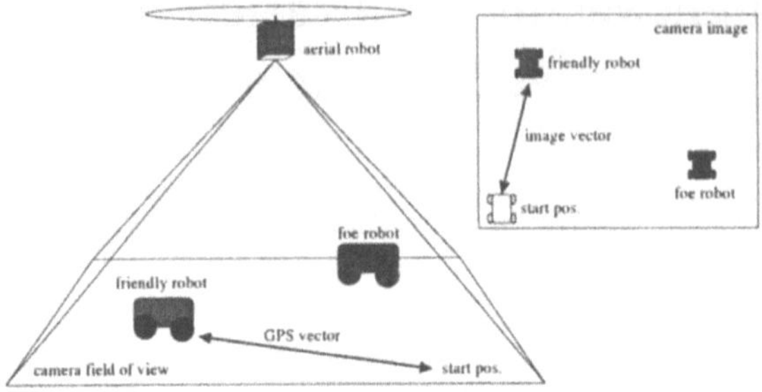


**Fig. 1.** Cooperating MAV/AGV scenario

mostly used linear control. Feedback linearization, a design technique for developing invariant controllers capable of operating over all flight modes, was used in [2] and [3]. In [2], the nonlinear controller robustness was increased using sliding mode and Lyapunov-based control. An approach using output tracking based on approximate linearization is discussed in [4]. Neural networks were combined with feedback linearization in [3] to create an adaptive controller with the same goal in mind.

Various techniques to *learn* a control system for helicopters have been proposed. These include a genetic algorithm-based approach [5] to discover fuzzy rules for helicopter control; a combination of expert knowledge and training data to generate and adjust the fuzzy rule base [6].

We have successfully implemented [7] a behavior-based control approach to the helicopter control problem, described in Section 4.1. Our past work on the subject has included a model-free method for generating a helicopter control system which also has an online tuning capability using teaching by showing.

## 3 Approach

If an MAV can observe the relative positions of itself and two objects with known location on the ground below it, it can localize itself completely by triangulation. If the objects on the ground were friendly robots ('Unmanned Ground Vehicles' - UGVs) with on-board localization, they could inform the MAV of their positions and even send updates as they moved. This requires only a camera, modest computation and communications. All these resources are likely to be required on the MAV for other tasks; this localization technique requires no dedicated resources.

Alternatively, a single cooperating robot can move over time to establish a baseline on the ground and on the image plane of the MAV's camera, as shown in Figure 1. This is subject to some assumptions about the behaviour of the MAV, but it is the minimal configuration for such a system and is the scenario investigated in the rest of this paper.

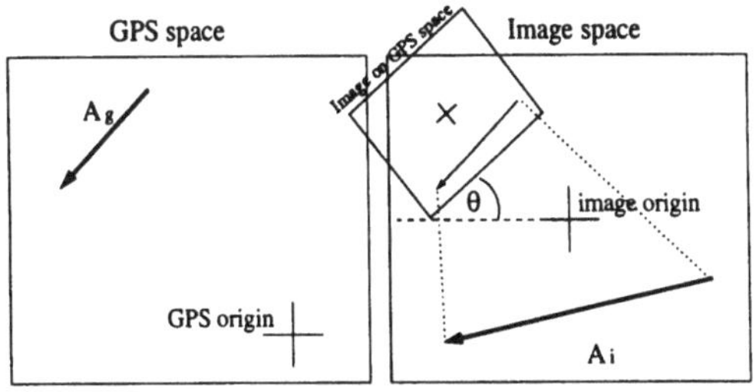


**Fig. 2.** Mapping a vector in image space $\boldsymbol{A_i}$ to its correlate in GPS space $\boldsymbol{A_g}$

Additional advantages of such a scheme are (i) scalability: a single ground robot can cooperatively localize many MAVs, effectively sharing its expensive and bulky GPS sensor among a swarm of aerial vehicles; (ii) the localization estimate is not subject to drift, and as such could be used to correct parallel high-resolution but drifting inertial sensors.

### 3.1 Assumptions

We assume these simplifying assumptions and constraints:

1. the camera is mounted perpendicular to the plane of the ground, pointing straight down;
2. the ground in the field of view is essentially flat;
3. the camera is mounted directly under the center of rotation of the MAV (though an off-center mounting can easily be compensated for by a rigid body transform);
4. the MAV is in the same place and pose when the two vector-end samples are taken.
5. over short distances (<100m) GPS space is approximately Cartesian; GPS latitude and longitude coordinates can be converted to meters by a scaling factor that is a function of latitude). The error introduced by this approximation is small compared to errors elsewhere in the system.

### 3.2 Heading

As the GPS coordinate system is aligned with north and the camera coordinate system is aligned with the helicopter, the vehicle's heading is the rotation angle $\theta$ of the coordinate systems, shown in Figure 2.

Given that $(x_{i0}, y_{i0})$ are the image coordinates of the UGV at $t = 0$ and $x_{g0}, y_{g0})$ are the GPS coordinates of the UGV at $t = 0$:

$$\boldsymbol{A_i} = \begin{pmatrix} x_i - x_{i0} \\ y_i - y_{i0} \end{pmatrix} \quad \boldsymbol{A_g} = \begin{pmatrix} x_g - x_{g0} \\ y_g - y_{g0} \end{pmatrix} \qquad \theta = \angle \boldsymbol{A_g} - \angle \boldsymbol{A_i}$$

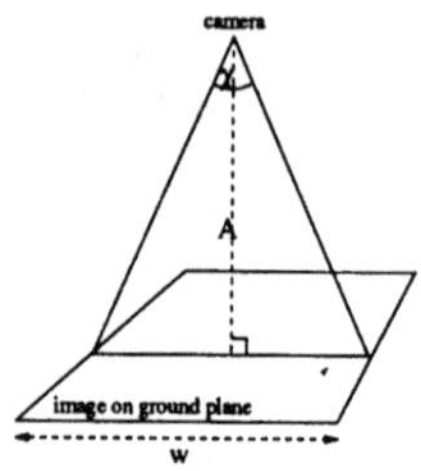


**Fig. 3.** The altitude calculation

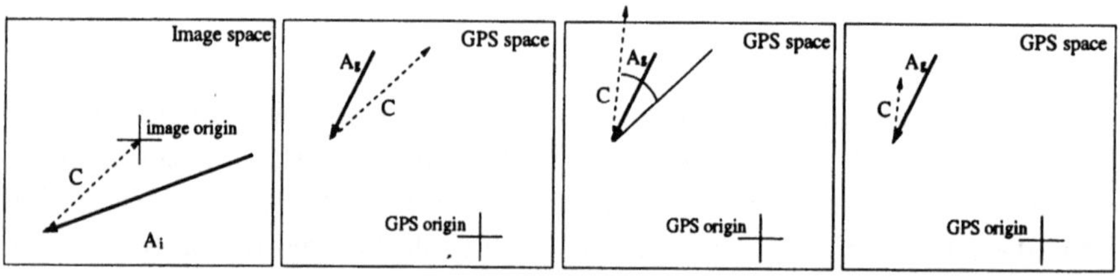


**Fig. 4.** Mapping the origin of image space into GPS space by translating (left) rotating (middle) and scaling (right) vector $C$.

### 3.3 Altitude

The scaling between image and GPS systems is the ratio of the lengths of the vectors in each coordinate frame $S = \frac{|\boldsymbol{A_i}|}{|\boldsymbol{A_g}|}$.

Once the scaling is found, given the camera aperture angle $\alpha$ of the camera and the width $W$ of the image in pixels, the altitude estimate $a$ can be obtained: $a = S\frac{\frac{W}{2}}{\tan\frac{\alpha}{2}}$ (Figure 3).

Now $a$ is the height above the ground in meters. If instead we want to know the absolute altitude above sea level, we can add $a$ to the GPS altitude measured by the UGV.

### 3.4 Position

The MAV's (x,y) position is assumed to be at the center of the image i.e. at the origin. To find the GPS location of the MAV we find the GPS location of the image origin. To do this we construct the vector $\boldsymbol{C}$ from the position of the current image sample to the image origin (Figure 4 left) and transform it into GPS space. First we translate $\boldsymbol{C}$ to the position of the current GPS sample $(x_g, y_g)$ (Figure 4 middle left), then rotate it by $\theta$ (Figure 4 middle right), then scale by $S$ to complete the coordinate transformation (Figure 4 right). The transformed $\boldsymbol{C}$ now points to the corresponding GPS location of the image origin, giving us the MAV's position estimate $(x_u, y_u)$.

$$\begin{pmatrix} x_u \\ y_u \end{pmatrix} = S\left[\begin{pmatrix} \cos\theta & -\sin\theta \\ \sin\theta & \cos\theta \end{pmatrix}\begin{pmatrix} x_g - x_i \\ y_g - y_i \end{pmatrix}\right]$$

### 3.5 Target localization

With the parameters of the image-to-GPS coordinate transform completely known, any pixel in the image can be mapped to GPS space. Thus we can estimate the position of any object tracked in the image. This has application for aerial mapping and reconnaissance, friend/foe identification and target interception.

We demonstrate this in the next Section by estimating the GPS location of a 'foe' robot placed arbitrarily in the MAV's field of view.

### 3.6 Accuracy constraints

The accuracy of the MAV and target localization estimates are limited by four main factors. Of these, the first two are directly related to the assumptions described above; the last two are due to limitations in the accuracy of the sensors.

1. **Camera movement:** the method requires that the two coordinate frames to not move relative to one another in the time interval between the two samples. In particular, the method is sensitive to rotation of the coordinate frames. However, camera movement can be compensated for if an estimate of the movement is available. This estimate in turn is subject to inaccuracy, but very accurate, very small inertial sensors are available that might afford movement compensation.
2. **Camera misalignment:** any displacement of the camera from vertical will induce error. This also can be compensated for if measurements of the offset angles are available.
3. **Image sampling resolution:** the resolution of the images puts a lower bound on the attainable accuracy of the system, even assuming a perfect object tracking system. For a MAV at high altitudes a ground vehicle may have to move a long way to produce a good image vector.
4. **GPS measurement error:** over small distances, the effect of GPS measurement error is significant. The maximum accuracy attainable is a function of the measurement error and the baseline, the length of the vector between the two samples. To achieve good accuracy, the length of a GPS vector must be significantly longer than the variance in the error of its measured end points.

## 4 Demonstration

The localization method described above is attractively simple, but has some fairly strong constraints which could make it impractical. In order to assess its utility for real world applications we implemented the system on our real MAV and UGV systems. This section describes the implementation and compares the localization estimates generated by the method with those obtained from an independent combined GPS/IMU sensor.

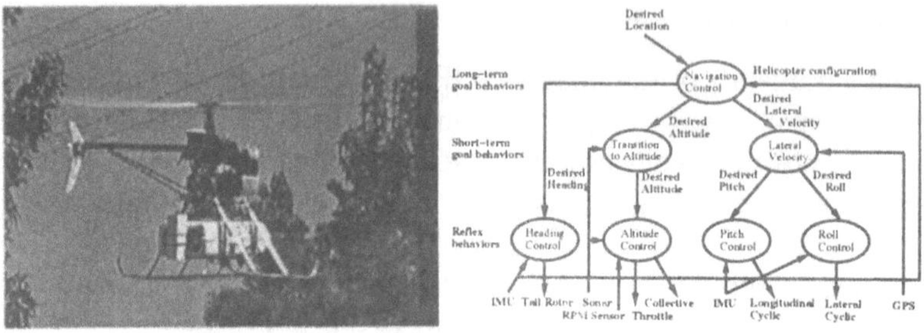


**Fig. 5.** The AVATAR Unmanned Aerial Vehicle (left) and its control system (right)

## 4.1 MAV platform

The USC AVATAR (Autonomous Vehicle Aerial Tracking And Reconnaissance) MAV is a gas-powered model helicopter fitted with a PC104 computer and custom control electronics (Figure 5 left) [7,8]. It has been developed through three generations over the last 9 years in our lab. It carries a high-quality Inertial Measurement Unit (IMU), a Novatel RT20 GPS receiver/decoder, an engine RPM sensor and a color video camera. Communication is via 2.4GHz wireless Ethernet and 2.3GHz wireless video.

AVATAR can autonomously servo to GPS locations and orient itself to GPS headings. The AVATAR control system is implemented using a hierarchical behavior-based control system architecture [9]. The behavior-based control approach partitions the control problem into a set of loosely coupled computing modules called 'behaviors'. Each behavior is responsible for a specific task and they act in parallel to achieve overall robot goals. For example a low-level behaviour controls the vehicle's pitch angle by changing the tilt of the rotor head; a higher-level behaviour controls the forward speed by generating desired pitch angles; a top-level behavior controls the vehicles position by generating desired forward and lateral speeds. A schematic of the controller architecture is shown in Figure 5 (right).

In this experiment AVATAR was required to hover and maintain the same 3D location and heading as accurately as possible. At the time of the experiments the altitude control loop had not been closed, so our pilot adjusted throttle and collective manually to maintain as constant an altitude as possible.

## 4.2 UGV platform

Our UGV platform is an ActivMedia Pioneer AT robot, augmented with a PC104 stack as shown in Figure 6 (left). It has five forward and two side facing sonars, a compass and wireless Ethernet. It carries a Novatel GPS receiver/decoder which is older and less accurate than that carried by AVATAR.

**Fig. 6.** The Pioneer AT robot UGV (left); an example image from the helicopter camera, showing two Pioneer AT UGVs correctly identified (right)

In this experiment the UGV moves over the field at a constant speed and heading to establish the GPS and image vectors. The Pioneer broadcasts its GPS location on the local network's broadcast channel at 5Hz. The AVATAR monitors these messages to obtain the GPS vector.

### 4.3 Object tracking

A simple blob tracking algorithm is used to track the position of high contrast regions in the image. The experiments were carried out over an open grassy field so the black and white UGVs stand out clearly from the background. The images are sampled from the MAV's wireless video stream by a low-cost framegrabber on a Pentium II workstation. An update rate of 25Hz was achieved while reliably tracking two objects. Figure 6 (right) shows an example frame from the MAV video camera with the locations of the two UGVs correctly tracked by the software.

### 4.4 Procedure

The AVATAR was used to represent a modestly capable MAV. It can hold its position for a few seconds, capture video images of the ground below it, receive data from a friendly ground vehicle, and perform some computation. The localization geometry requires relatively little computation; the object tracker is more greedy, but several systems exist that perform this task in hardware.

AVATAR's onboard GPS/IMU systems were used to maintain the hover and their position and pose measurements logged. These logs were used as the ground truth to compare with the estimates produced by the geometric method.

AVATAR was commanded to hover above the center of our test field. Two Pioneer UGVs were placed in arbitrary start positions below it. Once we determined that the UGVs were being tracked by the vision system, the initial image positions were recorded and the trial began. The friend UGV then moves approximately 10m to the west and stops for a few seconds, then

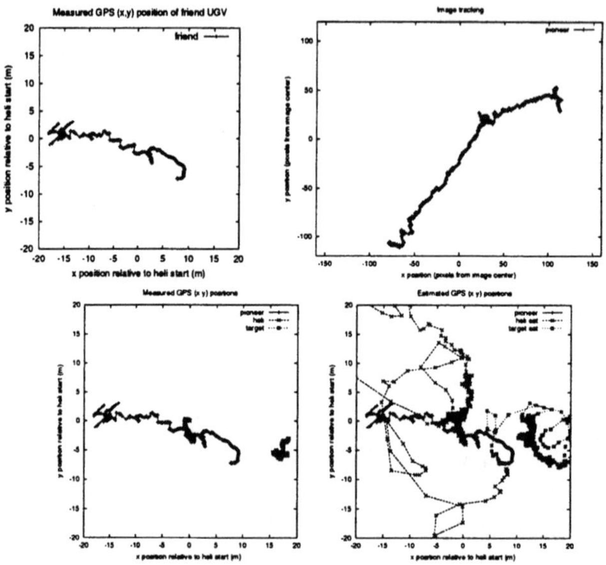


**Fig. 7.** Top: Inputs to the estimator: Friend UGV GPS path (left) and tracked image positions of friend and foe (right). Bottom: Helicopter and foe locations recorded by on-board GPS (left) and estimated (right).

moves west another 15m, by which time it has disappeared from the camera's field of view. Each time AVATAR receives a GPS location from the friend it generates an $(x, y, z, \theta)$ localization estimate for itself and an $(x, y)$ estimate for the foe UGV based on the vectors from the initial to current positions in GPS and image space. Data were recorded for 80 seconds.

### 4.5 Results

Figure 7 shows the inputs to the estimator. The GPS plot (left) shows the locations reported by the friend's GPS as it traces a path from $(8, -7)$ to $(-17, 0)$. The scale is in meters with the origin at the initial helicopter position. The image plot (right) shows the friend's path through the helicopter camera image as detected by the tracker.

Figure 7 (bottom left) shows the measured GPS positions of the helicopter and foe UGV. The helicopter drifts slightly around its initial position and the foe UGV is stationary at approximately $(17, -5)$ but the apparent variation in position shows the varying error in its GPS measurement.

The estimated positions of the helicopter and foe are shown in Figure 7 (bottom right). Comparing this plot with the previous one, both traces appear to converge reasonably close to the corresponding measured points. Figure 8 (left) shows the distance error in helicopter $(x, y)$ localization over time. The estimate starts out poor at $t = 0s$, but quickly begins to approximate the measured value.

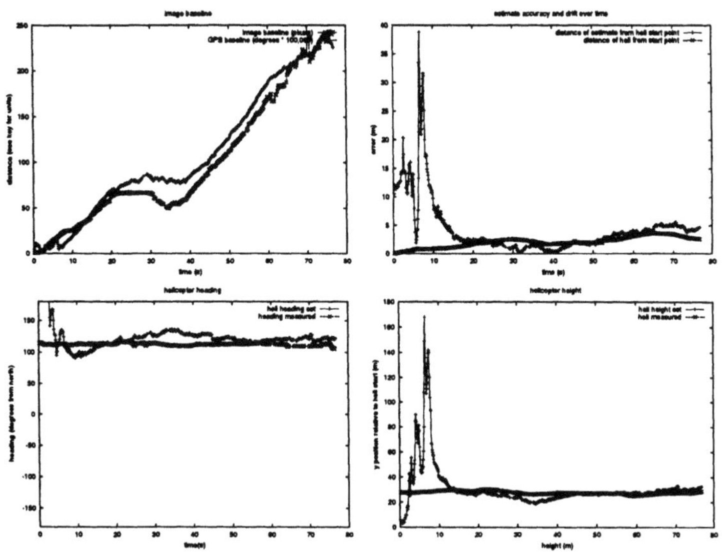

**Fig. 8.** Top: MAV drift over time (left) and MAV localization error (right). Bottom: Measured and estimated MAV heading (left) and altitude (right).

The length of the GPS and image baselines are plotted in Figure 8 (top left). The slope of these curves is approximately constant, indicating a constant translation speed for the UGV, except between $t = 20s$ and $t = 40s$, when the UGV was stationary. We would expect the estimate to improve as a function of the GPS baseline. However in the error plot Figure 8 (top right) we see that the error reaches a minimum at around $t = 40s$, then increases slowly over time. This is due to the lateral drift in the helicopter's position, violating assumption 3 in Section 3.1. The amount of drift is also plotted. It can be seen that the drift is of a similar order to the error it induces. We conclude that a small lateral drift in MAV position may be tolerable in some applications, though we suspect that this may not be true of a change in heading.

Figure 8 (bottom left) shows the heading of the helicopter as measured by its onboard GPS/IMU hardware and that estimated by our method. Again the error is large for the first few seconds but decreases rapidly; from t=10s until the end of the trial the estimate remains within 10% of the measured value. Note that the measured helicopter heading varied very little, allowing generation of a good estimate.

Figure 8 (bottom right) shows the helicopter's measured and estimated height above the ground. Again the estimate is initially poor, but begins to approximate the measured value after a few seconds. We omit error plots for heading and altitude for lack of space, but as can be seen from the estimate plots, the errors do not increase over time. We conclude that these estimates are also rather insensitive to the lateral drift in helicopter position.

## 5 Conclusions and further work

We have demonstrated a simple geometric method that approximately localizes an aerial robot using minimal sensing. We suggest it may be suitable for use in micro air vehicles, as it exploits general-purpose sensors that are likely to be carried already and requires only modest computation. It also allows multiple MAVs to share a single GPS receiver carried by a cooperating ground robot. We aim to further evaluate this method and to increase the interaction between air and ground vehicles. In particular we aim to direct the friendly ground robot to intercept the foe robot using the estimate calculated by the MAV. The localization accuracy observed in these experiments suggests that this is feasible.

## Acknowledgements

Thanks to Kale Harbick, Gerry Bohne, the AVATAR team, Brian Gerkey and Goksel Dedeoglu. This work is supported by DARPA grant DABT63-99-1-0015, contract F04701-97-C-0021 and contract DAAE07-98-C-L028.

## References

1. O. Amidi, T. Kanade, and K. Fujita. A visual odometer for autonomous helicopter flight. In *Proceedings of IAS-5*, 1998.
2. D. Y. Maharaj. *The Application of Nonlinear Control Theory to Robust Helicopter Flight Control.* PhD thesis, Department of Aeronautics, Imperial College of Science, Technology, and Medicine, 1994.
3. B. S. Kim. *Nonlinear Flight Control Using Neural Networks.* PhD thesis, Department of Aerospace Engineering, Georgia Institute of Technology, 1993.
4. T. J. Koo and S. Sastry. Output tracking control design of a helicopter model based on approximate linearization. In *Proceedings of IEEE Conference on Decision and Control*, Orlando, Florida, 1998.
5. C. Phillips, C. L. Karr, and G. Walker. Helicopter flight control with fuzzy-logic and genetic algorithms. *Engineering Applications of Artificial Intelligence*, 9(2):175–184, April 1996.
6. M. Sugeno, M. F. Griffin, and A. Bastian. Fuzzy hierarchical control of an unmanned helicopter. In *17th IFSA World Congress*, pages 179–182, 1993.
7. J. F. Montgomery, A. H. Fagg, and G. A. Bekey. The USC AFV-I: A behavior-based entry in the 1994 International Aerial Robotics Competition. *IEEE Expert*, 10(2):16–22, April 1995.
8. J. F. Montgomery and G. A. Bekey. Learning helicopter control through 'teaching by showing'. In *1998 IEEE Conference on Decision and Control*, December 1998.
9. R. A. Brooks. A robust layered control system for a mobile robot. *IEEE Journal Robotics and Automation*, 2(1):14–23, March 1986.

# Dynamic Robot-Target Assignment – Dependence of Recovering from Breakdowns on the Speed of the Selection Process

Tomoyuki Kaga[1], Jens Starke[2], Péter Molnár[3], Michael Schanz[4], and Toshio Fukuda[5]

[1] Graduate School of Engineering, Nagoya University
[2] Institute of Applied Mathematics, University of Heidelberg
[3] Center for Theoretical Studies of Physical Systems, Clark Atlanta University
[4] Institute of Parallel and Distributed High Performance Systems, University of Stuttgart
[5] Center for Cooperative Research in Advanced Science and Technology, Nagoya University

**Abstract.** The self-organized and fault tolerant behavior of a novel control method for the dynamic assignment of robots to targets using an approach proposed by Starke and Molnár is investigated in detail. Concerning the robot-target assignments the method shows an excellent error resistivity and robustness by using only the local information of each robot. Experimental results verify the dynamic assignment of the mobile robots to the targets and the capability to cope with sudden changes like a breakdown of one of the robots. The dependence of the assignment on the speed of the target-selection dynamics is shown by both experiments and numerical simulations. The results suggest the existence of an optimal value for the speed of the target-selection dynamics.

## 1 Introduction

Recently, a production with high varieties has been required which can be realized by flexible manufacturing systems (FMS). In systems based on conventional centralized control methods, it is difficult to cope with errors which usually occur in proportion to the number of changing conditions. Therefore, a realization of decentralized and error resistant flexible production systems is required. As example, distributed autonomous robotic systems are noticed as systems which allow fault tolerance and flexibility [2], [3], [4].

Many problems have been still left before being able to apply distributed autonomous robotic systems to actual production fields, and among the problems, this paper addresses the assignment of mobile robots to some targets which is a general problem in many production scenarios. For example in transportation systems, automatic guided vehicles (AGV) [9] have to be assigned to loading and delivering stations.

Though it is possible that solutions of assignment problems are obtained by using standard techniques of combinatorial optimization, some breakdowns or interruptions by tasks with higher priority like engine breakdowns or necessary battery charging, lower the flexibility of the systems, since replanning is required on all such occasions. To cope with this problem a self-organized control based on [10], [11] and suggested in [7] is used in the following. This approach is motivated by pattern formation in physical, chemical and biological systems [5], [6].

In the present paper, the experimental results for three mobile robots show that the robots can be assigned dynamically. The algorithm is based only on local target recognition using vision sensors and can cope with sudden breakdowns of robotic units. Furthermore, the experiments and numerical simulations suggest the existence of an optimal evolution speed of the decision dynamics for selecting the targets. The criterion for the optimality is chosen as the necessary time for the robots to reach the targets.

## 2 Dynamic Assignment using Self-organized Control

In the present work, the well-known assignment problem [8] is extended to a dynamic assignment problem to formalize a manufacturing process. This is defined as follows. In a planar working area, mobile robots with e.g. manipulators or welding machines have to perform tasks at several targets (work sites) which are spatially distributed. Each target has to be served by one and only one robotic unit, and each robotic unit has to be assigned to one and only one target. This assignment has to be done such that the sum of the covered distance of movements (or working costs) becomes minimal.

Figure 1 indicates the used experimental environment which is an example for the above description, where three robots have to be assigned to three targets represented as cones. Furthermore, collisions with obstacles represented as white boxes have to be avoided.

The two-index assignment problem, known in graph theory as weighted bipartite matching problem, can be solved optimal by the Hungarian method [8]. However, this method cannot be used for controlling robots including error resistivity. In contrast to that, the self-organizing control method presented here can cope with errors, since the robotic units choose their targets dynamically through a competition process among the robots.

### 2.1 Dynamic Selection of Targets

As already mentioned, in the used self-organized approach, the final assignment of robots to targets emerges by the dynamic selection of targets by each robotic unit. The normalized vector $\mathbf{e}_i^0(t) \in \mathbf{R}^2$ denotes the destination direction of robot $i$ at time $t$ and is defined by

$$\mathbf{e}_i^0(t) = \mathbf{\Phi}_{\gamma\delta}\left(\sum_j \xi_{ij}(t)\mathbf{\Phi}_{\gamma'\delta'}\left(\mathbf{g}_j - \mathbf{r}_i\right)\right), \tag{1}$$

where the function $\mathbf{\Phi}_{\gamma\delta}(\mathbf{x}) = \frac{1}{\|\mathbf{x}\|+1/(\gamma\|\mathbf{x}\|+\delta)} \cdot \mathbf{x}$ with $\gamma, \delta > 0$ is a kind of normalization operator which can avoid the singularity at $\mathbf{x} = \mathbf{0}$ and $\mathbf{r}_i$ and

**Fig. 1.** Experimental environment composed of robotic units, targets (cones) and obstacles (white boxes).

$\mathbf{g}_j$ are positions of the robotic unit $i$ and the target $j$ respectively. In this definition the destination direction $\mathbf{e}_i^0(t)$ represents a kind of time-dependent linear combination of all normalized direction vectors $(\mathbf{g}_j - \mathbf{r}_i)$ of the robotic unit $i$ to the targets $j$. Hereby $\xi_{ij}(t)$ works as a preference coefficient of robotic unit $i$ for target $j$. In order to guarantee that a target is served by one and only one robotic unit, a competition process among the preference coefficients $\xi_{ij}(t)$ governed by coupled selection equations [10] is introduced by

$$\frac{d}{dt}\xi_{ij} = \kappa\xi_{ij}\left(1 - \xi_{ij}^2 - \beta\sum_{i'\neq i}\xi_{i'j}^2 - \beta\sum_{j'\neq j}\xi_{ij'}^2\right) \tag{2}$$

with $\beta > 1/2$. Hereby, non-negative initial values of eq. (2) ensure for square matrices $(\xi_{ij})$ that the asymptotic dynamics always converge to stable fixed points which correspond to permutation matrices (proved in [10]). These coupled selection equations work similar like standard selection equations which are used to describe pattern formation in self-organization processes like the formation of rolls or hexagonal patterns in the so-called Bénard problem.

To understand the optimization heuristics behind the coupled selection equations, they can be considered as a *winner takes all* strategy, which selects the element with the largest initial value. In order to minimize the moving costs of the robotic units, the initial values are therefore defined as

$$\xi_{ij}(0) = 1 - \frac{\|\mathbf{r}_i(0) - \mathbf{g}_j\|}{\max\limits_{i',j'}(\|\mathbf{r}_{i'}(0) - \mathbf{g}_{j'}\|)}. \tag{3}$$

### 2.2 Movement of Robots and Collision Avoidance

The robotic units move according to their destination direction $\mathbf{e}_i^0$ which is governed by the coupled selection equation (2). On the way to their destination, the robotic units have to avoid collisions with obstacles and other units. For this purpose, a behavioral force model [7] based on specific repulsive forces is used for motion control of the units in the experiments.

The dynamics of the velocity $\mathbf{v}_i(t) \in \mathbf{R}^2$ of the robotic unit $i$ is defined by the equation of motion

$$\frac{d}{dt}\mathbf{v}_i(t) = \frac{1}{\tau}\left(v_i^0\mathbf{e}_i^0(t) - \mathbf{v}_i(t)\right) + \alpha\sum_{k=1}^{n^s}\hat{\mathbf{f}}_{ik}(\mathbf{d}_{ik}), \tag{4}$$

where $\tau$ is a relaxation parameter which describes how fast the velocity $\mathbf{v}_i(t)$ tends to the default value $v_i^0\mathbf{e}_i^0(t)$, $\alpha$ is a scaling factor, $\mathbf{d}_{ik}$ is a difference vector from the sensor $k$ of the robotic unit $i$ to the boundary of the obstacle sensed by this sensor as shown in Fig. 2, and $n^s$ is the number of installed distance sensors. The repulsive forces are defined by

$$\hat{\mathbf{f}}_{ik}(\mathbf{d}_{ik}) = \begin{cases} \left(\tan g(\tilde{d}) - g(\tilde{d})\right)\frac{\mathbf{d}_{ik}}{\|\mathbf{d}_{ik}\|} & \text{for} \quad 0 < \tilde{d} \leq \sigma^o \\ 0 & \text{for} \quad \tilde{d} > \sigma^o \end{cases} \tag{5}$$

where $\tilde{d} = \|\mathbf{d}_{ik}\| - d^o/2$, $g(\tilde{d}) = \frac{\pi}{2}\left(\frac{\tilde{d}}{\sigma^o} - 1\right)$, and $d^o$ is used to keep a certain distance to the obstacles. The range of the repulsive force is given by $\sigma^o$.

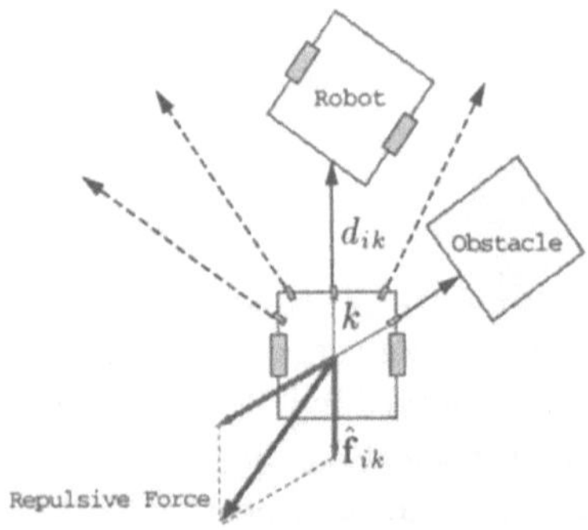


**Fig. 2.** The linear combination of several repulsive forces determines the resulting force.

Since the robotic unit $i$ requires in eq. (4) only the normalized difference vectors $\mathbf{\Phi}_{\gamma\delta}(\mathbf{g}_j - \mathbf{r}_i)$ to the targets, the dynamic assignment can be performed with local vision sensing by each robot in the experiments.

## 3 Experimental Setup

For the experiments the autonomous mobile robots named CEBOT Mark V [1] were used. They have two wheels which are driven independently by DC motors and are equipped with batteries so that they are able to move without any external power supply. In order to demonstrate the decentralized task-solving properties of the proposed control method, the calculation of eq. (2) is distributed among the robotic units. Using an on-board CPU, each robot solves its part of the coupled selection equations and its own equation of motion separately and communicates the result to the other robotic units by using a wireless LAN with a transmission rate of 2 Mbps.

### 3.1 Detection of Targets and Obstacles

For the implementation of the distributed control mechanism the robotic units must be able to recognize directions to given targets and these targets have to be distinguished from each other in order to ensure a unique solution to the underlying selection process. This is achieved by using blue, red and green colored cones as targets and color CCD cameras as detection devices. Fig. 3 shows the system configuration for the target detection and an example image of a CCD camera in which a target is visible. If a target is detected

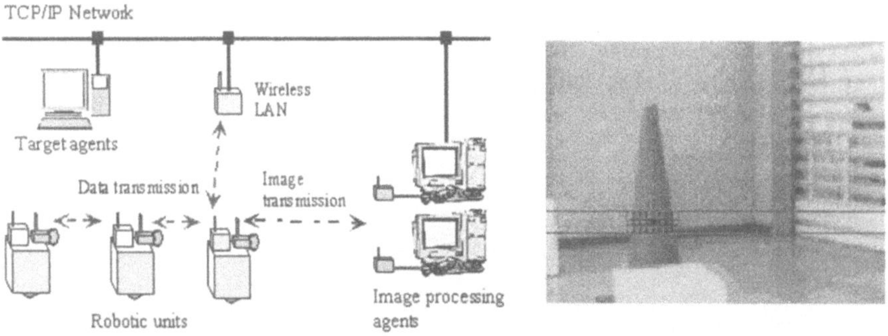


**Fig. 3.** System configuration for the target detection and an image of a CCD camera during the scanning process.

in the image, the image processing agent calculates the horizontal position of the latticed area which is restricted between parallel lines so as to reduce the calculation time. After the image processing it transmits the result to the robot who requested the calculation. With this information, the robotic unit can calculate the relative direction to the detected target. If no target is detected in the current image, the robot rotates about 32 degrees which corresponds to the angle of the camera's view, and starts the scanning process over. This procedure is continued until all targets have been detected.

One robot (R3) has five ultrasonic sensors which can measure the distance up to 2000 mm and the other robotic units (R1, R2) uses eleven infrared sensors with a scanning range of up to 500 mm. Due to the discrete arrangement of the distance sensors the robotic units are not able to detect precise positions of the obstacles. However, the obtained information is sufficient for the implemented collision avoidance algorithm.

### 3.2 Distributed Control

In the distributed environment each robotic unit $i$ maintains its own set of preferences $(\xi_{i1}, \xi_{i2}, \cdots, \xi_{in})$ to the targets. These preferences are updated using the following explicit Euler forward scheme

$$\xi_{ij}(t + \Delta t_{\xi}) = \xi_{ij}(t) + \Delta t_{\xi} \frac{d}{dt} \xi_{ij}(t) \tag{6}$$

for the integration of the coupled selection equations (2) with a fixed time step $\Delta t_\xi$. This integration method is sufficient here, because the numerical integration of the coupled selection equations is not crucial. Since the calculation is performed independently on each robotic unit, the updated preferences have to be shared among them. Therefore we use for each target one separate agent which runs as a computational process on a host computer. If the targets are also equipped with their own computer hardware these agents can also be implemented on the targets directly to achieve a totally distributed architecture. Each agent corresponding to one target collects the calculated preferences from the robots. When the robotic unit $i$ has calculated the values of $\xi_{ij}(t+\Delta t_\xi)$, it reports these preferences to all target agents. After the collection of all preferences, the target agents report these back to the robotic units. The scheme of this distributed control is shown in Fig. 4.

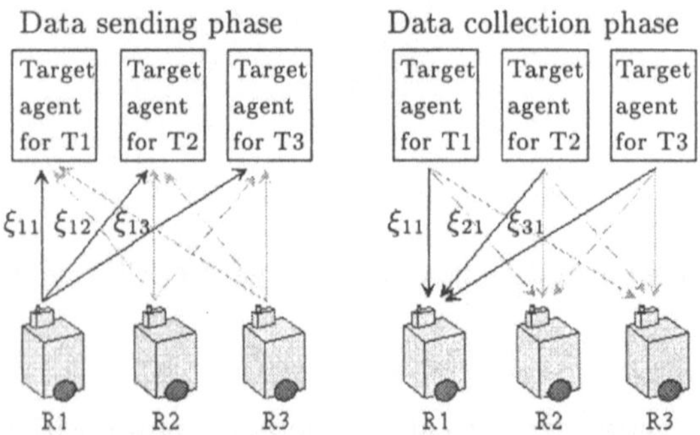


**Fig. 4.** Scheme of the distributed control.

# 4 Experimental Results

To demonstrate the dynamic assignment capability of the proposed distributed control method we used three robots and three targets in our experiments, although the approach works very well in the case of more robotic units and targets.

## 4.1 Dynamic Assignment

The initial arrangement of the robotic units, targets and obstacles is shown in Fig. 5. For this arrangement we obtain from eq. (3) the following initial values for the coupled selection equations (2):

$$(\xi_{ij}(0)) = \begin{pmatrix} 0.525 & 0.0 & 0.390 \\ 0.488 & 0.408 & 0.414 \\ 0.169 & 0.393 & 0.359 \end{pmatrix} . \tag{7}$$

The equation of motion for the velocity of each robotic unit is calculated also by the explicit Euler forward method:

$$\mathbf{v}_i(t+\Delta t_\mathbf{v}) = \mathbf{v}_i(t) + \Delta t_\mathbf{v} \frac{d}{dt}\mathbf{v}_i(t) \tag{8}$$

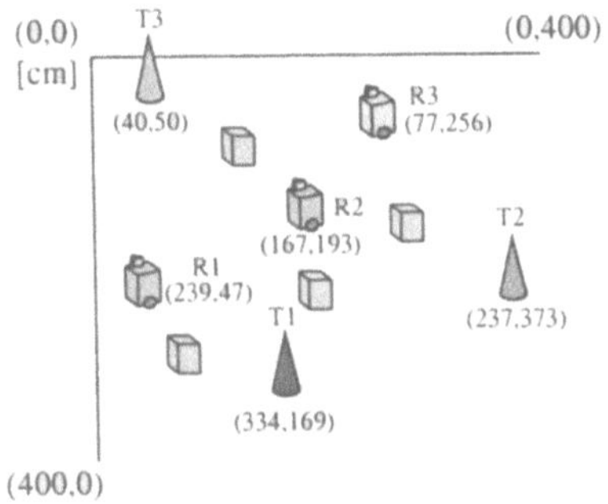


**Fig. 5.** Initial arrangement of robotic units, targets and obstacles. The positions of the obstacles are (306,79), (103,122), (245,190) and (181,272) respectively.

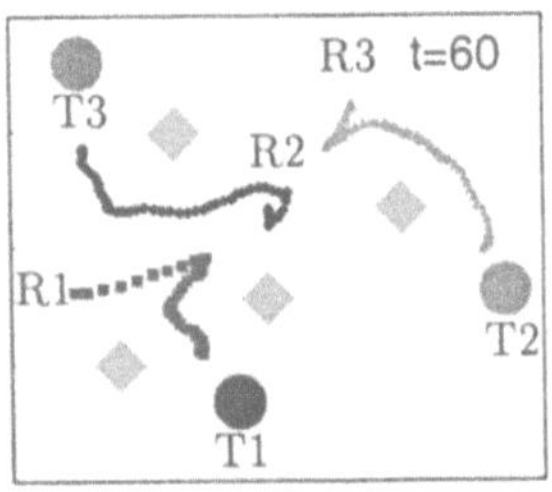


**Fig. 6.** Experimental trajectories of the robotic units in the case of the assignment of three robots to three targets. Finally the robotic unit R1 is assigned to target T1, R2 is assigned to T3 and R3 is assigned to T2.

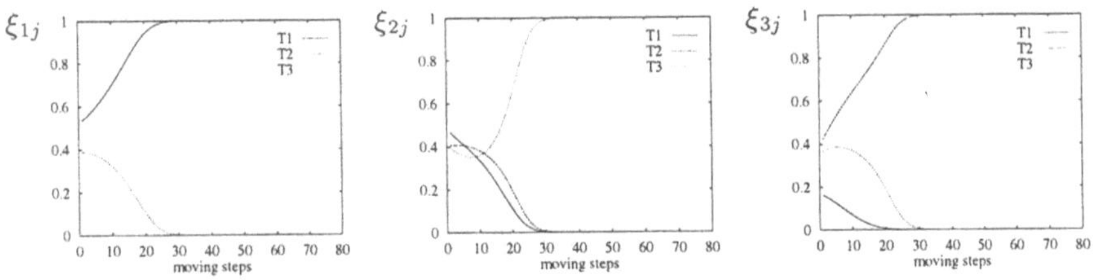


**Fig. 7.** Evolution of the preferences $\xi_{ij}$ in the case of the assignment of three robots to three targets.

Eqs. (6) and (8) are updated every moving step of the corresponding robotic unit and after eight steps a new target scanning process is initiated again which finally results in an update of the destination vectors $\mathbf{e}_i^0$. The calculation of both the coupled selection equations (6) and the equation of motion for the velocity (8) is suspended during the target scanning process. The parameters used for the calculations are shown in Tab. 1.

Fig. 6 shows the trajectories of the robotic units obtained by using the dead-reckoning method. The evolution of the preferences is depicted in Fig. 7 and Fig. 8 shows the experimental realization, i.e., the total assignment process as a sequence of pictures. As one can see the robotic units are dynamically assigned to the targets avoiding hereby collisions with obstacles or other robotic units.

**Table 1.** Parameter settings for the experiments.

| $\Delta t_\xi, \Delta t_v$ [sec] | $\beta$ | $\tau$ | $\sigma^o$ [mm] | $d^o$ [mm] | $v_i^0$ [mm/sec] | $\alpha$ |
|---|---|---|---|---|---|---|
| 0.5 | 1.5 | 2.0 | 500.0 | 100.0 | 200.0 | 60.0 |

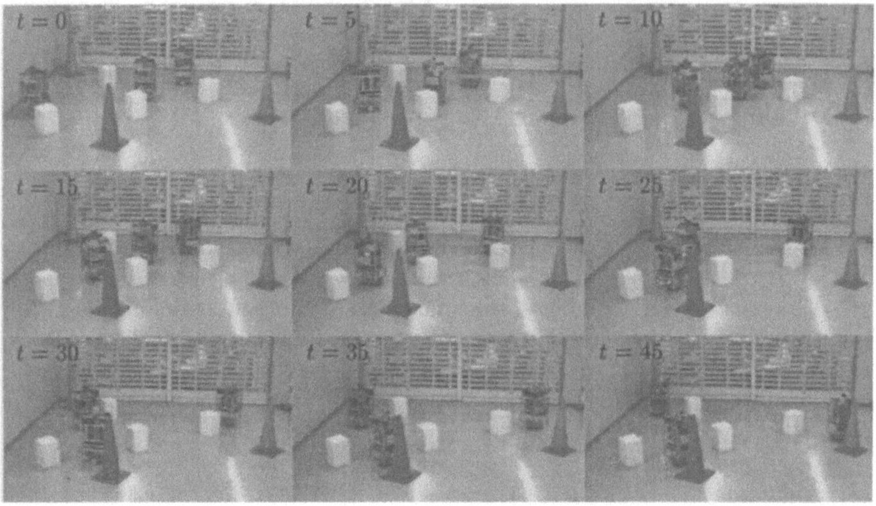


**Fig. 8.** Experimental realization of the assignment of three robots to three targets.

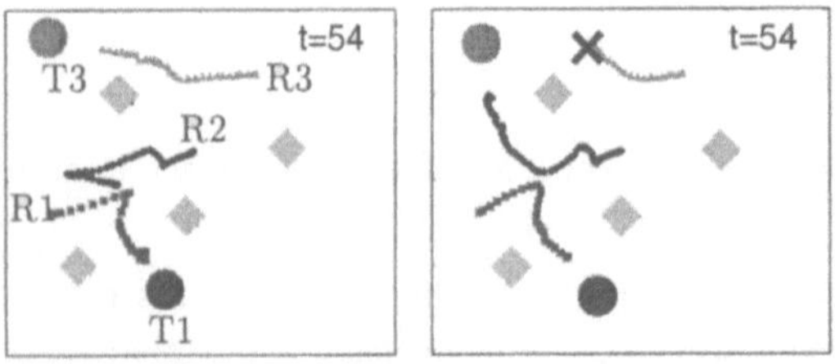


**Fig. 9.** Experimental trajectories of the robotic units in the case of two targets (left: without breakdown, right: with breakdown of robotic unit R3)

### 4.2 Verification of the Error Resistivity

The proposed control method is capable of making use of spare units in case of a breakdown of one of the assigned robotic units. The spare robot can take over the assignment from the broken robot as shown in Fig. 9.

The time scaling parameter $\kappa$ of eq. (2) adjusts the intrinsic time-scale of the coupled selection equations to that of the equation of motion for the velocity (4). If it is guaranteed, that breakdowns of robotic units or interruptions by tasks with high priority can not occur, the assignment task can be solved with integer programming techniques and the robots can head off directly for the targets without any detours. A similar behavior can be achieved by the proposed approach with a large $\kappa$. In this case the assignment is done very fast and hence seems to be static and the spare unit has to start moving from the initial point in the case of breakdowns. To minimize the assignment

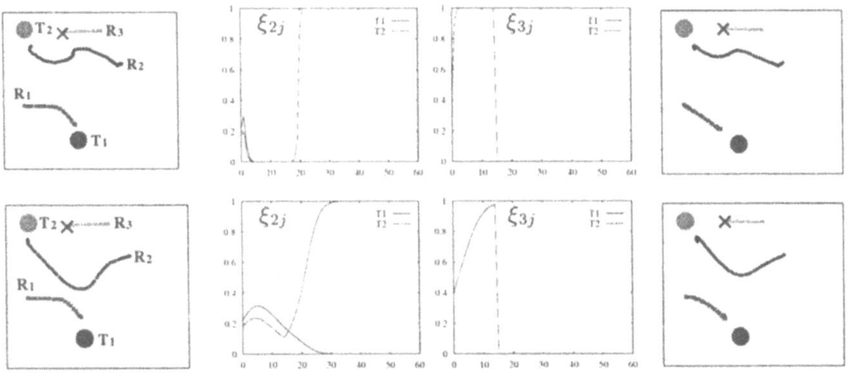


**Fig. 10.** Experimental trajectories (left), the evolution of the preferences of robot R2 and the simulated trajectories (right). The upper row was obtained for $\kappa = 1.0$, the lower row for $\kappa = 0.1$.

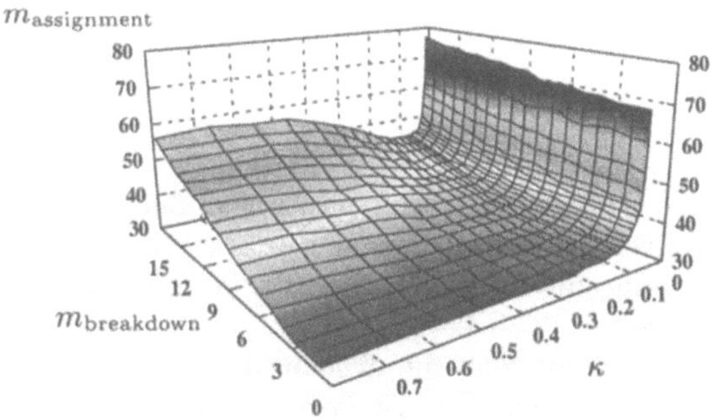


**Fig. 11.** Number $m_{\mathrm{assignment}}$ of steps of robot R2 necessary to reach the target T2 in dependence on the time scaling parameter $\kappa$ and $m_{\mathrm{breakdown}}$, the occurence of the breakdown of robot R3.

time the spare robot should move to a location where it can cope with sudden breakdowns immediately. This can be achieved by an appropriate chosen value of the parameter $\kappa$.

Fig. 10 shows results of the experiments and the numerical simulations with $\kappa = 1.0$ and 0.1 respectively. The robot R3 stops moving after 15 steps, setting hereby its preferences $\xi_{3j}$ to 0 which indicates the breakdown. Obstacles are removed in this experiment in order to clarify the relation between the preferences and the movement of the robotic units. Without breakdown the robot R1 moves to the target T1 and robot R3 moves to target T2 and the preferences $\xi_{21}$ and $\xi_{22}$ of the spare robot R2, converge immediately to 0 and this robot stops moving. After the breakdown of the robot R3, the robot R2 starts moving again and reaches finally the target T2. In the case of $\kappa = 1.0$, robot R2 reaches the target after 63 steps whereas in the case of $\kappa = 0.1$ it needs only 48 steps, since the speed of the selection dynamics is more appropriate (compare Fig. 10).

Figure 11 shows the number of steps of robot R2 which are necessary to reach the target T2 in dependence on the time scaling parameter $\kappa$ on the one hand and the occurrence of the breakdown of robot R3 on the other hand. As one can see this number of steps is minimal at the value $\kappa = 0.1$.

## 5 Conclusion

A novel control method using an approach based on self-organization principles is proposed and real-world experiments demonstrate its effectiveness and fault tolerance. Autonomous mobile robots were assigned to targets dynamically and can cope with sudden breakdowns of some of the robotic units. Furthermore, flexibility against errors was analyzed regarding the parameter which decides the evolution speed of the preference coefficients. The proposed control method ensures the unique assignment of the robotic units to the targets by using only local sensor information. However a global positioning system will enable the robots to reach also targets which are far away and which can't be detected by the local sensors. Furthermore, other sensing devices such as laser scanners or omni-directional cameras will improve detection quality and reduce the time needed for the target detection.

As a future work, it is planned to investigate also time dependent assignment processes. This is necessary to cope with more realistic problems for instance when new targets appear over time or when the targets themselves are mobile.

## References


1. A. Cai: "Behavior Decision and Distributed Sensing on Cellular Robotic System", Ph.D. thesis, Nagoya University, 1983, in Japanese
2. T. Lueth, R. Dillmann, P. Dario and H. Wörn: "Distributed Autonomous Robotic System 3", Springer, 1998
3. T. Fukuda and S. Nakagawa : "Approach to the dynamically reconfigurable robotic system", Journal of Intelligent and Robotic Systems, pp. 55–72, 1988 pp. 239–248, 1991
4. T. Fukuda and T. Ueyama: "Cellular Robotics and Micro Robotic Systems", World Scientific Series in Robotics and Automated Systems, World Scientific, 1994
5. H. Haken: "Synergetics, An Introduction", Springer Series in Synergetics, Springer-Verlag, Heidelberg, Berlin, New York, 1983
6. H. Haken: "Advanced Synergetics", Springer Series in Synergetics, Springer-Verlag, Heidelberg, Berlin, New York, 1983
7. P. Molnár and J. Starke: "Control of distributed autonomous robotic systems using principles of pattern formation in nature and pedestrian behaviour", submitted, 1999
8. C. Papadimitriou and K. Steiglitz: "Combinatorial Optimization - Algorithms and Complexity", Prentice-Hall.Englewood Cliffs, New Jersey, 1982
9. P. Pu and J. Hughes: "Integrative AGV schedules in a scheduling system for a flexible manufacturing environment", in Proceedings of the IEEE International Conference on Robotics and Automation, pp. 3149–3154, 1994
10. J. Starke: "Kombinatorische Optimierung auf der Basis gekoppelter Selektionsgleichungen", Ph.D. thesis, Universität Stuttgart, Verlag Shaker, Aachen, 1997
11. J. Starke, M. Schanz and H. Haken: "Self-Organized Behaviour of Distributed Autonomous Mobile Robotic Systems by Pattern Formation Principles", Distributed Autonomous Systems 3, Springer, pp. 90–100, 1998 and T. Fukuda:

# Fuzzy Control of Distance and Motion Tracking Applied to Cooperative Autonomous Robots

M. R. Hitchings. Lj. B. Vlacic, Z. O'Sullivan

Intelligent Control Systems Laboratory (ICSL), Griffith University
Queensland, Australia 4111.
M.Hitchings@me.gu.edu.au, L.Vlacic@me.gu.edu.au, Z.Osullivan@me.gu.edu.au

**Abstract:** An approach to distance and motion tracking control investigated from the viewpoint of the cooperative behaviour of autonomous robots is presented in this paper. A fuzzy control-based solution using laser alignment and ultrasonic distance ranging techniques as inputs to a two input, two output, non-linear controller is the major focus of this paper. The solution uses a fuzzy-controller to control the orientation and speed of a differential drive robot to maintain distance and motion tracking control with respect to a robot in front of it. The fuzzy controller has been fully implemented on an 8-bit microcontroller, using S-type, Z-type and Lambda-type membership functions, MAX/MIN inference and CoM defuzzification methods.

## 1 Introduction

Much of the research effort by others [1] [2] [3] [4] [5] has focused on distance control only - without the use of automatic motion tracking control to explore cooperation among robots. The solution presented in this paper combines distance control with leader motion tracking control to form one complete solution requiring no intervention from humans or supervisory systems. By combining both distance and motion tracking control, the solution is targeted towards *totally autonomous robot* applications.

The *Distance and Tracking Control (DTC)* problem, from the viewpoint of the cooperative behaviour of autonomous robots, is the cooperative behaviour of two successive robots travelling in the same direction in the same environment. This behaviour requires one robot (designated the leader) to move autonomously around it's environment with other robots (designated followers) maintaining a coincident travel path and desired longitudinal distance with respect to the leader. The followers must perform DTC without any intervention from external supervisory systems or humans.

Other work on both distance and tracking control [6] uses path planning and odometry for the follower to directly trace the path of the leader. However, this was not the goal of the solution presented in this paper. The path planning method is also not essentially a real-time, event driven solution in that the follower does not immediately adapt to changes in the leader's trajectory. The method detailed in this paper is based on real-time distance and motion tracking control, whereby the follower tracks the leader's motion - thus performing the same manoeuvres as the leader while maintaining a set longitudinal distance to it. It is the authors opinion that an exact traced-out follower path with respect to the leader is not essential for many applications.

An efficient navigation strategy for a group of multiple ground vehicles, such as Autonomous Guided Vehicles (AGVs) in factories, would be to group the vehicles into two classes, *leaders* and *followers* [6]. The leader robots may have the intelligence and navigational capabilities to guide a number of followers to a task. The leader takes no part in the task performance, but is used to lead the followers to and from the work-site. The followers perform application specific tasks and do not have the navigational ability of the leader. This approach lowers the overall complexity and cost of service robots. One specific application of this solution could be in the case where two or more robots cooperate together to carry a heavy or large sized load. An example of this application is robots cooperating to carry a long pipe [7]. The pipe is of a size that precludes it from being carried by one robot only. In this case the robots work together to pick up and transport the long pipe from one place to another. Only one robot is required to have the navigational capabilities and assumes the role of leader. The other robot/s only require a passive manipulator to support the load as well as perform distance and tracking control such as that presented in this solution.

This fuzzy control-based DTC solution features two inputs, which are the distance and alignment of the follower with respect to the leader. The fuzzy controller asserts two outputs, which are left and right wheel velocities to control the speed and trajectory of a differential drive robot. Each of the input and output fuzzy membership functions has seven terms based around lambda, Z-type and S-type functions. The fuzzy rule base consists of 49 rules and the fuzzy inference stage is based on the MAX/MIN method. A Centre of Maximum (CoM) defuzzification method is used to provide the two crisp valued outputs to the robot motion control. The methods chosen for the fuzzy control of distance and tracking for autonomous robots were selected based on a compromise between their computational complexity and performance characteristics. This compromise was necessary in order to implement the chosen controller structure on pre-existing robots based on 8-bit microcontrollers with limited memory and processing resources.

## 2 Fuzzy Controller

The fuzzy controller on the following robot measures its alignment and distance with respect to the position of the leader and outputs left and right wheel velocities in order to maintain these values within set-point values. The leader emits a laser-based signal behind it to the robot immediately following which uses the position of this signal on an array of photo-diodes to measure it's alignment with respect to that of the leader. The follower measures the distance to the leader by using ultrasonic ranging techniques. Pre-processing of the sensor data occurs prior to its use in the fuzzy controller. This involves the following:

- Normalising the distance sensor reading to the set-point distance. This yields an input universe for the distance variable of [0, 2], where 0 represents no separation distance between the two robots and 2 represents twice the set-point distance between them. These input limits were found experimentally to be a sufficient universe for the distance input, but may easily be extended in software should other applications warrant this.
- Deriving a single alignment result. This involves reading the input laser signal strength from all 39 sensors, determining a threshold value for ambient light effects and normalising each sensor reading to the threshold value. More than one sensor may have a non-zero, normalised reading and therefore a single alignment reading must be resolved from all non-zero inputs using a Centre of Gravity (COG) algorithm.

The DTC fuzzy controller takes inputs from the distance and alignment sensors and fuzzifies these input crisp values. The definitions of the input and output linguistic variables were based on the following conditions:

- Only Z-type, Lambda-type and S-type basis functions be used to define the linguistic variable terms
- The input and output linguistic variables contain at most seven terms which is considered quite sufficient for the application [7]. This maximum number of terms is based on a restriction imposed by the memory capacity of the test-beds.
- The linguistic variables are symmetric. This means that the number of terms in the variables must be odd with one term describing the middle of the upper and lower extremes of the variable input space.

Equally spaced linguistic variable terms were used in the initial definitions, with a simulation stage used to refine their size and shape in order to obtain satisfactory controller performance. The membership functions of the input and output fuzzy variables are as follows:

***Input Linguistic Variables:***

Distance $\in$ {Too_Far, Far, Bit_Far, OK, Bit_Close, Close, Too_Close}

Alignment $\in$ {Far_Left, Left, Bit_Left, Centre, Bit_Right, Right, Far_Right}

***Output Linguistic Variables:***

Left_Wheel_Speed $\in$ {Very_Slow, Slow, Bit_Slow, Medium, Bit_Fast, Fast, Very_Fast }

Right_Wheel_Speed $\in$ {Very_Slow, Slow, Bit_Slow, Medium, Bit_Fast, Fast,

(a)

| ALIGNMENT \ DISTANCE | VERY_FAR | FAR | BIT_FAR | OK | BIT_CLOSE | CLOSE | VERY_CLOSE |
|---|---|---|---|---|---|---|---|
| FAR_LEFT | VF | VF | VF | VF | F | BF | M |
| LEFT | VF | VF | VF | VF | F | BS | BS |
| BIT_LEFT | VF | VF | VF | VF | BF | BS | S |
| CENTRE | VF | VF | VF | M | VS | VS | VS |
| BIT_RIGHT | F | BF | BS | VS | VS | VS | VS |
| RIGHT | BS | BS | S | VS | VS | VS | VS |
| FAR_RIGHT | M | M | M | BS | BS | S | VS |

(b)

| ALIGNMENT \ DISTANCE | VERY_FAR | FAR | BIT_FAR | OK | BIT_CLOSE | CLOSE | VERY_CLOSE |
|---|---|---|---|---|---|---|---|
| FAR_LEFT | M | M | M | BS | BS | S | VS |
| LEFT | BS | BS | S | VS | VS | VS | VS |
| BIT_LEFT | F | BF | BS | VS | VS | VS | VS |
| CENTRE | VF | VF | VF | M | VS | VS | VS |
| BIT_RIGHT | VF | VF | VF | VF | BF | BS | S |
| RIGHT | VF | VF | VF | VF | F | BS | BS |
| FAR_RIGHT | VF | VF | VF | VF | F | BF | M |

**Fig. 1.** Fuzzy rule matrices for each output variable (**a**) Left_Wheel_Speed and (**b**) Right_Wheel_Speed

Very_Fast }

The designed fuzzy rule base consists of 49 equally weighted rules as shown in the rule matrix in figure 1.

The two outputs from the fuzzy controller are used as set-point values by a motion control module, which has two separate PI controllers - one for each wheel velocity control. Each PI wheel velocity controller performs the single task of keeping the velocity parameter at a constant level quite well. However the operating points derived from the overall process of distance and tracking control represent a non-linear, multi-variable control task. The fuzzy controller serves this task.

# 3 Hardware Description

## 3.1 Hardware Test-beds

Hardware test-beds in the form of autonomous mobile robots were utilised in order to implement the fuzzy DTC solution. These test-beds are shown as figure 2.

The test-beds were built for the use of research staff of the Intelligent Control Systems Laboratory to implement algorithms and sensor designs relevant to autonomous mobile robots and vehicles. The robots are capable of full autonomy, being equipped with batteries and microcontroller-based navigation.

**Fig. 2.** Hardware test-beds used in the fuzzy control of distance and tracking solution

## 3.2 Distance and Tracking Controller Module

A hardware module was developed specifically for the fuzzy control of distance and tracking solution and was integrated onto the existing test beds. An 8-bit microcontroller stores the fuzzy controller software and provides an interface to the peripheral hardware.

An alignment sensor array, made up of 39 silicon photodiodes arranged in a linear array on the vertical plane of the DTC hardware module is used to measure the alignment of the follower with respect to the leader. The contribution from any background sources of light are removed by performing the following:

- An A/D converter and memory storage are used to determine the voltage level of each sensor element and store its value
- A threshold value across the entire array is found using an averaging method
- If the contribution from each sensor is below the threshold it is treated as zero
- If it is above the threshold its actual input value is stored.
- The solution uses a Centre of Gravity method to determine the centre of the laser light "spot" on the alignment array by treating the array as continuous rather than a discretised sensor array.

A laser diode module is mounted on the DTC hardware module so that it faces toward the rear of the robot on which it is mounted. The laser light that it emits is utilised by a following robot to perform tracking control based on the laser light incident on its alignment sensor array. The laser light emitted is of a wavelength of 670 nanometres and power of less than five milliwatts. The laser beam width is adjustable using a collimator. For the purposes of DTC, the beam is deliberatelde-focussed to give a larger laser spot incident on the alignment sensor array of the following robot. This is done in order to maintain the spot on the array irrespective of minor variations in the surface on which the robots travel.

Ultrasonic ranging hardware and a time-of-flight measurement technique are used to calculate the distance from the follower to the leader robot. The ultrasonic ranging hardware was not developed specifically for the DTC solution and existed on the test-beds prior to the solution. The ultrasonic ranging hardware is physically attached to a sensor interface module and the distance data measured by this sub-system is obtained using the $I^2C$ data communications bus linking the distance and tracking controller to the sensor interface module.

# 4 Experimental Results and Discussion

The performance of the DTC solution applied to the hardware test-beds was evaluated by experimental measurement of the system's response to various initial errors and disturbances imposed on it. In evaluating the operation of the solution, there are two parameters that require measurement - distance and alignment of the follower with respect to the leader. It is desirable to measure these parameters against their desired (or set point) value over a range of situations. This was performed using two methods as follows:

- The error in the follower's trajectory with respect to that of the leader at any point in time (termed the *alignment*) is measured by tracing out the trajectories of both leader and follower robots. The trajectories of the leader and follower are traced out using a pen-plotter attached to the underside of the robot that contacts the ground and hence traces a line beneath it as it moves along.
- The distance from the leader to the follower is measured at regular time intervals using ultrasonic ranging techniques

Typical results obtained for distance error (figure 3) and alignment error (figure 4) of the follower with respect to the leader as a function of forward distance that the leader travels are shown below. In all cases the leader's trajectory was a straight line, with the experiments repeated for increasing leader speeds. The follower in each case was started at an initial alignment error of 60 millimetres and an initial distance error of 95 millimetres.

Overall, the reliability of the distance and tracking control can be summarised as follows:

- The response of the distance control of the system is mainly affected by the leader speed
- The rate of change of orientation of the leader affects the reliability of the tracking control
- The unevenness of the surface on which the robots traverse affects the tracking control

## 4.1 Straight Line Motion

From the results it can be seen that the response of the distance control is heavily dependent on the speed of the leader. If the actual distance from the follower to the leader lags behind the set-point distance due to some sort of disturbance, the follower takes a longer time to correct the distance error for faster leader speeds when compared to the same conditions for slower leader speeds. Conversely, if the distance from the follower to the leader is closer than the set-point distance, the follower takes a shorter time to correct the distance error for fast leader speeds when compared to the same conditions for slow leader speeds.

The tracking control performance is not largely affected by the speed of the leader. Only the magnitude of the second overshoot increases slightly with increasing leader speed. However, the time to reach the steady state of tracking control is not largely affected. It can be seen that there is typically two overshoots in the alignment error when the follower distance and tracking controller is correcting large alignment errors. These overshoots are due to the dependence of the solution on the differential configuration of the robots. Even though the follower test-bed may be in line with the leader, its orientation with respect to the leader may be in error.

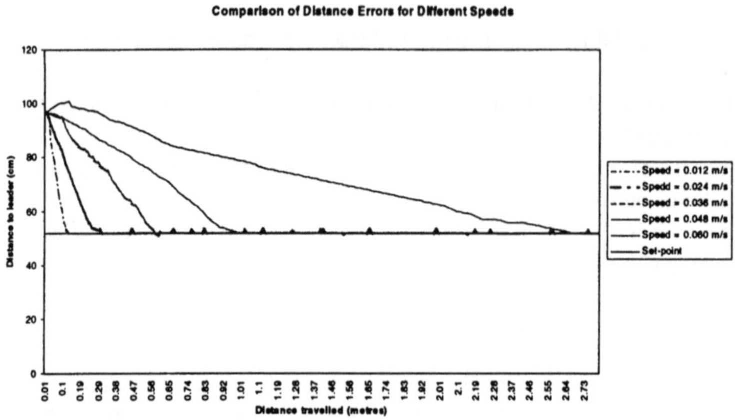


**Fig. 3.** Experimental results achieved for distance control

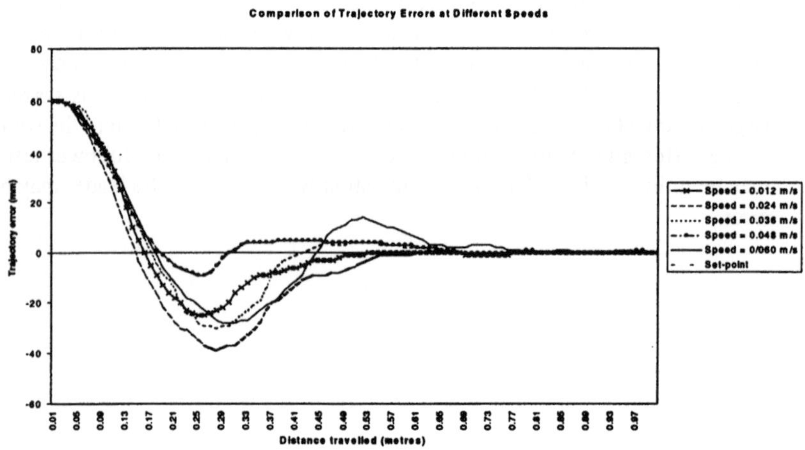


**Fig. 4.** Experimental results achieved for tracking

The laser-based alignment measuring system used in the solution is line-of-sight and hence the orientation of the follower with respect to the leader effects the alignment reading. When the follower is correcting large alignment errors, it typically turns inward towards the leader and hence has a large difference in orientation with respect to the leader when it crosses the zero-error point - leading to overshoots. This problem was reduced somewhat by placing the follower's alignment array forward from its central axis so that readings are slightly dependant on its orientation.

It may also be observed that there are slight and random disturbances in the followers trajectory during steady state motion. The cause of these disturbances was attributed to the motion of the leader test-bed and its effect on the laser signal it emits to the follower. The leader motion was observed to jump slightly due to slight floor and wheel imperfections. While the magnitude of these jumps was very small, they had an observable effect on the position of the laser signal on the followers alignment sensor array. This problem is due to the physical construction of the test-beds and may be overcome using different hardware configurations. Future work, using these test-beds, will address a solution by way of filtering of the alignment input.

## 4.2 Curved Motion

Experiments on the DTC solution were conducted in order to observe its performance during curved leader trajectories. Curved leader trajectories essentially represent a dynamic set-point condition for the alignment parameter. During curved (or any other) motion of the leader, the follower uses line-of-sight methods to measure its distance and alignment with respect to that of the leader. The follower, in performing DTC, will tend to overshoot any curved motion of the leader to some degree as shown in figure 5.

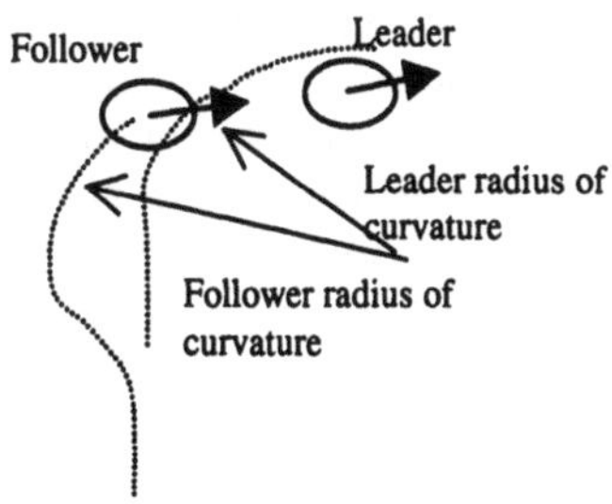


**Fig 5.** Behaviour of the follower in response to leader curved trajectory

By using these methods, the follower cannot exactly mimic the travel path of the leader [6]. Figure 5 shows how the follower's trajectory around the curved section of motion follows a larger radius of curvature than that of the leader. The trajectory traced out by the follower, when viewed *after* both robots have performed the motion, appears in error to that of the leader. However, the distance and alignment errors at any moment in time *during* the motion are within acceptable limits. This phenomenon is due solely to the methods used to measure these parameters, which are based on line-of-sight techniques. In other words, the follower measures its distance and alignment with respect to the leader at any point in time irrespective of the path taken by the leader. It is not the aim of the solution presented in this paper to enable the follower to mimic, in an absolute coordinate sense, the path traced out by the leader exactly, but to track the motion of the leader in a real time sense. Others [6] have specified path tracking as an important goal of their solution and have developed solutions based on odometry and path planning. The follower must obtain exact position data from the leader and plan a path to trace out the leader trajectory. The follower in the solution presented in this paper measures its distance and alignment with respect to the leader's motion and attempts to control any errors in these parameters in a real-time sense. The follower in this case adjusts its orientation and speed to try to achieve a set-point distance and alignment coincident with the leader at any point in time. This does not necessarily mean its path will be traced out exactly as that of the leader. Both these approaches have their merits and limitations depending on the application intended on the solution.

# 5 Conclusions and Future Work

This paper has outlined a novel solution and identified relevant issues relating to DTC investigated from the viewpoint of the cooperative behaviour of autonomous robots. A fuzzy control-based solution using laser alignment and distance ranging techniques as inputs to a two input, two output, non-linear controller is the major result of the research. The fuzzy-controller is used to control the orientation and speed of a differential drive robot to maintain distance and motion tracking control with respect to a robot in front of it.

The solutions presented in this paper were implemented on differential drive test-beds. This configuration uses the speed of the left and right driving wheels to determine both the speed and orientation of the platform. The experiments on the tracking control performance found that this configuration was responsible for overshoots in the alignment parameter. The configuration was also found to be partially responsible for slight disturbances in the steady state motion of the test-beds. Future work on the solution will address these steady state disturbances. Subsequent modifications to the solution may require some filtering of the alignment input in order to remove its sensitivity to slight movements in the laser-based signal emitted by the leader.

As discussed previously, it was found that the reliability of the fuzzy control-based distance and tracking solution for distance errors is dependent on the leader speed. A method of reducing this dependence is for the follower to signal to the leader when there is a large distance error, so that the leader may speed up or slow down as appropriate. This will have the effect of reducing the settling time for large distance errors. Future work should investigate this method.

Perhaps the major limiting factor of the solution presented in this paper relates to the line-of-sight alignment measuring technique and the resulting difference in radii of curvature between the leader and follower trajectories. It was asserted that in many applications this is not necessarily a disadvantage. However, it is accepted that it does preclude the solution's application to environments where larger separation distances are required in lane-type environments. A method identified as likely to overcome this limitation is a radio communications link between the leader and follower.

Overall the fuzzy control-based solution presented in this paper effectively solves the DTC problem. The solution was applied to differential drive hardware test-beds and was tested to verify performance. The solution is a feasible solution to service robotics to enable two or more robots to cooperate together to perform a task. One robot may act as a leader and guide one or more followers to a work-site where they can perform application specific tasks. This approach reduces the cost and complexity of service and manufacturing robotics applications.

## References


1. J.A. Dickerson, H. Kim, B. Kosko. (1997) Fuzzy Control for Platoons of Smart Cars. Chapter from Fuzzy Engineering. Prentice-Hall Publishing, New Jersey U.S.A., 77-209.
2. R. Holve, P. Protzel. (1995) Adaptive Fuzzy Control for Driver Assistance in Car-Following, Proceedings of the 3rd European Congress in Intelligent Techniques and Soft Computing, Aachen, Germany, August 1995, 1149-1153.
3. R. Holve, P. Protzel. (1995) Generating Fuzzy Rules for the Acceleration Control of an adaptive Cruise Control System, Proceedings of the NAFIPS Conference, Berkeley, California U.S.A., June 1995.
4. H. Raza, P. Ioannou. (1996) Vehicle Following Control Design for Automated Highway Systems, IEEE Control Systems, December 1996,43-60.
5. A. Schone, X. Sun. (1995) Different In-Vehicle Control Concepts for the Distance Control of Cars Moving in Convoys, Proceedings of the 2nd International Conference on Road Vehicle Automation, Bolton, UK, September 1995, 50-69.
6. T. Tsumura, H. Okubu, N. Komatsu. (1992) Directly Follow-up and/or Traced Control System of Multiple Ground Vehicles, Proceedings of the 1992 IEEE/RSJ International Conference on Intelligent Robots and Systems, 1769-1776.
7. A. Ming, V. Masek, C. Kanamori, M. Kajitani. (1996) Cooperative Operation of Two Mobile Robots, Proceedings of Distributed Autonomous Robotic Systems 2 Springer-Verlag, 339-349
8. C. Von Altrock. (1995) Fuzzy Logic and Neuro-Fuzzy Applications Explained, Prentice-Hall Publishing, New Jersey

# Broadcast of Local Eligibility for Multi-Target Observation

Barry Brian Werger and Maja J Matarić

USC Robotics Research Lab, Los Angeles, CA 90089-0781

**Abstract.** While various researchers have investigated group behavior of robots which are each controlled in a behavior-based manner, none have yet thoroughly investigated the possibilities of extending the *port-arbitrated behavior* (PAB) paradigm across networks of robots. We present an extension to the well-defined PAB techniques of behavioral interaction which provides standard abstractions for messaging, inhibition, and suppression over IP networks. The Broadcast of Local Eligibility is a general technique built from these abstractions that allows fully-distributed, flexible team coordination. We present a BLE approach to the CMOMMT multi-target observation problem, implemented on a team of physical robots.

## 1 Introduction

Although several lines of research have investigated group behavior of robots which are each controlled in a behavior-based manner [1,2,6,7,11], none have yet thoroughly investigated the possibilities of extending the port-arbitrated behavior [12] paradigm across networks of robots. While it has often been hypothesized that there need be no distinction between inter-robot and inter-behavior communication (e.g., [2]), to our knowledge no previous system has provided standard tools that allow port-based messaging, suppression, and inhibition between behaviors on separate networked robots.

Our intention is to demonstrate that behavior-based systems restricted to well-defined port-arbitrated interactions can scale to higher levels of competence than is generally assumed. Specifically, we show that when the port-arbitration paradigm is extended across networks, the resulting systems are able to dynamically reconfigure themselves in order to efficiently allocate resources in response to task constraints, environmental conditions, and system resources. We introduce the Broadcast of Local Eligibility (BLE) as a general tool for coordination between robots, and then demonstrate its application to the CMOMMT [8] multi-target observation task.

### 1.1 PAB: Port-Arbitrated Behavior-Based Control

In PAB systems, controllers are written in terms of *behaviors*, which are groups of concurrent processes that share a public interface. This interface is composed of *ports*, which are registers that each hold a single data item (e.g., an integer, float, string, or complex data structure).

Ports in different behaviors are linked together by *connections*, which are unidirectional data paths between a *source port* and a *destination port.* A port can have any number of incoming and outgoing connections. When a data item is written to a port, either directly from a process within the behavior or indirectly through a connection, it is generally propagated along all of that port's outgoing connections. We say "generally," because data flow can be modified by connections specified to be *suppressive*, *inhibitory*, or *overriding*, as described in this volume in [12].

It is through these mechanisms of suppression and inhibition that subsumption [2] hierarchies, as well as other forms of arbitration, can be efficiently and intuitively implemented. Since connections are external to the behaviors, behavior code is easily re-usable, and interaction between behaviors can be modified dynamically. The port abstraction enforces a data-driven approach to programming that "grounds" computation in sensor readings and effector actions. The PAB approach allows a clean, uniform interface between system components (behaviors) at all levels that abstracts away many issues of timing and communication; the "black boxes" of behaviors may contain reactive mappings or deliberative planners. While our research focuses on non-deliberative approaches, we believe that PAB interaction between system components can help reduce the complexity of the components themselves, whatever their type.

We developed Ayllu, a C-based language for behavior-based control (described in this issue in [12]), to facilitate implementation of distributed PAB algorithms. Connections between ports (normal, inhibitory, or suppressive) can be made either locally or over IP, and can be specified to broadcast. In addition, Ayllu adds specialized port types, such as *max ports* which filter arriving messages for maximum values. The Broadcast of Local Eligibility technique combines broadcast connections and Max ports to build systems that are arbitrarily scalable.

## 2 Broadcast of Local Eligibility

We now introduce our Broadcast of Local Eligibility (BLE) approach to multi-robot coordination. The BLE mechanism involves a comparison of locally determined eligibility with the best eligibility calculated by a peer behavior on another robot. When a robot's local eligibility is best for some behavior $B_n$ which performs task $T_n$, it inhibits the peer behaviors (that is, behaviors $B_n$) on all other robots, thereby "claiming" task $T_n$. Since this inhibition is an active process, failure of a robot which has claimed a task results in the task being immediately "freed" for potential takeover by another robot.

Since BLE is based on broadcast messages and receiving ports that filter their input for the "best" eligibility, BLE systems are inherently scalable. Up to the limit of communication bandwidth, any number of BLE-enabled

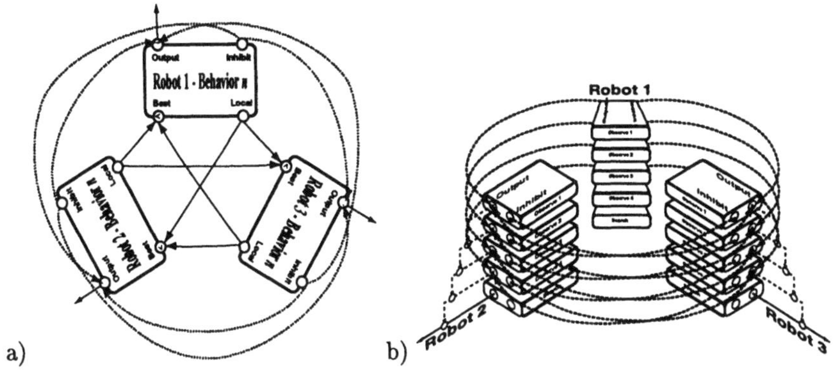


**Fig. 1.** *a) Cross-Inhibition*: A cross-inhibited peer group. *b) Cross-Subsumption*: The structure of a cross-subsumptive system. Some lines are omitted for clarity; each "layer" is connected as a).

robots can be added to a system. As we will also see in Sec. 2.2, BLE allows heterogeneous robots to efficiently allocate themselves to appropriate tasks.

## 2.1 BLE-Enabled Behaviors

BLE action selection requires that each BLE-arbitrated behavior include three ports named *Local*, *Best*, and *Inhibit* (see Fig. 1a). Useful behaviors will generally have additional ports for task-related input and output. We generically refer to the BLE-arbitrated output of a behavior as *Output*, though the actual output may be through any number of ports of arbitrary name. The *Best* port is a *max port*, accepting only values that are larger than its current value.

## 2.2 Cross-Inhibition of Behaviors

Cross-inhibition refers to the process of arbitration between *peer behaviors*, instances of the same BLE behavior on different robots. Given a behavior instance $B_n$ (which performs task $T_n$) on each robot, cross-inhibition results in the selection of at most one robot (the most eligible) to perform $T_n$. There may be multiple groups of cross-inhibiting behaviors active at the same time; below we discuss one manner in which local arbitration between different cross-inhibited behaviors can take place.

As illustrated in Fig. 1a, the *Local* port of each robot's behavior $B_n$ broadcasts a locally-computed eligibility estimate to the *Best* port of each other robot's behavior $B_n$. Each *Best* port maintains the maximum of the eligibility messages it has received in the current cycle. Whichever robot has a local eligibility better than or equal to the *Best* it receives writes to its *Inhibit* port, causing inhibition of behavior $B_n$ in the other robots.

Cross inhibition is particularly well-suited to heterogeneous systems, in which not all robots are able to perform all tasks. Robots in which some behavior $B_n$ is not instantiated will naturally never inhibit $B_n$ in other robots and claim $T_n$; thus if robots locally instantiate only behaviors appropriate to their capabilities nothing more needs to be done in order to assign heterogeneous robots to appropriate tasks. If the local arbitration gives priority to the tasks each robot is specialized for, then this assignment of robots to tasks should be very efficient.

### 2.3 Cross-Subsumption

Cross-inhibition arbitrates only between peer behaviors on different robots; some local mechanism must arbitrate between different behaviors on the same robot. We believe that simple subsumption, when combined with BLE, is sufficient for flexible, scalable, and robust team cooperation in many tasks. We call the combination of cross-inhibition and local subsumption *cross-subsumption.*

In cross-subsumption, each robot has a local subsumption hierarchy. Each layer of this hierarchy may be cross-inhibited (Fig. 1a), resulting in a system similar to the one in Fig. 1b. As a result, each robot is controlled by its behavior $B_n$ which has the highest priority of any behavior which is generating output. A behavior that is not generating output fails to do so either because its current input is unsuitable for the task (e.g., some necessary object is not in the field of view), or because its output is cross-inhibited. Thus, each robot will claim the highest-priority task for which it is most suitable.

## 3 The CMOMMT Task

We have tested our BLE approach on a multi-target observation task known as CMOMMT (Cooperative Multi-robot Observation of Multiple Moving Targets) [8], and a prioritized variation. CMOMMT is an NP-hard problem that requires strong cooperation [7] for good performance. It has the benefit of simple formulation and analysis, and implemented systems for comparison.

### 3.1 Definition of CMOMMT

Our version of the CMOMMT problem is defined as follows. Given $S$ : a bounded, enclosed region; $R$ : a team of $m$ robots with noisy, limited sensors; and $O(t)$ : a set of $n$ targets $o_j(t)$ such that $In(o_j(t), S)$ is true, where $In(o_j(t), S)$ means that target $o_j(t)$ is within $S$ at time $t$, define an $m \times n$ matrix $A(t)$ where

$$a_{ij}(t) = \begin{cases} 1, \text{ if robot } r_i \text{ is observing target } o_j \\ \quad \text{at time } t \\ 0, \text{ otherwise} \end{cases} \tag{1}$$

(a robot is *observing* a target when the target is in the robot's field of view and within a certain distance)

and define a logical $OR$ operator over a vector $H$ :

$$\bigvee_{i=1}^{k} h_i = \begin{cases} 1, \text{if there exists an } i \text{ such that } h_i = 1 \\ 0, \text{otherwise} \end{cases} \tag{2}$$

The goal of the CMOMMT is then to maximize

$$Observation = \frac{\sum_{t=1}^{T} \sum_{j=1}^{m} \bigvee_{i=1}^{r} a_{ij}(t)}{t \times m} \tag{3}$$

that is, to maximize the time during which each target in $S$ is being observed by at least one robot. We assume that the area covered by the sensors of the robots is much smaller than the total area to be monitored and that targets move slower than the robots. The original formulation of the problem [8] assumes that robots share a known global coordinate system; we replace this with the assumption that the robots can visually distinguish each target from the others. Thus our formulation focuses on *task space* where Parker's formulation [8], predictive tracking, and *local force-vector* algorithm [8] tend to be more oriented towards physical space.

We also introduce a prioritized version of the problem which we call Weighted CMOMMT, or W-CMOMMT. Given: $W$ : a vector of weights such that $w_i$ reflects the priority of target $o_i$, the goal of W-CMOMMT is to maximize

$$W\text{-}Observation = \frac{\sum_{t=1}^{T} \sum_{j=1}^{m} w_j \bigvee_{i=1}^{r} a_{ij}(t)}{t \times m \times \sum_{v=1}^{m} w_v} \tag{4}$$

# 4 Experiments

We implemented controllers for CMOMMT on a team of three ActivMedia Pioneer 2DX robots. These are differentially-steered wheeled bases with on-board sonar (for obstacle avoidance) and vision (for identifying and tracking targets). The video cameras have a 45-degree field of view. Each robot is connected to a wireless Ethernet LAN, and programmed using Ayllu [12].

## 4.1 The Experimental Environment

Current experiments take place in an 18×22 ft. corral. Targets are colored paper cylinders which the experimenter moves by hand in a fixed pattern at an average speed of about 2 ft/min. The targets begin at one end of the corral, and move in a criss-cross pattern that varies from a very dispersed to a very condensed formation (see Fig. 2a). Trials are run with three robots and four targets.

a) b)

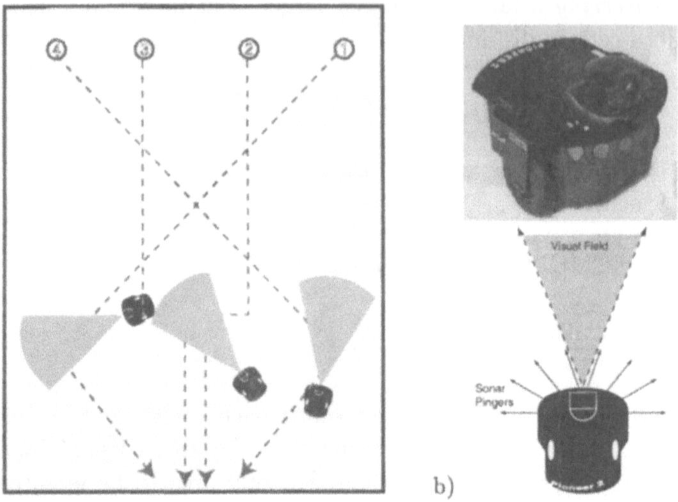


**Fig. 2.** a) *Experimental Environment*: The 18×22 ft. corral. Robots are shown with observation ranges; targets are numbered circles. Light grey targets and dashed lines indicate initial positions and paths of targets. b) *Robots*: Pioneer 2DXs.

### 4.2 Robot Behaviors

Four controllers were implemented for comparison in CMOMMT: a BLE controller, a local greedy controller, a local subsumption controller, and a random controller. Each of these controllers is implemented using the same behaviors, with differences in behavior arbitration.

**Common Behaviors** A single behavior on each robot controls translational motion to maintain a safe velocity based on the distance to sonar-detected obstacles. The behaviors described below only control rotational motion of the robots. Two classes of task-achieving behavior were implemented:

*Observer* behaviors: target-observing behaviors rotate the robot towards a specific target in its field of view. This, combined with the common velocity control behavior, causes the robot to approach a specific target and maintain a distance of approximately 1 foot. One observer behavior is instantiated for each target. The observation range of the robots is approximately four feet, and the robots are able to perceive targets up to fifteen feet away, depending on lighting conditions in different parts of the corral.

*Search* behavior: search behavior is a random wander.

**BLE Coordination:** The BLE controller is a subsumption hierarchy of *Observer* behaviors, with the Target 1 observer having highest priority. Each is then joined into a cross-inhibiting peer group which consists of *Observers* of the same target on each robot (Fig. 1a), such that the controller becomes a cross-subsumption hierarchy (Fig. 1b). The highest-priority behavior that is not cross-inhibited controls the robot - that is, each robot approaches and tracks the highest-priority target it sees that is not being observed by another robot.

The local eligibility evaluation function for each *Observer* behavior is proportional to the width of its associated target in the visual field (an approximation of distance). It favors observation of multiple targets by increasing for each additional target viewed in observation range:

$$LE_{ij}(t) = \begin{cases} \sum_{k=1}^{n} a_{ik}(t) \times width_{ik}(t), \text{ if } a_{ij}(t) = 1 \\ -\infty, \qquad\qquad\qquad\qquad otherwise \end{cases} \tag{5}$$

The wander behavior is active when all other behaviors are either cross-inhibited or unable to perceive any targets.

**Local Subsumption Only** The Local Subsumption (LS) controller is the same as the BLE controller, but connections are not made across the peer groups so that no cross-inhibition takes place. The robot approaches and tracks the highest-priority target it sees, or wanders if it sees no targets.

**Local Greedy** The Local Greedy (Greedy) controller has neither cross-inhibition nor local subsumption; instead, the behavior with the highest evaluation function controls the robot. The robot approaches and tracks whatever target is most salient (as influenced by perceptual uncertainty) in its field of view, or wanders if no target is perceived.

**Random** In the Random controller, only the random wander behavior is active, at all times.

### 4.3 Results

Five trials of approximately 12 minutes each were run for each of the BLE, Greedy, and LS controllers; two trials of the Random controller were run for a baseline.

The most important measures are of course the *Observation* and *W-Observation* metrics of Sec. 3.1. As seen in Fig. 3a, on CMOMMT the BLE approach, averaging 0.7963, scored significantly higher ($p = 0.0017$ on a pairwise t-test) than the Greedy approach at 0.69940 and the LS approach at 0.51995 ($p < .0001$). Using the W-CMOMMT metric, the relative performance of BLE was even better, scoring 0.860984 to the Greedy score of 0.717251 ($p < 0.00001$) and the LS score of 0.630928.

The distribution of the robots across the targets can be clarified with information on simultaneous target coverage, illustrated in Fig. 3b. On average, the BLE approach observed all four targets 41.68% of the time (compared to

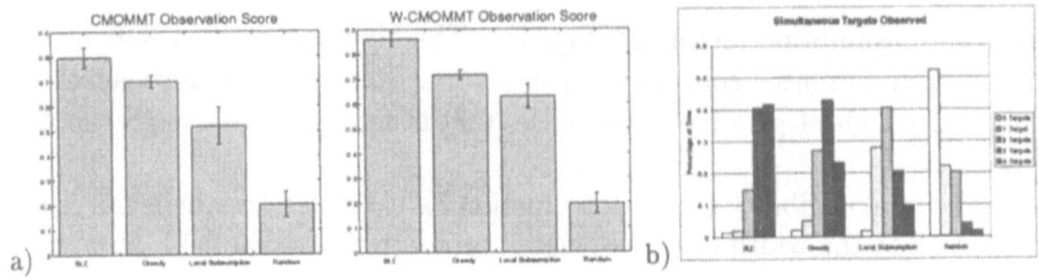


**Fig. 3.** a) *Average Observation and W-Observation scores by algorithm.* Error bars span 2 standard deviations. b) *Simultaneous Observation* The percentage of time at which 4, 3, 2, 1, or no targets were observed, averaged over trials.

Greedy: 23.27% and LS: 9.1%), and observed at least three targets 82.17% of the time (Greedy: 66.12%, LS: 30.38%).

It can be seen from the target motion patterns of Fig. 2a that during the last third of each trial, targets 2 and 3 were consistently close enough to be observed by a single robot. While the BLE approach resulted in a stable configuration of all four targets being observed for the majority of the final third of *every trial*, neither the Greedy nor LS approaches maintained this stable 4-target observation in even a single trial. This can be seen in Fig. 4, which illustrates typical patterns of observation. In the BLE approach, the three highest-priority targets are covered fairly constantly, although the particular observing robots change, and overlap of observation is minimal, demonstrating effective team cooperation in task assumption. In the Greedy and LS traces, there are both larger periods of overlap and larger periods in which some targets are not covered at all.

In all trials, periods in which a particular robot seems not to be observing anything often reflect a blocked robot which is tracking a target, but not close enough to observe. This situation was common to Greedy and LS trials where robots often "queued up" behind other robots observing the same salient target. Further, our collision avoidance, described in Sec. 4.2, did not deal effectively with robots approaching each other from the side, as when two were trying to get close to the same target; this resulted in occasional collisions during LS and Greedy trials. The *task-space* separation of the BLE approach effectively prevented both of these *physical-space* problems of interference.

Observation of the different approaches in action led to the realization that the BLE approach was effective in overcoming perceptual limitations of the robots. While in Greedy and LS trials robots tended to cluster around targets that were "better perceived" (due to details of the color-tracking implementation and environment), exacerbating the physical-space problems described above, in the BLE trials, highly visible targets were quickly and consistently observed, freeing robots to pursue less-salient targets.

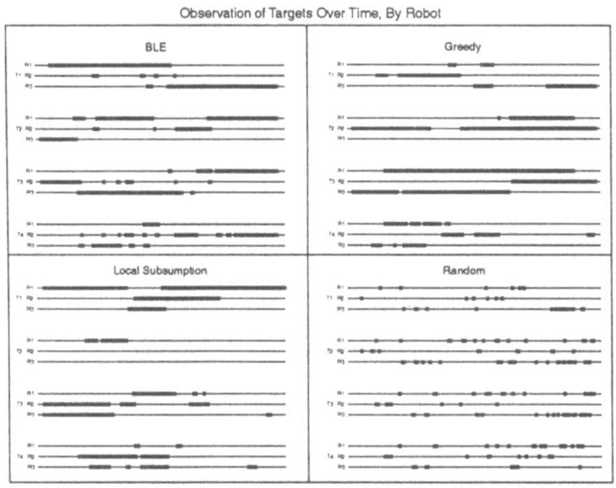


**Fig. 4.** *Observation Over Time, by Algorithm.* For each algorithm we have chosen a trace of the trial which scored closest to the average.

# 5 Related Work

Pioneering work with groups of behavior-based robots involved *weakly cooperative* tasks such as flocking and foraging [6] and clustering [5]. [3] modified such foraging strategies to include coordinated territorial division, and [4] investigated role-based task division in foraging systems, with and without explicit coordination. [1] discuss how communication is "foreign" to their behavior-selection methods. [9] discuss levels of abstraction that allow entire robots to be seen as single behaviors in a behavioral system, but rely on "observer" behaviors which come close to a centralized solution. The BeRoSH system [10] extends a subsumption approach to allow inter-robot communication as input to behaviors (but not for inhibition/suppression). Task performance is fully distributed, but one robot "host" must coordinate.

The ALLIANCE architecture [7] comes closest to our approach. Differences include the need in ALLIANCE for motivational behaviors to store information about other individual robots, the lack of uniform inter-behavior communication (inter-robot communication only takes place between motivational behaviors, and cannot be arbitrated as other behavioral communication), and ALLIANCE's monitoring of time other robots have spent performing behaviors rather than BLE's local eligibility estimates. Motivational behaviors must be re-written (rather than re-connected) to reflect changes in system structure, and restrict the ways in which robots can interact. Direct comparison of the two systems is not quite fruitful, as BLE is aimed at a lower level of abstraction than ALLIANCE. It is clear that ALLIANCE could be implemented using Ayllu and BLE, and that for certain tasks it would be convenient, but our interest is in examining the range of tasks that can be covered by our small set of clean, standard, "stateless" [11] language-level abstractions.

## 6 Conclusion

Our experiments have shown that the PAB paradigm, and BLE in particular, is able to support fully distributed, efficient coordination of teams of robots using simple and general low-level components. The resulting systems are scalable, robust, and flexible, adapting to changing environmental conditions and resource availability. Cross-subsumption can assign heterogeneous robots to tasks appropriately with no need for explicit negotiation or recognition. PAB is a principled approach, providing standard, well-defined abstractions for behavior coordination. Behavior are fully encapsulated, facilitating "bottom up" system design and testing.

## Acknowledgements

The research described here was supported by the Office of Naval Research Grants N00014-95-1-0759 and N0014-99-1-0162, and the Sandia National Labs Grant 53-4509-3060.

## References

1. T. Balch and R. Arkin. Behavior-based formation control for multi-robot teams. *IEEE Transactions on Robotics and Automation*, December 1998.
2. R. A. Brooks. *Cambrian Intelligence.* MIT Press, 1999.
3. Miguel Schneider Fontán and Maja J. Matarić. A study of territoriality: the role of critical mass in adaptive task division. In *Proceedings of the Fourth International Conference on Simulation of Adaptive Behavior*, Cape Cod, September 1996.
4. Dani Goldberg and Maja J Matarić. Interference as a tool for designing and evaluating multi-robot controllers. In *Proceedings of AAAI-97*, pages 637–642, Providence, RI, July 1996.
5. Owen Holland and Chris Melhuish. Stigmergy, self-organisation, and sorting in collective robotics. *Artificial Life*, 5:2, 2000.
6. Maja J Matarić. Designing and understanding adaptive group behavior. *Adaptive Behavior*, 4:1:51–80, December 1995.
7. L. E. Parker. Alliance: An architecture for fault tolerant multi-robot cooperation. *IEEE Transactions on Robotics and Automation*, 14, 1998.
8. L. E. Parker. Cooperative robotics for multi-target observation. *Intelligent Automation and Soft Computing*, 5:5–19, 1999.
9. J. Salido, J. M. Dolan, J. Hampshire, and P. Khosla. A modified reactive control framework for cooperative mobile robots. In *Sensor Fusion and Decentralized Control in Autonomous Robotic Systems, Proceedings of SPIE 3209*, 1997.
10. Z.-D. Wang, E. Nakano, and T. Matsukawa. Designing behavior of a mult-robot system for cooperative object manipulation. In *Proceedings of the International Symposium on Microsystems, Intelligent Materials, and Robots*, 1995.
11. Barry B. Werger. Cooperation without deliberation: A minimal behavior-based approach to multi-robot teams. *Artificial Intelligence*, 110:293–320, 1999.
12. Barry Brian Werger. Ayllu: Distributed port-arbitrated behavior-based control. In *Proceedings of DARS*, 2000.

# A Communication-free Behavior for Docking Mobile Robots

Brian W. Minten, Robin R. Murphy, Jeff Hyams, Mark Micire

University of South Florida, Dept. of Computer Science and Engineering, Tampa FL 33620, USA; minten, hyams, murphy, micire @csee.usf.edu

**Abstract.** Physical cooperation between robotic agents frequently requires the ability to dock. This paper reports on a reactive docking behavior which uses a vision algorithm that grows linearly with image pixels. The docking behavior was implemented on a marsupial team of robots, where a daughter micro-rover had to re-enter the mother robot from an approach zone with a 2 meter radius and 140° angular width. Experiments showed that the docking behavior had a similar success rate and was faster than 22 teleoperators.

*Keywords:* docking, control, multi-agents

## 1 Introduction

Much of the initial focus with multi-agent teams has been on strategies for the deployment and cascaded control of members over large distances, ignoring the specific issue of how to collect the agents back into one unit. Docking the agents, however, is essential for recovery, recharging, and sample processing. The use of human teleoperation for final docking maneuvers is an undesirable solution in that it results in cognitive strain for the operator and an open (and therefore, potentially detectable and corruptible) communication channel. The ability of agents to autonomously dock to one another extends the duration and types of missions possible.

Marsupial robots are a type of cascaded heterogeneous team where a "mother" agent transports one or more "daughter" micro-rover agents to a mission site [4]. The marsupial concept has been discussed in the USA for a variety of military applications, including the use of "mother ships" to carry and release micro-aerial vehicles. Due to the decreased transportation time and distributed sensing capabilities, such teams have received much attention for their potential in tactical situations [8]. The work reported here may serve to enable many applications involving multi-agents, such as space exploration, urban search and rescue, and reconnaissance.

Docking is a challenging problem for several reasons, most notably the need to accurately extract relative position and orientation [1]. Marsupial daughters typically have very limited on-board computational and sensing resources, as well as possessing restrictive "mouse eye" viewpoints. Nonholonomic vehicles and those that are not identical front and back complicate the situation further, typically necessitating limited path planning and periods of operation using only dead reckoning.

This paper reports on a reactive vision-based docking behavior that requires no communication and can be used under a variety of lighting conditions. It also presents the results of comparative experiments with 22 teleoperators, showing that the docking behavior is as reliable as and faster than a human teleoperator.

## 2 Related Work

Docking of a single agent was originally explored by Arkin and Murphy, who produced a reactive behavior for docking [1]. Much of the subsequent docking research [5], [9], [10], [6] either duplicates these results or produces more computationally complex methods. Spofford et al [8] also deals with docking, but is primarily concerned with agents' vision abilities, rather than their motion control.

Our work is an extension of earlier work done by Arkin and Murphy. Their approach used combinations of potential fields encapsulated in a motor schema to dynamically control docking without explicit path planning. This motor schema had low computational complexity, making it suitable for a daughter, as long as a low-order vision algorithm could be found. The robot's approach is computed instantaneously as a function of moving to the correct orientation (tangential to the dock) and depth (attraction). Two later treatments of docking ([7] and [9]) independently reached the same control scheme, although they use more computationally complex methods of extracting relative position and orientation, such as optic flow.

The work presented in this paper differs in two ways: it uses a direct perception algorithm for visual guidance, using the change in perspective between two colored regions to guide the robot. The resulting system is similar to the optomotor control system inspired by the visual orientation of flies [2]. Also, it is different in scope. Arkin and Murphy did not consider extensions to cooperative behavior, where the mother can actively place herself to aid docking. (Our lab is currently working on this.)

## 3 Approach

The general approach to the docking behavior follows [1]. This paper deals primarily with the motor schema controlling movement within the wedge leading to the dock, an area termed the approach zone. (See Fig. 2) The angular coverage of the approach zone is a function of the optics and fiducial; in this implementation it is approximately 140°. The behavior is divided into two schema: a *perceptual schema*, which supplies the stimulus, and a *motor schema*, or observable actions that the daughter takes to dock. The behavior is reflexive or stimulus-response. The perceptual schema supplies perspective and size cues using a color segmentation algorithm with complexity linear with the number of image pixels. The motor schema uses a potential fields methodology following Arkin and Murphy.

### 3.1 Perceptual Schema

The behavior is based on perspective and looming. To utilize these effects, a fiducial consisting of two blocks of solid color was placed on a flap over the entrance of the dock. The blocks were of equal size and similar brightness, and of colors chosen such that they would be distinct from the environment. No external landmarks were necessary. In our implementation, the fiducial is concealed when the gate is closed, thus allowing the vehicle to remain camouflaged when the daughter is not docking. (See Figs. 5 and 6. Notice that the dock is at the rear of the mother, with the daughter tilting her camera so as to look backwards.) From the perceived relative size of the two blocks (in number of segmented pixels), it is possible to roughly determine the daughter's position on the ground plane behind the mother. Thus, perspective influences the relative sizes such that they may be used to determine in which direction to steer to come inline with the dock. By also considering the total size of both blocks, distance can also be approximated.

The color segmentation algorithm selected for this application used a *Spherical Coordinate Transform* (SCT). Segmentation using this technique has been shown to perform well, even when subjected to fluctuations of image intensity caused by varying lighting conditions [3]. Using this transform, the daughter was able to dock using only the light from its own headlamps, though the range from which it was able to do so was reduced to approximately 1.5 m. To eliminate unlikely regions, only dilation and erosion are performed, allowing the algorithm to run in linear (with number of pixels) time. In Fig. 1, (a) shows a video frame from the camera, and (b) highlights the pixels that were segmented for each block.

a) b)

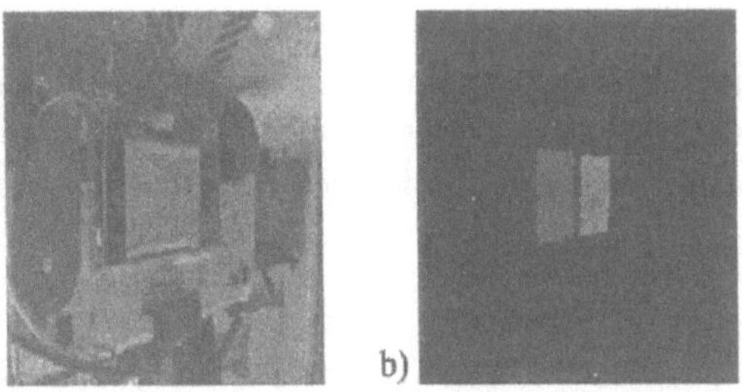

**Fig. 1.** Fiducial a) as seen, upside down, through the daughter's camera during docking, and b) after segmentation

### 3.2 Motor Schema

The motion control scheme might be best understood with a potential fields diagram approximating our method. In Fig. 2, the field over which the daughter can be expected to dock successfully is left white. Outside of this sector, the blocks of the fiducial appear too small to be properly segmented. Within this sector, vectors are computed by summing an attractive and tangential component. The attractive component is computed as growing in inverse proportion to the total number of pixels segmented as belonging to either block. The tangential component is influenced by the relative sizes of the two blocks. For example, if the vision algorithm segments more pixels as belonging to the right block, the correct action is to turn left. When the robot reaches a point within the innermost sector (darker area), it must either proceed into the dock or move back for another attempt. Though the path dictated by this field may not be the shortest possible, it assures good alignment when the robot encounters the dock. Running this code on a platform that did not offer zero-radius turning capability would almost certainly require such an approach.

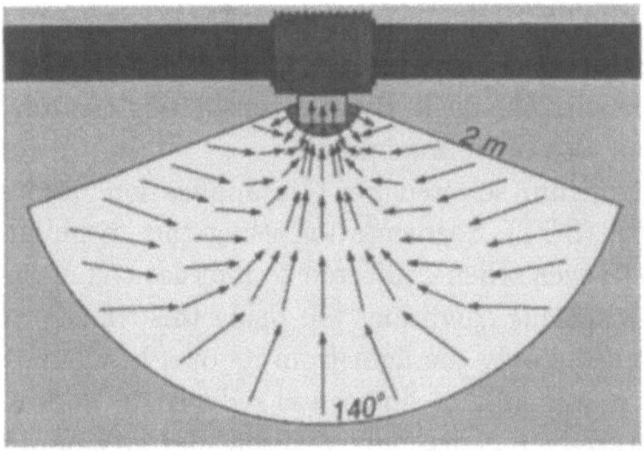


**Fig. 2.** Potential field within the daughter's approach zone.

Our deviation from a strict potential fields methodology is dictated by the need to keep the fiducial in view. When the fiducial has been lost, the robot turns in the direction that it was last seen. Thus, the daughter will tend to weave as it trades off between steering in a direction that will bring it into alignment behind the mother and turning to reaquire a lost fiducial. Accordingly, the fields shown on the diagram may be interpreted as the average of forces felt by the robot as it encountered a given position over multiple runs.

# 4 Implementation

The close dock behavior was implemented and tested on a marsupial team consisting of Silver Bullet, a modified Power Wheels Jeep, and Bujold, an Inuktun VGTV MicroRover. The pair is shown in Fig. 5. Bujold carries no power source or processing capability of her own, but instead uses Silver Bullet's resources via a tether. Although Bujold's processing was done on board Silver Bullet, the code could equally well be run on the docking agent if this capability existed. The design requires no communication between team members.

One complication of this configuration is that Bujold's tether requires that she back into Silver Bullet. Bujold's abilities to shape-shift (see Fig. 6) and tilt her camera did, however, make it possible to view Silver Bullet for the majority of the docking process, though the image was upside down. When she finally maneuvers into position at the dock, she must lower herself and back in based only on dead reckoning.

Within Silver Bullet, the dock size allows for approximately 1 cm of space on each side of Bujold. These tight tolerances are somewhat mitigated, however, by a funneled opening, which can allow for an approach which is up to 15° off-axis. In this implementation, however, the tolerances were set much closer, as this "cushion" was frequently destroyed by uneven slippage of Bujold's tracks on the gate. This tighter tolerance was also beneficial in cases where the two colors of the fiducial were being unevenly segmented. After some experimentation, parameters were chosen that would only allow Bujold to dock if she was within about $\pm 2$ cm laterally and $\pm 5°$ of center.

On board Silver Bullet, images were acquired with a Matrox Meteor framegrabber, and processing was performed on a 400 MHz AMD K6-II. Running the segmentation code, Bujold's $256 \times 240$ pixel images could be processed at a rate of about 3.25 Hz. The lens on Bujold's camera has a full angle field of view of approximately 70°, allowing the entire fiducial to be in view up to the point Bujold encounters Silver Bullet's ramp. These camera parameters also served to limit the docking region to about 2 m from the fiducial.

In addition to the pixel count of each block, the perceptual schema also extracts `xBar`, the average horizontal position of all segmented pixels. This value is stored within the motor schema from frame to frame, in case the daughter needs to reacquire the fiducial. It also is used when the total pixel count is above a given threshold, indicating that the robot is at the dock. In this situation, if the current `xBar` is well-centered in the image, the dead reckoning phase of docking begins. If it is not, the daughter drives forward (away from the dock) for another attempt. In all cases, the motor schema produces a vector. The heading is scaled according to the ratio of left and right pixel counts, though the code purposely understeers to avoid weaving. The speed is governed by the daughter's proximity to the mother, slowing on approach. As a further result of the time required to process the video frame, when the total pixel count indicates that the daughter is at the dock, she will stop and wait for the percepts from the next frame to insure that the ensuing action is appropriate.

Figure 3 shows a survey of the possible states that are encountered within the code. The daughter's field of view is within the wedge that extends backward from her camera. In (a), the daughter is at the dock, the ratio is close to one, and `xBar` is within tolerance. At this point, she should lower and back in. Panel (b) presents a similar situation, but the ratio is not within tolerance. The daughter will go forward, away from the mother, to try again. In (c), the daughter is at the dock and the left fiducial block is partially out of view. This may result in the pixel counts being nearly equal, but **xBar** will not be within tolerance so the prescribed action is to go forward for another attempt. In (d) the fiducial is out of view. Very likely, **xBar** will have last been seen in the left side of the inverted image, and so the correct action is to turn right in an attempt to reaquire the fiducial. Panel (e) is representative of the daughter's state for the majority of the docking operation. In this case, the pixel count of the right block will be greater than that of the left block, so the daughter will turn left according to their ratio. Because of situations such as the one portrayed in (f), sufficient frame rate and drive control precision result in an infrequent loss of the fiducial, as shown in (d). In this case, since the left portion of the fiducial is partially out of view, its pixel count will be lower and the daughter will turn left, even though she is on left side of the mother.

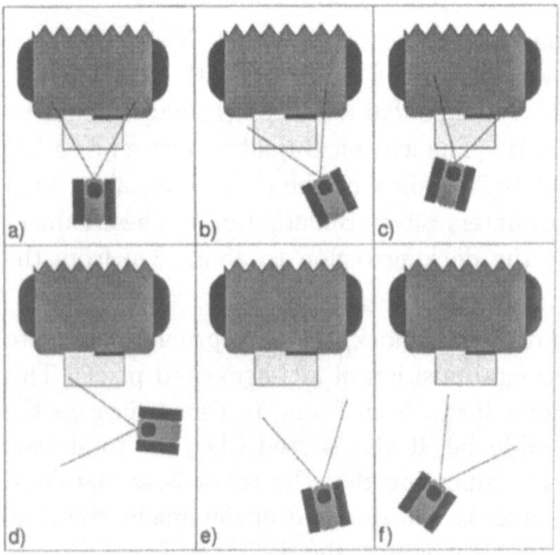


**Fig. 3.** Representative states of the docking vehicle

## 5 Experiments

To evaluate the behavior's performance, the daughter's docking time from 16 locations distributed over a field behind the mother was recorded. This field extended to 2 m, and in 60° either direction off center, as shown in Fig. 4. In all cases, the robot was placed directly facing the fiducial on Silver Bullet. Since the robot rotates until it sees the fiducial, placing it where did not face the fiducial would simply add time to the beginning of the nominal docking process.

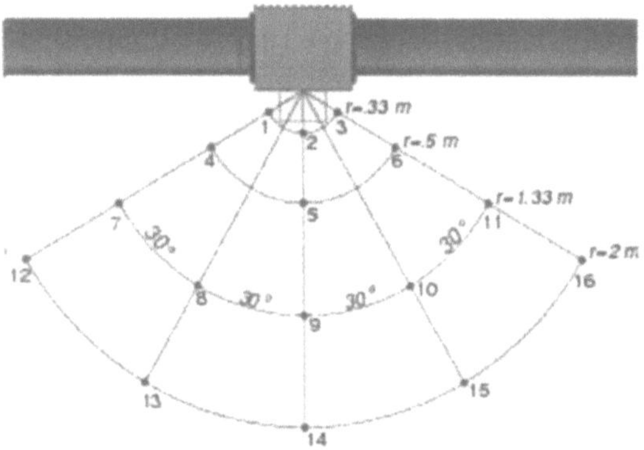


**Fig. 4.** Autonomous vs. teleoperated docking experiment starting positions

As a comparison, the times of twenty-two teleoperators attempting the same task were used. Sixteen were non-experts, undergraduate volunteers, and six were experts who routinely teleoperated Bujold or another robot at least once a week. Each was given up to fifteen minutes of practice docking the robot while permitted to view the robot as well as the display. The display consisted of a direct video feed from the robot. Once again, since the robot is backing in, the camera images are actually upside-down, as Bujold's camera is tilted 180°. Operators used the same five-function control system (forward, back, right, left, stop) as the system. During the trials, the teleoperator's view of the robot was blocked. Figure 5 illustrates the experimental set up. Each teleoperator performed three trials, each from one of the sixteen positions, randomly selected by the recording program. Timing was automatic, starting with the presentation of the video signal and ending upon Bujold's contact with pressure plates installed within Silver Bullet's bay.

**Fig. 5.** Experimental setup with Silver Bullet and Bujold (the daughter). The teleoperator sits behind the blind on the right

While attempting to dock autonomously, the system missed the dock twice. In such instances, the daughter got stuck on the edge of the entrance, and stalled. In the interest of robot preservation, such a run was stopped immediately. Three human operators also missed the dock during the dead reckoning phase. When this happened, it necessitated driving forward, changing geometry to bring the mother robot into view, adjusting as necessary, changing the geometry again, and reattempting the final dead reckoning phase. The run was allowed to continue, but for comparison purposes, these times were not used to compute the mean.

| | Mean time to Dock (s) | |
|---|---|---|
| Position | Teleoperator | Autonomous |
| 1,3 | 59.3 | 89.2 |
| 2 | 21.5 | 25.6 |
| 4,6 | 70.2 | 61.3 |
| 5 | 46.5 | 39.6 |
| 7,11 | 80.7 | 64.0 |
| 8,10 | 86.7 | 59.8 |
| 9 | 52.5 | 54.3 |
| 12,16 | 96.9 | 81.6 |
| 13,15 | 86.6 | 67.5 |
| 14 | 52.5 | 66.6 |

**Table 1.** Mean time to dock as a function of starting position

| | Mean Time to Dock (s) | |
|---|---|---|
| Position | Teleoperator | Autonomous |
| 2,5,9,14 | 43.2 | 46.5 |
| 8,10,13,15 | 86.7 | 63.6 |
| 1,3,4,6,7,11,12,16 | 79.4 | 73.8 |

**Table 2.** Mean time to dock as a function of angle

| | Mean Time to Dock (s) | |
|---|---|---|
| Position | Teleoperator | Autonomous |
| 1,2,3 | 44.2 | 65.3 |
| 4,5,6 | 61.6 | 54.1 |
| 7,8,9,10,11 | 78.5 | 60.1 |
| 12,13,14,15,16 | 84.9 | 73.0 |

**Table 3.** Mean time to dock as a function of distance

The mean times for autonomous and teleoperated docking are shown in Table 1. Note that positions are grouped by symmetry. Table 2 presents these times as a function of angle, and Table 3 presents them as a function of distance. Not surprisingly, docking time increased with both angle and distance. Interestingly, however, teleoperators took longer to dock from 30° than from 60° off- center.

The teleoperators were able to successfully dock in 95% of the 66 trials, with an average velocity of 2.0 cm/s. Under autonomous operation, the robot successfully docked in 96% of its 48 trials, with an average velocity of 2.4 cm/s. On average, the robot took 17% less time to successfully dock than did a teleoperator.

Comparing the data with a paired-sample t-test reveals a 22% chance that human operators can, on average, dock the daughter as quickly as can be done autonomously. It was not, however, originally intended for the robot to be able to dock from positions 1 and 3. Experimenting with the robot revealed that it would eventually work its way around the edge of the gate by repeatedly moving back and forth. If these positions are dismissed from consideration, the t-test result is .047, less than a .05 significance level.

With the six experts, the number of commands required to successfully dock was also recorded. Once again, each expert teleoperated the robot from three randomly selected positions, including 1 and 3. The average number of commands issued per meter was 29, issued at the average rate of 24 per minute. In contrast, the autonomous system required no commands to be transmitted, though the motor schema provided a new vector for each percept, at a rate of 3.25 Hz for the duration of the nominal docking process.

## 6 Conclusion

This paper has shown that reactive docking is faster than teleoperated docking, with a statistical confidence greater than 95%. It also has a similar success rate to teleoperation (96% vs. 95%). The docking behavior requires no communication with a teleoperator, while a teleoperator has to give an average of 29 commands per meter to dock a robot. This docking behavior uses direct perception of an unobtrusive artificial landmark which can be visually segmented in time that grows linearly with image pixel count. While this paper considers the application only to marsupial robots, the technique should be extensible to any team of physically cooperating robots, including reconfigurable robots.

Comparing the data with and without positions 1 and 3 gives us some important performance data. Clearly, this is not the best algorithm for docking from these locations. It might be worthwhile to devise another tactic for removing the robot from these positions. As a last resort, the robot might even defer control to a human operator.

Admittedly, the human/computer interface was a direct view through Bujold's upside-down camera, and the data collected might reflect this. However, this can be seen as one of the benefits of automation. When the robot

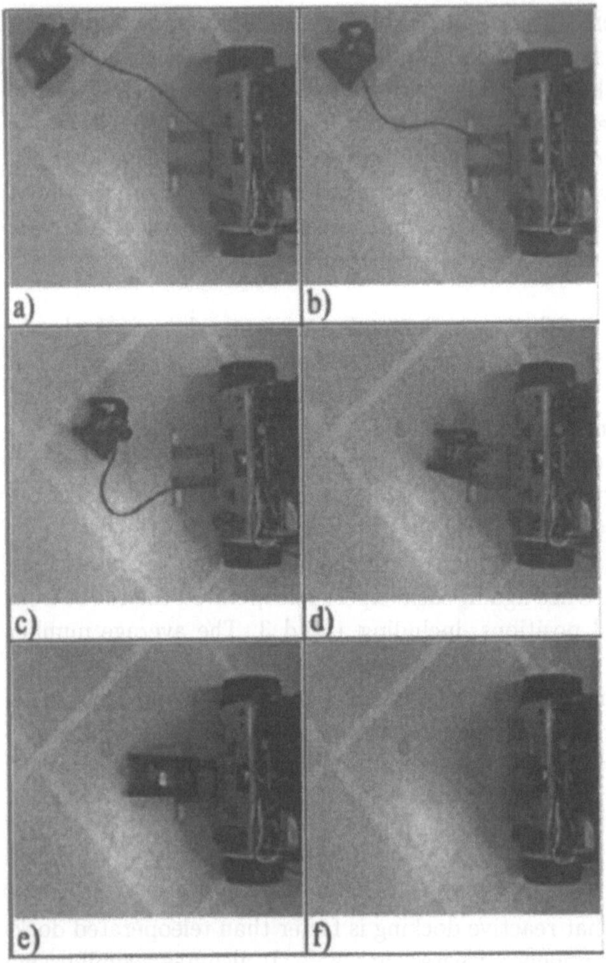


**Fig. 6.** Bujold docking. She a) begins turned away from the camera; b) turns to acquire the fiducial; c) approaches; d) stops when at the dock; e) changes geometery and backs in using dead reckoning; and f) triggers Silver Bullet's gate to close

is under its own control, it is not necessary to develop an optimal control mechanism, make special efforts for displays to be non-disorienting, or suffer the communications bandwidth to involve an operator.

Though the behavior can resume normal docking after an obstacle temporarily obscures the fiducial, it does not avoid static obstacles. On the contrary, the tendency is for the robot to steer toward an object that is positioned between it and the fiducial. If it were required for the robot to dock under such circumstances, an obstacle avoidance behavior running as a separate motor schema would be required.

Terrain types and lighting conditions were also not considered in this experimentation, since it was intended to establish the baseline characteristics of the motor schema. In more recent work, the perceptual schema has been made adaptive and capable of sampling, thereby making it even more robust in varying lighting conditions.

## 7 Acknowledgements

This work was supported by DARPA award F04701-99-C-0317. The authors would like to thank SAIC for their feedback and support.

## References

1. R. C. Arkin and R. R. Murphy. Autonomous navigation in a manufacturing environment. *IEEE Transactions on Robotics and Automation*, 6(4):445–454, 1990.
2. S. A. Huber, M. O. Franz, and H. H. Bülthoff. On robots and flies: Modeling the visual orientation behavior of flies. *Robotics and Autonomous Systems*, 29:227–242, 1999.
3. J. Hyams, M. Powell, and R. Murphy. Cooperative navigation of micro-rovers using color segmentation. *To appear in Autonomous Robots*, 1999.
4. R. R. Murphy. Marsupial robots for urban search and rescue. *To appear in IEEE Intelligent Systems*, 2000.
5. P. Questa, E. Grossmann, and G. Sandini. Camera self orientation and docking maneuver using normal flow. In *Proceedings SPIE*, volume 2488, pages 274–283, Orlando, FL, April 17-18 1995.
6. H. Roth and K. Schilling. Navigation and docking maneuvers of mobile robots in industrial environments. In *IECON 98*, pages 2458–2462, Aachen Germany, Aug. 31-Sept. 4 1998.
7. J. Santos-Victor and G. Sandini. Visual behaviors for docking. *Computer Vision and Image Understanding*, 67(3):223–238, 1997.
8. J. Spofford, J. Blitch, W. Klarquist, and R. Murphy. Vision-guided heterogeneous mobile robot docking. In *Proceedings SPIE*, volume 3839, pages 112–121, 1999.
9. M.C. Tse and R. Luo. Integrated multi-behavioral mobile robot navigation using decentralized control. In *IROS 98*, pages 564–569, Victoria, BC Canada, Oct. 13-17 1998.
10. J. Vandorpe, H. Xu, and H. Van Brussel. Dynamic docking integrated in a navigation architecture for the intelligent mobile robot lias. In *IAS 95*, pages 143–149, Karlsruhe Germany, March 27-30 1995.

# Chapter 10

# Multi-Robot Learning

# On Behavior Classification in Adversarial Environments

Patrick Riley[1] and Manuela Veloso[1,2]

[1] {pfr,veloso}@cs.cmu.edu, Computer Science Department, Carnegie Mellon University, Pittsburgh, PA 15213
[2] Currently a visiting professor at Massachusetts Institute of Technology, Boston, MA

**Abstract.** In order for robotic systems to be successful in domains with other agents possibly interfering with the accomplishing of goals, the agents must be able to adapt to the opponents' behavior. The more quickly the agents can respond to a new situation, the better they will perform. We present an approach to doing adaptation which relies on classification of the current adversary into predefined adversary classes. For feature extraction, we present a windowing technique to abstract useful but not overly complicated features. The feature extraction and classification steps are fully implemented in the domain of simulated robotic soccer, and experimental results are presented.

## 1 Introduction

In order for robotic systems to be successful in complex domains, they must be able to adapt to the environment, especially to the current behavior of other agents. This adaptation should occur at all levels of strategy, from individual reactive behaviors to team strategy. In highly complex domains with no clear optimal policy, agents which can adapt more quickly and effectively will perform better.

We consider "complex domains" to be those with enormous, or possibly continuous, state and action spaces. Further, we consider dynamic environments, i.e. those with limited time to reason before each agent must choose an action. We are further interested in domains with noisy perceptions and actions.

Reinforcement learning is a common technique for having an agent automatically discover its optimal policy. However, as discussed in [3], reinforcement learning often results in poor performance in the types of domains in which we are interested. Further, reinforcement learning will generally require trying every possible strategy of the agent many times. We would like to be able to adapt much faster than this.

Systems such as Carmel and Markovitch's [1] succesfully adapt by building up models of the opponents (in this case finite automata) and then computing the best strategy given that opponent. However, the environment they consider is a simple two-person repeated matrix game. It is not clear how this sort of technique will scale to the sort of complex environments mentioned above, and we propose that other techniques will be needed for these environments.

Another approach by Sen and Sekaran[8] allows agents to learn to cooperate with agents who are willing to cooperate, and simultaneously avoid exploitation from hostile agents. However, in their environment, it is clear what action to take when another agent is hostile. In more complex domains, there may be many possible strategies to deal with opposing agents.

The system we propose is quite different. The hypothesis underlying this research is that effective adaptation can be done by analogy to previous opponents. By noting similarities in "style of behavior," the agent team can employ the most effective strategy against that behavior.

This is partially inspired by human behavior. Human beings are often quite good at the sort of adaptation in which we are interested. Good sports players will carefully watch how their opponents play and change strategy based on their opponents' actions. We would like to better understand this ability and be able to map it onto autonomous agents.

The long-term goal of this research has several aspects:

1. Abstracting useful features from the raw sensor data and world state available to the agents. An approach to feature extraction in complex domains is introduced in Section 3
2. Constructing "adversary classes" from the information in the features. The adversary classes should embody strategic information about how this type of adversary behaves. Closely related to this is the work Ramoni, *et al.* have done on unsupervised Bayesian clustering of world dynamics[6]. Using this technique, clusters are formed which describe general features about what is going on in the world. For example, they describe one cluster as cases "in which the blue team won despite a large mass deficit." In this research, however, creating adversary classes is done by averaging over many observations with *a priori* distinctions (see Section 4.2).
3. Matching what is observed of a current adversary to the adversary classes. This gives a "classification" of the current adversary into the predefined classes. This step is addressed in Section 4.
4. Affecting an effective change in behavior based on the classification. There should be some mapping from adversary classes onto strategies that our agents' may use. This final change in behavior is the goal of the classification process. However, the other aspects may be analyzed in the absence of this change in strategy, so this step is not addressed here.

This research focuses mostly on abstracting features and classifying the adversary, numbers 1 and 3 above.

## 2 The Test Domain and Software

The test domain for this research is simulated robotic soccer, specifically, the Soccer Server System[4] as used in the Robot World Cup Initiative, an international AI and robotics research initiative. The Soccer Server System is a server-client system which simulates soccer in a discrete fashion. Clients communicate using a standard network protocol with well-defined actions. The server keeps track of the current state of the world, executes the actions which the clients request, and periodically sends each agent noisy, incomplete information about the world. Agents receive noisy information about the direction and distance of objects on the field (the ball, players, goals, etc.); information is provided only for objects in the field of vision of the agent. The Soccer Server System also allows an agent to connect as a "coach" client who has a global view of the world, but whose only action is to send short messages to the players while the ball is out of play. This is ideal for the sort of classification described here. The coach can classify based on global information, then communicate that information to the more active players. The classification thus provides a very limited but effective language for strategy communication. We have previously worked on classification without the coach agent by fusing the partial information of the agents in order to classify the adversary[7].

Each of the agents has an opportunity to act 10 times a second. Each of these action opportunities is known as a "cycle." Visual information is sent 6 or 7 times per second. Over a standard 10 minute game, this gives 6000 action opportunities and 4000 receipts of visual information. It would be impractical to attempt to classify the adversary every time new visual information arrives. Also, intuitively, some features which capture an adversary's performance are time-dependent. Therefore, we need to consider multiple cycles at once when classifying the adversary.

Further, learning from just the raw world data would also be infeasible. $10^{17}$ is an estimate for the size of the state space of the Soccer Server. We cannot expect to adapt as quickly as we would like without providing some abstractions to our learning algorithm.

The algorithms described here are implemented based on the successful simulated robotic soccer team CMUnited-98[9] which was the champion of RoboCup-98. All decision tree learning was done using C4.5[5].

## 3 Feature Extraction

When trying to learn in complex domains, some features must usually be abstracted from the raw data. It would be impractical and not very revealing to expect a learning algorithm to effectively decode all sensor information if a good model for doing so is already known. Extracting relevant features from sensor data can make learning feasible.

```
time := 0;
do {
    W := get_new_world_state;
    (D,I) := f(D,I,W);
    time += 1;
  }until (time == epoch_length);
return D;
```

**Fig. 1.** Pseudo-code for the use of an observation

Often, this step of feature extraction is taken somewhat lightly. However, we are working in a complex domain and trying to learn high-level strategy, so feature extraction is both more critical and complicated.

### 3.1 Windowing

To reduce the complexity of our features, we remove time sequencing from the feature's data. Removing time sequencing from data has been effective in other complex domains such as text classification with the "bag-of-words" approach, and music analysis and generation[10]. We refer to features of this type as "observations." An observation occurs over a fixed length of time, known as the window length.

Formally, an observation $A$ is a 3-tuple $\langle D, I, f\rangle$. $D$ is the data that results from the observation. $I$ is the internal state used in the updating function $f : D \times I \times W \rightarrow D \times I$, where $W$ represents the state of the world. The internal state $I$ is used to capture time dependent features (see below). Every discrete cycle, the function $f$ is called to update $D$ and $I$. Pseudo-code indicating how such an observation would be used is given in Fig. 1.

Further, the definition of an observation is slightly more strict than given above. The updating function $f$ is an indicator for some event. If that event occurs, the fact that it occurred is recorded into the data structure $D$; the time of the occurrence is *not* recorded. For example, say $E_1$ and $E_2$ are events of the type for which $f$ is looking. If $E_1$ occurs at time 25 and $E_2$ at time 50, the resulting data structure $D$ would be identical to the data structure produced if $E_2$ occurred at time 1 and $E_1$ at time 2.

Note that the observation keeps some internal state $I$. This allows $f$ to detect features that involve sequences of events. For example, an observation might be for an event $E$ which consists of an event of type $E_1$ followed immediately by an event of type $E_2$. Passing and shooting in simulated robotic soccer require this sort of description. While the internal state $I$ may record time information, the data structure $D$ may not. At the end of the time window, only the data structure $D$ is returned; the internal state $I$ is discarded.

### 3.2 The Observations Used

**RectGrid** The data structures for the observations used here are built on a data structure called a RectGrid. A RectGrid is a division of the field into small geographic regions. The RectGrid is designed to record the occurrence of certain events in specific geographic areas. An observation is recorded by adding a count to some of these geographic areas. The counts in the geographic areas represent how many times the specified type of event (like a pass or shot) occurred in that region.

In all the experiments performed here, the RectGrid dimensions used were 8×15, with the longer dimension in the longer dimension of the field.[1]

**Detailed Observations** The following observations were used in all of the classification experiments.

- **Ball Position**: Every cycle, the ball's position is stored in a RectGrid
- **Opponent Position**: Every cycle, each opponent's position is stored in a RectGrid, one RectGrid per opponent number
- **Opponent Passing**: This observation records all opponent passes in a RectGrid. All bins that a pass goes through are marked. A pass is defined as follows: Some opponent is in control of the ball at some cycle. Within 50 cycles, a different opponent controls the ball, with no significantly long control by another player in the middle.
- **Opponent Dribbling**: This observation records all opponent dribbles in a RectGrid. All bins which an opponent goes through while dribbling are marked. A dribble is defined as follows: Some opponent is in control of the ball continuously (i.e. no more than 4 cycles in a row where it does not control the ball) and his position changes by at least 3m.
- **Opponent Shooting**: This observation records in a RectGrid from where all shots are taken by the opponents.

**An Example** To further demonstrate the structure and use of observations, we will examine the Ball Position observation more closely. The data structure $D$ is just a RectGrid as described in Section 3.2. There is no internal state $I$. The updating function $f$ takes the current RectGrid $D$, adds a count to whichever geographic region the ball is currently in, and then returns the updated RectGrid $D'$.

A result of the ball position observation is pictured in Fig. 2. The field is divided into 6 rectangular regions. As the ball moves from left to right over 10 cycles (part (a)), the agent adds counts to the geographic regions of the RectGrid, resulting in the counts shown in part (b).

## 4 Construction and Classification

The goal here is to use the observations to *classify* the current adversary. That is, given a predefined set of adversary types, $C_1, \dots, C_n$, determine into

[1] A variety of numerical values are provided for the observations described here in order to allow others to reproduce these results. Further, all software developed for this research is available at http://www.cs.cmu.edu/~pfr/adv_class.html

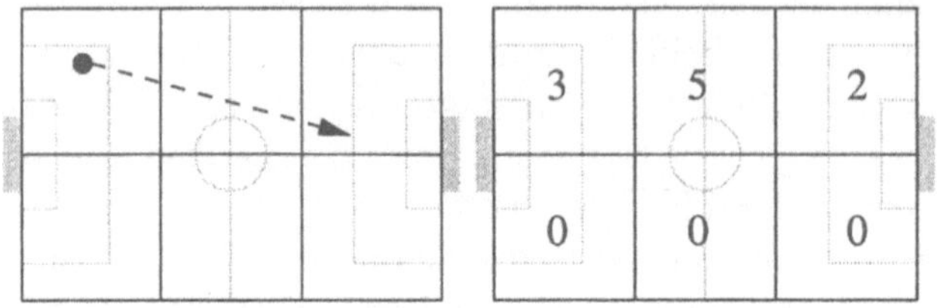


(a) (b)

**Fig. 2.** An Example of the Ball Position Observation

which class the current adversary fits. On a conceptual level, these adversary classes should represent some strategic qualities about the adversary, which can then be used to improve the agent team's performance. Also, these classes provide a compact way for the agents to describe this adversary to each other.

In this framework, an adversary class is a set of target configurations for observations. For example, one adversary class may consist partly of a target configuration for "Opponent Dribbling" which indicates that the opponent only dribbles on the sides of the midfield.

### 4.1 Experimental Setup

The issue of creating an effective strategy change was not addressed in this research. However, it is important to verify that some useful information can be obtained from this sort of windowing approach to feature extraction.

As discussed in Section 2, we used a "coach" agent who has a global view, but whose actions are limited to communication. In this version of the software, there is no strategic change based on the classification. Consequently, the coach is entirely passive. This allows the coach to look at a logfile of a previously played game exactly as if the game was currently in progress.

Hence, the logfiles from all of the games at RoboCup-98 and RoboCup-99 were used as the data-sets. The algorithms and data structures were developed while analyzing the RoboCup-98 logfiles, and once completed, the analysis of the RoboCup-99 logfiles was done.

### 4.2 Generating Adversary Classes

In each of the data sets, one adversary class was created for each team. The class was created by observing the team on the data set. There is an underlying assumption that the strategy of a team does not vary significantly over the course of the competition. There are several known exceptions to this, especially in RoboCup-99[2]. However, most teams did not make significant

[2] Specifically, the team HCIII from Linköping University has a nice strategy editor which they used to design/select strategies for each game. Also, some teams made significant revisions to their software during the competition after seeing how opponents played

changes during the competition. The instances of the observation were then averaged together. A RectGrid is averaged by averaging all of the counts for each geographic region.

After creating these adversary classes, the goal was to correctly identify which teams were playing based on the observations. In RoboCup-98, there were 34 teams, making the accuracy of random guessing to be 2.9%. In RoboCup-99, there were 37 teams, making the accuracy of random guessing to be 2.7%. The window length varied between 25 cycles and 1000 cycles. With 6000 cycles in a regular length 10-minute game, this gives between 6 and 240 windows in a game. The difference in the amounts of data at these extremes can be seen in the results (Section 5).

In order to compare observations of the current adversary to the adversary classes, there must be a notion of similarity between different values for an observation. Since all of the observations are built on the RectGrid data structure, we need a similarity metric for a RectGrid. We previously developed such a metric to take into account the spatial locality of differences in observations[7].

### 4.3 Classification

Our first approach to classification was a simple nearest neighbor approach. This performed quite poorly, not doing signifigantly better than random. One problem we perceived with the nearest-neighbor approach was the small amount of information available for making the classification decision. In generating a similarity (distance) between instances of observations and an adversary class, you have a single real number in $[0, 1]$ for each type of observation (5 types in this case).

This led us to two changes: using decision trees to deal with the large amount of data, and disassociating the target configurations with a specific adversary class. Consequently, even though the target configurations were generated for a specific team, classification would occur based on all similarities to all target configurations. All of the similarities are fed into the decision tree learning as features on which to split.

Therefore, when distinguishing between class $C_1$ and $C_2$, the best feature to distinguish on may be the similarity to a target configuration from class $C_3$. Why should the similarity to a target configuration in $C_3$ affect whether the adversary is of type $C_1$ or $C_2$? Consider the classes to be two rectangular regions on the plane (not necessarily axis-parallel), with their centers on the x-axis. Just measuring the similarity between the centers of the two regions allows you to split the plane only with lines parallel to the y-axis. Using other centers can allow you the split the plane in other directions.

Decision trees also allow us to capture disjunctive concepts. For example, a team may have several modes of behavior, like "defensive", "offensive" or "goal kicks". Comparing to a single target configuration does not allow us to effectively capture these different modes. Having a greater variety of

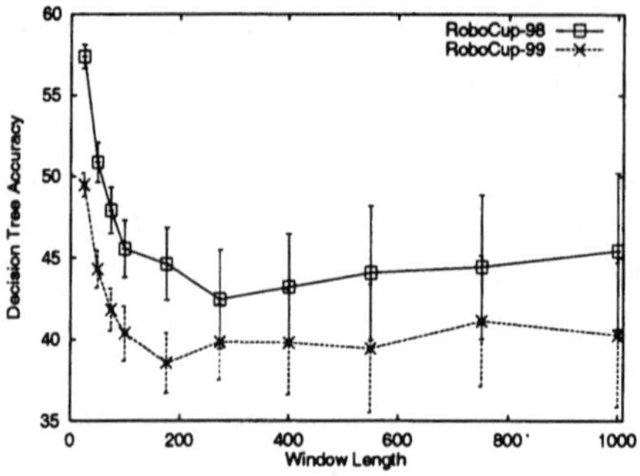


(a) All Data

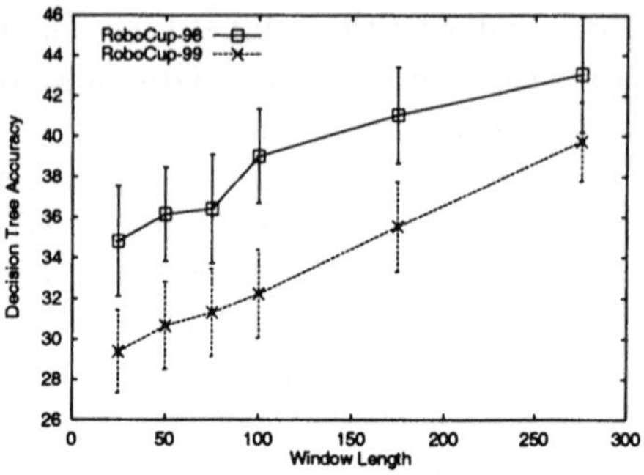


(b) Normalized Amount of Data

**Fig. 3.** Classification Accuracy (note that accuracy of random guessing < 3%)

configurations allows us to express that adversary type $A$ may have different distinct modes of behavior.

We are also interested in how varying the window length affects classification accuracy. Too short of a window presumably does not allow the agent time to capture the occurrence of important events. Using too long of a window negatively affects the goal of quick adaptation to an adversary, as well as causing unnecessary smoothing of the features.

# 5 Results

In the experiments, all of the data was used to generate the target configurations. Then, each time window was observed again, and the similarities to all of the target configurations were recorded. 90% of the data was then used for training of the decision tree and 10% was reserved for testing. All accuracy rates reported are for the testing data. In order to better ascertain the accuracy, 50 decision trees were generated for each window length, using different randomly selected training and test data sets.

The results are shown in Fig. 3(a). The results are somewhat surprising: the smallest windows have the highest accuracy rates. It was expected that too small a window would not allow many important features to be captured.

Notice that as window length decreases, the amount of data available for decision tree training increases like $1/x$. Increasing the data available for training often greatly increases accuracy in decision tree training.

Therefore, we trained decision trees for all window lengths less than 275 with reduced data and testing sets. That is, the data sets for all of these window lengths were reduced to the size of the data set for a window length of 275. The data sets were reduced uniformly in the sense that if the data needed to be reduced by a factor of $n$, 1 out of every $n$ consecutive windows was kept.

The results for these trials is shown in Fig. 3(b). This error curve is a bit more as expected, with the smaller epochs performing slightly worse.

# 6 Conclusions and Future Work

Accuracies around 40% may not seem all that successful, but there are several points to note:

- There are 34-37 classes to be distinguished, and the accuracy of guessing is less than 3%. Therefore, our accuracy is on order of magnitude better than guessing.
- The data sets are "real" data sets. The logfiles from the RoboCup competitions are a result of many people working to produce interesting teams, and reflect a great variety of techniques and strategies.
- One class was made for each named team in the data sets. If two teams play similarly, we are still requiring that the classification methods distinguish them. It is likely that a more refined method of creating adversary classes would have higher classification accuracy.

The difference in the accuracy trends on epoch length points out an interesting tradeoff in this windowing approach. The smaller the windows used, the more training data one will get, probably improving accuracy. However, a small window also means that some necessarily time consuming events will rarely be observed because they will not occur perfectly inside of a window. Also, a small window is likely to give higher random fluctuation in behaviors.

Given the importance of choosing the correct window length, the ability to dynamically change the window length would probably improve the system. Alternatively, running several different window lengths in parallel and using a meta-learner to combine the classifications is also an interesting future direction.

The main contributions of this research are:

- A windowing approach to extracting useful features in complex domains.
- A successful classification method using decision tree learning.
- A demonstration that this approach to classification is effective in a particular complex, dynamic domain.

As indicated in the introduction, this research addresses only part of the whole problem of adversary adaptation. For example, in this research, an adversary class was generated for each named team of agents. However, creation of adversary classes should be based on playing style and not solely on *a priori* team distinctions, suggesting a clustering approach. Further, understanding how to map adversary classes to strategy changes requires a greater understanding of the different strategies our own agent team may employ.

The use of Hidden Markov Models in the updating function $f$ of observations also seems promising in creating more complex and reliable observations. Hidden Markov Models have been used successfully in recognizing simple behaviors in robotic soccer[2].

While the use of decision trees to produce classifications gave fairly good results, having a classification method which indicates how *well* this adversary matches to all classes would be helpful. Other statistically based machine learning algorithms could produce this kind of information. That way, truly novel adversaries would be recognized, as would adversaries who have features from several adversary classes.

We have described a general method for adapting to adversaries in complex domains. We further explored some pieces of this approach, using decision trees in the final classification step. We also presented experimental results for this system in the domain of simulated robotic soccer. We hope that this approach to adversary adaptation will further developed, tested, and improved.

## References


1. D. Carmel and S. Markovitch. Opponent modeling in multi-agent systems. In G. Weiss and S. Sen, editors, *Adaptation and Learning in Multi-Agent Systems*, Lecture Notes in Artificial Intelligence, pages 40–52. Springer, 1995.
2. K. Han and M. Veloso. Automated robot behavior recognition applied to robotic soccer. In *Proceedings of IJCAI-99 Workshop on Team Behaviors and Plan Recognition*, 1999.
3. M. J. Matatić. Learning in multi-robot systems. In G. Weiss and S. Sen, editors, *Adaptation and Learning in Multi-Agent Systems*, Lecture Notes in Artificial Intelligence, pages 206–217. IJCAI'95 Workshop, Springer, 1995.
4. I. Noda, H. Matsubara, K. Hiraki, and I. Frank. Soccer server: A tool for research on multiagent systems. *Applied Artificial Intelligence*, 12:233–250, 1998.
5. J. R. Quinlan. *C4.5: Programs for Machine Learning.* Morgan Kaufmann, San Mateo, CA, 1993.
6. M. Ramoni, P. Sebastiani, P. Cohen, J. Warwick, and J. Davis. Bayesian clustering by dynamics. Technical Report KMi-TR-78, Knowledge Media Institute, The Open University, United Kindgdom MK7 6AA, February 1999.
7. P. Riley. Classifying adversarial behaviors in a dynamic, inaccessible, multi-agent environment. Technical Report CMU-CS-99-175, Carnegie Mellon University, 1999.
8. S. Sen and M. Sekaran. Using reciprocity to adapt to others. In G. Weiss and S. Sen, editors, *Adaptation and Learning in Multi-Agent Systems*, Lecture Notes in Artificial Intelligence, pages 206–217. IJCAI'95 Workshop, Springer, 1995.
9. P. Stone, M. Veloso, and P. Riley. The CMUnited-98 champion simulator team. In M. Asada and H. Kitano, editors, *RoboCup-98: Robot Soccer World Cup II*, pages 61–76. Springer Verlag, Berlin, 1999.
10. B. Thom. Learning models for interactive melodic improvisation. In *International Conference on Computer Music*, China, October 1999.

# Learning-Based Task Allocation in Decentralized Multirobot Systems

Poj Tangamchit(poj@andrew.cmu.edu)[1], John M. Dolan(jmd@ri.cmu.edu)[2], and Pradeep K. Khosla(pkk@cs.cmu.edu)[1,2]

[1] Dept. of Electrical and Computer Engineering, Carnegie Mellon Univ. , Pittsburgh PA, USA
[2] The Robotics Institute, Carnegie Mellon Univ. , Pittsburgh PA, USA

**Abstract.** This paper presents an efficient decentralized multirobot system through the use of reinforcement learning coupled with heuristics to accelerate the learning process. Our heuristic functions are not reinforcement signals, but are biases to the learning space that help accelerate the exploration process and have no effect on the robot's knowledge. Two kinds of heuristics are used: local and global. The local heuristic is derived from the local knowledge of the robot, whereas the global heuristic is derived from the data obtained from other robots. Both heuristics can help accelerating the learning process.

## 1 Introduction

A basic design choice in multirobot systems is that between centralized control, in which a single entity plans and directs all system actions, and decentralized control, in which system behavior emerges from the collaboration of multiple entities. Centralized systems are easier to implement and optimize, whereas decentralized systems are superior in terms of robustness and scalability. Robustness is the ability of a system to gracefully degrade when some units in the system fail. A decentralized system does not rely on a single central unit whose failure will cause the whole system to die. Scalability is the ability of a system to handle a large number of robots, a computationally intensive task for centralized systems. Decentralized systems are worthy of consideration due to these advantages, but their lack of centralized planning can lead to inefficiency and suboptimality. This paper presents a decentralized system with improved efficiency through the use of reinforcement learning coupled with heuristics to speed up the learning rate. Unlike related work in reinforcement learning, our heuristics are not reinforcement signals, which generate rules in the robots' knowledge base. Instead, we use a simple reinforcement signal and add heuristics only to generate a bias in the exploration of the state-action space.

This paper is organized as follows. Section 2 presents related work in decentralized systems and reinforcement learning. Section 3 poses the multirobot-patrolling problem. Section 4 describes our approach to this problem, discussing dynamic task allocation, reinforcement learning and heuristics to improve the task-allocation efficiency. Section 5 presents results and Section 6 gives the conclusion.

## 2 Related Work

Parker [1] used the idea of dynamic task allocation in a decentralized multirobot architecture by maintaining a list of tasks that robots can choose from and allowing those robots to choose tasks independently. Parker has extended dynamic task allocation to create a new architecture called ALLIANCE, which has a mechanism of impatience and acquiescence, promoting robustness and fault tolerance. She used reinforcement learning to optimize the parameters of impatience and acquiescence motivation. Mataric [2] used reinforcement learning for a group of robots performing a foraging task. In her algorithm, she introduced the progress estimator, which was used to increase the performance of the reinforcement signal. The progress estimator is a special reinforcement signal, added to the normal reinforcement signal in order to give an additional guide to robots about their actions. Balch [3] showed that different types of reinforcement signals could generate different behaviors on robots.

The work presented here differs from previous work in that a conventional, rather than a special, reinforcement signal is used. Reinforcement signals normally generate rules in the robots' knowledge. The use of a special reinforcement signal may generate rules that are undesirable in that they may distract from the main goal, which is supported by the conventional reinforcement signal. In this paper, we use heuristic functions to generate biases in exploration of the state-action space. As a result, our heuristic functions are decoupled from the robots' knowledge. This method maintains the original purpose of the learning and eliminates the side effect caused by a special reinforcement signal.

## 3 Problem Statement

We chose the multirobot-patrolling problem as a testbed. A group of robots with short-range vision sensing must patrol a large area by continuously traversing a set of checkpoints, like real security guards. The robots need to decide how best to divide up the patrolling points. When a robot reaches a point, it selects and goes to another point. After all the points are reached, a new cycle begins. Figure 1 shows a rectangular area with six patrolling points and two robots. Each robot has its own trajectory for a cycle. Every point needs to be reached to complete a cycle. We assume that the robots already have a navigation algorithm to avoid obstacles and move to goal points (see, e.g., [6,7]).

Because the system is fully decentralized, each robot will not know ahead of time what other robots are doing. To eliminate repeated actions, each

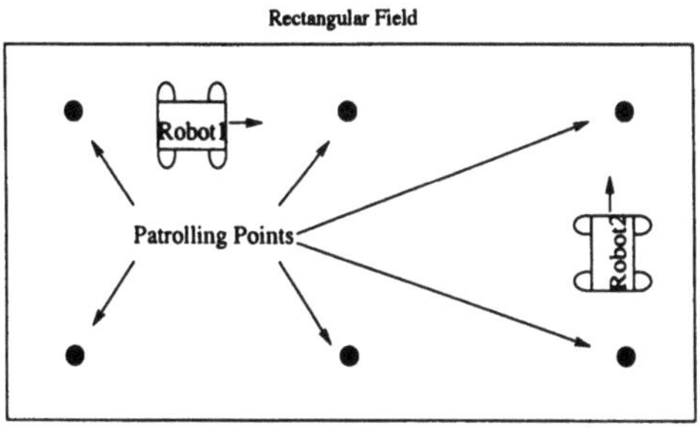


**Fig. 1.** Multirobot-Patrolling Problem

robot needs to broadcast messages to all other robots telling which task it is doing. The other robots will then know which patrolling points have not been reached. They will consider those points for the next task. This mechanism is called dynamic task allocation, in which each robot maintains a synchronized list of points (tasks) which have not been visited and chooses the next point from that list. The choice of the point sequence is based on minimizing the time used in each patrolling cycle. Choosing a different order of patrolling points will result in a different cycle time. Although this problem may look like the traveling salesman problem (TSP), it is different, due to its dynamic nature. TSP has a global planner, so the environment is deterministic. In TSP, one can plan the path before doing the task, whereas in the multirobot patrolling problem, one cannot (see [4] for details about TSP).

# 4 Approach

We begin this section by introducing dynamic task allocation, which is the task management mechanism in our system. We then describe how reinforcement learning is applied to dynamic task allocation. Next, we discuss the knowledge that we use to construct the heuristic functions. Finally, we present two types of heuristic functions.

## 4.1 Dynamic Task Allocation

Our decentralized implementation has no mechanism of instantaneous central management (like a master/slave or leader vote) which dynamically promotes one unit to be a central manager. In our system, each robot manages tasks by itself using communications with other robots in order to achieve the ultimate goal of the group. This method yields a higher level of robustness and scalability. This mechanism is called dynamic task allocation or dynamic task selection [5] (see Figure 2). It maintains a local synchronized task pool on each robot and assumes that the task pool is unified. Each robot needs to actively choose tasks from the task pool by itself.

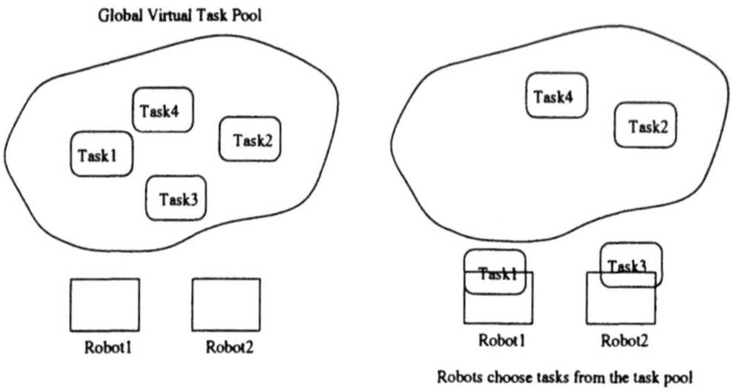


**Fig. 2.** Dynamic Task Allocation

The robots may choose tasks randomly from the task pool or by following some rules. Choosing tasks in different orders may result in differing system performance. In the next subsection, we discuss the use of reinforcement learning to generate rules for choosing tasks in order to optimize system performance.

### 4.2 Reinforcement Learning

Reinforcement learning has proven to be a successful learning method in robotic applications. It requires only a feedback signal (called a reinforcement signal) to tell a robot how good the behavior of the robot is. If the reinforcement signal is good for a given state-action pair, then the probability of the robot choosing this action given the same state is higher. If the reinforcement signal is bad, the robot will tend to choose another action instead the next time.

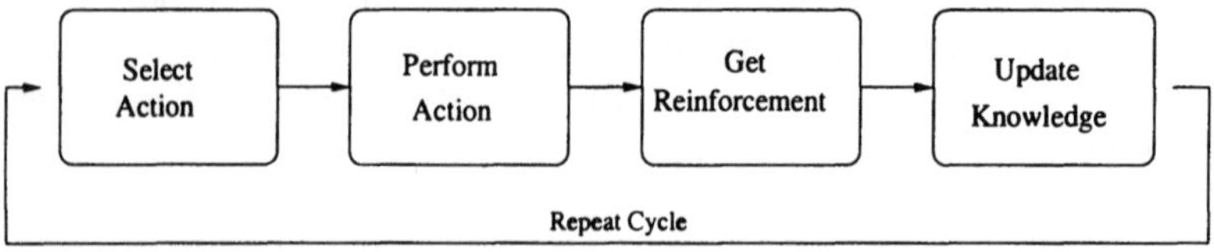


**Fig. 3.** Procedure of typical Reinforcement Learning

In order to decide which point to choose to optimize the patrolling cycle time in the multirobot-patrolling problem, we implement reinforcement learning on every robot using the travelling distance of each robot in each cycle as a reinforcement signal. The distance is measured as a closed-loop path to form a cycle (start and end at the same point). The state of the robot is the last point that it visits. The action of the robot is the next point that it will go to. The robot that goes to the first and the last patrolling points in the cycle needs to send messages to tell other robots about these events so that every robot can measure the cycle time. In the learning experiment with two robots, learning progress was very slow. The learning process required more than 125 cycles to become stable (see section 5 for more details).

### 4.3 Local Knowledge and Global Knowledge

Robots can use their knowledge to decide which task they should perform. There are two types of knowledge: local knowledge and global knowledge. Local knowledge is the information within the robot itself. Other robots will not have this information. Global knowledge is the information shared among robots. Global knowledge needs to be exactly the same for all robots with an acceptable delay. An example of local knowledge of a robot is its position or its current task, whereas an example of global knowledge is another robot's position or another robot's current task. Usually, global knowledge is important and necessary for multirobot systems to exhibit cooperative behavior. A robot can make local knowledge global via communication. For example, if a robot broadcasts its current position to its teammates, this position data will become global knowledge to all robots. If all robots can share all information they have, the system will act like a single entity. However, the bandwidth of the communication channel is always limited. Therefore, robots need to communicate only the most useful information.

### 4.4 Heuristic Functions

Due to the dynamic environment, the learning process in the multirobot patrolling problem is unacceptably slow. We therefore tried to accelerate the learning process by introducing heuristics to guide the patrolling-point selection. A heuristic function is a set of rules used to prejudge the value of an action. However, these rules are sometimes inadequate to accurately judge the value of an action. An example of a heuristic is the distance to the goal in a robot navigation task. In order to navigate to a goal point, there are several choices of via points that the robot can visit. A point that has the smallest distance to the goal is preferred by this heuristic. However, this point could be a bad choice if it is blocked by an obstacle so that the robot cannot really go to the goal via this point. The heuristic does not have information about the obstacle and provides a bad result in this situation.

Heuristic functions can be divided into two types: local and global. A local heuristic function uses information from local knowledge, whereas a global heuristic function is constructed from both global and local knowledge. We use heuristic functions to produce bias in learning exploration. They are not a part of the reinforcement signal, which generates rules in the robot's knowledge. Because heuristics are not guaranteed to choose good actions, they may generate bad rules in the robot's knowledge if used as a reinforcement signal. Instead, heuristic functions are used in the action-selection step (see figure 4).

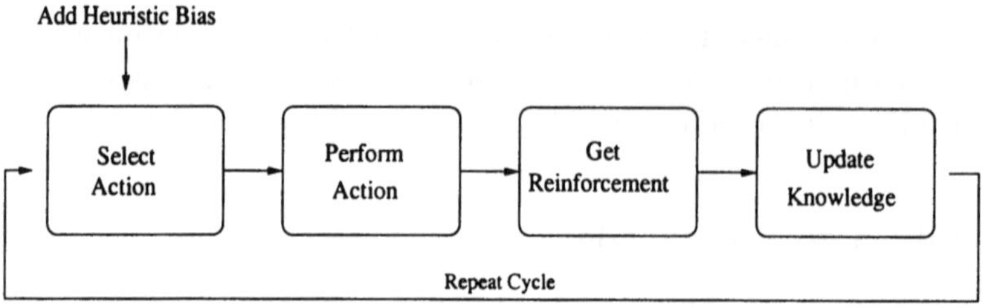


**Fig. 4.** Heuristic Function on Reinforcement Learning

For a local heuristic in the multirobot patrolling problem, we use the same heuristic as is common in the TSP, called the "nearest-neighbor" heuristic (see [4] for more details). The nearest-neighbor heuristic encourages the robot to choose the nearest point that has not been visited. This heuristic is simple, well known, and effective. Let $h$ denote the heuristic function, where $i$ denote the $i^{th}$ patrolling point that has not been visited. Then,

$$h_{local}(i) = \texttt{distance from the robot to point i}$$

For the global heuristic, we used the fact that the robots need to be maximally spatially distributed for efficient patrolling, so that each robot is encouraged to choose the patrolling point that is farthest from other robots:

$$h_{global}(i) = \texttt{min distance from the other robots to point i}$$

Note that the robots need to choose the point with a large $h_{global}$ and a small $h_{local}$. These heuristics are not only suitable for the multirobot-patrolling problem. They are also suitable for the mission that requires the robots to navigate to different work sites located far away from each other. However, for missions that require little navigation, we may need other forms of heuristics that do not depend on the distance.

## 5 Experiment and Results

We tested the performance of learning with heuristics in simulation. Two simulated robots patrolled a rectangular field with six patrolling points located at the edges and corners (see Figure 5). The optimal solution for this problem is to have the points divided equally among the robots, resulting in two closed-loop triangular paths.

We define a set of actions and states equal to the number of the patrolling points. All transitions to and from any state are valid. However, the points that are already visited are removed. That is, each robot can move from its current point to any point in the area that has not been visited. In our experiment, we define

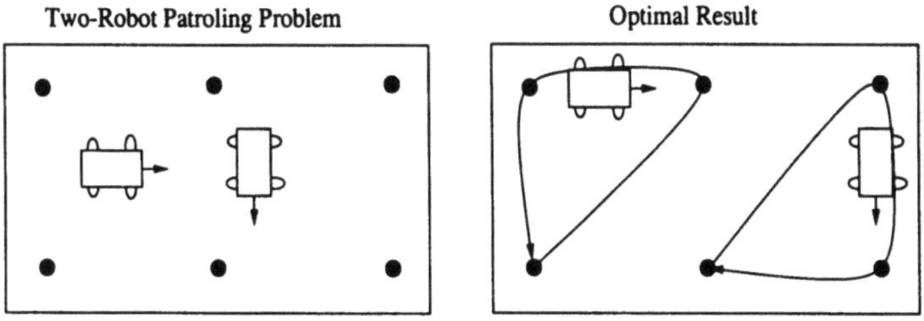


**Fig. 5.** Two-robot Patrolling Problem Experiment

$$State = (point1, point2, point3, point4, point5, point6)$$
$$Action = (gotopoint1, gotopoint2, gotopoint3, gotopoint4, gotopoint5, gotopoint6)$$

Each robot's knowledge base is represented with a 6x6 matrix, which keeps scores of state-action pairs. The rows of the matrix represent six states and the columns of the matrix represent six actions. The smaller the value of the $ij^{th}$ element in the matrix, the better it is for the robot at point $i$ to move to point $j$ in the next step. For example, if the robot is at point 3, it will look at the third column of the matrix and find the smallest value. If the smallest value is at column 5, the robot will choose to go to the fifth point in the next step. At the beginning, when the robot has not learnt anything yet, all the values in the matrix are initially set to zero. When the robot learns, it keeps track of actions that it does in a cycle and measures the time for a complete cycle. The time is used as the reinforcement signal to modify the matrix based on state-action pairs.

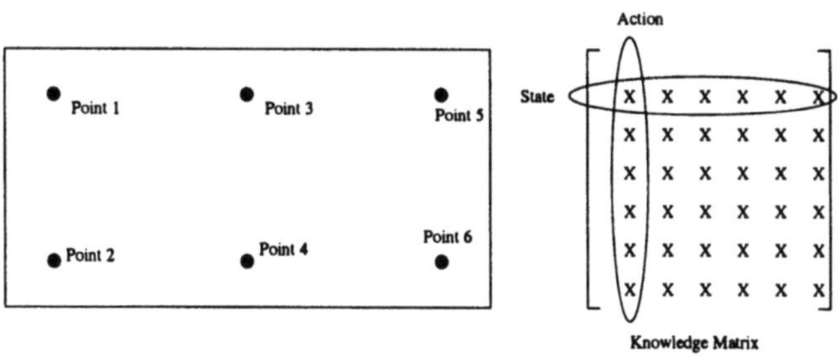


**Fig. 6.** Patrolling Points and Knowledge Matrix in the Experiment

We have tested the learning process with four scenarios: without heuristic, with local heuristic, with global heuristic, and with both local and global heuristics. The number of cycles needed for the learning process to become stable is recorded. The system is considered to be stable if the path that the robots take is unchanged, meaning that there is no improvement in the learning process. Figure 7 shows the learning progress on one robot using different heuristics. The X-axis represents the number of learning cycles. The Y-axis represents the total travelling distance in a cycle (this value is the reinforcement signal). This value fluctuates in the beginning of the learning process and becomes stable when the robots learn the optimal path. The table shows the number of learning cycles the robots need to stably follow the optimal path.

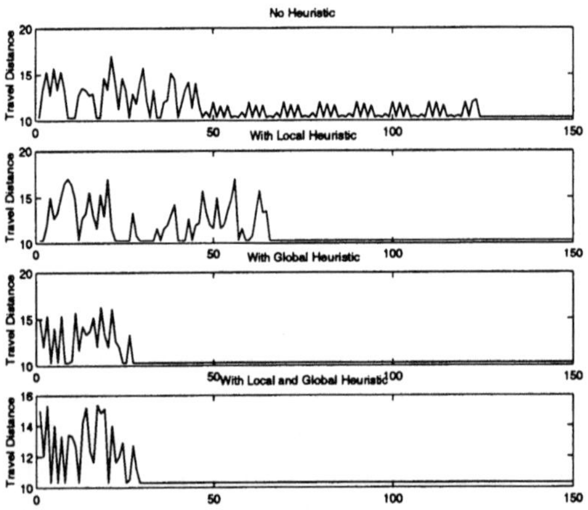


**Fig. 7.** Plot of Learning Cycle and the Cycle Distance with Different Heuristics

**Table 1.** Learning Result

| Heuristic/Learning Cycles | 2 Robots 6 Points | 3 Robots 6 Points | 3 Robots 9 Points |
|---|---|---|---|
| No Heuristic | 125 | 95 | 310 |
| Local Heuristic | 65 | 70 | 200 |
| Global Heuristic | 27 | 78 | 180 |
| Local & Global Heuristics | 30 | 115 | 210 |

The result shows a long learning time with no heuristic. After we add the local heuristic, the learning time is reduced by more than half. Using the global heuristic, the learning time is even shorter. When we combine both local and global heuristics, the learning process is somewhat worse than with the global heuristic alone. This is because the local heuristic and the global heuristic generate conflicting decisions. When we combine the bias generated from both heuristics, it provides a guide to bad directions. This combination of heuristics can even decelerate the learning process. Using the global heuristic alone is sufficient. However, it requires direct information from other robots, which requires significant communications among robots.

In additional experiments, we used more robots and added more points. We considered the cases of 3 robots/6 points (Figure 8) and 3 robots/9 points (Figure 9). The optimal paths that the learning algorithm found are also shown. The points are divided equally among the three robots. The heuristics cannot produce much improvement when the number of points per robot is small (two points per robot in Figure 8). In the third experiment (Figure 9), increasing the number of points increases the percentage of improvement. This is because when the robots have more points to choose from, they can benefit more from the heuristic guidance. When there are only two points for each robot, the heuristics do not give much useful information more than the information that the robots have.

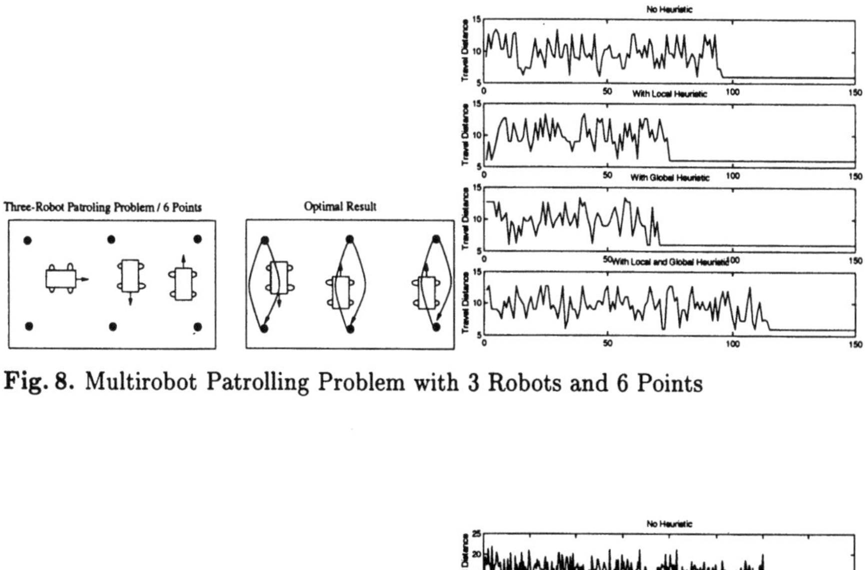


**Fig. 8.** Multirobot Patrolling Problem with 3 Robots and 6 Points

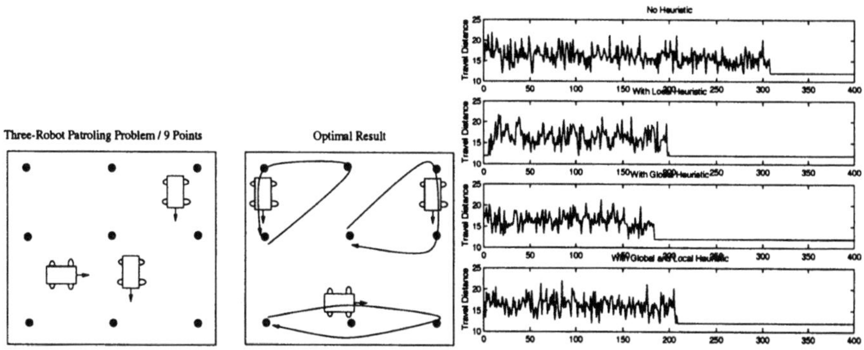


**Fig. 9.** Multirobot Patrolling Problem with 3 Robots and 9 Points

## 6 Conclusion

Decentralized multirobot systems are potentially more robust and scalable than centralized systems, but they have no inherent means of organization. The use of dynamic task allocation and reinforcement learning allows the system to self-adapt and organize the mission, but it is unacceptably slow. To deal with this problem, we used heuristic functions to generate bias in the learning process and achieve improvement of the learning speed. Different heuristics resulted in different performance. The global heuristic alone performed better than either a local heuristic alone or the combination of local and global heuristics.

## References

1. L. Parker (1994) Heterogeneous Multi-Robot Cooperation PhD. Thesis, Dept. of Electrical Eng. and Computer Science MIT.
2. M. Mataric (1998) Learning in Multi-Robot Systems, Technical Report, CS Dept. Brandeis University.
3. T. Balch (1998) Behavioral Diversity in Learning Robot Teams, PhD. Thesis, Computer Science Dept., Georgia Tech.
4. E.L. Lawler (1985) The Traveling Salesman Problem, John Wiley & Sons.
5. P. Tangamchit et al. (2000) Dynamic Task Selection: A Simple Structure for Multirobot Systems, DARS 4, Springer Verlag.
6. R.A. Brooks (1985) A Robust Layered Control System for A Mobile Robot, MIT AI-LAB Technical Report.
7. R.C. Arkin (1989) Motor schema-based mobile robot navigation, The International Journal of Robotics Research 8(4):92-112.
8. Y.U. Cao, A.S. Fukunaga, A.B. Khang (1997) Cooperative Mobile Robotics: Antecedents and Directions, Autonomous Robots, 4, 1-23 Kluwer Academic Publishers.
9. P. Stone, M. Veloso (1997) Multiagent System: A Survey from a Machine Learning Perspective, CMU CS Technical Report.

# Multi-Robot Learning in a Cooperative Observation Task

Lynne E. Parker and Claude Touzet

Center for Engineering Science Advanced Research
Computer Science and Mathematics Division
Oak Ridge National Laboratory, Oak Ridge TN 37831-6355, USA
email: ParkerLE@ornl.gov

**Abstract.** An important need in multi-robot systems is the development of mechanisms that enable robot teams to autonomously generate cooperative behaviors. This paper first briefly presents the Cooperative Multi-robot Observation of Multiple Moving Targets (CMOMMT) application as a rich domain for studying the issues of multi-robot learning of new behaviors. We discuss the results of our hand-generated algorithm for CMOMMT, and then describe our research in generating multi-robot learning techniques for the CMOMMT application, comparing the results to the hand-generated solutions. Our results show that, while the learning approach performs better than random, naive approaches, much room still remains to match the results obtained from the hand-generated approach. The ultimate goal of this research is to develop techniques for multi-robot learning and adaptation that will generalize to cooperative robot applications in many domains, thus facilitating the practical use of multi-robot teams in a wide variety of real-world applications.

## 1 Introduction

Research in multi-robot cooperation has grown significantly in recent years. Before multi-robot teams will ever become widely used in practice, however, we believe that advances must be made in the development of mechanisms that enable the robot teams to autonomously generate cooperative behaviors and techniques. With the current state of the art, the implementation of cooperative behaviors on physical robot teams requires expert behavior programming and experimentation, followed by extensive tuning and revision of the cooperative control algorithms. It is unlikely that a significant real-world impact of cooperative robot teams will occur as long as the current level of effort is required to implement these systems.

Researchers have recognized that an approach with more potential for the development of cooperative control mechanisms is autonomous learning. Hence, much current work is ongoing in the field of multi-agent learning (e.g., [12]). This paper discusses our research in the learning of new behaviors in multi-robot teams. The types of applications that are typically studied for this area of multi-robot learning vary considerably in their characteristics. Some of the applications include air fleet control [9], predator/prey [2,3], box pushing [4], foraging [6], and multi-robot soccer [10,5]. Particularly challenging domains for multi-robot learning are those tasks that are *inherently* cooperative. By this, we mean that the utility of the action of one robot is dependent upon the current actions of the other team members. Inherently cooperative tasks cannot be decomposed into independent subtasks to be solved by a distributed robot team. Instead, the success of the team throughout its execution is measured by the combined actions of the robot team, rather than the individual robot actions. This type of task is a particular challenge in multi-robot learning, due to the difficulty of assigning credit for the individual actions of the robot team members.

Of these previous application domains that have been studied in the context of multi-robot learning, only the multi-robot soccer domain addresses inherently cooperative tasks with more than two robots while also addressing the real-world complexities of embodied robotics, such as noisy and inaccurate sensors and effectors in a dynamic environment that is poorly modeled. To add to the field of challenging application domains for multi-robot learning, we have defined and have been studying a new application domain – the Cooperative Multi-robot Observation of Multiple Moving Targets (CMOMMT) – that is not only an inherently cooperative task, but, unlike the multi-robot soccer domain, is also a domain that must deal with issues of scalability to large numbers of robots. In [7] we presented the motivation for using this domain for multi-robot learning. In this paper, we briefly review this motivation, and then describe a hand-generated solution to this problem, along with the results we obtained with this approach. We then define a learning approach to enable robot teams to autonomously generate viable solutions to the CMOMMT application and compare the results to the hand-generated approach. The final section of the paper concludes with some summary remarks.

## 2 The CMOMMT Application

The application domain that we are studying for use as a multi-robot learning testbed is the problem we entitle *Cooperative Multi-robot Observation of Multiple Moving Targets* (CMOMMT). This problem is defined as follows. Given:

- $\mathcal{S}$ : a two-dimensional, bounded, enclosed spatial region
- $\mathcal{V}$ : a team of $m$ robot vehicles, $v_i, i = 1, 2, ...m$, with $360^0$ field of view observation sensors that are noisy and of limited range
- $\mathcal{O}(t)$ : a set of $n$ targets, $o_j(t)$, $j = 1, 2, ..., n$, such that target $o_j(t)$ is located within region $\mathcal{S}$ at time $t$

We say that a robot, $v_i$, is *observing* a target when the target is within $v_i$'s sensing range. Define an $m \times n$ matrix $B(t)$, as follows:

$$B(t) = [b_{ij}(t)]_{m \times n} \text{ such that } b_{ij}(t) = \begin{cases} 1 \text{ if robot } v_i \text{ is } \textit{observing} \text{ target} \\ \quad o_j(t) \text{ in } \mathcal{S} \text{ at time } t \\ 0 \text{ otherwise} \end{cases}$$

Then, the goal is to develop an algorithm, which we call *A-CMOMMT*, that maximizes the following metric $A$:

$$A = \sum_{t=1}^{T} \sum_{j=1}^{n} \frac{g(B(t), j)}{T}$$

where:

$$g(B(t), j) = \begin{cases} 1 \text{ if there exists an } i \text{ such that } b_{ij}(t) = 1 \\ 0 \text{ otherwise} \end{cases}$$

That is, the goal of the robots is to maximize the average number of targets in $\mathcal{S}$ that are being observed by at least one robot throughout the mission that is of length $T$ time units. Additionally, we define $sensor\_coverage(v_i)$ as the region visible to robot $v_i$'s observation sensors, for $v_i \in \mathcal{V}$. Then we assume that, in general, the maximum region covered by the observation sensors of the robot team is much less than the total region to be observed. That is, $\bigcup_{v_i \in \mathcal{V}} sensor\_coverage(v_i) \ll \mathcal{S}$. This implies that fixed robot sensing locations or sensing paths will not be adequate in general, and instead, the robots must move dynamically as targets appear in order to maintain their target observations and to maximize the coverage.

The CMOMMT application is an excellent domain for embodied multi-robot learning and adaptation. CMOMMT offers a rich testbed for research in multi-robot cooperation, learning, and adaptation because it is an inherently cooperative task. In addition, many variations on the dynamic, distributed sensory coverage problem are possible, making the CMOMMT problem arbitrarily more difficult. For example, the relative numbers and speeds of the robots and the targets to be tracked can vary, the availability of inter-robot communication can vary, the robots can differ in their sensing and movement capabilities, the terrain may be either enclosed or have entrances that allow objects to enter and exit the area of interest, and so forth. Many other sub-problems can also be addressed, including the physical tracking of targets (e.g. using vision, sonar, IR, or laser range), prediction of target movements, multi-sensor fusion, and so forth.

## 3 A Hand-Generated Solution to CMOMMT

We have developed a hand-generated solution to the CMOMMT problem that performs well when compared to various control approaches. This solution has been implemented on both physical and simulated robots to demonstrate its effectiveness. The hand-generated solution, which we call *A-CMOMMT*, is described briefly as follows. Robots use weighted local force vectors that attract them to nearby targets and repel them from nearby robots. The weights are computed in real time by a higher-level reasoning system in each robot, and are based on the relative locations of the nearby robots and targets. The weights are aimed at generating an improved collective behavior across robots when utilized by all robot team members.

The local force vectors are calculated as follows. The magnitude of the force vector attraction of robot $v_l$ relative to target $o_k$, denoted $| \mathbf{f_{lk}} |$, for parameters $0 < do_1 < do_2 < do_3$, is:

$$| \mathbf{f_{lk}} |= \begin{cases} \frac{-1}{do_1} & \text{for } d(v_l, o_k) \leq do_1 \\ \frac{1}{do_2 - do_1} & \text{for } do_1 < d(v_l, o_k) \leq do_2 \\ \frac{-do_2}{do_3 - do_2} & \text{for } do_2 < d(v_l, o_k) \leq do_3 \\ 0 & \text{otherwise} \end{cases}$$

where $d(a, b)$ returns the distance between two entities (i.e., robots and/or targets). The magnitude of the force vector repulsion of robot $v_l$ relative to robot $v_i$, denoted $| \mathbf{g_{li}} |$, for parameters $0 < dr_1 < dr_2$, is:

$$| \mathbf{g_{li}} |= \begin{cases} -1 & \text{for } d(v_l, v_i) \leq dr_1 \\ \frac{1}{dr_2 - dr_1} & \text{for } dr_1 < d(v_l, v_i) \leq dr_2 \\ 0 & \text{otherwise} \end{cases}$$

Determining the proper setting of the parameters $do_1, do_2, do_3, dr_1$, and $dr_2$ is one approach to solving the CMOMMT multi-robot learning task using an ***a priori*** model-based technique.

Using only local force vectors for this problem neglects higher-level information that may be used to improve the team performance. Thus, the hand-generated approach enhances the control approach by including higher-level control to weight the contributions of each target's force field on the total computed field. This higher-level knowledge can express any information or heuristics that are known to result in more effective global control when used by each robot team member locally. The hand-generated approach expresses this higher-level knowledge in the form of a weight, $w_{lk}$, that reduces robot $r_l$'s attraction to a nearby target $o_k$ if that target is within the field of view of another nearby robot. Using these weights helps reduce the overlap of robot sensory areas toward the goal of minimizing the likelihood of a target escaping detection.

The higher-level weight information is combined with the local force vectors to generate the commanded direction of robot movement. This direction of movement for robot $v_l$ is given by: $\sum_{k=1}^{n} w_{lk} \mathbf{f_{lk}} + \sum_{i=1, i \neq l}^{m} \mathbf{g_{li}}$, where $\mathbf{f_{lk}}$ is the force vector attributed to target $o_k$ by robot $v_l$ and $\mathbf{g_{li}}$ is the force vector attributed to robot $v_i$ by robot $v_l$. To generate an $(x, y)$ coordinate indicating the desired location of the robot corresponding to the resultant force vector, we scale the resultant force vector based upon the size of the robot. The robot's speed and steering commands are then computed to move the robot in the direction of that desired location.

## 4 Results from Hand-Generated Solution

Figure 1 shows two of the simulation runs of the hand-generated algorithm (out of over 1,000,000 simulation test runs). Figure 2 shows snapshots of two of the physical robot experiments (out of over 800) in which the the robots perform the task either with no obstacles in the work area or with randomly distributed obstacles.

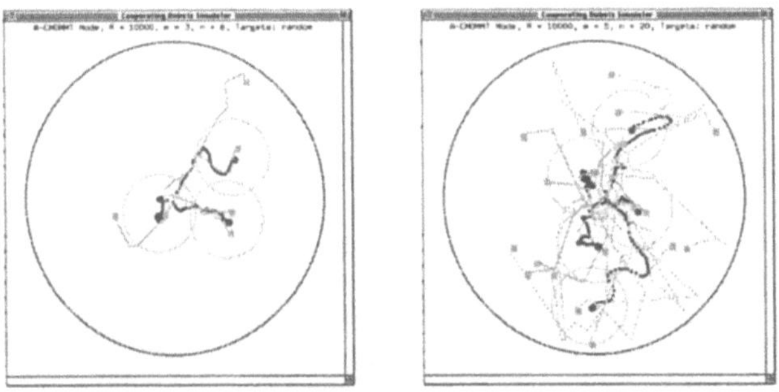

**Fig. 1.** Simulation results of three robots and six targets (first image), and five robots and twenty targets (second image), with the robots using the hand-generated solution to CMOMMT, and the targets moving randomly.

The results of the hand-generated approach to CMOMMT vary depending upon a number of factors, including the relative numbers of robots and targets, the size of the work area, the motions of the targets (i.e., whether random or evasive), and the setting of the weights. In general, the *A-CMOMMT* algorithm performed best for a ratio of targets to robots greater than 1/2. We compared the hand-generated *A-CMOMMT* approach with a non-weighted local force vector approach, as well as two control cases in which robots either maintained fixed positions or are moved randomly. Figure 3 gives a typical set of these comparative results.

## 5 Learning in the CMOMMT Application

We have studied the CMOMMT problem from a learning perspective without the assumption of an *a priori* model. This approach uses a combination of reinforcement learning, lazy learning, and a Pessimistic algorithm able to compute for each team member a lower bound on the utility of executing an action in a given situation. The challenges in this multi-robot learning

**Fig. 2.** Robot team executing hand-generated solution to CMOMMT. The first photo shows robots operating in an area with no obstacles. The second photo shows the robots amidst randomly distributed obstacles.

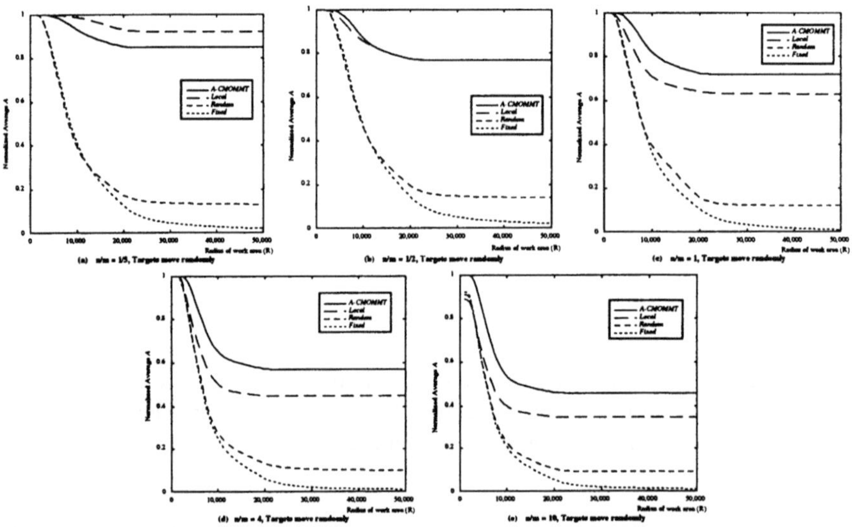

**Fig. 3.** Simulation results of four distributed approaches to cooperative observation, for random/linear target movements, for various ratios of number of targets ($n$) to number of robots ($m$).

problem include a very large search space, the need for communication or awareness of robot team members, and the difficulty of assigning credit in an inherently cooperative problem.

In this learning approach, lazy learning [1] is used to enable robot team members to build a memory of situation-action pairs through random exploration of the CMOMMT problem. A reinforcement function gives the utility of a given situation. The pessimistic algorithm for each robot then uses the utility values to select the action that maximizes the lower bound on utility. The resulting algorithm is able to perform considerably better than a random action policy, although it is still significantly inferior to the hand-generated algorithm described in the previous section. However, even with a performance less than that of the hand-generated solution, this approach makes an important contribution because it does not assume the existence of a model (as is the case in the Partially Observable Markov Decision Process (POMDP) domain), the existence of local indicators that help individual robots perform their tasks, nor the use of symbolic representations. The following subsections describe this approach and its results in more detail.

## 5.1 Lazy learning and Q-learning

Lazy learning [1] – also called instance-based learning – promotes the principle of delaying the use of the gathered information until the necessity arises (see Fig. 4). The same pool of information (i.e., memory) is used for different behavior syntheses. The lazy memory provides a good way of reducing the duration of any robotic learning application. In the context of reinforcement learning, lazy learning provides an instantaneous set of situation-action pairs (after the initial and unique sampling phase). Lazy learning samples the situation-action space according to a random action selection policy, storing the succession of events in memory and, when needed, probes the memory for the best action. The exploration phase is performed only once. By storing situation-action pairs, a lazy memory builds a model of the situation transition function.

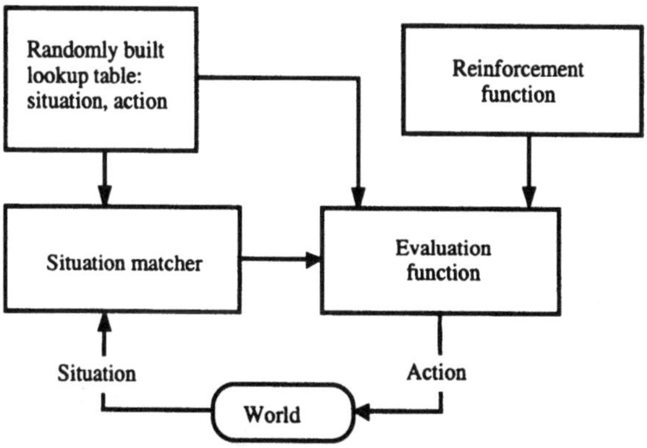


**Fig. 4.** Lazy learning: randomly sampled situation-action pairs in the lookup table are used by the situation matcher to select the action to execute in the current situation. The reinforcement function qualifies the actions proposed, helping to select the best one.

In order to express a behavior, the memory must be probed. To do this probing, we use a modified version of the technique proposed in [8]. In [8] the objective is to provide a method for predicting the rewards for state-action pairs without explicitly generating them. For the current real world situation, a situation matcher locates all the states in the memory that are within a given distance. If the situation matcher has failed to find any nearby situations, the action comparator selects an action at random. Otherwise, the action comparator examines the expected rewards associated with each of these situations and selects the action with the highest expected reward. This action is then executed, resulting in a new situation. There is a fixed probability (0.3) of generating a random action regardless of the outcome of the situation matcher. New situation-action pairs are added to the memory, along with a Q-value computed in the classical way [11]. Among similar situation-action pairs in the memory, an update of the stored Q-values is made. However, there is a limit to the generality of this lazy memory because the Q-values associated with the situation-action pairs only apply for a particular behavior. With the desire of reducing the learning time as much as possible, as well as preserving the generality of the lazy memory, we modified the algorithm as follows: (1) the situation matcher always proposes the set of nearest situations – no maximum distance is involved, (2) there is no random selection of actions by the action comparator, and (3) the Q-values are not stored with the situation-action pairs, but are computed dynamically as the need arises.

### 5.2 The Pessimistic Algorithm

We define a Pessimistic Algorithm for the selection of the best action to execute for a given robot in its current local situation as follows: find the lower bounds of the utility value associated with the various potential actions that may be conducted in the current situation, then choose the action with the greatest utility. A lower bound is defined as the lowest utility value associated with a set of similar situations.

The idea behind the Pessimistic Algorithm is that a local robot situation is an incomplete observation of the true state of the system. Thus, instead of trying to solve the observation problem by completing the observation (usual POMDP approach), we are only interested in ranking the utility of the actions. If we use a unique instance of the memory to obtain the utility of the situation, then chances are that the utility attributed to this local situation is due in fact to other robot's actions. This probability decreases proportionally with the number of similar situations that are taken into account. If a large number of situations are considered, then there must be at least one for which the reward directly depends on the local situation. By taking the minimum utility value of the set of similar situations, we are guaranteed that, if the value is null, then the situation achieved does not imply loosing target(s).

The Pessimistic Algorithm is then given as follows:

- `Let M be the memory, a lookup table of situation-action pairs gathered during an exploration phase --` $M = [(s(1), a(1)), ..., (s(t), a(t)), (s(t+1), a(t+1)), ...]$.
- `Let` *sit* `be the current situation.`

- Find $S(sit)$, the set of $n$ situations of $M$ similar to $sit$.
- Let $S_{follow}(sit)$ be the set of the situations that directly follows each situation of $S(sit)$.
- Compute the lower bound (LB) of the utility value (U) associated with each situation $s(k) \in S_{follow}(sit)$:
  - $LB(s(k)) = \min(U(s(m)))$, for $s(m) \in S(s(k))$, the set of situations similar to $s(k)$.
- Execute the action that should take the robot to the new situation $s^*$: $s^* = \max(LB(s))$ and $s \in S_{follow}(sit)$.

The utility U associated with a given situation can be computed in many ways. It can be the exact value of the reinforcement function for this particular situation-action pair, or it can be a more elaborate variable. For example, in our experience we store the situation-action pairs, plus the number of targets under observation in the lookup table (M). However, the value that is used as utility is +1 if one or more targets have been acquired compared to the previous situation, -1 if one or more targets have been lost, or 0 otherwise. An exact Q value requires running the Q-learning algorithm with the samples stored in the memory.

### 5.3 Results of Learning Approach

We studied the efficiency of the Pessimistic Algorithm by comparing the performance of a team of robots with a purely random action selection policy, a user-defined non-cooperative policy and *A-CMOMMT*. In these experiments, each robot situation is a vector of two times 16 components. The first 16 components code the position and orientation of the targets. It simulates a ring of 16 sensors uniformly distributed around the robot body. Each sensor measures the distance to the nearest target. The sensor position around the body gives the orientation. The second ring of 16 components code in the same manner the position and orientation of neighboring robots. The maximum range for a target or a robot to be seen is 1, for an arena radius of 5. The actions of each robot are rotation and forward movement. The measure of performance is the mean observation time of all targets.

Figure 5 shows the performance of a Pessimistic lazy Q-learning policy versus the size of the lazy memory, from 100 to 900 situation-action pairs. Each point is the average of 10 experiments. The lazy memories are obtained through an initial exploration involving from 15 to 25 targets and a single robot. During the sampling, the targets are fixed and the robot's policy is random action selection (with 5% chance of direction and orientation changes). The reinforcement function returns +1 if the total number of targets under observation increases, -1 if this number decreases, or 0 otherwise.

As we see there is an important performance gain associated with the Pessimistic lazy Q-learning over a purely random selection policy. This clearly demonstrates the importance of lazy Q-learning as a learning technique. Even more interestingly, lazy Q-learning performs much better than the user-defined non-cooperative policy (*Local*). It is important to note that neither policy is aware of the existence of the other robots. Both policies use the same sensory information – i.e., the distance and orientation of nearby targets. It is our opinion that the variation of performance is due to the fact that the lazy Q-learned behavior is somewhat less rigid than the user-defined policy. A lazy Q-learning guided robot will follow a target further than it could be, and, in doing so, will exhibit an erratic path, moving from one side of the target to another, back and forth without losing the target. In doing so, the surface under observation per unit of time is larger than the covered surface by the more rigid center-of-gravity-oriented robot. On the other hand, because it does not take into account the neighboring robots, it is easy to understand why the lazy Q-learned behavior performance cannot reach the level of the *A-CMOMMT* performance.

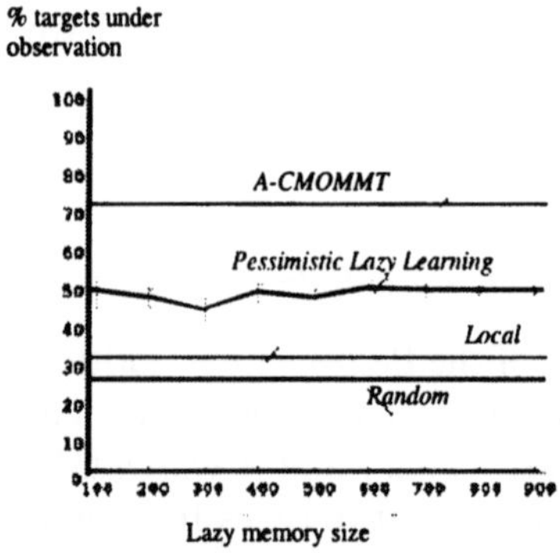


**Fig. 5.** Performances of the Pessimistic lazy Q-learning approach compared to a random action selection policy, a user-defined non-cooperative policy and the hand-generated solution *A-CMOMMT*, for 10 robots and 10 randomly moving targets. The results are the mean of 10 different experiments per point for lazy learning policy, and 100 experiments for the other 3 policies. Each experiment duration is 1000 iterations.

## 6 Conclusions

In this paper, we have proposed that the Cooperative Multi-robot Observation of Multiple Moving Targets (CMOMMT) application domain provides a rich testbed for learning and adaptation in multi-robot cooperative teams. We have described the need for learning and adaptation in multi-robot teams, and have defined the CMOMMT application, along with the characteristics that make it an interesting testbed for learning and adaptation. We reported on a hand-generated solution to the CMOMMT problem and discussed how the results from the implementation of this solution reveal the need for learning and adaptation in this domain. We discussed our work that uses the CMOMMT problem as a learning domain. The ultimate objective is to develop learning techniques using the CMOMMT domain that will generalize to other real-world domains, and will thus help realize the ultimate goal of enabling the widespread, practical use of multi-robot teams.

**Acknowledgments**

This research is sponsored by the Engineering Research Program of the Office of Basic Energy Sciences, U. S. Department of Energy. Accordingly, the U.S. Government retains a nonexclusive, royalty-free license to publish or reproduce the published form of this contribution, or allow others to do so, for U. S. Government purposes. Oak Ridge National Laboratory is managed by UT-Battelle, LLC for the U.S. Dept. of Energy under contract DE-AC05-00OR22725.

## References

1. D. Aha, editor. *Lazy Learning.* Kluwer Academic Publishers, 1997.
2. M. Benda, V. Jagannathan, and R. Dodhiawalla. On optimal cooperation of knowledge sources. Technical Report BCS-G2010-28, Boeing AI Center, August 1985.
3. Thomas Haynes and Sandip Sen. Evolving behavioral strategies in predators and prey. In Gerard Weiss and Sandip Sen, editors, *Adaptation and Learning in Multi-Agent Systems*, pages 113–126. Springer, 1986.
4. S. Mahadevan and J. Connell. Automatic programming of behavior-based robots using reinforcement learning. In *Proceedings of AAAI-91*, pages 8–14, 1991.
5. S. Marsella, J. Adibi, Y. Al-Onaizan, G. Kaminka, I. Muslea, and M. Tambe. On being a teammate: Experiences acquired in the design of RoboCup teams. In O. Etzioni, J. Muller, and J. Bradshaw, editors, *Proceedings of the Third Annual Conference on Autonomous Agents*, pages 221–227, 1999.
6. Maja Mataric. *Interaction and Intelligent Behavior.* PhD thesis, Massachusetts Institute of Technology, 1994.
7. L. E. Parker. A case study for life-long learning and adaptation in cooperative robot teams. In *Proceedings of SPIE Sensor Fusion and Decentralized Control in Robotic Systems II*, volume 3839, pages 92–101, 1999.
8. J.W. Sheppard and S.L. Salzberg. A teaching strategy for memory-based control. In D. Aha, editor, *Lazy Learning*, pages 343–370. Kluwer Academic Publishers, 1997.
9. Randall Steeb, Stephanie Cammarata, Frederick Hayes-Roth, Perry Thorndyke, and Robert Wesson. Distributed intelligence for air fleet control. Technical Report R-2728-AFPA, Rand Corp., 1981.
10. P. Stone and M. Veloso. A layered approach to learning client behaviors in the robocup soccer server. *Applied Artificial Intelligence*, 12:165–188, 1998.
11. C. Watkins. *Learning from Delayed Rewards.* PhD thesis, King's College, Cambridge, 1989.
12. Gerhard Weiss and Sandip Sen, editors. *Adaption and Learning in Multi-Agent Systems.* Springer, 1996.

# Ultrafast Neural Network Training for Robot Learning from Uncertain Data

J. Barhen and V. Protopopescu

*Center for Engineering Science Advanced Research*
Computer Science and Mathematics Division
Oak Ridge National Laboratory
*Oak Ridge, TN 37831-6355, USA*
barhenj@ornl.gov, vvp@ornl.gov

**Abstract.** A capability for learning from uncertain data has been a major and perennial requirement for many real-life robotic applications. In that context, a new methodology for ultrafast learning using neural networks is presented. It requires only a single iteration to train a feed-forward network with near-optimal results. Uncertainty reduction algorithms are also incorporated in a natural and optimal fashion. As such, this methodology is intended to become an essential building block for future architectures of intelligent systems. Its application to multi-robot observation of multiple moving targets is illustrated.

## 1. Introduction

Engineered intelligent systems such as autonomous robots behave with remarkable rigidity when compared to their biological counterparts, especially in their ability to recognize objects or speech, to manipulate and adapt in an unstructured environment, and to learn from past experience. They lack common sense reasoning and knowledge structures for processing complex patterns, i.e., they fail to recognize their own limitations. On the other hand, the biological machinery is capable of providing satisfactory solutions to such hard problems with remarkable ease and flexibility [1-3]. Hence, a cornerstone underlying any paradigmatic development for intelligent systems today is to understand and replicate the aforementioned unstructured computations. In particular, the spontaneous emergent ability of biological systems to self-organize and adapt their structure and function is of great interest.

Many unstructured robotics problems involve complex issues such as "emergent behavior" [4], sub-symbolic cognition [5], and/or pattern directed inference [6]. In recent years, the quest for innovative approaches to robot learning using neural networks has received considerable attention [7-11]. Early on, the emphasis was primarily on intelligent control of autonomous robots. The proven ability of neuromorphic algorithms to deal with uncertain information and to interact with dynamic environments is currently providing a strong incentive to extend their application to more complex problems in robotics. Of particular interest is the topic of cooperative learning, which is of crucial importance when attempting, for example, to control a team of robots [12]. A successful approach to this formidable challenge requires revolutionary advances in fast, efficient, and robust methodologies for training a neural network. Even though very promising techniques, such as distributed lazy Q-learning [13] and tropism-based cognition [14], have recently been proposed, conventional algorithms (including backpropagation or reinforcement learning) remain at the heart of state-of-the art methodologies. Hence, the purpose of this paper is to introduce to the robotics community the framework of a new, ultrafast and robust learning paradigm that could constitute an essential building block in future architectures of intelligent systems.

The remainder of this paper is organized along the following lines. In Section 2 we review available learning concepts for neural networks. A new ultrafast (single iteration) learning algorithm is presented in Section 3. The need for incorporating explicitly the uncertainties of the training data in the learning process is discussed in Section 4. The methodology is developed in detail in Section 5 for the neural network model introduced in Section 3. The highlights of our approach and its potential application to the problem of multi-robot observation of multiple moving targets are summarized in Section 6.

## 2. Learning and Neural Networks

Artificial neural networks are adaptive systems that process information by means of their response to discrete or continuous input [15]. Neural networks can provide practical solutions to a variety of artificial intelligence problems, including pattern recognition [16], autonomous knowledge acquisition from observations of correlated activities [17], real-time control of complex systems [18], and fast adaptive optimization [19]. At the heart of such advances lies the development of efficient computational methodologies for "learning" [20]. One typically categorizes learning into three paradigms: supervised, unsupervised, and reinforcement learning. Supervised learning requires a set of training data, which includes "correct" answers. It can be interpreted as a kind of function approximation. Unsupervised learning seeks appropriate representations of input signals to the network according to a specified criterion. In reinforcement learning, an agent tries to maximize the total amount of reward it receives when interacting with a complex, uncertain environment. In principle, no explicit set of answers is required. For certain classes of applications involving dynamical systems driven by non-stationary signals, the reformulation of reinforcement learning as supervised temporal learning with "teacher forcing" has been shown to result in considerable benefits in terms of computational cost and accuracy [21,22].

The development of neural learning algorithms has generally been based upon the minimization of an energy-like neuromorphic error function or functional [23]. Gradient-based techniques have typically provided the main computational mechanism for carrying out the minimization process, often resulting in excessive training times for the large-scale networks needed to address real-life applications. Consequently, to date, considerable efforts have been devoted to: (1) speeding up the rate of convergence [23,24] and (2) designing more efficient methodologies for deriving the gradients of these functions or functionals with respect to the parameters of the network [25,26]. The primary focus of such efforts has been on recurrent architectures. However, the use of gradient methods presents challenges even for the less demanding multi-layer feed-forward architectures. For instance, entrapment in local minima has remained one of the fundamental limitations of most currently available learning paradigms. The recent successful development of the innovative global optimization algorithm TRUST [27] has been suggested [28] as a promising new avenue for addressing such difficulties.

In a major departure from the above paradigms, Biegler-König and Bärman proposed a learning approach solely based on linear algebraic methods [29]. In their seminal paper, they observed that it is possible to separate the linear (inter-layer propagation) and nonlinear (individual neuron activation function) aspects of information propagation within a neural network. Using linear least squares, they computed the synaptic weights between each pair of layers. The inverted activation function enabled the accurate propagation of each remaining error back into the preceding layer of the network. The essence of their approach was to minimize the learning error at each layer separately, rather than globally, i.e. for the entire network.

Based on these ideas, we recently introduced [30] a single iteration training algorithm for a generalized feedforward neural network. In such a network architecture, the nodes (or neurons) are organized in layers, namely: (i) *input*, (ii) (one or several) *hidden*, and (iii) *output*. In addition to these traditional layers, we also introduced a *virtual* layer between the input layer and the (first) hidden layer. The virtual layer acts as a preprocessor of the input patterns, and replaces a highly overdetermined linear system with an invertible one. This method, implemented in the *DeepNet* code, was successfully demonstrated [31] in the accurate characterization of an oil field using low precision data from seismic sensors.

In this paper we further evolve the *DeepNet* architecture in view of potential robotics applications requiring real-time performance. We consider a network architecture having $N_0 = I$ input nodes, $N_1 = V$ virtual nodes, and $N_2 = O$ output nodes. The numbers $I$ and $O$ are equal to the dimensionalities of the input and output data and, for a given application, are in general fixed. The number $V$ is related to the number of examples, $K$, and may take different values for the same application. The goal of the learning process is to minimize the discrepancy between *DeepNet* predictions and measurements of the responses of the robot's "environment" for the same inputs. In particular, we wish to determine the synaptic interconnections, while incorporating explicitly the uncertainties associated with the training data.

The idea of our learning paradigm is the following. A two step algorithm is envisioned. In the first step, a novel architecture and ultrafast training procedure are introduced to determine the *nominal values* of the network parameters assuming no uncertainties in the data. In the second step, best estimates of these parameters are obtained by minimizing a generalized Bayesian loss function in a space where the inverse of a generalized covariance matrix (which captures all uncertainties) serves as metric of the computational manifold. As result of the minimization process, all uncertainties of interest are considerably reduced.

## 3. Ultrafast Computation of Nominal Network Parameters

Two sets of $K$ vectors are used for training. They are stored as rows of the matrices $\boldsymbol{\Omega}_{KI}$ and $\mathbf{R}_{KO}$ respectively, which represent the input signal patterns and the target outputs. The number of columns of each matrix equals the number of nodes of the corresponding processing layer. Typically, $K >> I$. For convenience, the matrix dimensions are explicitly indicated as subscripts.

Two successive *nonlinear* transformations map $\boldsymbol{\Omega}_{KI}$ into the $K$ x $V$ presynaptic matrix, $\mathbf{H}_{KV}$, output by the virtual layer. We construct these transformations such that $\mathbf{H}_{KV}$ becomes a *nonsingular square* matrix, which requires, in particular, that $V = K$ be chosen. We also decouple the nonlinearity of the transfer function at the output layer from the linear interlayer pattern propagation mediated by the synaptic weights $\mathbf{W}_{VO}$. This transformation is used to compute the postsynaptic input to the output layer as a $K$ x $O$ rectangular matrix. Since the latter is connected via a bijective sigmoid mapping to the output training examples, the synaptic interconnection matrices $\mathbf{W}_{VO}$ can be exactly determined by solving a system of linear equations [32]. Biases to the output layer can be included as needed.

The processing between the input and virtual layers is specified as follows. For a given set of training vectors, we assume that there exists a particular nonlinear transfer function $\psi$ that maps row vectors from the input pattern matrix $\boldsymbol{\Omega}_{KI}$ to row vectors of the postsynaptic matrix $\Xi_{KK}$. The usual sigmoid transform $\varphi$ is applied to each element of $\Xi$ to produce the presynaptic matrix $\mathbf{H}_{KK}$ output by the virtual layer. The function $\psi$ is not altered during the learning process. Note also that the number of nodes in the virtual layer is being set to $K$. We have

$$\mathbf{H}_{KK} = \varphi(\, \Xi_{KK}\,) = \varphi(\, \psi(\, \boldsymbol{\Omega}_{KI})\,). \tag{1}$$

The above condition illustrates the critical role played by the virtual layer, which differentiates it drastically from a conventional hidden layer. To better understand the difference, consider a network with a single hidden layer having $N = H$ nodes. Let $\Xi_{KH}$ denote the postsynaptic input to the hidden layer. It is common for the number $K$ of training examples to exceed the number of input nodes $I$. If we then attempt to use a conventional algebraic training process [29] with $H$ fixed, the system $\Xi_{KH} = \boldsymbol{\Omega}_{KI}\, \mathbf{W}_{IH}$ is overdetermined for $\mathbf{W}_{IH}$, and the resulting minimum norm solution may be a poor approximation. Furthermore, as indicated in [30,31], $\Xi_{KH}$ itself would be approximate because of the assumptions used to calculate it in the backward sweeping process.

In the following, we present a mapping $\psi$ which always produces a *nonsingular square matrix*, $\Xi_{KK}$. We first consider the special case of scalar inputs, so $I = 1$. Let $D$ denote an upper bound for the maximum distance between two distinct training input values. For any input $\omega_i$, let $\psi(\omega_i)$ be the $K$-dimensional vector whose $k$-th component is

$$\psi_k(\omega_i) = 1 - \frac{|\omega_i - \omega_k|}{D} \quad , \tag{2}$$

where $\omega_k$ is the $k$-th training input value, $k = 1,\ldots, K$. Clearly, $\psi(\omega_k)$ denotes the $k$-th row of $\Xi_{KK}$.

We may assume, without loss of generality, that the training input values are ordered from smallest to largest. If not, we can interchange rows of the matrix, two at a time, and interchange the corresponding two columns. The net effect on the determinant is multiplication by $(-1)^2$. Let $u(k)$ denote the distance between $\omega_k$ and $\omega_{k+1}$. We have proven [30] the following theorem:

$$\det(\Xi_{KK}) = 2^{K-2}\left(\prod_{k=1}^{K-1}\frac{u(k)}{D}\right)\left(2-\sum_{k=1}^{K-1}\frac{u(k)}{D}\right) = \frac{1}{2}\prod_{k=1}^{K-1}\frac{2u(k)}{D} > 0 \quad . \tag{3}$$

If the input vectors have dimension $I > 1$, the one-dimensional procedure above is applied one component at a time. For each component $i = 1, 2, \ldots I$, construct a $K \times K$ matrix $\Xi^{(i)}$:

$$\Xi^{(i)}_{KK}(k,l) = 1 - \frac{|\omega_k^{(i)} - \omega_l^{(i)}|}{D^{(i)}} \tag{4}$$

with $k, l = 1, \ldots K$, and $i = 1, \ldots I$. Here $\omega_k^{(i)}$ denotes the $i$-th component of the $k$-th training vector $\boldsymbol{\omega}_k$, and $D^{(i)}$ is the maximum of $|\omega_k^{(i)} - \omega_l^{(i)}|$ over all $k, l = 1, \ldots, K$.

Let $\Xi_{KK}$ be the block diagonal matrix whose $i$-th block is given by Eq. (4). The determinant of the full matrix is

$$\det(\Xi_{KK}) = \prod_{i=1}^{I}\left(\frac{1}{2}\prod_{k=1}^{K-1}\frac{2u^{(i)}(k)}{D^{(i)}}\right) > 0 . \tag{5}$$

The above block implementation of $\Xi_{KK}$ for $I > 1$ guarantees that the matrix is nonsingular, but significantly increases its size. As an alternative, the mapping $\psi$ used in the one-dimensional case can be generalized to higher dimensions by defining the $(k, l)$-th component to be

$$[\Xi_{KK}]_{kl} = 1 - \frac{\|\omega_k - \omega_l\|}{D} \tag{6}$$

where $\omega_k$ denotes the $k$-th training input vector, and $D$ is the maximum distance between all input vectors in the training set. The distance is the usual Euclidean distance in the $I$-dimensional space.

Each network node implements a nonlinear transfer function $\varphi : \Re \rightarrow (0,1)$, typically a sigmoid. As result of applying $\varphi$, the presynaptic matrix, $\mathbf{H}_{KK}$, output by the virtual layer is obtained. Since $\varphi$ is bijective, the inverse $\varphi^{-1}$ is well defined. Then, the postsynaptic inputs $\mathbf{T}$ to the output layer corresponding to the given target outputs $\mathbf{R}$ are

$$\mathbf{T}_{KO} = \varphi^{-1}(\mathbf{R}_{KO}) . \tag{7}$$

The postsynaptic inputs to the output layer computed by the network are obtained from

$$\mathbf{P}_{KO} = \mathbf{H}_{KK}\,\mathbf{W}_{KO} . \tag{8}$$

The next phase of the learning algorithm minimizes the norm $\|\mathbf{T}_{KO} - \mathbf{P}_{KO}\|$ by solving the system

$$\mathbf{T}_{KO} = \mathbf{H}_{KK}\,\mathbf{W}_{KO} \tag{9}$$

for $\mathbf{W}_{KO}$. Here, $\mathbf{T}_{KO}$ and $\mathbf{H}_{KK}$ are known. We compute $\mathbf{W}_{KO}$ using a singular-value decomposition (SVD) of $\mathbf{H}_{KK}$ from the left. In general, if the matrix $\mathbf{H}$ is obtained via Eq. (4), it is non singular. For some training sets, the matrix $\mathbf{H}_{KK}$ computed by using the simpler transformation (6) may, however, be singular. Nevertheless, this does not present a problem, since the SVD method does not require a nonsingular matrix to perform the inversion.

The new *DeepNet* algorithm represents an ultrafast (single iteration) approach to neural network learning for applications involving static patterns. As such, it has considerable advantages in efficiency and accuracy over backpropagation. It is interesting to note that network retraining, typically associated with an excessive cost when using conventional learning, becomes trivial. Another important aspect of neural learning is the network's ability to generalize beyond the training sets. We address this issue in the sequel from the perspective of uncertainties associated with the network parameters and with the data in the training sets.

## 4. Sensitivity and Uncertainty Analysis for Robot Learning

Many novel sensors (such as imaging micro-spectrometers), which are expected to play an ever growing role in future robotic applications, produce large data sets (e.g., hyperspectral data cubes). With such sensors, even relatively simple tasks may involve an ensemble of often-complex models embedded in sophisticated computer codes. How much confidence should then be placed in decisions made by a robot on the basis of predictions obtained from these models, when it is known that they are driven by sensory data possibly corrupted by uncertainty? It is clear that answers to such a question based solely on physical intuition or engineering judgement are precluded.

Here, our proposed objective is twofold. First, we will show how to determine the confidence limits of results predicted by model codes. In terms of robotic learning, we will illustrate this concept using the *DeepNet* model discussed in Section 3 above. Then, we will present a methodology that allows a robot to consistently combine sensor measurements with model-predicted results. Our goal is to obtain best estimates for the model parameters, and to drastically reduce the uncertainties underlying decision processes based on learning, such as target detection and/or classification. Concepts and tools from *sensitivity* and *uncertainty analysis* provide the enabling means for achieving these objectives.

Sensitivities can be used to determine and rank the importance of model parameters and input data to computed quantities of interest (also referred to as *system responses*), and to assess model uncertainties due to uncertainties in parameters and data. They are defined as the derivatives of the system responses with respect to parameters and inputs. In past years, a number of application areas have immensely benefited from theoretical and computational advances in sensitivity and uncertainty analysis. Nuclear reactor core physics and shielding have involved models with time-independent linear equations, and relatively simple linear response functionals [33]. Nuclear fuel cycle studies encountered mildly nonlinear, time dependent model equations, and simple linear responses. Thermal-hydraulic problems were much more challenging [34]. Here the models were coupled nonlinear equations, and the responses could be maxima or minima of nonlinear functionals. Furthermore, the location of the maxima or minima could be variable. Sensitivity and uncertainty analyses have also been applied to energy-economic models [35]. The mathematical complexity of the models was similar to that encountered in other fields, with perhaps the added difficulty of handling discontinuities and the need to incorporate the results in a decision-theoretic framework. Most recently, seismic exploration projects have benefited from this methodology, particularly in attempting to fuse information from multiple sensor arrays to help analyze methane hydrate concentrations [36].

To enable *reliable decisions*, uncertainty analysis methods must possess five *key capabilities*. First, there should be a guarantee that no *important effects* are overlooked, i.e., a *full set of sensitivities* should be available. A full set means that the sensitivities with respect to all parameters are needed, without making an *a-priori* judgement as to which one is important. Second, we require an *efficient computation of the sensitivities*, since we must process large data sets fast. Thus, adjoint operator methods and/or automated differentiation preprocessors are essential. Third, proposed methods should allow for a systematic treatment of *nonlinearities*. The fourth criterion addresses the rigorous treatment, where relevant, of *time dependence*. This includes model inputs, parameters, and responses. Finally, one requires a coherent method for combining experimental (i.e., measured) data and model results, the primary goal being to *reduce the uncertainties*.

There are several potential approaches to uncertainty analysis [37]. Response surface methods [38] are often a preferred paradigm because of their intrinsic conceptual simplicity. Other techniques frequently used in the neurocomputing community include fuzzy logic [39] and cross validation [40,41]. The methodology we propose here for robot learning applications is based on the constrained minimization of a generalized Bayesian loss function [42]. When combined to the ultrafast *DeepNet* algorithm, it provides the basis for robust learning. It incorporates all the requirements listed above, although the time dependent formalism will not be discussed in this paper.

## 5. Learning under Uncertainty

The incorporation of uncertainty information into the *DeepNet* learning mechanism is essential for enabling proper generalization. For illustrative purposes, we present here a simple time-independent framework. We begin by specifying the assumptions and the notation. As result of the single-iteration training process in *DeepNet*, a set of *nominal* values for the intrinsic network parameters (e.g., $\mathbf{W}_{KO}$) has been determined. There is uncertainty associated with $\mathbf{W}_{KO}$, since there is uncertainty in the training sets. Let $\omega$ denote an $I$-dimensional input pattern. It may be selected from the input training set $\Omega_{KI}$, or be a new pattern for which a measured $O$-dimensional response pattern $\mathbf{r}$ is available. The intrinsic network parameters $\mathbf{W}$ are concatenated (by rows) with the inputs $\omega$ as a vector $\mathbf{a}$ of *system* parameters. The dimension of $\mathbf{a}$ is of order $KO + I$. The responses calculated by *DeepNet* as function of $\mathbf{a}$ are denoted by $\mathbf{q}$. The nominal uncertainties in the parameters are quantified by specifying covariance matrices, e.g.,

$$\mathbf{C}_{\mathbf{aa}} = \left\langle \Delta\mathbf{a}\,\Delta\mathbf{a}^{t} \right\rangle \quad . \tag{10}$$

In Eq. (10), the brackets denote integration over a joint probability density function (PDF). Many uncertainty analysis methods choose a form for the PDF. We will be more general, and need only to specify the first few moments of the PDF: e.g., mean value and covariance matrix. Initially, $\mathbf{C}_{\mathbf{aa}}$ is block diagonal, each block corresponding to the covariance matrices associated with $\mathbf{W}$ and $\omega$.

Sensitivities provide a systematic way to propagate uncertainties in complex, non-stationary, nonlinear models. For example, to first order in a stationary system, the sensitivity of the calculated response $n$ with respect to parameter $i$ evaluated at the nominal values $\mathbf{a}$ is

$$S_{ni} = \partial q_n / \partial a_i \quad . \tag{11}$$

When neural networks are implemented as dynamical systems, sensitivities can be efficiently computed using an adjoint operator formalism [25], or existing automated differentiation preprocessors [43]. Since only a feed-forward multilayer architecture is considered in *DeepNet*, they can be calculated analytically in a straightforward manner, and the detailed formulae are not included here.

Using the sensitivity matrix $\mathbf{S}$, we can calculate the nominal covariance matrix of the *DeepNet* responses. By expanding about the centroid of the joint PDF of the system parameters, we obtain, again, to first order

$$\mathbf{C}_{\mathbf{qq}} = \left\langle \Delta\mathbf{q}\,\Delta\mathbf{q}^{t} \right\rangle = \mathbf{S}\,\mathbf{C}_{\mathbf{aa}}\,\mathbf{S}^{t} \quad . \tag{12}$$

We seek *best estimates* for the parameters and responses, denoted by $\hat{\mathbf{a}}$ and $\hat{\mathbf{q}}$. These values are related to the current estimates by the sensitivities:

$$\hat{\mathbf{q}} = \mathbf{q} + \mathbf{S}\left(\hat{\mathbf{a}} - \mathbf{a}\right) \quad . \tag{13}$$

To obtain the best estimates, we must *consistently combine* computational results and experimental measurements. We achieve this by optimizing a generalized Bayesian loss function, which simultaneously minimizes (*i*) the differences between the best estimate and the measured responses and (*ii*) the best estimate and the nominal values of the system parameters. Our optimization process uses the inverse of a generalized total covariance matrix as the natural metric of the calculational manifold. In particular, we write:

$$Q=\left[\hat{\mathbf{q}}-\mathbf{r}\,\middle|\,\hat{\mathbf{a}}-\mathbf{a}\right]\begin{bmatrix} C_{\mathbf{rr}} & C_{\mathbf{ra}} \\ C_{\mathbf{ar}} & C_{\mathbf{aa}} \end{bmatrix}^{-1}\begin{bmatrix} \hat{\mathbf{q}}-\mathbf{r} \\ \hat{\mathbf{a}}-\mathbf{a} \end{bmatrix}. \tag{14}$$

Additional potential contributions to the covariance matrix such as method biases may also be included in the above expression. To capture the constraints between parameters and responses, it is convenient to define new variables:

$$\mathbf{x}=\hat{\mathbf{a}}-\mathbf{a} \quad , \tag{15}$$

$$\mathbf{y}=\hat{\mathbf{q}}-\mathbf{r} \quad . \tag{16}$$

The discrepancy between calculations and measurements is then expressed as

$$\mathbf{e}=\mathbf{q}-\mathbf{r} \quad . \tag{17}$$

Using the new variables, the constraints become

$$\mathbf{y}=\mathbf{S}\,\mathbf{x}+\mathbf{e} \quad . \tag{18}$$

For simplicity, we have illustrated here this relationship to first order only. One can now construct an augmented Lagrangian, *L*, given by

$$L=Q+\boldsymbol{\lambda}^{T}\left[\mathbf{S}\,\mathbf{x}-\mathbf{y}+\mathbf{e}\right] \quad . \tag{19}$$

The necessary conditions for an optimal solution are obtained by requiring that the partial derivatives of the augmented Lagrangian with respect to **x** and **y** be zero, in conjunction with the constraint satisfaction. Thus, we seek:

$$\frac{\partial L}{\partial \mathbf{x}}=\mathbf{0}_{\mathrm{a}} \quad , \qquad \frac{\partial L}{\partial \mathbf{y}}=\mathbf{0}_{r} \quad . \tag{20-21}$$

We can solve these equations for the optimal values of **x** and **y**, which determine the best estimates for both parameters and responses. For example, assuming no bias contribution to the optimization metric, we obtain

$$\hat{\mathbf{a}}=\mathbf{a}+(\mathbf{C}_{\mathbf{ra}}^{T}-\mathbf{C}_{\mathbf{aa}}\mathbf{S}^{T})\boldsymbol{\lambda} \quad , \tag{22}$$

where

$$\boldsymbol{\lambda}=(\mathbf{C}_{\mathbf{rr}}-\mathbf{S}\mathbf{C}_{\mathbf{ra}}^{T}-\mathbf{C}_{\mathbf{ra}}\mathbf{S}^{T}+\mathbf{S}\mathbf{C}_{\mathbf{aa}}\mathbf{S}^{T})^{-1}\mathbf{e} \quad . \tag{23}$$

Then, we can show that the best estimates of the covariance matrices yield reduced uncertainties. After some lengthy algebra, we find for instance that the covariance matrix corresponding to the best estimates of the parameters is

$$\mathbf{C}_{\hat{\mathbf{a}}\hat{\mathbf{a}}} = \mathbf{C}_{\mathbf{aa}} - (\mathbf{C}_{\mathbf{ra}}^T - \mathbf{C}_{\mathbf{aa}}\mathbf{S}^T)(\mathbf{C}_{\mathbf{rr}} - \mathbf{S}\mathbf{C}_{\mathbf{ra}}^T - \mathbf{C}_{\mathbf{ra}}\mathbf{S}^T + \mathbf{S}\mathbf{C}_{\mathbf{aa}}\mathbf{S}^T)^{-1}(\mathbf{C}_{\mathbf{ra}} - \mathbf{S}\mathbf{C}_{\mathbf{aa}}). \quad (24)$$

It should be noted that the above expression reduces the uncertainties around a nominal set of values of the input data and the model parameters. If *global uncertainties* over the entire domain of these variables are desired, two approaches are possible. One could propagate the uncertainties with the help of generalized derivatives, which are defined in terms of *interval arithmetic*. It is also possible to combine the estimation based on local minimization described above with the TRUST global optimization algorithm.

## 6. Multi-Robot Observation of Multiple Moving Targets

Our new learning methodology can readily be applied to the problem of cooperative multi-robot observation of multiple moving targets (CMOMMT). This problem was proposed by Parker as a benchmark for on-line distributed control strategies [12]. The reader should refer to [12] for further details. We note that the concepts underlying CMOMMT are very similar in nature to the multi-task scheduling and load-balancing problem encountered in multiprocessor computer systems. An elegant and efficient solution to the latter, based on a perturbation technique applied to neural networks has already been demonstrated [44] and could be adapted to CMOMMT. Here, however, we propose an approach that exploits the ultrafast learning algorithm discussed in Section 3.

In our formulation, in order to determine its direction of motion and velocity at each time step, every robot $i$ may seek to minimize a local objective function

$$E_i(t+\delta_t) = \sum_{\tau \in N_\pi} \eta_\tau \Delta_{i\tau}^2(t+\delta_t) + \sum_{j \neq i} \xi_j / \Delta_{ij}^2(t+\delta_t). \quad (25)$$

In the above equation, $\eta_\tau$ and $\xi_\tau$ are weights, function of the predictive-tracking and sensing distance system parameters, and $N_\pi$ denotes the set of targets within the predictive tracking range of robot $i$. The distances $\Delta$ (which include a small positive offset to avoid collisions) at time $t + \delta_t$ are computed from the known positions and velocities of targets and other robots at time $t$ and from the known position of $i$. They are function of the desired velocity magnitude and direction to be assigned to $i$ at time $t$. The latter can be obtained by solving, for each robot and at each time step, the nonlinear optimization problem specified by Eq. (25). Since $E_i$ possesses many local minima, a global optimization algorithm such as TRUST [27] may typically be required.

One can avoid this costly alternative by training a single neural network, which can then be uploaded to each robot. To illustrate the concept, consider the following simple case. Let $\{E^+_k, E^-_k, \psi_k, \zeta_k\}$ denote the vector constituting the $k$-th row of the matrix $\boldsymbol{\Omega}_{KI}$ (defined in Section 3) representing a set of $K$ $I$-dimensional (with $I = 4$) input patterns. The terms $E^+_k$ and $E^-_k$ correspond to discrete values (selected in a range $[E_{min}, E_{max}]$) of the two right- hand-side terms of Eq. (25). The angles $\psi_k$ and $\zeta_k$ are computed from the directions of Parker's target and robot local force vectors [12]. The set of $K$ output training patterns $\{\upsilon_k, \theta_k\}$ is obtained by solving Eq. (25) in terms of the given input. They are stored as rows in the matrix $\mathbf{R}_{KO}$ (where $O$ = 2). The methodology developed in Section 3 then applies directly. Note that the training set does not need to be excessively large, since our neural net inherently provides a robust tool for nonlinear interpolation. Alternative patterns may also be considered, particularly ones involving time-delayed information.

In summary, we have presented a novel training methodology for feedforward neural networks that introduces several innovations of major relevance to multi-robot learning. First, nominal values of the network parameters are obtained in a single iteration; in fact, we eliminate all the hidden layers and connect the input and output layers only through the specially constructed virtual layer. Second, contrary to backpropagation, our method does not rely on calculating gradients, so it is not affected by local minima. Third, excellent generalization to signal patterns outside the training set is observed. Finally, if uncertainty information for the training data is available (in the form of covariance matrices), it can be explicitly accounted for. This may require a few additional iterations to obtain best estimates for all network parameters while reducing the uncertainties in the network predicted results. As a result, a new paradigm for cooperative multi-robot observation of multiple moving targets was proposed. In this paradigm, each robot uses a neural network with virtual layer technology to rapidly compute its own direction of motion (and velocity along this direction) using a multidimensional input pattern directly derived from sensory data.

## Acknowledgments

This research was performed at the Center for Engineering Science Advanced Research, Computer Science and Mathematics Division, Oak Ridge National Laboratory. Primary support for the effort was granted by the Basic Energy Sciences Program of the DOE Office of Science, under contract DE-AC05-96OR22464 with Lockheed Martin Energy Research Corporation.

## References

1. S.-I. Amari and M. Arbib, *Competition and Cooperation in Neural Networks*, Springer (1982).
2. S. Grossberg, *The Adaptive Brain, I: Cognition, Learning, Reinforcement, and Rhythm*, Elsevier (1987).
3. S. Grossberg, *The Adaptive Brain, II: Vision, Speech, Language, and Motor Control*, Elsevier (1987).
4. R. Arkin, *Behavior-Based Robotics*, MIT Press (1998).
5. D. Rumelhart and J. McLelland, *Parallel Distributed Processing*, MIT Press (1986).
6. D. Waterman and F. Hayes-Roth, *Pattern-Directed Inference Systems*, Academic Press (1978).
7. J. Barhen, W. Dress, and C. Jorgensen, "Application of Concurrent Neuromorphic Algorithms to Autonomous Robots", in *Neural Computers*, NATO *ASI*, **F41**, pp. 321-334, Springer (1988).; ***see also***: J. Barhen, M. Zak, and N. Toomarian, "Non-Lipschitzian Neural Dynamics", in *Advanced Neural Computers*, R. Eckmiller, ed., pp. 103-112, North Holland (1990); ***see also***: N. Toomarian, J. Barhen, and S. Gulati, "Neural Networks for Real-Time Robotic Applications", in *Parallel Computation Systems for Robotics*, A. Fijany and A. Bejczy eds., pp. 213-247, World Scientific (1992).
8. G. Bekey and K. Goldberg, *Neural Networks in Robotics*, Kluwer (1993).
9. O. Omidvar and P. van der Smagt, *Neural Systems for Robotics*, Academic Press (1997).
10. R. Brooks, S. Grossberg, and L. Optican, eds, "Neural Control and Robotics", special issue of *Neural Networks*, **11**, no. 7-8 (1998).
11. T. Ziemke and N. Sharkey, eds, "Artificial Neural Networks for Robot Learning", special issue of *Autonomous Robots*, **7**, no. 7 (1999).
12. L. Parker and C. Touzet, "Multi-Robot Learning in a Cooperative Observation Task", in *Distributed Autonomous Robotic Systems 4*, L. Parker, G. Bekey, and J. Barhen eds., Springer (in press, 2000).
13. C. Touzet, "Distributed Lazy Q-learning for Cooperative Mobile Robots", *Autonomous Robots* (in press, 2000).
14. A. Agah and G. Bekey, "Cognitive Architecture for Robust Adaptive Control of Robots in a Team", *Journal of Intelligent and Robotic Systems*, **20**, 251-273 (1997).
15. M. Hassoun, *Fundamentals of Artificial Neural Networks*, MIT Press (1995).
16. M. Bishop, *Neural Networks for Pattern Recognition*, Oxford University Press (1997).

17. M. Beckerman, *Adaptive Cooperative Systems*, Wiley Interscience (1997).
18. D. White and D. Sofge, *Handbook of Intelligent Control: Neural, Fuzzy and Adaptive Approaches*, Van Nostrand Reinhold (1992).
19. A. Cichocki and R. Unbenhauen, *Neural Networks for Optimization and Signal Processing*, Wiley (1993).
20. P. Mars, J. Chen, and R. Nambiar, *Learning Algorithms*, CRC Press (1996).
21. N. Toomarian and J. Barhen, "Learning a Trajectory using Adjoint Functions and Teacher Forcing", *Neural Networks*, **5**, 473-484 (1992).
22. N. Toomarian and J. Barhen, "Fast Temporal Neural Learning Using Teacher Forcing", *U.S. Patent No. 5,428,710* (March 28, 1995).
23. Y. Chauvin and D. Rumelhart, *Backpropagation: Theory, Architectures, and Applications*, Lawrence Erlbaum (1995).
24. J. Barhen, S. Gulati, and M. Zak, "Neural Learning of Constrained Nonlinear Transformations", *IEEE Computer*, **22**(6), 67-76 (1989).
25. J. Barhen, N. Toomarian, and S. Gulati, "Applications of Adjoint Operators to Neural Networks", *Appl. Math. Lett.*, **3**(3), 13-18 (1990).
26. N. Toomarian and J. Barhen, "Neural Network Training by Integration of Adjoint Systems of Equations Forward in Time", *U.S. Patent No.* 5,930,781 (July 27, 1999).
27. J. Barhen, V. Protopopescu, and D. Reister, "TRUST: A Deterministic Algorithm for Global Optimization", *Science*, **276**, 1094-1097 (1997).
28. A. Shepherd, *Second-Order Methods for Neural Networks*, Springer (1997).
29. F. Biegler-König and F. Bärman, "A Learning Algorithm for Multilayered Neural Networks based on Linear Least Squares Problems", *Neural Networks*, **6**, 127-131 (1993).
30. J. Barhen, R. Cogswell, and V. Protopopescu, "Single Iteration Training Algorithm for Multi-layer Feed-forward Neural Networks," *Neural Proc. Lett*, **11**(2) (in press, 2000).
31. J. Barhen, D.Reister, and V. Protopopescu, "*DeepNet:* An Ultrafast Neural Learning Code for Seismic Imaging", *Procs. IJCNN'99*, CD-ROM, IEEE Press (1999).
32. G. Golub and C. Van Loan, *Matrix Computations*, Johns Hopkins University Press (1983).
33. C. Weisbin *et al*, "Sensitivity and Uncertainty of Reactor Performance Parameters", in *Advances in Nuclear Science and Technology*, **14**, Plenum Press, New York (1981).
34. D. Cacuci *et al*, "Sensitivity Theory for General Systems of Non-linear Equations", *Nucl. Sci. Eng.*, **75**, 88-110 (1980).
35. R. Alsmiller, J. Barhen, and J. Horwedel, "The Application of Adjoint Sensitivity Theory to a Liquid Fuels Supply Model", *Energy*, **9**, 239-253 (1984).
36. D. Reister and J. Barhen, "Application of Sensitivity and Uncertainty Analysis to the Estimation of Methane Hydrate Concentration", *Jour. of Petroleum Science and Engineering* (in preparation, 2000).
37. D. Cox and P. Baybutt, "Methods for Uncertainty Analysis: A Comparative Survey", *Risk Analysis*, **1**(4), 251-258 (1981).
38. R. Myers, *Response Surface Methodology*, Edwards Bros. (1976).
39. B. Kosko, *Fuzzy Engineering*, Prentice Hall (1996).
40. A. Dubrawski, "Stochastic Validation for Automated Tuning of Neural Network's Hyper Parameters", *Robotics and Autonomous Systems*, **21**, 83-93 (1997).
41. F. Aminzadeh, J. Barhen, C. Glover, and N. Toomarian, "Estimation of Reservoir Parameters using a Hybrid Neural Network", *Jour. of Petroleum Science & Engineering*, **24**, 49-56 (1999).
42. J. Barhen, D. Cacuci, J. Wagschal, and M. Bjerke, "Uncertainty Analysis of Time-Dependent Nonlinear Systems", *Nuclear Science & Engineering*, **81**(1), 23-44 (1982).
43. J. Tolma and P. Barton, "On Computational Differentiation", *Comp. Chem. Engin.*, **22**(4/5), 475-490 (1998).
44. J. Barhen, N. Toomarian and V. Protopopescu, "Optimization of the Computational Load of a Hypercube Supercomputer Onboard a Mobile Robot", *Applied Optics*, **26** (23), 5007-5014 (1987).

# Chapter 11

# Cooperative Object Transport

# Object Transport Using Multiple Mobile Robots with Pin Joint Endeffectors

Chad K. Humberstone and Dr. Kevin B. Smith

Brigham Young University, Provo UT 84602, USA. (email: ckh2@email.byu.edu and kbsmith@ee.byu.edu)

**Abstract.** This paper presents the mechanical design and control policy for a set of robots used to carry a solid object. Contributions in our approach include a mechanical design that does not require compliance and is simple and inexpensive to manufacture. All the robots receive the same commands from a supervisor, and no communication is required between robots. Feedforward and feedback linearization using force and acceleration inputs is employed to produce a point with linear motion on the nonholonomic robots. Acceleration feedback is used on the wheel accelerations. The control has been validated using simulations, and the mechanical design and control policy have been validated experimentally.

## 1 Introduction

### 1.1 Context

This paper can be categorized with those that deal with moving a solid or non-deformable object in two dimensions over a flat surface using mobile robots, where unconstrained motions[1] of the object are possible. Some solutions to the problem of object transport with multiple mobile robots constrain object motion, such as leader/follower designs [1] [2], where the object motion is constrained to behave like a trailer. Presented here is a method to produce unconstrained motion of an object using robots with pin joint endeffectors.

### 1.2 Current Status

Many others have addressed the problem of moving an object with abitrary or unconstrained motions with multiple robots on a smooth floor. This review is limited to those which use an impedance control for high-level object-motion control. A multiple robot transfer system that allows arbitrary object motion is conceptualized in [3] and [4] where the robots connect to the object through a 1-DOF prismatic joint. This solution requires fast communication. In [5], a similar solution is proposed using a 2-DOF prismatic joint, which requires less dense communication. In [6], a robot design is used where the robot "hand" is placed at a distance ahead of the axle to avoid the nonholonomic

[1] The motion of the object is constrained to move in the plane, but its motion within the plane is unconstrained.

constraints at the center of the robot. The hand is built with compliance in the moving plane, and the robots are velocity controlled. The hands are capable of regrasping the object. None of the above solutions are presented with experimental results.

### 1.3 Our Approach

In our approach, all the robots connect to the object with simple free rotating pin joints. The joints are attached to the robots at a point off the wheel axles. Newly presented in this paper is 1) a simple robot design, and 2) a control policy for controlling the robot so that the desired acceleration of the object is achieved.

## 2 Robot Design: Point with Linear Motion on a Nonholonomic Base

The objective of the robot design was to make a simple and inexpensive mechanism for cooperative transport in the plane. The robot base has two wheels driven by two DC motors. The robot is able to spin in place. Two castors are used for support.

The primary difference between robot designs in other similar research and the robot design used in this research is that the robot hand has no compliance except for rotation; the object is free to rotate with respect to the robot. In [7], it was stated that "flexibility mechanisms were designed to solve the problems of nonholonomic constraints, excessive inner forces and vibrations." In [6], the problem of the nonholonomic constraint is solved by moving the free-rotating robot hand from the center of the robot (or over the axle) to a point offset from the axle; "the robots can then generate onmidimensional forces in a horizontal direction at every moment". However, compliance was still added to the hand in the x and y direction. The robot hand in this research has no compliance in the x and y direction; it is a pin joint.

Pomet [8] showed that the motion of a point offset from the axle is linearizable for the kinematic model. This is another reason for moving the robot's hand off of the axle. This method will be extended to the robot's hand in the dynamic model in Section 4.

## 3 Control Structure

### 3.1 Notation

The following notational conventions will be used in this paper:

- Bold letters represent vectors, and nonbold letters represent scalars.

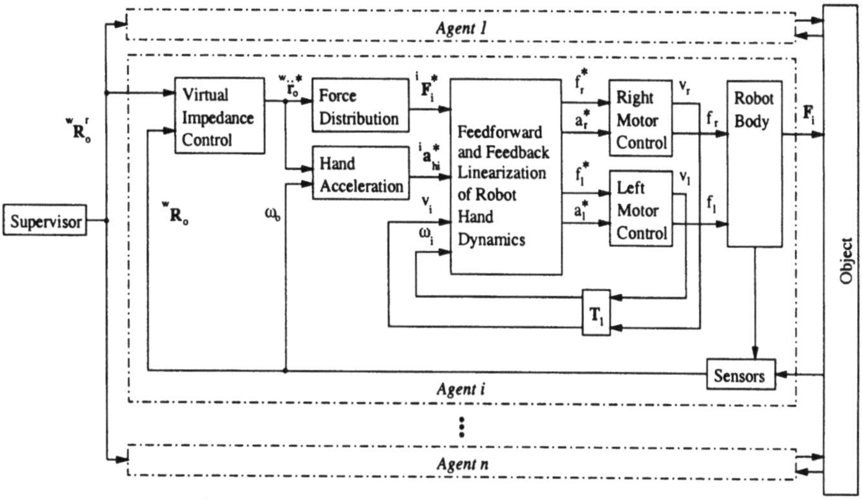


**Fig. 1.** Control Architecture.

- The subscript i, ($_i$), denotes the $i^{th}$ robot.
- The subscript o, ($_o$), denotes the object.
- A superscript preceding a variable or constant indicates the reference frame it is referenced to; i.e., $^o x_i$ refers to the x-position of the $i^{th}$ robot in the object reference frame. If no superscript precedes a variable, the assumed reference frame is world.
- A superscript r, ($^r$), following a variable denotes a reference value.
- An asterisk (*) following a variable denotes a desired value.

The following are two-dimensional reference frame definitions:

- $\sum_o \equiv$ the object reference frame
- $\sum_i \equiv$ the $i^{th}$ robot's reference frame
- $\sum_w \equiv$ the world reference frame

## 3.2 Structure

The control structure under which our contributions apply is flexible (see Figure 1). The supervisor generates the reference trajectory, $^w\mathbf{R}_o^r$, of the object. The supervisor could be a process separate from the robots, a process running on one of the robots, or a process instantiated on each of the robots. In Figure 1, the supervisor is shown as a process running independent of the robots.

The supervisor sends the reference trajectory of the object to the robots. Given the mass, and the rotational inertia of the object, the robots apply forces to make the object track the reference trajectory using "virtual impedance control" [6] [3], or "object impedance control" [9], which for one dimension is: $F^* = M(\ddot{x}_o^r) + k_v(\dot{x}_o^r - \dot{x}_o) + k_p(x_o^r - x_o)$.

Here, $\ddot{x}_o^r$, $\dot{x}_o^r$, and $x_o^r$ are the reference trajectory. $M$, $k_v$, and $k_p$ are the virtual mass, spring force, and damping force for the impedance control. The force F is divided by the virtual mass to get the desired acceleration, $\ddot{x}_o^*$. From the virtual impedance model comes the overall desired object acceleration ${}^w\ddot{\mathbf{r}}_o^*$, which contains $\ddot{x}$, $\ddot{y}$ and $\ddot{\theta}$. Knowing the configuration of the robots with respect to the object, each robot is able to solve for the forces it must apply. The method the robots use to solve for the forces they each must apply is shown in both [3] and [6] and is represented in Figure 1 as the block labeled "Force Distribution".

Our contributions in the area of the control begin after a robot solves for the force it must apply to the object. The robot is made to also solve for the desired acceleration ${}^i\mathbf{a}_{hi}^*$ of its hand. Using the desired force and acceleration, feedforward and feedback linearization are used to solve for the forces the motors must apply to the wheels in order to get the desired force applied to the object. (See the block labeled "Feedforward and Feedback Linearization of Robot Hand Dynamics.") Because the effectiveness of feedforward and feedback linearization depends on accurate modeling, acceleration feedback is applied to the acceleration of each robot wheel to reduce the control's sensitivity to modeling errors. This feedback is implemented in the blocks labeled "Left Motor Control" and "Right Motor Control." The "motor control" blocks and the "Feedforward and Feedback Linearization of Robot Hand Dynamics" block make up what is called the robot's "control policy."

# 4 Feedback Linearization with Force and Acceleration Input

## 4.1 Variable Definitions

The state of the $i^{th}$ robot is given by:

- $x_i \equiv$ position of robot center along the world $x$-axis
- $y_i \equiv$ position of robot center along the world $y$-axis
- $\theta_i \equiv$ robot orientation
- $v_i \equiv$ velocity of robot center along the robot's $x$-axis
- $\omega_i \equiv$ angular velocity of the robot

The position and orientation of the object are given by:

- $x_o \equiv$ position of object center along the world $x$-axis
- $y_o \equiv$ position of object center along the world $y$-axis
- $\theta_o \equiv$ object orientation

The point on the robot that connects to the object is referred to as the robot's hand. The notation for the $i^{th}$ robot's hand location is:

- $x_{hi} \equiv$ hand position along the world $x$-axis

- $y_{hi} \equiv$ hand position along the world $y$-axis

The location of the $i^{th}$ robot's hand in the object frame is given in polar coordinates where

- ${}^o\phi_i \equiv$ angle to the robot hand
- $d_i \equiv$ distance from object center to hand

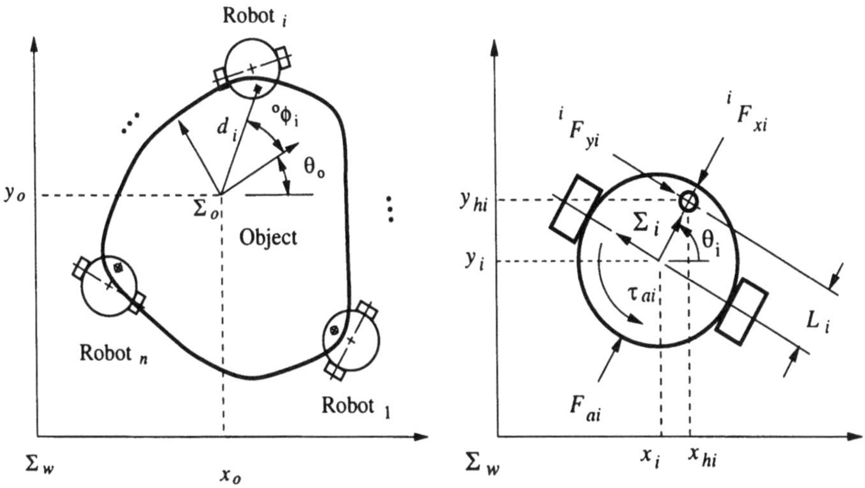


**Fig. 2.** Parameters describing the geometry of the object and robots.

## 4.2 Robot Calculations

The following calculations are for the $i^{th}$ robot. The desired force the robot is to apply in the robot's reference frame (Figure 2) is ${}^i\mathbf{F}_i^* = [{}^iF_{xi}^*\ {}^iF_{yi}^*]^T$.

The desired hand accelerations are calculated by

$$\ddot{x}_{hi}^* = \ddot{x}_o^* - d_i\ddot{\theta}_o^* sin(\theta_o + {}^o\phi_i) - d_i\dot{\theta}_o^2 cos(\theta_o + {}^o\phi_i) \tag{1}$$

$$\ddot{y}_{hi}^* = \ddot{y}_o^* + d_i\ddot{\theta}_o^* cos(\theta_o + {}^o\phi_i) - d_i\dot{\theta}_o^2 sin(\theta_o + {}^o\phi_i), \tag{2}$$

where $\ddot{x}_o^*$, $\ddot{y}_o^*$, and $\ddot{\theta}_o^*$ are the desired accelerations of the object, and where $\theta_o$ and $\dot{\theta}_o$ are the current angle and angular velocity of the object. The robot must accelerate its hand according to the desired object acceleration while generating the desired force. The forces applied by the wheels can be represented by a force in the $x$ direction of the robot, $F_{ai}$, and a torque, $\tau_{ai}$. The forces exerted on the robot hand from the object will be equal and opposite to what the robot hand exerts on the object (see Figure 2). The force and torque to be applied on the robot by the wheels is calculated using the desired force and accelerations in $\sum_i$:

$$F_{ai}^* = m_i({}^i\ddot{x}_{hi}^* + L_i\omega_i^2) + {}^iF_{xi}^* \tag{3}$$

$$\tau_{ai}^* = (J_i/L_i)({}^i\ddot{y}_{hi}^* - v_i\omega_i) + L_i\,({}^iF_{yi}^*). \tag{4}$$

A derivation of these two last equations follows.

### 4.3 Derivation

The model for the $i^{th}$ robot is

$$\dot{x}_i = {}^i v_i(cos\theta_i)$$
$$\dot{y}_i = {}^i v_i(sin\theta_i)$$
$$\dot{\theta}_i = \omega_i$$
$${}^i\dot{v}_i = (F_{ai} - {}^iF_{xi})/m_i$$
$$\dot{\omega}_i = (\tau_{ai} - L_i\, {}^iF_y i)/J_i.$$

The motion of the robot's hand is described by

$$\ddot{x}_{hi} = {}^i\dot{v}_i cos\theta_i - {}^iv_i\omega_i sin\theta_i - L_i\dot{\omega}_i sin\theta_i - L_i\omega_i^2 cos\theta_i \tag{5}$$
$$\ddot{y}_{hi} = {}^i\dot{v}_i sin\theta_i + {}^iv_i\omega_i cos\theta_i + L_i\dot{\omega}_i cos\theta_i - L_i\omega_i^2 sin\theta_i. \tag{6}$$

Substituting from the robot model (here $c \equiv cos$ and $s \equiv sin$):

$$\ddot{x}_{hi} = \frac{F_{ai}}{m_i}c\theta_i - \frac{{}^iF^*_{xi}}{m_i}c\theta_i - {}^iv_i\omega_i s\theta_i - L_i\frac{\tau_{ai}}{J_i}s\theta_i + L_i^2\frac{{}^iF^*_{yi}}{J_i}s\theta_i - L_i\omega_i^2 c\theta_i \tag{7}$$
$$\ddot{y}_{hi} = \frac{F_{ai}}{m_i}s\theta_i - \frac{{}^iF^*_{xi}}{m_i}s\theta_i + {}^iv_i\omega_i c\theta_i - L_i\frac{\tau_{ai}}{J_i}c\theta_i + L_i^2\frac{{}^iF^*_{yi}}{J_i}c\theta_i - L_i\omega_i^2 s\theta_i. \tag{8}$$

The previous two equations describe the motion of the robot's hand as a function of the force and torque generated by the wheels and the desired force to apply to the object. These equations can be linearized by following the example of [8]. Using the feedback transformation similar to [8], let

$$F_{ai} = u_1 m_i sin\theta_i + u_2 m_i cos\theta_i \tag{9}$$
$$\tau_{ai} = u_1\frac{J_i}{L_i}cos\theta_i - u_2\frac{J_i}{L_i}sin\theta_i, \tag{10}$$

and substitute them into $\ddot{x}_{hi}$ and $\ddot{y}_{hi}$. This reduces the equations of hand motion to

$$\ddot{x}_{hi} = u_2 - \frac{{}^iF^*_{xi}}{m_i}cos\theta_i - {}^iv_i\omega_i sin\theta_i + L_i^2\frac{{}^iF^*_{yi}}{J_i}sin\theta_i - L_i\omega_i^2 cos\theta_i \tag{11}$$
$$\ddot{y}_{hi} = u_1 - \frac{{}^iF^*_{xi}}{m_i}sin\theta_i + {}^iv_i\omega_i cos\theta_i + L_i^2\frac{{}^iF^*_{yi}}{J_i}cos\theta_i - L_i\omega_i^2 sin\theta_i. \tag{12}$$

The remaining nonlinear terms can be cancelled through feedback linearization by letting

$$u_2 = u_3 + \frac{{}^iF^*_{xi}}{m_i}cos\theta_i + {}^iv_i\omega_i sin\theta_i - L_i^2\frac{{}^iF^*_{yi}}{J_i}sin\theta_i + L_i\omega_i^2 cos\theta_i \tag{13}$$
$$u_1 = u4 + \frac{{}^iF^*_{xi}}{m_i}sin\theta_i - {}^iv_i\omega_i cos\theta_i - L_i^2\frac{{}^iF^*_{yi}}{J_i}cos\theta_i + L_i\omega_i^2 sin\theta_i, \tag{14}$$

and after making this substitution the result is that $\ddot{x}_{hi} = u_3$ and $\ddot{y}_{hi} = u_4$. We then let $u_3 = \ddot{x}^*_{hi}$ and $u_4 = \ddot{y}^*_{hi}$. Substituting $u_3$ and $u_4$ back into $u_2$ and $u_1$, then substituting those back into $F_{ai}$ and $\tau_{ai}$ and simplifying, we have the desired force and torque to apply to the robot as a function of the desired hand acceleration and the desired force to apply to the object (Equations 3 and 4). The hand accelerations were rotated to the robot's reference frame as part of the simplification.

## 5 Motor Control

The robot motors must be controlled so that the desired force, torque and hand acceleration are achieved. The force to be exerted on the robot by the right and left wheels (the force is the wheel torque divided by the wheel radius) is calculated by $f_r^* = 0.5F^*_{ai} + (1/2D)\tau^*_{ai}$, and $f_l^* = 0.5F^*_{ai} - (1/2D)\tau^*_{ai}$, where D is the distance of half of the wheel base. The desired acceleration of the center of the right and left wheels (in the direction the wheels are pointed) is calculated by $a_r^* = {}^i a^*_{hx} + (D/L)^i a^*_{hy}$, and $a_l^* = {}^i a_{hx^*} - (D/L)^i a^*_{hy}$. The acceleration here is the angular acceleration of the wheel multiplied by the wheel radius.

The idea of the motor control is to get the actual force and acceleration to be exactly the desired force and acceleration. Feedforward and feedback linearization are used in an effort to cancel the dynamics of the motor and wheels. Using a dynamic model of the motors and wheels, the desired force and acceleration are used in feedforward control The model for the motors is $f = K_1 * u - K_v * v - friction(v)$, where $f$ is the wheel torque divided by the wheel radius, $u$ is the input voltage, and $v$ is the velocity of the center of the wheel. Friction is a nonlinear function of $v$. $K_1$ and $K_v$ are motor constants. Solving for $u$: $u = (f + K_v * v + friction(v))(K_1)$. Feedback linearization is

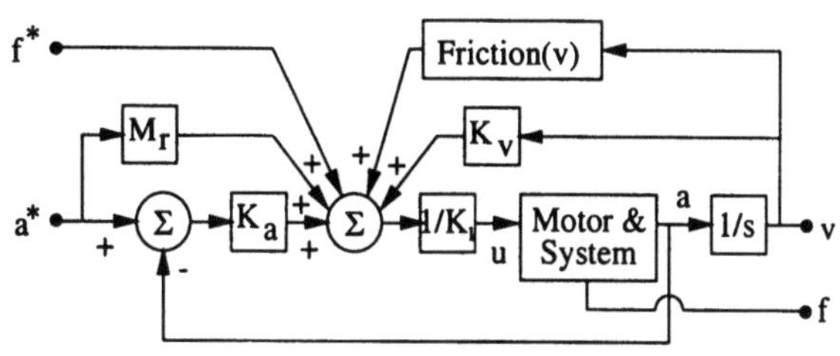


**Fig. 3.** Block diagram for individual motor control.

used to cancel the motor dynamics and friction. The desired acceleration is fed forward through a term $M_r$ to account for accelerating the motor's rotor. Because feedforward and feedback linearization depend on accurate models of the motors, a simple proportional feedback control is used with the desired accelerations. Figure 3 shows the inputs to the motor control as being the desired force the wheel is to apply to the base and the desired acceleration of the center of the wheel. The outputs are the velocity of the wheel. The actual force applied by the wheel to the robot base. If the modeling is perfect and there is no noise present (as within a perfect simulation), the acceleration feedback loop does nothing.

# 6 Simulated and Experimental Results

This section presents simulated and experimental results of multiple robots carrying a solid object using the method previously described. The simulations are done with two and three robots, and the experiment is done with two.

In the simulations and the experiment, the objective is to drive the state vector of the object to zero. The object is started at nonzero initial conditions. The desired forces to be applied to the object are calculated by a simple PD control on position and orientation errors in meters and radians respectively. This simple PD control is what the virtual impedance model turns into if there is only a desired position and no desired trajectory. The virtual object's mass and rotational inertia are set to be the same as the actual object's.

## 6.1 Simulations

The connection between the robots and the object is a pin joint. In the simulations, the reaction forces between each robot and the object through the pin joint need to be solved for. The reaction forces are solved for given the force applied to the robots by the motors and the states of the robots and the object. The derivation for solving for the reaction forces involves making free body diagrams for each robot and the object, and creating a set of equations from the free body diagrams where the unknowns are the reaction forces. Using linear algebra, the reaction forces are solved for. The derivation is tedious and lengthy but straightforward, so it is omitted for brevity.

On the right of Figure 4 is a dynamic simulation showing the trajectory taken by three robots moving an object. The robots and triangular object shown in dashed lines are the initial starting points. The robots and object shown in solid lines are the final positions.

The simulation shown in left of Figure 4 is done with two robots. This simulation is for comparison to a similar experiment, as shown in the next section. This simulation has the same initial conditions, desired position, and parameters as the experiment which follows.

## 6.2 Experimental Results

Two robots of the Multi-Agent Intelligent Coordinated Control (MAGICC) Lab at BYU [10] were used for the experiment. The object was a simple rod.

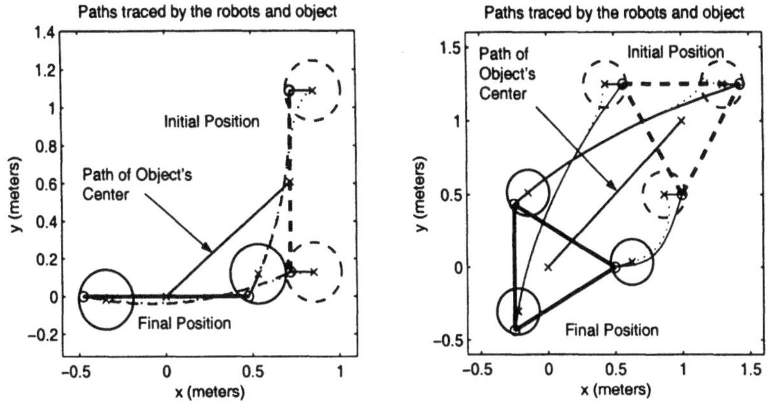


**Fig. 4.** Simulation showing the trajectory taken by three robots moving an object.

A picture of two of the robots transporting a rod is shown on the right of Figure 5. Each robot is controlled by a PC with a 400MHz processor over a wireless modem at a sample rate of 14 Hz. Motor encoder information is read on the robot, and this information is sent to the PC. An overhead camera provides position information on each robot to each PC to simulate onboard position centers. The encoder and camera information are fused on the PC's. A sensor on each robot measures the angle of the bar with respect to itself. Each PC has information about only one robot and makes calculations independent of the other PCs. The supervisor is artificially implemented on each PC. The supervisor's output in this experiment is a desired position of the object at $x_o = 0$, $y_o = 0$, and $\theta_o = 0$. Since the two robots have independent sensor information, the lines representing where the two robots calculated the path that the object's center traveled differ slightly.

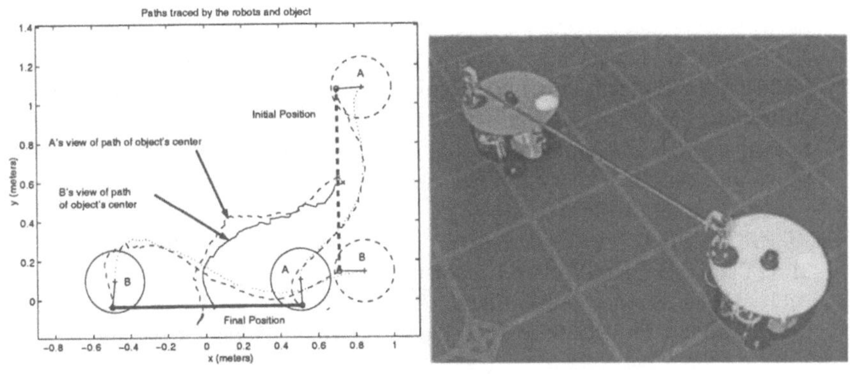


**Fig. 5.** Experimental results and picture of two robots moving a rod.

## 7 Conclusions

The main accomplishments of this paper are as follows. A simple and inexpensive robot-to-object connection is presented which uses a pin joint with no compliance in the plane. A control scheme using feedback linearization with force and acceleration inputs was derived, which included acceleration feedback. Simulation and experimental results show that the method realizes the desired motion of the object.

## References

1. D.J. Stilwell, J.S. Bay. (1993) Toward the development of a material transport system using swarms of ant-like robots. Proceedings - IEEE International Conference on Robotics and Automation, **1**, 766-771.
2. K. Kosuge, T. Oosumi, M. Satou, K. Chiba, K. Takeo. (1998) Transportation of a single object by two decentralized-controlled nonholonomic mobile robots. Proceedings - IEEE International Conference on Robotics and Automation, **4**, 2989-2994.
3. M. Hashimoto, F. Oba, H. Nakahara, K. Imamaki, T. Eguchi. (1991) Trajectory generation and tracking control methods for a multiple transfer robots system. International Workshop on Intelligent Robot and Systems", **2**, 799-804.
4. M. Hashimoto, F. Oba. (1993) Dynamic control approach for motion coordination of multiple wheeled mobile robots transporting a single object. IEEE/RSJ International Conference on Intelligent Robots and Systems, 1944-1951.
5. M. Hashimoto, F. Oba, S. Zenitani. (1995) Object-transportation control by multiple wheeled vehicle-planar Cartesian manipulator systems. IEEE International Conference on Robotics and Automation", 2267-2272.
6. N. Miyata, J. Ota, Y. Aiyama, J. Sasaki, T. Arai. (1997) Cooperative Transport System with Regrasping Car-like Mobile Robots. Proceedings - IEEE International Conference on Intelligent Robots and Systems, **3**, 1754-1761.
7. H. Osumi, T. Arai. (1994) Design of mechanisms for cooperation among multiple position-controlled robots. IEEE/RSJ/GI International Conference on Intelligent Robots and Systems, **2**, 792-799.
8. J.B. Pomet, B. Thuilot, G. Bastin, G. Campion. (1992) A hybrid strategy for the feedback stabilization of nonholonomic mobile robots. International Conference on Robotics and Automation, 129-134.
9. S. Schneider, R.H. Cannon, Jr. (1989) Object Impedance Control for Cooperative Manipulation: Theory and Experimental Results. Proc. IEEE Robotics and Automation, 1076-1083.
10. T.B. Gold. (1999) A Praxeic Utility Approach to Multi-Agent Coordination. M.S. Thesis. Brigham Young University.
11. C.K. Humberstone. (2000) Object Transport Using Multiple Mobile Robots with Pin Joint Endeffectors. M.S. Thesis. Brigham Young University.

# Distributed Motion Generation for Two Omni-Directional Robots Carrying a Ladder

Yuichi Asahiro[1], Eric Chung-Hui Chang[2], Amol Mali[2], Shunsuke Nagafuji[1], Ichiro Suzuki[2], and Masafumi Yamashita[1]

[1] Department of Computer Science and Communication Engineering
Kyushu University, Fukuoka 812-8581, Japan
[2] Department of Electrical Engineering and Computer Science
University of Wisconsin–Milwaukee, Milwaukee, WI 53201, U.S.A.

**Abstract.** The problem of efficiently moving a pair of omni-directional robots carrying a ladder using distributed control is discussed. We first consider the case in which two robots that may differ only in their maximum speeds are situated in an obstacle-free workspace, and present two distributed algorithms. Next, a distributed algorithm is presented for the case in which the workspace is a narrow corridor with a 90 degree corner and two robots are chosen from a large pool of robots having different characteristics in terms of the maximum speed, path generation strategy, sensitivity to the motion of the other robot, etc. The effectiveness of the algorithms is evaluated using computer simulation. Finally we present an outline of a distributed control scheme based on heuristic search that can potentially be used for more complex situations involving a larger number of robots in a workspace cluttered with obstacles.

## 1 Introduction

There are two general approaches for controlling multiple robots transporting an object. One is the centralized approach in which the motion of the robots is generated by an outside entity that can observe the global state of the system [7,12]. The other is the distributed approach, where every individual robot has to decide its motion based on the local information available to it [8,9]. This paper discusses the problem of transporting a ladder, or any long object such as a rocket and a bridge, using two omni-directional robots under distributed control.

Within the framework of the centralized approach, it is possible to compute a time-optimal motion of two such robots in an obstacle-free workspace using optimal control theory, under the assumption that the speed of the robots is either 0 or a given constant at any moment during a motion [5,10]. Fig. 1 shows a instance of this problem in which robots located at $A$ and $B$ must move to $A'$ and $B'$, respectively, as well as a time-optimal motion for the instance obtained by this method. Unfortunately, the applicability of this method for actually moving physical robots in optimal time is somewhat limited, because (i) the method uses complex calculation involving elliptic integral, and (ii) physical robots cannot always execute a computed motion precisely — they can neither accelerate to the maximum speed instantaneously, nor move along a given trajectory precisely due to mechanical imprecision and the unpredictability of the environment (such as a slight incline of the floor).

In contrast, in the distributed approach the robots can cope with unexpected perturbation by continuously monitoring their progress and dynamically adjusting their trajectories. The overall motion resulting from such a distributed strategy can be nearly as efficient as an optimal motion [1,2]. One should also keep in mind, however, that good distributed algorithms are usually much harder to design than centralized algorithms. For instance, the fact that the path of neither robot is straight in the optimal motion of Fig. 1 indicates that efficient motion may not be attainable distributively if either robot simply attempts to reach its destination as quickly as possible.

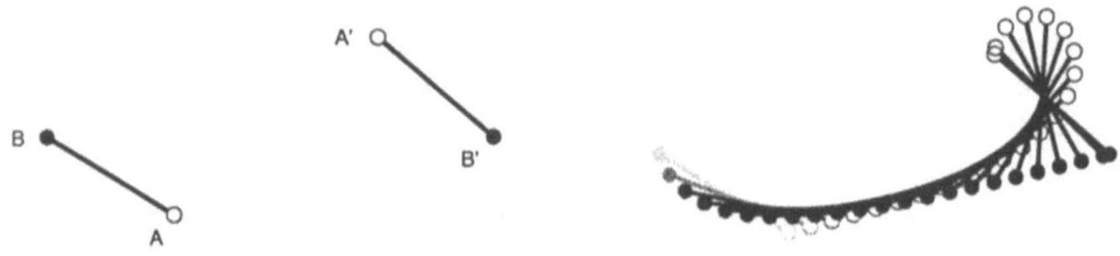


**Fig. 1.** An instance of the ladder carrying problem (left), and an optimal motion.

A number of distributed algorithms for carrying a ladder using two identical omni-directional robots (hence having the same maximum speed) have been reported for the case in which there are no obstacles in the workspace [1,2]. The goal of this paper is to consider the problem under the assumption that the robots are not necessarily identical and the workspace is not necessarily obstacle-free. Specifically:

1. We present two distributed algorithms for the case in which two robots possibly having different maximum speeds are situated in an obstacle-free workspace. The first algorithm, ALG1, assumes that each robot knows the maximum speeds of both robots, while the second algorithm, ALG2, is for the case this information is not available. The algorithms are evaluated using computer simulation.
2. We then present and evaluate an algorithm called ALG3 for the case in which two robots must transport a ladder through a corridor with a 90 degree corner. We assume that the two robots can have different characteristics in terms of the selection of a path through the corner, the maximum speed, "adherence" to stay on the intended path, and reaction to the other robot's motion.
3. Finally we give an outline of a more general framework for designing distributed algorithms for robots based on heuristic search, which is an AI

technique for finding a solution within a combinatorially large problem space [11]. In the new framework we propose, each robot's path planning process is viewed as real-time search for "most promising" motions interleaved with actual locomotion. This approach is expected to be effective even for more complex multi-robot motion coordination problems involving a larger number of robots in a workspace cluttered with obstacles.

Due to space limitation we are not able to present some of the details. The missing details may be found in the references, or will be reported in forthcoming papers.

## 2 The Model

The model of the robots we use is based on the omni-directional robots developed at RIKEN [3].

We represent each robot $R$ as a disk. One end of the ladder is attached to a force sensor that we model as an ideal spring located at the center of $R$. At any time during a motion the vector from the center of $R$ to the tip of the ladder attached to its force sensor is called the *offset vector*, and is denoted $\boldsymbol{o}$. The term *offset* refers to $|\boldsymbol{o}| = |(D-\ell)/2|$, where $D$ is the distance between (the centers of) the two robots and $\ell$ is the length of the ladder.

An ***algorithm*** for robot $R$ with maximum speed $V$ is any procedure that computes a ***velocity vector*** $\boldsymbol{v}$ with $|\boldsymbol{v}| \leq V$, using available information such as the robots' current and final positions, the offset vector $\boldsymbol{o}$, and the geometry of the workspace. We assume that $R$ repeatedly computes $\boldsymbol{v}$ and moves to a new position with velocity $\boldsymbol{v}$ for unit time.

We evaluate our algorithms by computer simulation. For simplicity we use discrete time and assume that both robots compute their respective velocity vectors and move to their new positions at time instants $0, 1, \ldots$.

## 3 Obstacle-Free Workspace

Consider two robots $A$ and $B$ in an obstacle-free workspace with respective goal positions $A'$ and $B'$, where $A$ and $B$ may have different maximum speeds. In the following we let $L_A = |AA'|$ and $L_B = |BB'|$, and describe instances of this problem using $L_B$ and two angles $\alpha$ and $\beta$, $-180° \leq \alpha, \beta \leq 180°$, that the ladder makes with $\overline{BB'}$ at the current and goal positions, respectively. See Fig. 2. The robots do not need to know $\alpha$ and $\beta$, but they do know the acute angle $\delta$, $0° \leq \delta \leq 180°$, that the ladder must rotate between the current and goal positions.

### 3.1 Distributed algorithms ALG1 and ALG2

We introduce two distributed algorithms ALG1 and ALG2. ALG1 is for the case where the two robots' maximum speeds are known to both robots, while

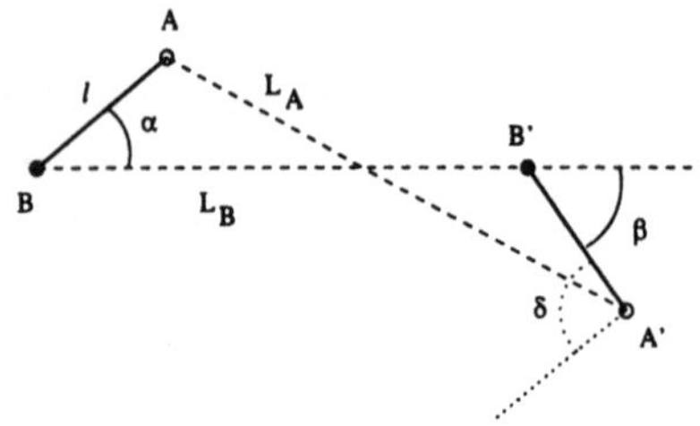


**Fig. 2.** The setup for ALG1 and ALG2.

ALG2 is for the case where neither robot knows the other robot's maximum speed. In either algorithm we assume that the robots' current and goal positions are known to both robots. (This assumption can be expensive to realize in practice.)

Both algorithms are memoryless in the sense that their output is a function of the current state (and is independent of the motions in the past). It is therefore sufficient to view $A$ and $B$ of Fig. 2 as the robots' current positions and specify how the velocity vector is computed from $A$, $B$, $A'$ and $B'$.

**Algorithm ALG1**

Both robots know the maximum speeds $v_A$ and $v_B$ of both, as well as the positions $A$, $B$, $A'$, $B'$, and hence $L_A$, $L_B$, angle $\delta$ and their offset vectors $\boldsymbol{o}_A$ and $\boldsymbol{o}_B$. $c_1$, $c_2$, and $s \geq 0$ are some constants. Since both robots have the same information we explicitly describe the procedure for both.

**Step 1:** Let $e_A = L_A/v_A$ and $e_B = L_B/v_B$. Let $\boldsymbol{t}_A$ and $\boldsymbol{t}_B$ be vectors directed from $A$ to $A'$ and from $B$ to $B'$, respectively, such that
- $|\boldsymbol{t}_A| = 1$ and $|\boldsymbol{t}_B| = (e_B/e_A)^{c_1}$ if $e_A \geq e_B$, and
- $|\boldsymbol{t}_A| = (e_A/e_B)^{c_1}$ and $|\boldsymbol{t}_B| = 1$ if $e_A < e_B$.

**Step 2:** Let $\boldsymbol{r}_A$ and $\boldsymbol{r}_B$ be the "rotational vectors" of length $c_2\delta$, perpendicular to the ladder, and in opposite directions to reduce $\delta$ (favoring a counterclockwise rotation if $\delta = 180°$).

**Step 3:** Scale the offset vectors as $\boldsymbol{h}_A = s\boldsymbol{o}_A$ and $\boldsymbol{h}_B = s\boldsymbol{o}_B$.

**Step 4:** $\boldsymbol{T}_A = \boldsymbol{t}_A + \boldsymbol{r}_A + \boldsymbol{h}_A$ and $\boldsymbol{T}_B = \boldsymbol{t}_B + \boldsymbol{r}_B + \boldsymbol{h}_B$.

**Step 5:** Output velocity vector
- $\boldsymbol{v}_A = v_A \boldsymbol{T}_A / \max\{|\boldsymbol{T}_A|, |\boldsymbol{T}_B|\}$ for robot $A$, and
- $\boldsymbol{v}_B = v_B \boldsymbol{T}_B / \max\{|\boldsymbol{T}_A|, |\boldsymbol{T}_B|\}$ for robot $B$.

In **Step 1** ALG1 tries to slow the robot that would otherwise reach its goal sooner than the other according to the estimates $e_A$ and $e_B$. The constant $s$ used in **Step 3** is a parameter indicating how a robot reacts to the offset.

**Algorithm ALG2**

The robots have all the information available in ALG1, except robot $A$ does not know $v_B$ and robot $B$ does not know $v_A$. We use additional constants $v'_A$ and $v'_B$.

**Step 1:** Robot $A$ runs ALG1 using $v_A$ and $v'_B$ in place of $v_A$ and $v_B$, respectively. Likewise robot $B$ runs ALG1 using $v'_A$ and $v_B$. Let $\boldsymbol{v}'_A$ and $\boldsymbol{v}'_B$ the velocity vectors obtained.

**Step 2:** Output velocity vector
- $\boldsymbol{v}_A = \boldsymbol{v}'_A$ if $|\boldsymbol{v}'_A| \leq v_A$, and $\boldsymbol{v}_A = v_A\boldsymbol{v}'_A/|\boldsymbol{v}'_A|$ otherwise, for robot $A$.
- $\boldsymbol{v}_B = \boldsymbol{v}'_B$ if $|\boldsymbol{v}'_B| \leq v_B$, and $\boldsymbol{v}_B = v_B\boldsymbol{v}'_B/|\boldsymbol{v}'_B|$ otherwise, for robot $B$.

In ALG2 we use $v'_A$ and $v'_B$ as an estimate of unknown $v_A$ and $v_B$, and the output of ALG2 coincides with that of ALG1 if $v'_A \leq v_A$ and $v'_B \leq v_B$. We assume that $v'_A$ and $v'_B$ are constants supplied to ALG2, since it is one of our basic goals to keep the algorithms memoryless (memoryless algorithms can tolerate a finite number of transient errors). It would be interesting, however, to modify ALG2 so that the robots will choose suitable values for $v'_A$ and $v'_B$ based on their recent history.

### 3.2 Experimental results by computer simulation

Using the setup of Fig. 2 we evaluate the performance of the algorithms in terms of the time necessary for the robots to reach their goals. The length $\ell$ of the ladder is 100, and the radius of the disks representing the robots is 10. We use $v_A = 1$ and $v_B = k$ (so $k$ is the ratio of $v_B$ to $v_A$), and in the following discuss mainly the results for the case $k \geq 1$. The parameters $c_1, c_2$ and $s$ of ALG1 are set to 3, 0.5, and 0.1, respectively, that have been found to work well when $v_A = v_B$ [1]. To reduce the number of instances to examine, we experiment with only two values of $L_B$, $L_B = 200$ $(= 2\ell)$ and 400 $(= 4\ell)$, while changing $\alpha$ in the range from 0° to 90° and setting $\beta = 180° - \alpha$.

Fig. 3 shows the motions generated by ALG1 for $k = 1$ and $\alpha = 30°$, for $L_B = 200$ (left) and $L_B = 400$ (right). Fig. 4 shows the same, for $k = 3$ instead of 1. The finish times for $L_B = 200$ are 205 ($k = 1$) and 201 ($k = 3$), and for $L_B = 400$ they are 404 ($k = 1$) and 401 ($k = 3$).

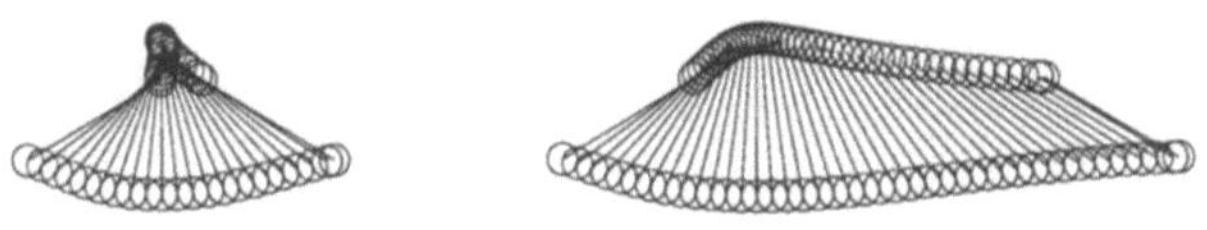

**Fig. 3.** ALG1 for $k = 1$, $\alpha = 30°$, $L_B = 200$ and 400.

Fig. 5 shows the finish times for $k$ = 1, 2, 3, 4, 5 and $0 \leq \alpha \leq 90°$, for $L_B$ = 200 (left) and 400 (right). We observe that the time decreases as $k$ increases when $\alpha \leq 60°$. This phenomenon is quite natural, since smaller $\alpha$ implies more necessary rotation, and larger $k$ causes $B$ to slow down, thus allowing $A$ to rotate more quickly.

**Fig. 4.** ALG1 for $k = 3$, $\alpha = 30°$, $L_B = 200$ and 400.

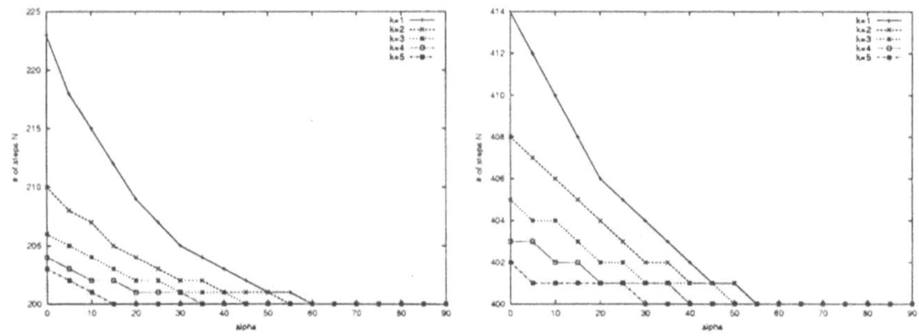


**Fig. 5.** Finish times of ALG1 for $k = 1$, 2, 3, 4 and 5, for $L_B = 200$ (left) and 400 (right).

Fig. 6 shows the motions generated by ALG1 (left) and ALG2 (right) for an instance with $k = 1/3$, $\alpha = 30°$ and $L_B = 200$, where in ALG2 both robots use an estimate of $k^* = 3$ in place of the unknown $k$. (That is, $A$ uses $v_A = 1$ and $v'_B = 3$, while $B$ uses $v'_A = 1/9$ and $v_B = 1/3$. The estimates of $k$ by $A$ and $B$ may differ in real situations.) Note that the robots successfully complete the task using ALG2 even though their estimate $k^*$ is not at all close to actual $k$. However, the large discrepancy between $k^*$ and $k$ has resulted in a noticeable decline in the performance in terms of the finish time — the finish time of ALG2 in Fig. 6 is almost 26% greater than that of ALG1.

**Fig. 6.** ALG1 (left), and ALG2 (right) using an estimate $k^* = 3$, for an instance with $k = 1/3$, $\alpha = 30°$, $L_B = 200$.

## 4 Corridor with a Corner

The distributed approach works well also for the case in which the robots must go through a 90 degree corner in a corridor avoiding both robot-to-wall and ladder-to-wall collision. The motion shown in Fig. 7 has been obtained by an algorithm that is similar in spirit to the ones presented in the preceding section, with an additional step in which each robot computes a target path through the corner before starting the motion. During the motion each robot attempts to move along the path while adjusting its positions based on both the motion of the other robot observed through the force sensor and the need to prevent collision. We assume that the robots can detect how close they and the ladder are to the walls, from their positions and the geometry of the workspace.

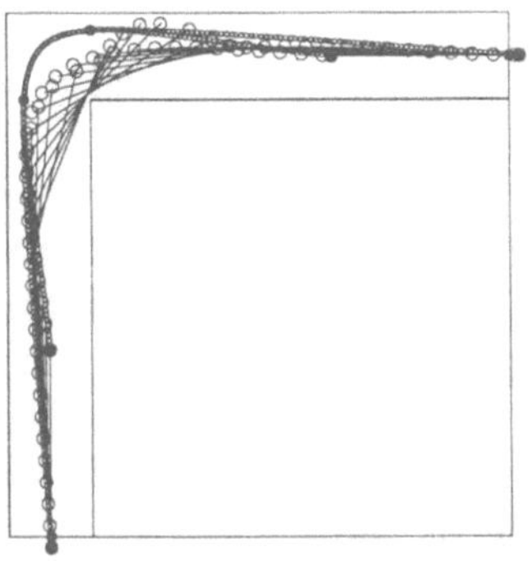

**Fig. 7.** Motion through a corner generated by ALG3. The robots start at the bottom of the figure and makes a right turn. The intended paths of the robots are shown in small circles.

While the robots in the preceding section are assumed to be identical in all aspects other than their maximum speed, in this section we assume that they may differ in a number of other characteristics as well — (i) selection of a path through the corner (in, middle or out), (ii) the maximum speed (high or low), (iii) adherence to the intended path (high or low), and (iv) reaction to the motion of the other robot (high, low, or nonlinear). These variations result in a total of 36 different types of robots from which two are chosen to carry out the task. We assume that the robots do *not* know the characteristics of the other robot.

First, robot $R$ generates a path $P$ through the corner as a Kochanek-Bartels spline [6] that travels either close to the inner walls of the corner (in), close to the outer walls of the corner (out), or somewhere in between the two (middle).

The remaining characteristics of $R$ are determined by the following algorithm ALG3 that $R$ uses to compute its velocity vector. It is assumed that the offset $|o|$ can be as large as the the radius 10 of the disk representing $R$.

**Algorithm ALG3**

**Step 1:** $\boldsymbol{u} = (1/10)(\boldsymbol{g} + f(\boldsymbol{o}) + \boldsymbol{c})$, where

- $\boldsymbol{g}$ is a vector of size 10 directed from the center of $R$ toward the "current target position",
- $\boldsymbol{o}$ is the offset vector and $f$ is a function that converts $\boldsymbol{o}$ into another vector such that $|f(\boldsymbol{o})| \leq 10$ (see details below), and
- $\boldsymbol{c}$ is a suitable "correction vector" needed to prevent collision. (We omit the details of $\boldsymbol{c}$.)

The factor 1/10 effectively reduces the sizes of $\boldsymbol{g}$ and $f(\boldsymbol{o})$ to within 1.

**Step 2:** $\boldsymbol{w} = \boldsymbol{u}$ if $|\boldsymbol{u}| \leq 1$, and $\boldsymbol{w} = \boldsymbol{u}/|\boldsymbol{u}|$ if $|\boldsymbol{u}| > 1$. That is, $\boldsymbol{w}$ is the result of "clamping" vector $\boldsymbol{u}$ at length 1 so that $|\boldsymbol{w}| \leq 1$.

**Step 3:** Output the velocity vector $\boldsymbol{v} = V\boldsymbol{w}$, where $V$ is the maximum speed of $R$.

In our experiments the maximum speed $V$ of $R$ can be either 2 (high) or 1 (low).

A robot with high adherence attempts to return to $P$ more quickly than one with low adherence when it deviates from $P$. We control the adherence of $R$ by choosing vector $\boldsymbol{g}$ in **Step 1** appropriately: If adherence is high then the "current target position" (at which $\boldsymbol{g}$ is aimed) is chosen to be a point on $P$ relatively close to the robot's current location. If adherence is low then $\boldsymbol{g}$ is directed toward a point on $P$ that is farther away. (We omit the details of how such points are actually chosen in our experiments.)

Function $f$ of **Step 1** determines how $R$ reacts to the motion of the other robot observed through $\boldsymbol{o}$. We use the following three variations.

1. high: $f(\boldsymbol{o}) = \boldsymbol{o}$. The robot is highly sensitive to $\boldsymbol{o}$.
2. low: $f(\boldsymbol{o}) = 0.5\boldsymbol{o}$. The robot's sensitivity is low.
3. nonlinear: $f(\boldsymbol{o})$ equals the zero vector $\mathbf{0}$ if $|\boldsymbol{o}| \leq 5$, and $2\boldsymbol{o} - \boldsymbol{o}/|\boldsymbol{o}|$ if $|\boldsymbol{o}| > 5$. The robot reacts to $\boldsymbol{o}$ only after its magnitude exceeds 5.

Note that $|f(\boldsymbol{o})| \leq |\boldsymbol{o}| \leq 10$ holds in all three cases.

The motion shown in Fig. 7 is obtained by ALG3 where both robots use the same out path and has the same low adherence. The robot in front has high maximum speed and high reaction ($f$), while the other robot has low maximum speed and nonlinear reaction.

To evaluate ALG3 we randomly generate 100 pairs of robots from the pool of 36 robots and examine the probability that the task is completed successfully (a motion is considered unsuccessful if the offset exceeds 10). The width of the corridor is 200 and the ladder length $\ell$ varies between 400 and 480. (The robots always fail when $\ell = 490$.) The success rate decreases from 100% for $\ell = 400$ to 75% for $\ell = 480$ if both robots are allowed to choose a path that is either in, middle or out, while the rate increases to 100% for

$\ell = 480$ if neither is allowed to choose an in path. However, if we allow the robots' adherence to be even higher (than high), then the rate drops again to 57% for $\ell = 480$ even without in paths. The reader is referred to [4] for additional results and a detailed analysis of the effect of these parameters to the overall performance.

## 5 Approach based on Heuristic Search

The distributed algorithms presented in the preceding sections use simple vector calculation to compute a velocity vector of a robot. As the simulation results indicate the approach is quite successful, at least for the cases we considered that involve two robots in a relatively simple workspace.

It is conceivable, however, that this simple approach may start to show its limitation as the problem complexity increases. For instance, if the workspace contains many obstacles (unlike just a few simple walls making up a corridor) then the target paths that the robots generate individually may not be mutually compatible, and resolving this conflict using a simple vector-based algorithm seems neither very effective nor intelligent. Coordinating the plans and motions of individual robots becomes even more difficult if the number of robots increases, as in the case of transporting a large object using several robots in a cluttered environment.

Based on the observation above, we are currently investigating a new general framework based on heuristic search for designing distributed algorithms for multiple robots. Heuristic search is an effective technique in AI for solving problems having a combinatorially large state space [11]. The basic idea of our approach is to view each robot's path planning process as real-time search for "most promising" motions interleaved with actual locomotion. One way to implement this idea is to suitably discretize the space of possible actions of the robot, and explore and evaluate some of the sequences of actions that look promising using an effective heuristic function that takes into account both the physical constraints of the robots and the quality of the resulting configurations.

Fig. 8 shows a motion of two robots obtained by this approach that successfully carry a ladder avoiding an obstacle. Again, details are omitted due to space limitation.

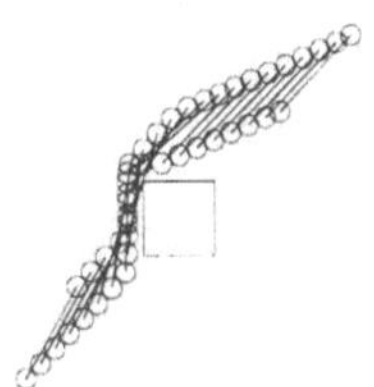

**Fig. 8.** Heuristic search-based motion avoids an obstacle.

## 6 Concluding Remarks

We have presented three distributed algorithms for two robots carrying a ladder under various conditions, and evaluated their performance using computer simulation. We are currently working on a detailed analysis of ALG3 and investigating the potential of the heuristic search-based approach. These results as well as the missing details will be reported elsewhere.

## References


1. Y. Asahiro, H. Asama, S. Fujita, I. Suzuki, M. Yamashita. (1999) Distributed algorithms for carrying a ladder by omnidirectional robots in near optimal time. *Sensor Based Intelligent Robots*, H.I. Christensen, H. Bunke and H. Noltemeier, Eds., Lecture Notes in Artificial Intelligence, Vol. 1724, Springer Verlag, Heidelberg, Germany, 240–254.
2. Y. Asahiro, H. Asama, I. Suzuki, M. Yamashita. (1999) Improvement of distributed control algorithms for robots carrying an object. *Proc. 1999 IEEE Int. Conf. on Systems, Man and Cybernetics*, VI 608–613.
3. H. Asama, M. Sato, L. Bogoni, H. Kaetsu, A. Matsumoto, I. Endo. (1995) Development of an omni-directional mobile robot with 3 DOF decoupling drive mechanism. *Proc. IEEE Int. Conf. on Robotics and Automation*, 1925–1930.
4. E.C.-H. Chang. (2000) Distributed motion coordination of two robots carrying a ladder in a corridor. MS Thesis, EECS Department, University of Wisconsin–Milwaukee.
5. Z. Chen, I. Suzuki, M. Yamashita. (1997) Time optimal motion of two robots carrying a ladder under a velocity constraint. *IEEE Trans. Robotics and Automation 13*, 5, 721–729.
6. D. Hearn, M.P. Baker. (1996) Computer Graphics C Version. Prentice Hall.
7. Y. Kawauchi, M. Inaba, T. Fukuda. (1993) A principle of distributed decision making of cellular robotic system (CEBOT). *Proc. IEEE Int. Conf. on Robotics and Automation*, 833-838.
8. K. Kosuge, T. Oosumi. (1996) Decentralized control of multiple robots handling and objects. *Proc. IEEE/RSJ Int. Conf. on Intelligent Robots and Systems*, 556-561.
9. N. Miyata, J. Ota, Y. Aiyama, J. Sasaki, T. Arai. (1997) Cooperative transport system with regrasping car-like mobile robots. *Proc. IEEE/RSJ Int. Conf. on Intelligent Robots and Systems*, 1754–1761.
10. S. Nagafuji, Y. Asahiro, M. Yamashita, I. Suzuki. (1999) Time-optimal motion of two heterogeneous robots carrying a large object. *Proc. 1999 Joint Conf. of Electrical and Electronics Engineers of Kyushu* (in Japanese), 742.
11. J. Pearl. (1984) Heuristics – Intelligent Search Strategies for Computer Problem Solving. Addison-Wesley.
12. Z. Wang, E. Nakano, T. Matsukawa. (1996) Realizing cooperative object manipulation using multiple behavior-based robots. *Proc. IEEE/RSJ Int. Conf. on Intelligent Robots and Systems*, 310–317.

# Experimental Manufacturing of Object Transfer System "Magic Carpet" Consisting of Actuator Array with Autonomous Decentralized Control

Hichirousai Oyobe, Ryo Marutani, and Yoichi Hori

Department of Electrical Engineering, The University of Tokyo,
7-3-1 Hongo Bunkyo-ku, Tokyo 113-8658, JAPAN
tel: +81-3-5841-6778
E-mail: oyobe@hori.t.u-tokyo.ac.jp, marutani@hori.t.u-tokyo.ac.jp,
hori@hori.t.u-tokyo.ac.jp

**Abstract.** The attempt to realize magic carpet driven by autonomous decentralized type control algorithm is proposed. The proposed system is manufactured as trial experimental setup and an experiment is done by itself. At last, a control algorithm for autonomous decentralized system is proposed. Due to the recent development of micromachine technology, we can integrate a lot of very small decentralized modules, each of which consists of sensor, actuator and electronic circuit. However, control algorithm of such systems have not yet been developed enough. To avoid complicated wiring problem of communication network, it is assumed that the module should exchange information only with neighboring several modules. The proposed system is manufactured for developing control algorithm of such system.

## 1 Introduction

Recent development of science and technology has realized a lot of things which have been drawn only in "SF" world. Micromachine is one of them. The submarine on the film "Fantastic Voyage" is often referred to as an example of future micromachine technology [1]. Micromachine is characterized by the following "3M" [2].

1. Miniaturization
2. Multiplicity
3. Microelectronics

Miniaturization means light and speedy action. Multiplicity means harmony of many micromachines. Microelectronics is important because semiconductor process technology is crucial in making micromachines. One micromachine module has actuator, sensor and electronic circuit like a microprocessor. As each micromachine module is very small and has poor ability, a lot of micromachine modules should be integrated to perform actual tasks [3] [4]. Micromachine technology has inherent problems as follows.

1. communication
   To avoid large scale and complicated wiring problem of communication network, it should be assumed that the module can exchange information only with neighboring several modules.
2. structure
   Semiconductor process technology is not good at making heterogeneous structure, but very suitable to make homogeneous structure.

As discussed above, micromachine modules must harmonize with each other. The proposed experimental setup is manufactured to develop a algorithm such that a micromachine module harmonizes with the other one [5].

## 2 Problem setting

For consideration of autonomous decentralized control system, we set an explicit problem. The problem is to move a ball put on the magic carpet to the target position. Each module of the magic carpet shown in Fig.1 can move along only $z$-direction [6].

<u>Move the Ball</u>

$$(x_{ini}, y_{ini}) \Rightarrow (x_c, y_c)$$

where, $(x_{ini}, y_{ini})$ is the initial position of the ball and $(x_c, y_c)$ is the target position.

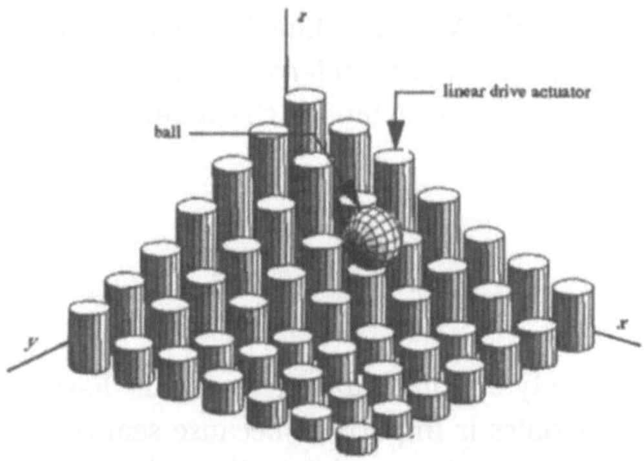


**Fig. 1.** Conceptual configuration of magic carpet. We try to realize this magic carpet in our real world.

We consider the micromachine array on the $xy$ plane $(x_{\min} \leq x \leq x_{\max}, y_{\min} \leq y \leq y_{\max})$, which consists of $m$ decentralized modules along $x$-direction and $n$ along $y$-direction [7]. The function of each decentralized module is given as follows. The sensor detects whether anything exists on the module or not. The actuator drives the module along $z$-direction and keeps it at any position. The microprocessor can perform very simple calculation and communicate only with several neighboring modules.

# 3 Manufacturing experimental setup for the proposed system

## 3.1 Configuration of the setup

Experimental setup is manufactured to develop autonomous decentralized control for micromachine modules. Micromachine researches are now in developing. Our final goal is the realization of magic carpet based on micromachine technology, but it is not so easy to develop micromachine linear drive actuator at present. Whereby, we decide to manufacture experimental setup by using real world scale actuators. Fig.2 shows the rough sketch of the experimental setup consists of 64 decentralized modules. Figs.3 and 4 are controllers and actuators of magic carpet under manufacturing. This setup is composed by 64 actuators, 64 designed variable structure controllers and "one" personal computer(PC). 64 microprocessors are virtually realized by a PC. In other words, hardware configuration of the setup is central control and decentralized control is realized by software in the PC.

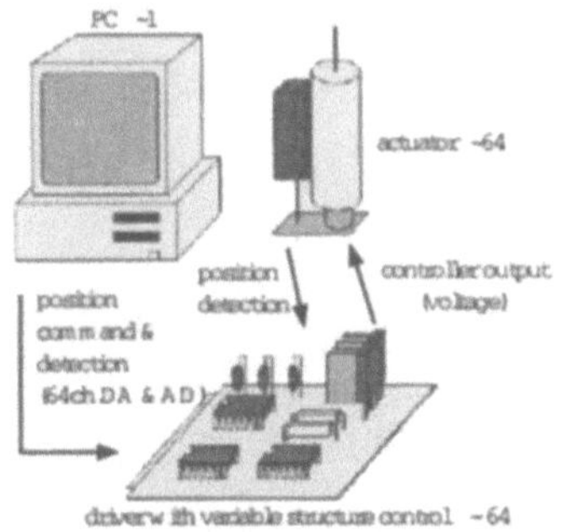


**Fig. 2.** Fundamental conceptual figure of experimental setup consisting of 64 variable structure controllers, 64 linear drive actuators and "one" PC.

**Fig. 3.** Overview of experimental setup including 64 controllers and main power supply.

## 3.2 Design of position control for each actuator

A solenoid plunger is adopted as a real world actuator. The position (or force) control are implemented in each adopted solenoid plunger to realize linear drive actuator. Fig.5 shows the solenoid plunger. At first, we try to design position controller for the solenoid plunger. The structure of this solenoid plunger is shown in Fig.6. Linear position sensor seen in Fig.6 is attached with

**Fig. 4.** A part of experimental setup, arrayed many actuators on 2-dimensional plane

**Fig. 5.** The real world actuator module consisting of solenoid plunger. We can easily see attached position sensor. This position sensor is putted to use position signal in feedback controller.

this solenoid plunger for position control. Electromagnetic force is calculated using the simple model in Fig.7 as electromagnetic model. The force of the

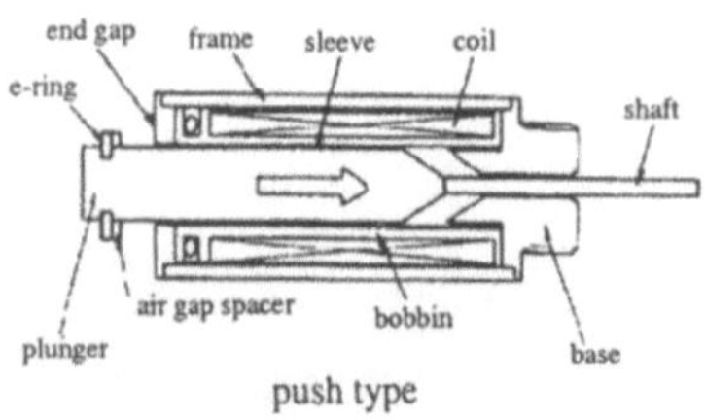


**Fig. 6.** Configuration of the solenoid plunger.

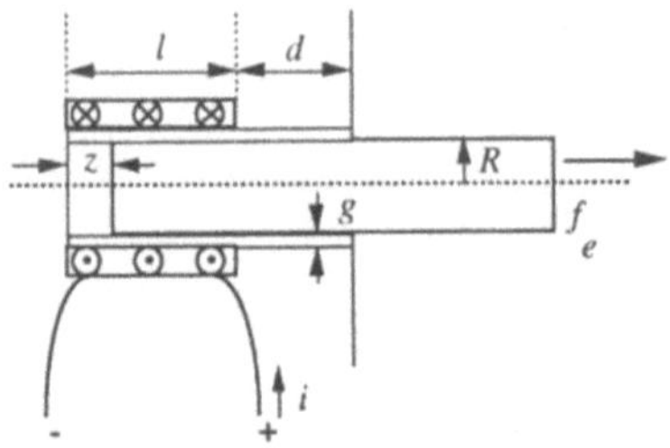


**Fig. 7.** Electromagnetic model of the solenoid plunger. Electromagnetic force generated in this solenoid plunger is calculated by using this simple model.

solenoid plunger takes the form of (1) using Ampere's integration law. It is inversely proportional to the square of position $z$ and proportional to the square of current $i$.

$$f_e = -\frac{2\mu_0 N^2 \pi R^2}{(2dz + gR)^2} i^2 \tag{1}$$

where, $\mu_0$, $N$, $R$, $g$, $\pi$ and $i$ are vacuum permeability, turn number, radius of plunger, gap length, the circle constant and solenoid current, respectively.

$$f_e = f_e(z^{-2}, i^2) \tag{2}$$

Because of high nonlinearity in force generation, position control of the solenoid is not so easy. We applied several robust control methods, and as the result, we selected variable structure control. Figs.8 and 9 are its block diagram and analog diagram respectively. The comparator represents variable structure and high pass filter suppresses chattering of position $z$ near desired position $z^*$ in Fig.8.

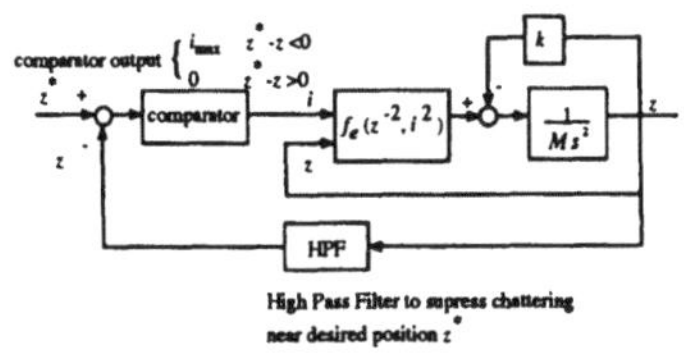


**Fig. 8.** Block diagram of solenoid plunger including variable structure controller and high pass filter.

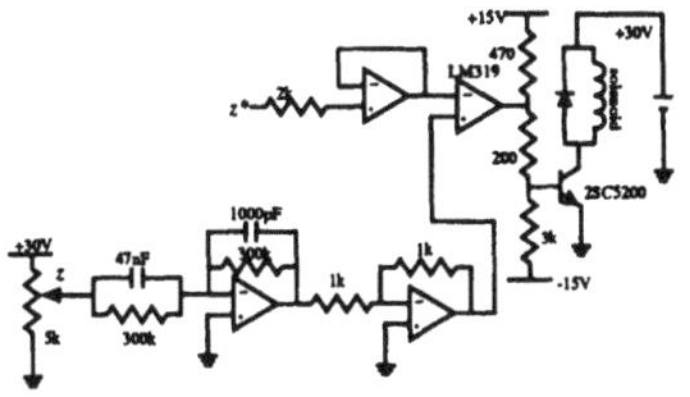


**Fig. 9.** Analog circuit of a variable structure controller. This variable structure controller actually used in the experiment. 64 sets of this analog circuit are prepared for realization of magic carpet. This circuit is very simple and adopted for position controller.

### 3.3 Position control for each actuator

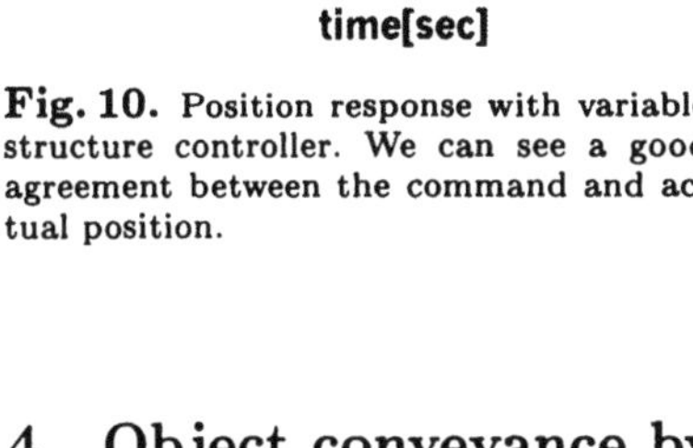


**Fig. 10.** Position response with variable structure controller. We can see a good agreement between the command and actual position.

In this section, the experiment results of solenoid plunger position control is shown. Fig.10 is the position response driven by the analog circuit of Fig.9. We can see a good agreement between the command and actual position and recognize to overcome nonlinearity of the solenoid plunger by using robust variable structure controller and high pass filter.

## 4 Object conveyance by the proposed system

We stand on the starting point of autonomous decentralized control research by completion of experimental setup. Experimental setup configuration is mainly reported in this paper. In this section, we consider to move the proposed system by autonomous decentralized scheme. At first, the proposed algorithm is divided into two steps: macro-scale and micro-scale, corresponding actuator array and actuator in itself. Micro-scale control is implemented in a robust manner by the variable structure controller. Macro-scale control is discussed in this chapter.

### 4.1 Concept of "field"

What is the concept of "field"? Generally speaking, autonomous decentralized system is a complex system composed of a lot of decentralized modules. However, its behavior can be divided into two steps. One is the micro-scale behavior of each module, and the other is the macro-scale behavior to rule many modules. For example, in our society, each human corresponds to a module and public opinion rules macro-scale behavior. Public opinion is a kind of "field". The micro behavior is relatively quick, and macro-scale behavior usually has a long time constant. Our proposal is to realize this "field" by central control.

### 4.2 Combined control of central / autonomous decentralized algorithms

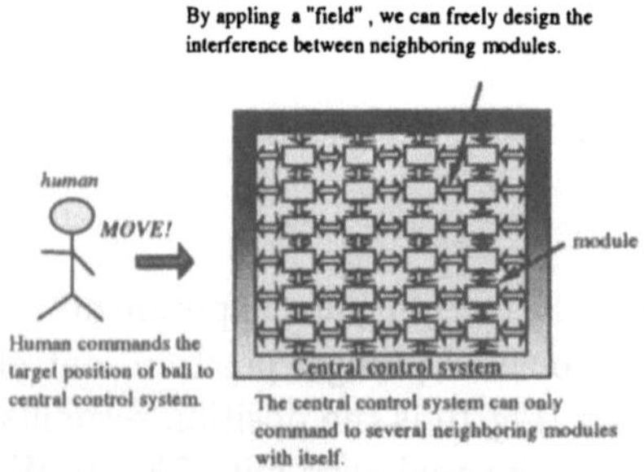


**Fig. 11.** Combined control of central / autonomous decentralized algorithms.

We apply the proposed "combined control of central / autonomous decentralized algorithms" to the magic carpet. Fig.11 shows its conceptual configuration. The central controller is allocated around decentralized modules. The modules have lattice structure. The operator commands only to the central controller. The central controller sends commands to the several neighboring modules close to the boundary. Each module is activated by communicating with neighboring modules.

Here, it is clear that the central controller is in charge of macro-scale control, and each module is of micro control.

### 4.3 Control of macro-scale behavior

To convey a ball to the target position, we hit upon the idea of constructing ball position feedback control based on estimation of ball position at first. That scheme was already proposed [8].

As next step, ball conveyance without estimation of ball position by the proposed system is considered. An algorithm is quite simple. To convey a ball to the target position, the carpet makes a hollow. To do this, "each" module should know the target position and its own position. How does each module get these important information? It is discussed in this section.

It is easily known from the configuration of the system, the central controller gives the boundary condition. At first, we design the field to tell each module of its own position. If $u^{xcd}$, the field value, has the solution given by (3), each module can know its own position as follows.

$$u^{xcd} = \frac{x_{\max} - x_{\min}}{x_{\max}} x + x_{\min} \qquad (x_{\min} \le x \le x_{\max}, y_{\min} \le y \le y_{\max}) \tag{3}$$

(4) is the field equation to give the solution (3). The field (4) can be realized only by local communication between modules.

$$\frac{\partial^2 u^{xcd}}{\partial x^2} + \frac{\partial^2 u^{xcd}}{\partial y^2} = 0 \tag{4}$$

To the central controller, (5) is given as the boundary condition.

$$u^{xcd}(x_{\min}, y) = x_{\min}, u^{xcd}(x_{\max}, y) = x_{\max}, \quad u^{xcd}(x, y_{\min}) = u^{xcd}(x, y_{\max}) = x \tag{5}$$

Next, the field to tell each module of the target position of the ball is designed. If $u^{xc}$, the field value, has the solution of (6), each module can know the target position.

$$u^{xc} = x_c \qquad (x_{\min} \le x \le x_{\max}, y_{\min} \le y \le y_{\max}) \tag{6}$$

The field (7) gives the solution (6). Here, (7) can be realized only by local communication, too.

$$\frac{\partial^2 u^{xc}}{\partial x^2} + \frac{\partial^2 u^{xc}}{\partial y^2} = 0 \tag{7}$$

In this case, (8) is given to the central controller as the boundary condition.

$$u^{xc}(x_{\min}, y) = u^{xc}(x_{\max}, y) = u^{xc}(x, y_{\min}) = u^{xc}(x, y_{\max}) = x_c \tag{8}$$

(9) is the function to make a hollow at the target position $(x_c, y_c)$. $(u^{xcd}, u^{ycd})$ shows the module position.

$$z(x,y) = f(\boldsymbol{u}) = f(u^{xc}, u^{xcd}, u^{yc}, u^{ycd}) = \frac{\alpha_1 \left\{(u^{xcd} - u^{xc})^2 + (u^{ycd} - u^{yc})^2\right\}}{\alpha_2 \left\{(u^{xcd} - u^{xc})^2 + (u^{ycd} - u^{yc})^2\right\} + 1} \tag{9}$$

The fields (4), (7) include no term concerning time domain. This means that each module acts independently.

## 5 Results

In this section, simulation and experimental results are reported. Parameters in Table 1 are used common in simulation and experiment. The hand-made experimental setup has capacity to perform an experiment by 64 (8x8) actuator arrays. Note that following experiment and simulation are done by 49 (7x7) actuators as shown in Table 1. This is caused by no large size rubber sheet coated on actuators shown in Fig. 15.

**Table 1.** Common parameters. The simulation and experiment are done with those parameters.

| | |
|---|---|
| $\Delta x$ | 0.065 [m] |
| $\Delta y$ | 0.065 [m] |
| numb. of module (x) | 7 |
| numb. of module (y) | 7 |
| $(x_{min}, y_{min})$ | (0, 0) |
| $(x_{max}, y_{max})$ | (0.39, 0.39) [m] |
| $(x_{ini}, y_{ini})$ | $(4.5\Delta x, 4.5\Delta y)$=(0.2925, 0.2925) [m] |
| $(x_c, y_c)$ | $(1.5\Delta x, 2.5\Delta y)$=(0.0975, 0.1625) [m] |
| $\alpha_1$ | 0.5 |
| $\alpha_2$ | 50 |

## 5.1 Simulation results

Behavior of the ball put on the magic carpet is simulated and shown in Figs.12 and 13. Fig.12 is time response of the ball and teaches that the ball position converges to the target position (0.0975, 0.1625) [m] with state error. This state error is caused by distance between actuators($\Delta x$, $\Delta y$). The shorter this distance is, the smaller the state error becomes. Also the ball locus with vibration is quite natural. The implemented algorithm moves the ball by gravity. Fig.13 is a trajectory of the ball during the simulation. Figs.14-a, 14-b and 14-c show the surface curves of the magic carpet at the start 0.00[s], midstream 1.39[s] and end 9.79[s], respectively.

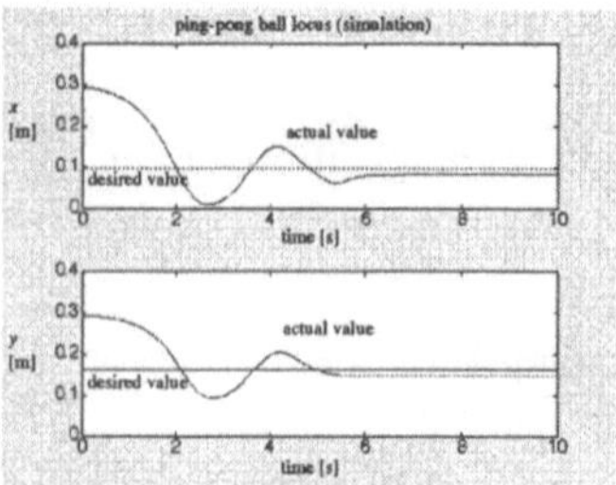


**Fig. 12.** Ping-pong ball locus (simulation). Upper figure is time response of ball position along x-direction. Lower one is about y-direction. The ball converges to the desired position with vibration.

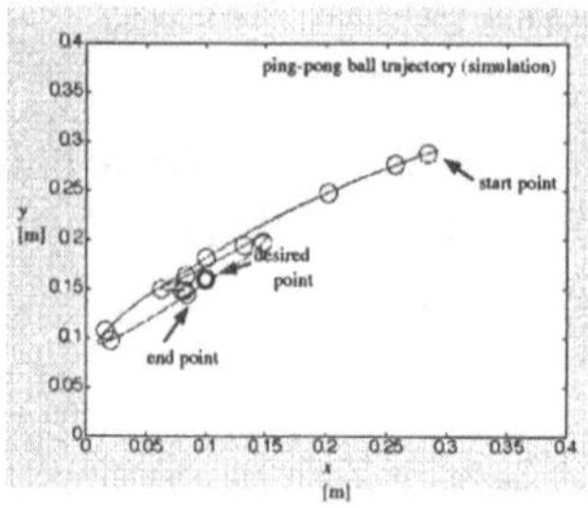


**Fig. 13.** Ping-pong ball trajectory (simulation). This figure is a ball trajectory obtained simulation results.

## 5.2 Experimental results

In this section, experimental results by the manufactured setup are reported. Parameters applied the experiment are same as those of the simulation shown

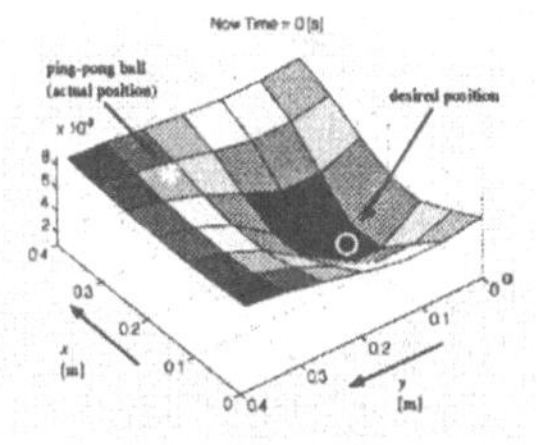


**Fig. 14-a.** Surface curve at 0.00 [s] (simulation)

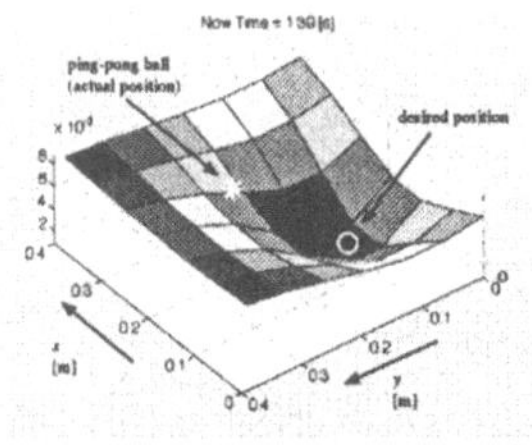


**Fig. 14-b.** Surface curve at 1.39 [s] (simulation)

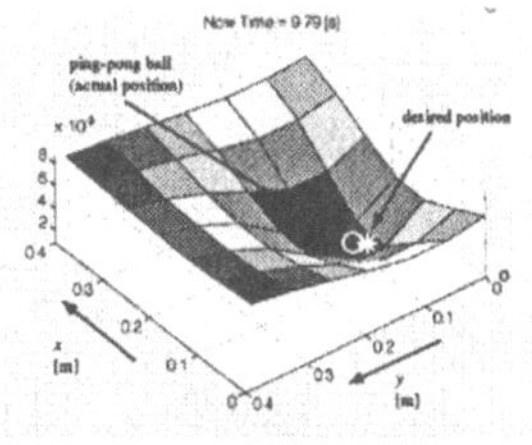


**Fig. 14-c.** Surface curve at 9.79 [s] (simulation)

**Fig. 14.** Magic carpet behavior (simulation) - interpolation curves between each actuator top

in Table 1. Experimental procedure is quite simple. At first, a ping-pong ball is put at the initial position (0.2925, 0.2925)[m] (Fig.15-c). Then, the desired position (0.0975, 0.1625)[m] is set on the PC. Fig.15-a is the position response of the ball. After 9 seconds, the ping-pong ball is carried to the desired position. The state error shown in Fig.15-a is caused by the same reason as that of the previous simulation. Fig.15-b is a trajectory of the ball. This experimental result corresponds to the simulation result of Fig.13.

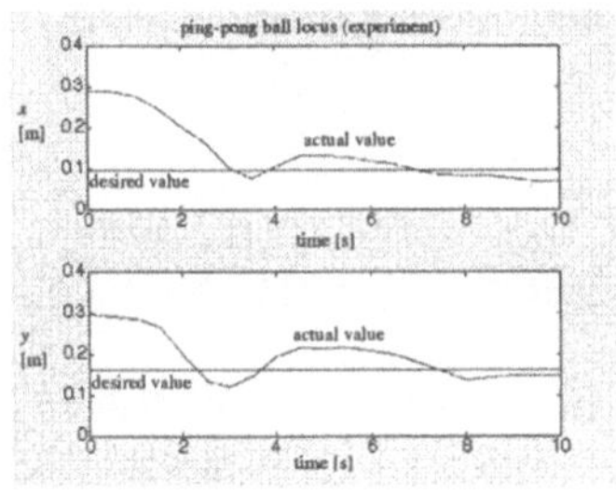


**Fig. 15-a.** Ping-pong ball locus (experiment). Upper figure is time response of ball position along x-direction. Lower one is about y-direction.

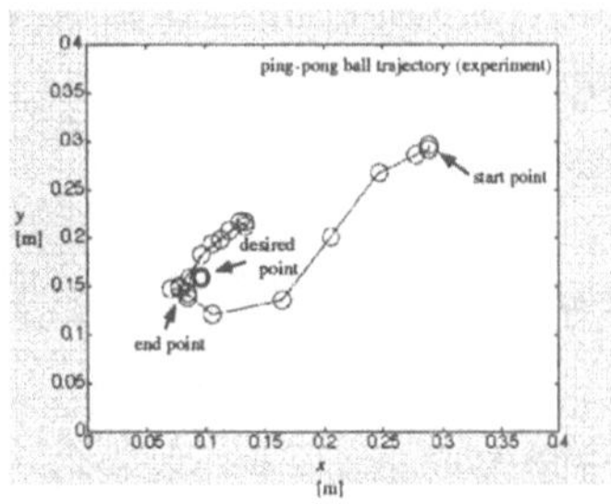


**Fig. 15-b.** Ping-pong ball trajectory (experiment).

# 6 Conclusion

We aim to construct object conveyance system composed by actuator arrays based on macromachine technology. In this paper, we mainly report

1. trial experimetanl manufacturing composed by actuator arrays
2. object conveyance algorithm development for such system
3. simulation and experimental results of the proposed system.

The experimental setup of the proposed system is manufactured in real size world. In order to realize this experiment, variable structure control of

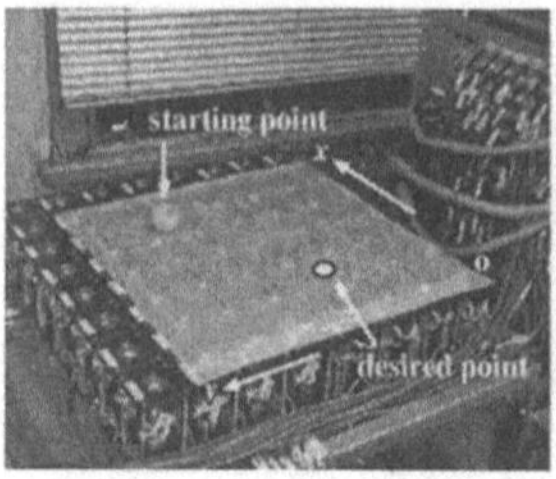


**Fig. 15-c.** Manufactured setup behavior before the experiment. The ping-pong ball is set at the initial point.

**Fig. 15-d.** Manufactured setup behavior during the experiment. The ping-pong ball is being carried to the target.

**Fig. 15-e.** Manufactured setup behavior after the experiment. The ping-pong ball has reached to the desired point.

**Fig. 15.** Magic carpet behavior (experiment) - An insufficient size rubber sheet covers the actuators.- .

solenoid plunger position was applied and this position control succeeded. Also, we proposed the combined control of central / autonomous decentralized algorithms in the realization of magic carpet. The proposed algorithm overcomes the problems concerning communication and structure. We have shown that the object can be conveyed to the target position by making a hollow in the "field", where each module estimates the target and its own positions only by communication with several neighboring modules. The simulation and experimental results prove that the ball are carried.

## References


1. Yasuhiro N.(1994) On the Publication of an Issue on "Micro Machine". JSME **97**, 252 (*in Japanese*)
2. Gabriel. K. J(1994) MEMS Research Project in U.S.A. JSME, **97**, 272-275 (*in Japanese*)
3. Hiroyuki F.(1993) Autonomous Distributed Systems for Micromachines, SICE, **32**, 848-853 (*in Japanese*)
4. Hiroyuki F.(1994) Autonomous Distributed Micro Systems. JSME, **97**, 298-301 (*in Japanese*)
5. Yoshiteru I.(1993) The Character of Information Processing in Autonomous Decentralized Systems. SICE, **32**, 830-836(*in Japanese*)
6. Hichirousai O., Hiroaki K., Yoichi H.(2000) Design and Realization of Autonomous Decentralized Object Transfer System:Magic Carpet. Proceedings of Advanced Motion Control, 25-29
7. Satoshi K., Yoshio M., Hiroyuki F.(1998) Two-dimensional conveyance system using cooperative motions of many fluidic microactuators. Advanced Robotics, **12**, 155-165
8. Hichirousai O., Yoichi H.(1998) Suggestion and Design of Autonomous Decentralized Type Magic Carpet Using Concept of Field. Proceedings of JIASC, **3**, 379-382 (*in Japanese*)

# A Control Method of a Multiple Non-holonomic Robot System for Cooperative Object Transportation

ZhiDong Wang, Yoshio Kimura, Takayuki Takahashi, and Eiji Nakano
Advanced Robotics Lab., GSIS, Tohoku University, Sendai-City, 980-8579, Japan

**Abstract.** This paper describes a multi-robot system which incorporates a behavior-based dynamic cooperation strategy for object transportation. This cooperation strategy is realized in two steps: designing the distributed robot's cooperative behavioral attributes according to robot's abilities, and organizing these behavioral attributes so that the team cooperation is realized. A method to control non-holonomic robots for cooperative object transportation which includes three control modes on realizing behavior attributes on each individual robot's ability direction is proposed. The switching among the three control modes is decided according to the position of the virtual target point and the orientation of the robot's body. An experimental robot system is designed to verify the proposed method and some experiment results are shown in this paper.

## 1 Introduction

Our main concern is studying teams of distributed cooperative robots performing real world tasks. We have chosen the object manipulation task with mobile robots for its well-known practical applications. For an object transferring task, several distributed robots system [1][6][11][10] have been proposed but most of them need common model of the whole system or need explicit communication among robots. Some protocols for box pushing was demonstrated [2][3][8], but they are short of guarantee for manipulation task performing. This paper addresses the control problem of a multiple robot system BeRoSH (**Be**havior-based Multiple **Ro**bot **S**ystem with **H**ost for Object Manipulation) incorporating a behavior-based dynamic cooperation strategy for multiple robots which have limited ability for object manipulation. The host acts as a manager which organizes robots' ability direction and generates final goals for the robots. A method to control non-holonomic robots for cooperative object transportation which include three control modes on realizing behavior attributes on each individual robot's ability direction is proposed. The switching among the three control modes is decided according to the position of the virtual target point and the posture of the robot's body. An experimental system is designed to verify the proposed control method and some experiment results are shown to support the main concept.

## 2 Cooperative System BeRoSH

In this study, rather than aiming at constructing a pure distributed robots system, we propose a multi-robot system BeRoSH [12] [13]. It is composed

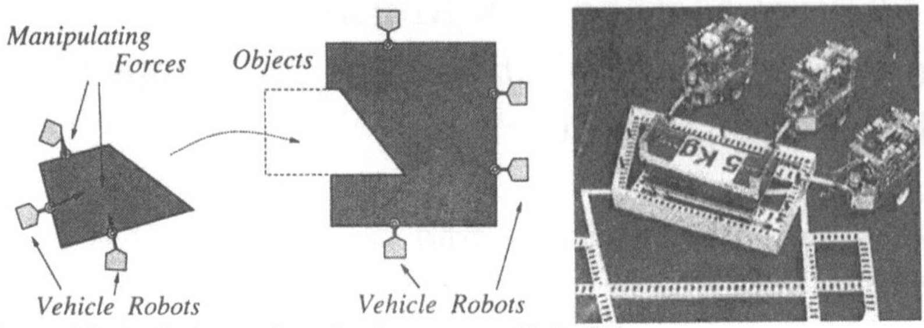

Fig. 1. Manipulation task and robot system BeRoSH

of a managing robot as its host, and incorporates homogeneous behavior-based robots, which have limited abilities for manipulation(Fig.1). Different from the centralized system, BeRoSH host acts as a leader and an organizer which coordinates the robots' behavior, generates goals for the robots, and offers some static but global position information for each robot. It does not perform any calculation of target objects' dynamics or force distribution for dynamic cooperation and does not do any path planning for each robot. Each robot has a 1DOF arm manipulation.

In general, there are two basic strategies to realize the transportation and assembly of large and heavy objects using multiple robot system. To push and pull the object which is supported on casters or free moving platform, or to carry the object directly. The latter is usually used in the research works [4] [6] of distributed manipulation. In this strategy, each robot must possess a strong mechanical parts to lift and carry the object. It also gives a very tight constraint among robots [10], e.g. the difficulty to reconfigure the position of the robot with object during the manipulation. The first strategy, which is used in this study, is easy to realize specially on handling heavy objects. It also leaves the possibility of robot's reorganization in the manipulation. Exerting large force or moment using an actuator at arm-mobile joint may not be possible, but a relative large force can be generated by the robot's mobile part. For the said implementation, a 1DOF arm direction is enough for each robot to manipulate the object. For these reasons, the robot BeRoSH has the same concept with a tow tractor which is used to pull a jumbo jet to move from the boarding bridge to the runway. The robot's arm is holding a contacting point on the object and can rotate around it freely.

## 2.1 Dynamics of the object manipulation

The dynamic equations of the two dimensional object manipulation by multiple robots(Fig.2), which can generate pushing and pulling force to the object, are represented from Eq.(1) $\sim$ Eq.(3).

$$\mathcal{F}^o = \mathcal{F}_0^o + \boldsymbol{I}_0 \, \ddot{\boldsymbol{\phi}}_0 \tag{1}$$

where

$$\ddot{\boldsymbol{\phi}}_0 = ( \ddot{\boldsymbol{r}}^{o\,T} \, , \, \dot{\omega}^T )^T, \qquad \boldsymbol{I}_0 = \begin{pmatrix} m\boldsymbol{E}_2 & \boldsymbol{0} \\ \boldsymbol{0} & I \end{pmatrix} \tag{2}$$

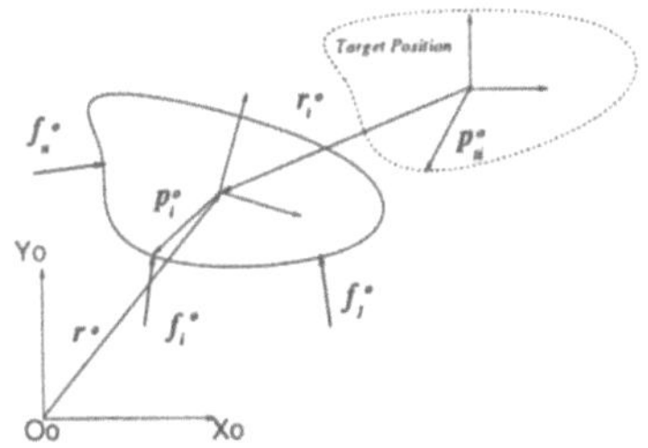


Fig. 2. The Coordination of Object Manipulation

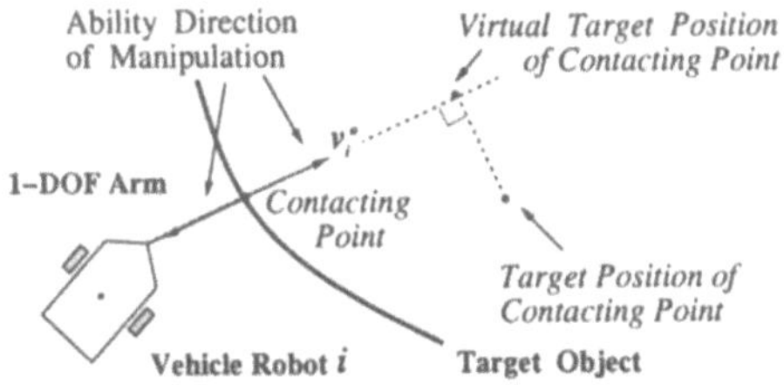


Fig. 3. Robot's ability direction and virtual target point in object manipulation

$$\mathcal{F}_0^o = ( \mathbf{0} \ , \ \omega \times I\omega \ )^T + \mathcal{F}_{fr}^o \tag{3}$$

. $\Sigma^o(O^o, X^o, Y^o)$ is the world reference. Here, $\mathcal{F}^o, \mathcal{F}_0^o, \mathcal{F}_{fr}^o \in R^3$ represent the general forces which includes the force component and moment applied on the target object. $\mathcal{F}_{fr}^o$ is the friction force and moment exerted on the object during the manipulation. $m$ and $I$ are the mass and mass moment of inertia of the target object. $\omega$ and $\ddot{\boldsymbol{r}}^o$ are the object's angular velocity and linear acceleration on its mass center. $\mathbf{0}$ is a zero vector for constructing the matrix. The general resultant force $\mathcal{F}^o$ on the target object is

$$\mathcal{F}^o = \begin{bmatrix} \boldsymbol{f}^o \\ \tau^o \end{bmatrix} = \begin{bmatrix} \sum \boldsymbol{f}_i^o \\ \sum \boldsymbol{p}_i^o \times \boldsymbol{f}_i^o \end{bmatrix} \tag{4}$$

Here $\boldsymbol{f}^o, \boldsymbol{f}_i^o, \boldsymbol{p}_i^o, \boldsymbol{r}^o \in R^2$. $\boldsymbol{f}_i^o$ is the manipulation force generated by robot $i$. $\boldsymbol{f}^o$ and $\tau^o$ are the resultant force and torque exerted on the mass center of the object. $\boldsymbol{r}^o$ is the current position of the center of mass of object and $\boldsymbol{r}_t^o$ is the vector from the center of mass in the target position to the current position. $\boldsymbol{p}_i^o$ is the vector to represent the position of each robot's contacting point. In the BeRoSH system, the contact points of all robots are decided by the managing robot and will not change in the object reference when an object manipulation task is performing.

### 2.2 Behavior-based cooperative strategy

In BeRoSH, a whole task can not be completed by one robot alone, because each robot can only perform a limited part of the action required to complete a task. The robot can only generate a 1-DOF force to the target object along the direction of the arm. Here, we define the *ability direction* of a robot in the manipulation space of an object, to show *both the directions in which the robot can act and the manipulating ability which the robot has in these directions* (Fig.3). We design the cooperative strategy in BeRoSH as:

(1) *every robot has the nature that it is active in its ability direction*
(2) *every robot has the nature that it is passive in its non-ability direction*
(3) *robots' ability directions are organized so that robots work together*

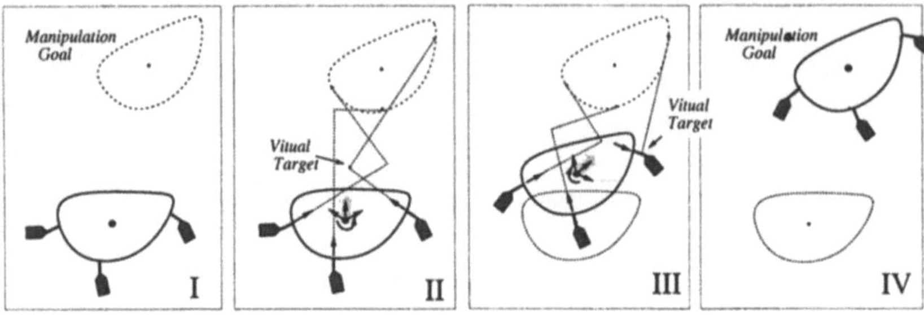


**Fig. 4.** Robots' active behaviors and the object motion toward the goal

In this cooperation, the passive nature means that a robot doesn't give negative influence as much as possible to the object's motion in its ***non-ability direction***. Then this passive nature prevents conflicts from occurring between robots. It is realized by some zero force feedback loops in the robot's behavior-based controller. The active nature in each robot's ***ability direction*** and the correct organization of each robot's ***ability direction*** enables all robots to work on performing the cooperative task manipulation. The host plays this organization role by using the concept of Form Closure [7].

The active nature in its ***ability direction*** is realized through defining a virtual target for its contact point. As shown in Fig.3, the virtual target is a point on the ***ability direction*** line which is projected from the real target of the contact point at each moment. Since the robot's ***ability direction*** has changed in the global reference system following the motion of the target object, the virtual target point is not fixed in either the object coordinate system or the global reference system during the manipulation. Variation in the virtual target will lead the robots toward achieving the manipulation task if the organization of all robots' behaviors is correct(Fig.4). An additional behavioral layer of each robot is that it attempts to have its arm perpendicular to the object's surface at the contact point. Therefore at all times during manipulation, the relation among ***ability directions*** does not change. This makes the organization, a task of managing robot, become static even for a dynamic cooperation task. Basically, only the goal is given to each robot and the target object's motion (path) results from the interaction of all robots' dynamic cooperative behavior outputs [12].

### 2.3 Manipulation force of each robot

Each robot's behavior-based controller will control the robot to generate its manipulation force $\boldsymbol{f}_i^o$ as follow: ($Q$ is positive scalar and $F_{pmax} = K_p Q$)

$$\boldsymbol{f}_i^o = (f_p(\boldsymbol{r}^o, \theta) + f_d(\dot{\boldsymbol{r}}^o, \omega))\boldsymbol{v}_i^o \tag{5}$$

where

$$f_p(\boldsymbol{r}^o, \theta) = \begin{cases} F_{pmax} & \boldsymbol{q}_{ti}^o \cdot \boldsymbol{v}_i^o > Q \\ -F_{pmax} & \boldsymbol{q}_{ti}^o \cdot \boldsymbol{v}_i^o < -Q \\ K_p\, \boldsymbol{q}_{ti}^o \cdot \boldsymbol{v}_i^o & else \end{cases} \tag{6}$$

$$f_d(\dot{\boldsymbol{r}}^o, \omega) = K_d\, \dot{\boldsymbol{r}}_i^o \cdot \boldsymbol{v}_i^o, \qquad \dot{\boldsymbol{r}}_i^o = \boldsymbol{0} - \dot{\boldsymbol{r}}^o - \boldsymbol{p}_i^o \cdot \omega \tag{7}$$

$$\boldsymbol{q}_{ti}^o = \boldsymbol{p}_{ti}^o - \boldsymbol{r}_t^o - \boldsymbol{p}_i^o \tag{8}$$

. $\boldsymbol{q}_{ti}^o$ is the vector from the target position to the current position of the contacting point. The $\boldsymbol{v}_i^o$ is a unit vector perpendicular to the surface of the target object at robot's contact point and it's direction is always the same with the robot's *ability direction.* Therefore, the $f_p(\boldsymbol{r}^o, \theta)\boldsymbol{v}_i^o$ is the factor in realizing the active nature in its *ability direction* and passive nature in its non-*ability direction.* For increasing the speed of convergence near the manipulation goal, a damping factor $f_d(\dot{\boldsymbol{r}}^o, \omega)\boldsymbol{v}_i^o$ is included on robot's *ability direction.*

# 3 Control of Non-holonomic Worker Robot

To realize the active nature on robot's ability direction and passive nature on its non-ability direction, the most simple and effective method is to set the force control on its ability direction to the joint position of the arm which has a certain length from the axle of the wheels. But, in some cases, such as a long distance transportation task, the problem of non-holonomic constraint and limitation of the rotation angle of robot's arm may bring to a difficult situation on performing the task. Then two additional control modes, Pseudo-Ability Control Mode and Posture Adjust Control Mode are proposed. The switching among the three control modes is decided according to the position of the virtual target point and orientation of the robot's body.

## 3.1 Ability Direction Control Mode

The Ability Direction Control Mode(Fig.5) is designed as Eq.(9).

$$\begin{cases} f_{ad} = K_{ap} d_{vir} + K_{ad}\, v_{vir} \\ \theta_{hd} = 0, \qquad \tau_{rd} = 0 \end{cases} \tag{9}$$

The first equation shows the desired force $f_{ad}$ applied on the *ability direction.* Here, $d_{vir}$ is the distance to the virtual target point and $v_{vir}$ is the velocity on the *ability direction* at contacting point. The second and third equations in Eq.9 show the passive nature on robot's *non-ability direction.* Then, the inputs of the robot's controller can be set in the following equation.

$$\begin{cases} f_a = f_{ad} + K_{af}(f_{ad} - \tilde{f}_a) \\ f_n = K_{np}(\theta_{hd} - \theta_h) - K_{nd}\dot{\theta}_h + K_{ni}\int(\theta_{hd} - \theta_h)dt \\ \tau_{rm} = K_{rm}(\theta_{ra} - \theta_{rm}) \end{cases} \tag{10}$$

$\tilde{f}_a$ is the force on arm direction measured from the sensor. $\theta_h$ is the angle between arm and object, and and $\theta_r m$ is angle from body to arm direction. $f_R, f_L$ are calculated according to the statics of the mobile robot from $f_a$, $f_n$.

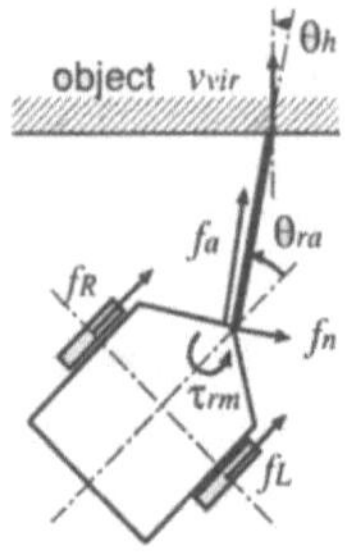


Fig. 5. Ability Direction Control

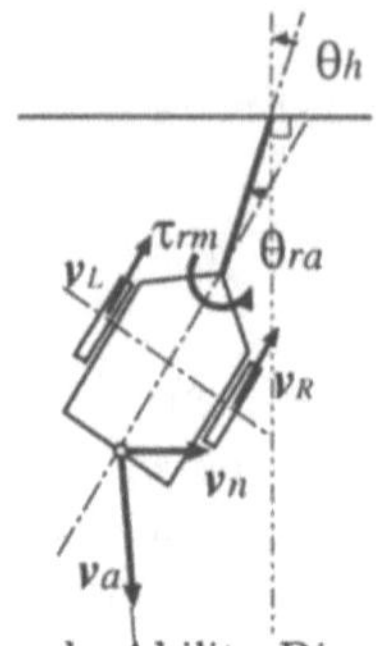


Fig. 6. Pseudo Ability Direction Control

### 3.2 Pseudo Ability Direction Control Mode

Pseudo Ability Direction Control Mode is used in situation where the virtual target point is behind the robot and has certain distance to the current contacting point. In this case, the control point is set to the rear part of the robot's body and $\boldsymbol{v}_a$ is the velocity on the *ability direction* of the control point. $\boldsymbol{v}_n$ is used to adjust the robot's posture(Fig.6).

$$\begin{cases} ||\boldsymbol{v}_{ad}|| = V_d \ (const) \\ \theta_{hd} \ = 0, \qquad \tau_{rd} = 0 \end{cases} \tag{11}$$

This pseudo active nature and passive nature is realized according to the following equation. $\boldsymbol{e}_a$ is a unit vector of pseudo *ability direction.*

$$\begin{cases} \boldsymbol{v}_a \ = V_d \, \boldsymbol{e}_a \\ \boldsymbol{v}_n \ = (K_{np}(\theta_{hd} - \theta_h) - K_{nd} \dot{\theta}_h + K_{ni} \int (\theta_{hd} - \theta_h) dt \,) \, \boldsymbol{e}_n \\ \tau_{rm} \ = K_{rm}(\theta_{ra} - \theta_{rm}) \end{cases} \tag{12}$$

### 3.3 Posture Adjust Control

There are two cases that will cause the robot's arm to rotate to arm's rotational limitation. The first is to pull the object in the Ability Direction Control Mode continually. Another is when a robot pulls an object after a continual motion in perpendicular direction. When the arm rotates to its limitation, any pulling action in the Ability Direction Control Mode will lead the robot to a deadlock since the non-holonomic constraint prevents the robot to adjust its orientation with the rigid arm connected to the object.

On solving this problem, one free joint which can be locked when the robot need to exert a pushing force to the object is designed in the middle of the robot's arm. By unlocking the free joint, the robot can adjust its posture freely under its non-holonomic constraint when it is necessary. But the time cost and path of the posture adjustment should be discussed since other robots may move the object toward the robot while it is adjusting the posture.

Here, we design the posture adjust motion by three arc sub-motions with different radiuses in the period of $[0 \sim \infty)$ (Fig.7), a most simple and basic posture adjustment method of a non-holonomic robot.

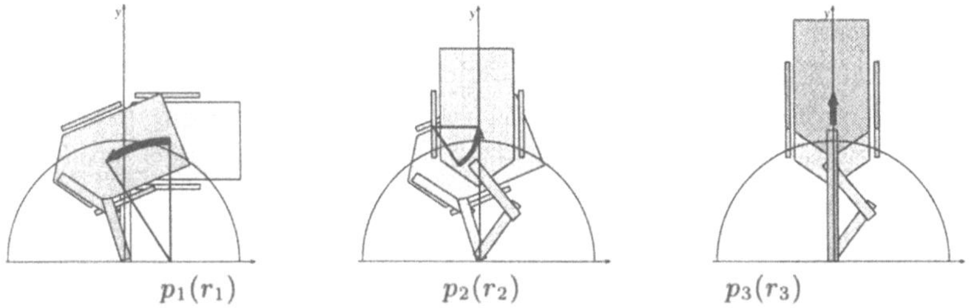


**Fig. 7.** Posture adjust is realized by three arc sub-motions($p_1(r_1)$, $p_2(r_2)$, $p_3(r_3)$)

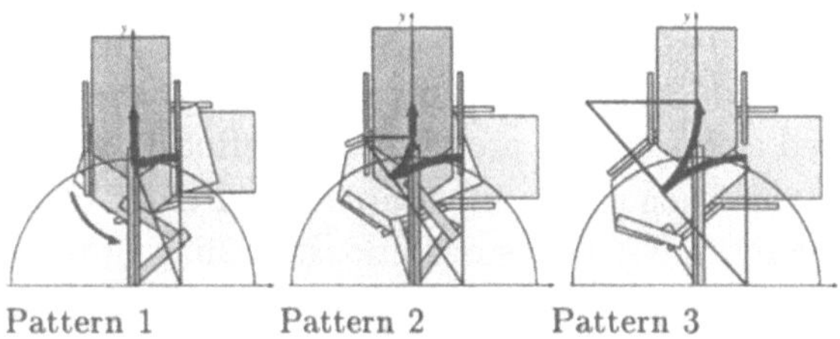


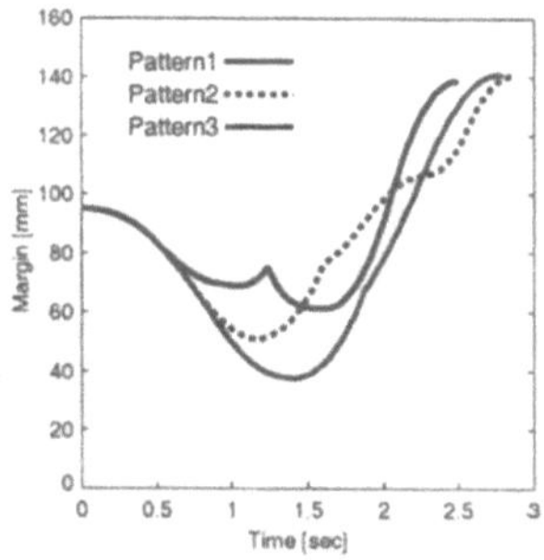


$r_1 = 160mm$

**Fig. 8.** Three different paths with same radius of $p_1$.

In Fig.8, we compare the three typical cases when the radius of $p_1$ is fixed in $160mm$. The robot is moving on a velocity continual trajectory under the physical limitation of the robot($v_{max} = 0.2m/s, \dot{v}_{max} = 0.4m/s^2$). In the view point of time cost and margin to the object, which is calculated from the shortest length between all points of robot body, Pattern $1(r_2 = 0, r_3 = \infty)$ is best even though Pattern 3 only need two arc sub-motions. By improving the path Pattern 1, we design the $p_1$ to be an arc which rotation center is exactly set on the contact point and the radius is the arm length(160mm). Then it is possible to generate force to the object during $p_1$ and just unlock the free joint of the arm from $p_2$ to $p_3$ as shown in Fig.9. This adjust motion can cope with the higher velocity of the object from collision.

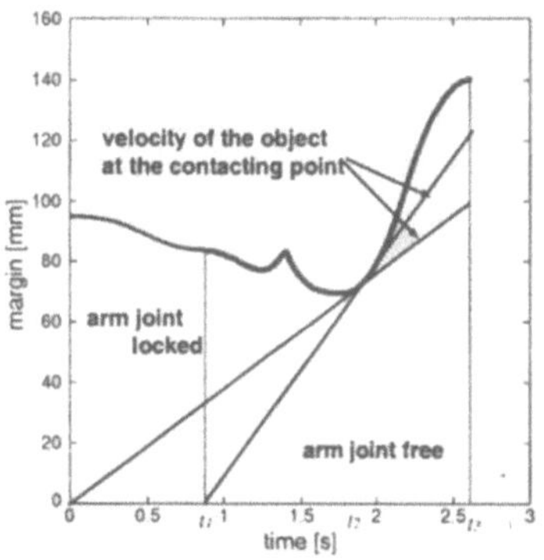


**Fig. 9.** Arm can generate force to the object during the period $p_1$.

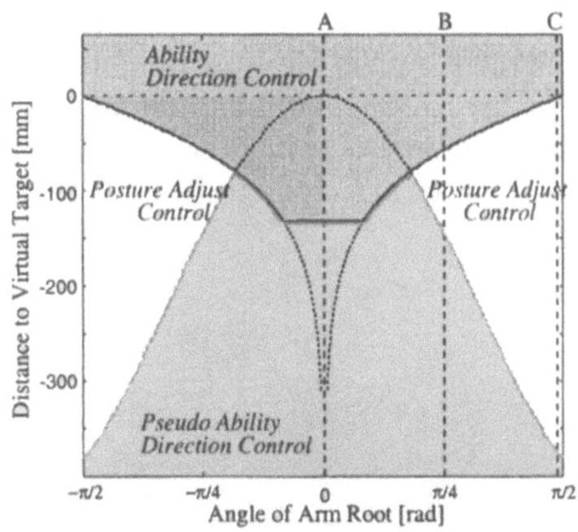


**Fig. 10.** Mode switching conditions

### 3.4 Switching of the Three Control Modes

Fig.10 shows the switching conditions on the distance to the virtual target point and orientation of the robot for the three control modes. The boundary of Ability Direction Control is that contact point can reach the goal before the arm reaches the rotational limitation, and is calculated according to the distance from the axle to the arm attachment. The boundary of the Pseudo Ability Direction Control is decided by the result of convergence experiments. Outside of these two regions, Posture Adjust Control should be used.

In the case that an object is pulled by a robot in Pseudo Ability Direction Control Mode, the control mode will switch back to the Ability Direction Control Mode when the object is moved around the goal position.

## 4 Experimental Cooperative Robots System

A experimental robot system, which includes a managing robot and four working robots, is constructed as the second prototype for BeRoSH.

A CCD camera, an on-board vision processing board and a PC-AT board computer(Pentumn MMX 200Hz) are implemented into the managing robot (Fig.11). A wide range of object manipulation is realizable due to the mobility of the managing robot. The mobile platform and camera motion are controlled by a board computer with an MC68332 CPU.

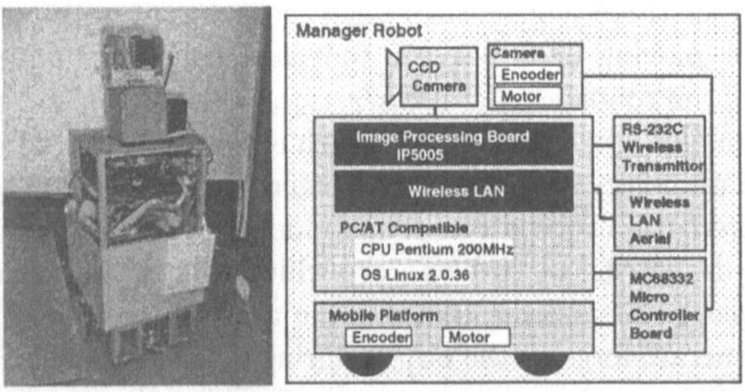


**Fig. 11.** The Managing Robot in BeRoSH II

A board computer with MC68332 CPU is installed on the worker robot as the main controller(Fig.12). A pair of the measuring wheels is included into the robot system for estimating its motion since measuring system fixed on the driving wheels will include slipping error during the pushing and pulling motion. Physical springs are installed in the arm's joint and arm directly to sense the force and torque for realizing the passive behavior on the non-*ability direction*.

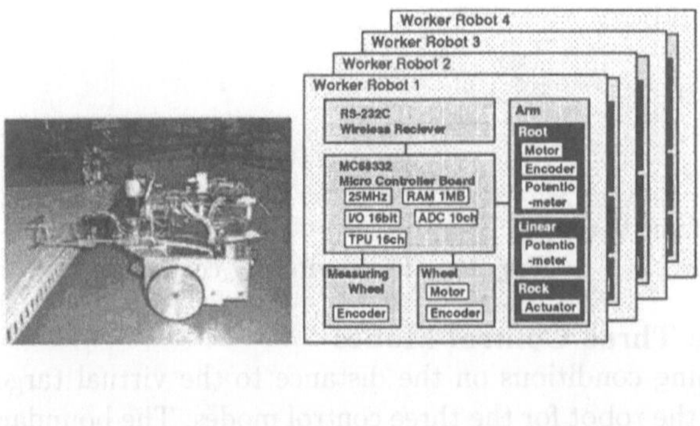


**Fig. 12.** The Worker Robot in BeRoSH II

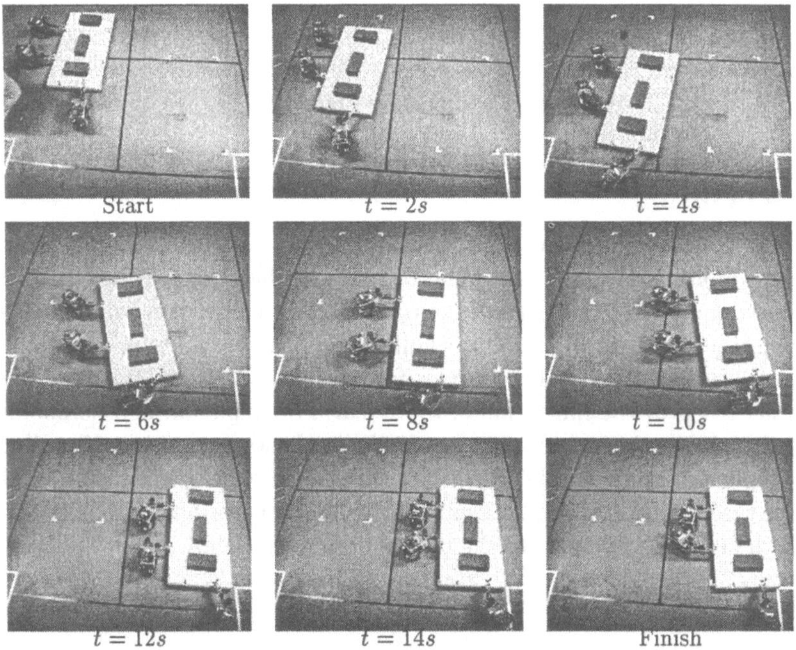


**Fig. 13.** Object is transferred from starting point(0,0,0) to the goal(1000,-500,0)

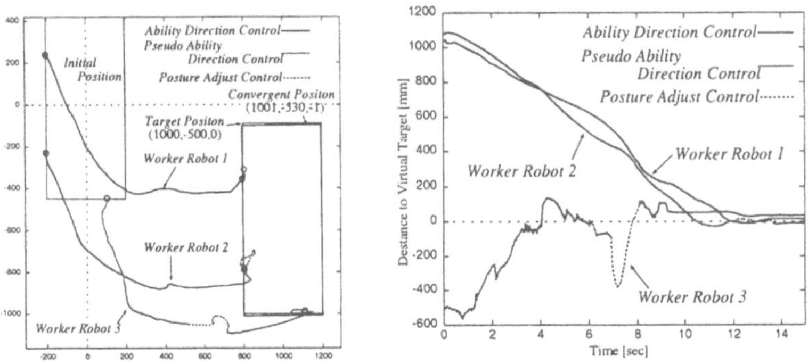


**Fig. 14.** Trajectories of contacting points and distances to the virtual target point

## 5 Experimental Result

An experiment result which an object is handled from $(0, 0, 0)$ to (1000,-500,0) by three Worker Robots is shown in Fig.13. The object is $12.5kg$ weight. The position and orientation of the object is represented as ($x[mm]$,$y[mm]$,$\theta[deg]$). Fig.14 shows the path of each contacting point according to each robot's odometry system and the distances to the virtual target point. The object converged to $(1010, -530, 2)$. The experiment error is cause by the odometry error of each robot and the friction between object and ground. During the transportation, Robot1 was in Ability Direction Control only but Robot2 changed to Posture Adjust Control in the later part of the manipulation. The Robot3 was in the Pseudo Ability Direction Control from the beginning, but after pulling the object around 3.2s second, it switched to Ability Direction

Control, and to Posture Adjust Control around 6 second from the start. After 10 second, it switched to Ability Direction Control.

With this result, it presents a smooth switching of the three control modes which avoided the deadlock motion of each robot.

## 6 Conclusion

The basic concept of system BeRoSH and a new experimental system which is designed for a wide range manipulation are described in this paper. Also we introduced a mechanical design of the robot arm and three control modes for realizing the active nature in robot's ability and passive nature in non-ability direction. By using these three modes for various conditions, the deadlock, which is caused by the limitation of robot's arm rotation and non-holonomic constraint of the robot wheels, can be avoided and wide range object handling by multiple autonomous robots can be realized. An experiment result shows that the three control modes are switching smoothly according to the distance to the virtual target point and the orientation of the robot.

## References


1. H.Asama, A.Matsumoto and T.Ishida, Design of an Autonomous and Distributed Robot System, *Proc. IROS'89*, pp.283-290, 1989.
2. R.G.Brown and J.S.Jennings, A Pusher/Steerer Model for Strongly Cooperative Mobile Robot Manipulation, *Proc. IROS'95*, Vol.3, pp.562-568, 1995.
3. B.Donald,J.Jennings,D.Rus, Analyzing Teams of Cooperating Mobile Robots, *Proc.ICRA'94*,pp.1896-1903.
4. P.J.Johnson and J.S.Bay, Distributed Control of Simulated Autonomous Mobile Robot Collectives in Payload Transportation, *Autonomous Robots*,Vol.2 No.1, pp.43-63, 1995.
5. C.R.Kube, H.Zhang and X.Wang, Controlling Collective Tasks with an ALN, *Proc. IROS'93*, pp.289-293.
6. K.Kosuge,T.Oosumi, Decentralized control of multiple robots handling and object, *Proc.IROS'96*,pp.318-323.
7. K.Lakshminarayana, Mechanics of Form Closure,*ASME Paper*,78-DET-32,1978.
8. M.J.Mataric,M.Nilsson and K.T.Simsarian, Cooperative Multi-Robot Box-Pushing, *Proc.IROS*, pp.556-561, 1995.
9. M.Nili Ahmadabadi and E.Nakano, Constrain and Move: A new concept to develop distributed object transferring protocols, *ICRA'97*, pp.2318-2325, 1997.
10. J.Ota, N.Miyata, T.Arai,*et.al*, Transferring and Regrasping a Large Object by Cooperation of Multiple Mobile Robots, *Proc. IROS'95*, Vol.3, pp.543-548, 1995.
11. L.E.Parker, ALLIANCE: An Architecture for Fault Tolerant, Cooperative Control of Heterogeneous Mobile Robots, *Proc. IROS'94*, pp.776-783, 1994.
12. Z.D.Wang, E.Nakano, T.Matsukawa, A New Approach to Multiple Robots' Behavior Design for Cooperative Object Manipulation,*Distributed Autonomous Robotic Systems2*, Springer-Verlag Press,pp.350-361,1996.
13. Z.D.Wang, M.Nili Ahmadabadi, E.Nakano, T.Takahashi, A Multiple Robot System for Cooperative Object Transportation with Various Requirements on Task Performing, *Proc.ICRA'99*, pp.1226-1233, 1999.

# Structural Characteristics of Intelligence and Embodiment in Two-Mobile Robots System

Tsutomu Sawada[1], Atsushi Sugai[1], Sumiaki Ichikawa[1], and Fumio Hara[1]

[1] Science University of Tokyo, Dept. of Mechanical Engineering, Kagurazaka 1-3, Shinjuku-ku, Tokyo 162-8601, Japan

E-mail: sawa@hafu0103.me.kagu.sut.ac.jp

**Abstract.** This paper deals with a quantitative investigation to the relationship between cooperative intelligence and embodiment. The robotic intelligence may be strongly influenced with the task, the environment, and the body morphology. In the task that two-mobile robots system, composed of a crawler and 1dof prismatic arm, carries a long bar horizontally along a compound road with keeping the bar-support length constant, the two pair of the robotic systems, with different body size and locomotion step width, are used in acquiring the cooperative intelligence by reinforcement learning. We investigated quantitatively the structural characteristics of relationship between the intelligence and the size of body, locomotion step width and found that the cooperative intelligence is much simpler in small-sized robotic systems than that in large-sized ones.

**Key words.** Intelligence, Learning, Embodiment, Size of Body, Locomotion step width

## 1. Introduction

Recently, robotists are more aware of the fact that embodiment is important in the design of autonomous intelligent robot [1].

We think a task-execution by autonomous intelligent robot in the following way: "The robot obtains information about its environment, that is, sensory input, through dynamical (kinematical) interaction of its body, and decides how to drive an actuator/actuators based on its intelligence. And an ability to execute the task emerges from the robot interaction with the environment through its intelligence, body, and dynamics (kinematics)" Note that it is hard to define intelligence, and no generally accepted definition exists [2]. In this paper, we regard "intelligence" as "a mapping from sensory input to motor-output" in robotics field as some researchers do [3].

From this point of view, we think that robotic intelligence, body, and dynamics (kinematics) are strongly interrelated each other under a given task and an environment. In these aspects, the designer can manipulate only its intelligence and body. Therefore in research works for autonomous intelligent robot, it is important to investigate the methodology to design coordinately the intelligence and the body morphology, *i.e.* material, mass, size of body, sensor and actuator arrangement.

There is a research work that the robots equipped with IR sensors and controlled by a very simple neural network can collect objects by interacting with environment, but the robots with different position of the IR sensors can no longer collect objects [4]. But it is not fully investigated what kind of characteristics of body is quantitatively related to robotic intelligence. Thus we think it is important to make clear the structural relationship between characteristics of intelligence and embodiment in the scope of robotic design in general. But this is said also for cooperation problem of multi-robotic systems.

In this paper, we employ a task that two-mobile robots carry a long bar horizontally with keeping the bar-support length constant along a compound road. Taking into account the "size of body" and "locomotion step width" as elements of body feature, we prepare several types of the robotic systems, which are different with respect to these parameters, and conduct the numerical experiments of learning. Then we compare the structural characteristics of the cooperative intelligence acquired by these robots, and investigate quantitatively characteristics of cooperative intelligence with respect to the embodiment parameters.

## 2. Learning Algorithm

In this paper, we use a learning method "STL (S-Temperature based on-line Learning) [5]". This is a kind of reinforcement learning based on $Q$-Learning for a robot to acquire intelligence autonomously. STL has two characteristics as follows:

- A robot has two action-modes (Learning-Mode and Task-execution-Mode). The robot transits its action-mode seamlessly between Learning-Mode at which the robot judges not have learned enough in the environment and Task-execution-Mode at which the robot judges already learned enough in the environment
- In Learning-Mode, the robot autonomously constructs a state space based on sensory data that the robot recognizes as same from the viewpoint of action

## 3. Simulation Experiment

### 3.1 Task and environment

Figure 1 shows a one-dimensional compound road composed of flat planes, slope and sinusoidal segments along which two-mobile robots locomote. Each robot is composed of crawler body and a single prismatic arm. The task is for the two robots to carry a long bar along the road with keeping the bar-support length constant (600 [mm]) and bar-posture horizontal (0 [deg]). But in simulation study this condition is realized by the fact that the bar-support length and bar-posture or support inclination are within the certain permissible ranges.

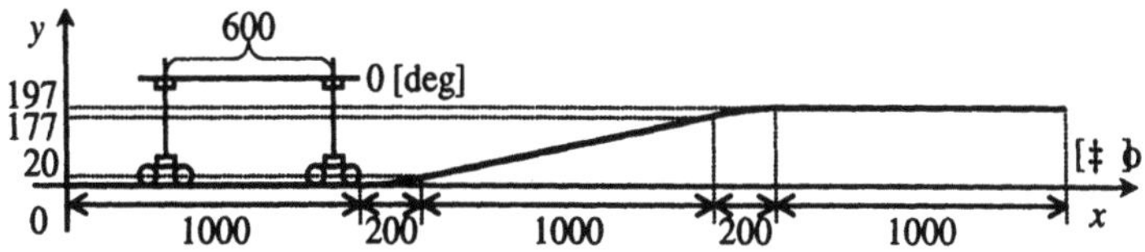


Fig.1 One dimensional compound road

### 3.2 Action space, initial state space and initial $Q$ value

We set action space $A=\{a: 0, 1,\ldots,8\}$ as shown in Fig.2 where thick arrows indicate the direction of motion of crawler and/or prismatic arm. Support length ($sl$) and support inclination ($\bullet_{si}$) are used to define the 2 dimensional sensory data. State space is thus represented by 2 dimensional plane ($sl$, $\bullet_{si}$). For initial state space, we set the permissible region, which is defined by the bar-support length ($580 \bullet sl \bullet 620$) and bar-posture or support inclination ($-1.5 \bullet \bullet_{si} \bullet 1.5$), and this region is considered as $s=1$, and the others as $s=0$. For initial $Q$ value, we set $Q(1,0)=1$ and the other $Q$ values are set 0 (See Fig.3). We represent cooperative intelligence of two-mobile robots on a 2 dimensional plane.

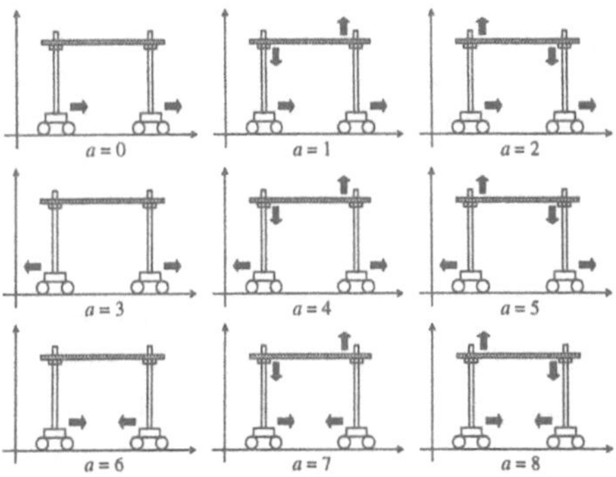


Fig.2 9 actions constituting action set $A$

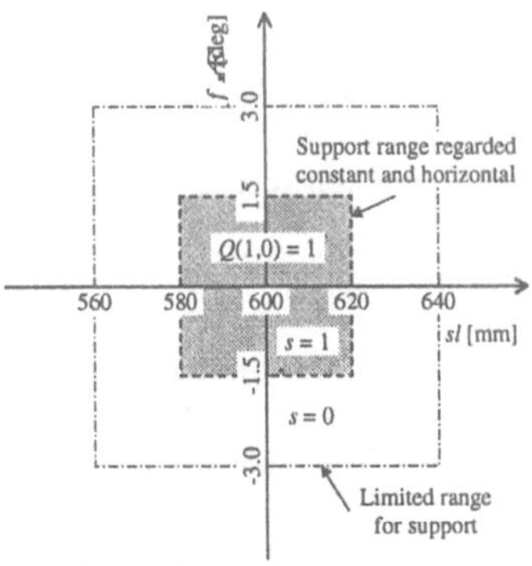


Fig.3 Initial state space and initial $Q$ value

### 3.3 Immediate reward

We assume that, when that the two-mobile robots proceed forward with holding the bar-support condition in the permissive range, the robotic system obtains the positive reward $r=1$. However, as the bar length and the length of both arms are finite, the robotic system might drop the bar from the arms in the course of action in learning-mode. We, thus, set the limitation to the range for the support length and the support inclination as shown in Fig.3. When the support condition is not kept within the limitation of the range, the robotic system is forced to receive the negative reward $r=-2$, which works as a penalty, and then the robots select the action not according to the action selection probability but the robotic system is to step-back at the next step.

For the other cases, the robotic system obtains the reward $r$=0.

Energy-consumption of the crawler actions such as (forward, backward) and (backward, forward) seems to be larger than that of the robots action (forward, forward), since the crawler actions that include backward locomotion seem not to be the ones for carrying the bar to the goal. It is obvious that the energy-consumption at the arm's actions such as (descent, ascent) and (ascent, descent) is larger than the one for (stillness, stillness).

Thus we can introduce the coefficients of energy-consumption rate for arm and crawler actions as $Ec_{arm}$, $Ec_{crw}$. At renewing $Q$ value for a set of ($s$, $a$), this kind of energy-consumption for each action can be reflected on the reward, when we use the eq.(1) for renewing $Q$ value.

$$Q_{t+1}(s,a)=(1-\alpha)Q_t(s,a)+\alpha(1-Ec_{arm})(1-Ec_{crw})\{r+\gamma\max_{a'\in A}Q_t(s',a')\} \qquad (1)$$

$\alpha$ : Learning rate, $\gamma$ : Discount rate, $r$ : Immediate reward,
$t$ : Time, $s'$ : State at $t$+1, $\alpha'$ : Action selected from $A$,
$Ec_{arm}$: Energy-consumption rate for arm's action,
$Ec_{crw}$: Energy-consumption rate for crawler's action

### 3.4 Characteristics of body

We employ the 16 cases of robot body parameters such as the size of body "$Sb$" and locomotion step width "$Lsw$". For the size of body $Sb$, $Sb$=50, 100, 150, 200 [mm] and locomotion step width $Lsw$, $Lsw$=1, 5, 10, 15 [mm] are used. Note that the size of crawler is not considered in this paper.

### 3.5 Parameters

Parameters used in the learning algorithm are shown in Table 1. We set $sl$=600, • $_{si}$=0.0, and $T_s$=$T_{min}$ as an initial condition for the robot task execution simulation experiments. The other details in the experiment are cited in the reference [5].

Table 1 Parameters for Learning

| | | | | | | | | | |
|---|---|---|---|---|---|---|---|---|---|
| • | 0.1 | • | 0.8 | $T_f$ | 20 | $R_d$ | 0.5 | $R_r$ | 0.1 |
| $T_{min}$ | 0.01 | $T_{max}$ | 5.00 | $T_{cr\,min}$ | 0.8 | $T_{cr\,max}$ | 0.1 | | |
| $Ec_{arm}$ | If $a$=0, 3, 6 then $Ec_{arm}=0.0$ else then $Ec_{arm}=0.4$ | | | | $Ec_{crw}$ | If $a$=0, 1, 2 then $Ec_{crw}=0.0$ Else then $Ec_{crw}=0.4$ | | | |

## 4. Experiment

According to the simulation set-up shown in chapter 3, each robotic system, with different characteristics of body, is set to start the learning of task execution. We define a trial as a series of robots actions by which the robots reach the goal. By repeating these trials, each robotic system progresses the learning.

By STL algorithm, the robot autonomously transits the action-mode between

Learning-Mode and Task-execution-Mode. Therefore, when the robot's action-mode is fully Task-execution-Mode in the trial, we judge the learning has converged.

From the simulation experiments for each robotic system among the 16 cases of the robot body characters, we collect 50 examples of the cooperative intelligence established in *Q*-table by changing the random number seed used in action selection.

## 5. Result and Discussion

### 5.1 Intelligence acquired in two-mobile robots

Figure 4 shows three examples of state space partition, obtained by the STL algorithm implemented, *i.e.* cooperative intelligence coordinating the sensor state with robots action, and also the locus of 2 dimensional sensory data *sl* and $\bullet_{si}$ is shown in the same figure for the case of *Sb*=200, *Lsw*=10.

Note here that the cooperative intelligence acquired by learning is built-up by the robots task execution experience through the kinematic interaction with the compound road geometry. Cooperation abilities are regarded almost same because the robots learned to put the bar back after or just before support condition violates the support range regarded constant and horizontal.

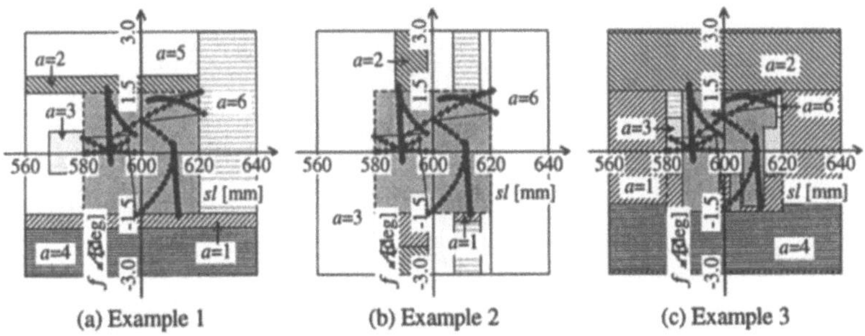


Fig.4 State space partition and the locus of 2 dimensional sensory data in the case of *Sb*=200, *Lsw*=10

As the robots action is selected based on probability, each robotic system that has different random seed number for action selection does not always acquire the same intelligence, even if the robot has the identical characteristics of body. For example, in Figs.4 (a) and (b), the sensory data locus are identical, but the state-space partition, *i.e.* cooperative intelligence, is different between them. Then for quantitative investigation how the characteristics of body affects the cooperative intelligence, we compare the structure of cooperative intelligence with respect to the rate of usage of each action in 50 examples of cooperative intelligence and the tendency of sensory state in action selection. From these we will make clear the structural characteristics between the cooperative intelligence and the size of body, as well as locomotion step width.

### 5.2 Characteristics of cooperative intelligence with respect to body size

Figure 5 shows the rate of usage of each action in each case of the robot body characteristics when the robots execute the task. Note here that the rate of usage 0.8 means that 40 cases of 50 intelligence examples used the action at least one time during the task execution. Note also that two numerical values in the parentheses indicate the (*Sb*-*Lsw*).

From this figure, for the case of $Sb$=50, 100, the rate of usage is very high for $a$=0, 1, 2, and for the case of $Sb$=150, 200, the rate of usage is also high for $a$=0, 1, 2, 3, 6. This tendency is independent of *Lsw*.

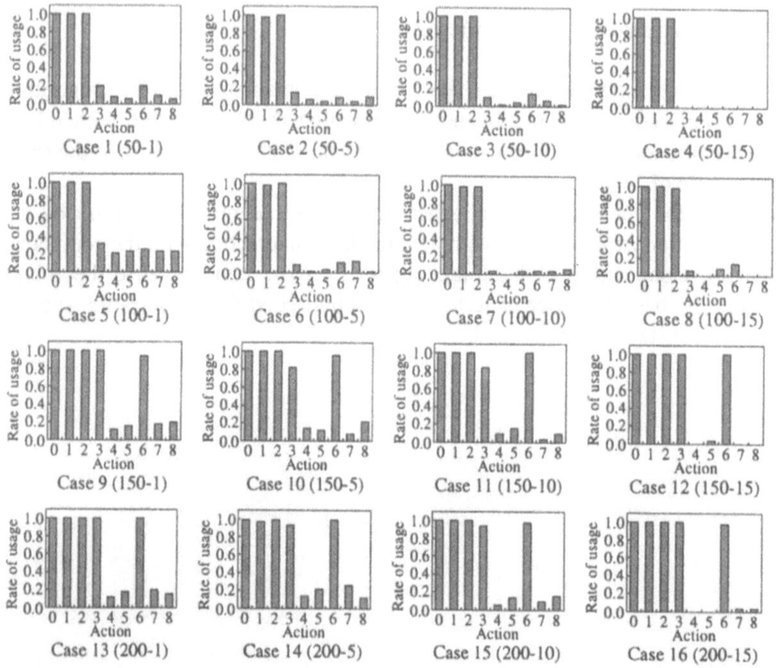


Fig.5 Rate of usage of each action

Figure 6 shows the locus of 2 dimensional sensory data ($sl$,$\bullet_{si}$), when the robot having the body size specified by $Sb$ takes only $a$=0 until it reaches the goal. From this figure it is said that the maximum error of $\bullet_{si}$ from $\bullet_{si}$=0 [deg] is constant ($\bullet_{si}$=8.9 [deg]) and independent of the size of body. On the contrary, the maximum error of $sl$ from $sl$=600 [mm] is affected by the size of body. The larger the size of body is, the larger the maximum error of $sl$ is.

Especially for the cases of $Sb$=50 and 100, $\bullet_{si}$ occasionally exceeds the range corresponding to the condition that the bar-posture is regarded as horizontal, but $sl$ never exceeds the range for the support length condition. Therefore the robots with body size of 50 or 100 could acquire the cooperative intelligence for executing the task to carry the bar without using $a$=3-8. On the other hand, for the cases of $Sb$=150 and

*Sb*=150 and 200, *sl* occasionally exceeds the range for the support length condition. Therefore such robots have a tendency to acquire the cooperative intelligence that uses the actions *a*=3-8, especially *a*=3 and 6.

The locus of 2 dimensional sensory data in Fig.6 is seen dependent on only the size of body and independent of locomotion step width. Thus it is said that only the size of body determines the action selection *a*=0, 1, 2 or *a*=0, 1, 2, 3, 6.

For the case of *Sb*=120, the maximum error of *sl* from *sl*=600 [mm] is coincided with the range for the support length condition and consequently it is said that the critical size of body is *Sb*=120 for which the robots acquire the cooperative intelligence of using *a*=0, 1, 2 or of using *a*=0, 1, 2, 3, 6. This critical body size is determined by the conditions associated with the environment and the task.

In this kind of environment and task, the robots, which have the size of body less than 120, can execute the task by much simpler cooperative intelligence that needs fewer kinds of action.

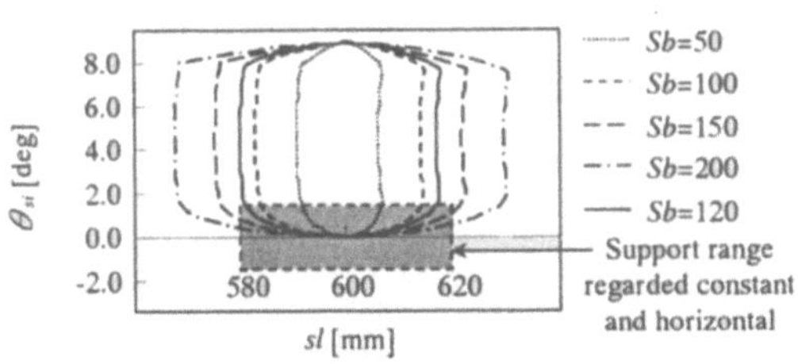


Fig.6 Locus of 2 dimensional sensory data for each *Sb*

## 5.3 Characteristics of cooperative intelligence with respect to locomotion step width

We employ a 2-dimensional ellipsoid calculated by Mahalanobis' generalized distance as shown in eq.(2) [6] to express the tendency of state-partition for the actin selection. We apply this ellipsoid to only the action whose rate of usage is higher than 0.8 except for the case *a*=0.

$$(\boldsymbol{x}-\boldsymbol{\mu})^{T}\,\Sigma^{-1}(\boldsymbol{x}-\boldsymbol{\mu})=\mathrm{m}+2 \qquad (2)$$

$\boldsymbol{x}$: $(sl,\ \theta_{si})$ $\boldsymbol{\mu}$: the mean vector of $\boldsymbol{x}$

$\Sigma$:variance covariance matrix of $\boldsymbol{x}$ m: sensory dimension (m=2)

Figure 7 shows the result. From this figure, we found that the larger locomotion step width is, the smaller the ellipsoid to select the action is. Figure 8 shows the root $\sqrt{S}$ of the ellipsoid area *S*, which indicates the mean radius of the ellipsoid for the action, and plotted against the body size *Sb*. Note that each sensory dimension is normalized by the range for the support length and inclination condition ($580 \leqq sl \leqq 620$ , $-1.5 \leqq \theta_{si} \leqq 1.5$).

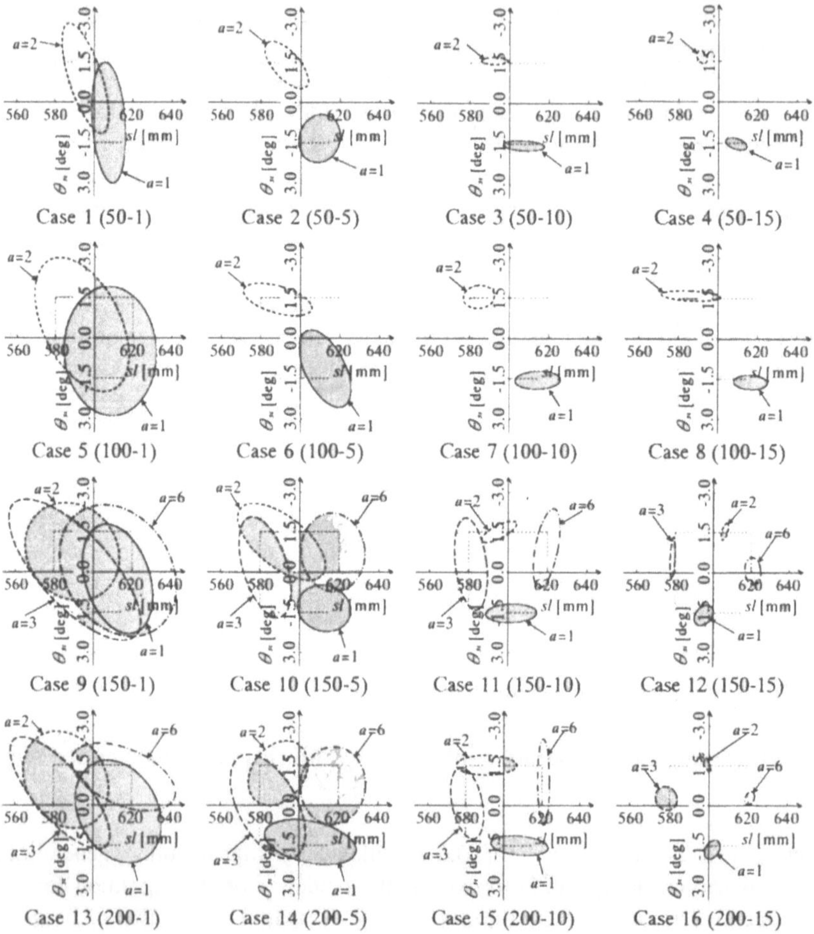


**Fig.7 Ellipsoid of state partition for each action**

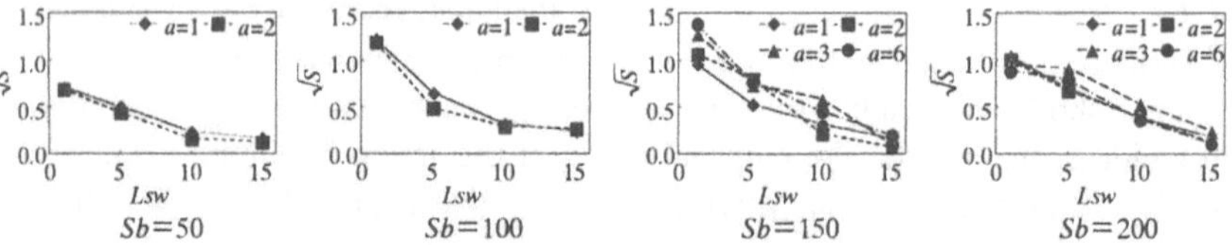


**Fig.8 Mean radius of ellipsoid of each action in each *Sb***

In this paper, robotic intelligence is regarded as a transformation rule to select an actuator input from sensory data, so it also can be regarded as a controller in control system. Then, from the viewpoint of control system, we analyze the structural relationship between the cooperative intelligence and the locomotion step width.

Our two-mobile robots system is considered as a sampled data control system because its locomotion step is fixed. As a mechanical dynamic system is generally represented as a second order system, we assume the sampled data control system to be a second order system $G(s)$ with a proportional feedback control. Note that the hold device $G_H(s)$ is assumed as zero-order hold. (See Fig.9)

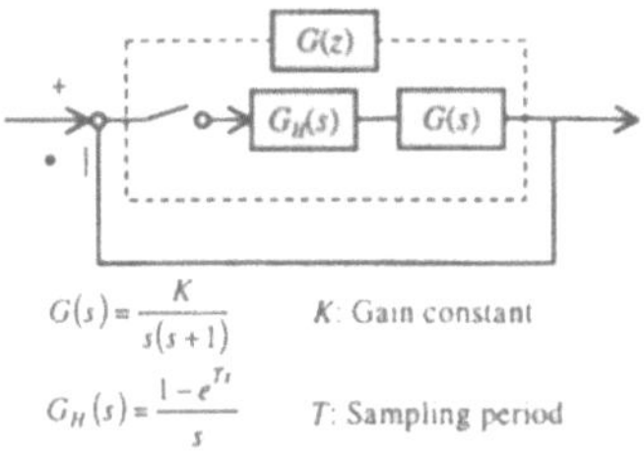

Fig.9 Sampled data control system

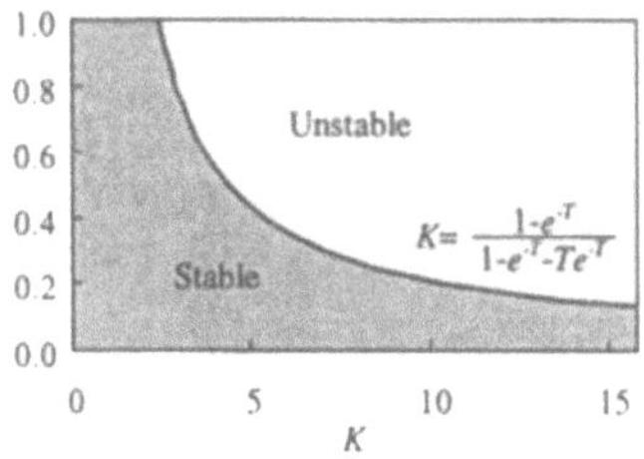

Fig.10 Stable region

Then, the $z$-transfer function is obtained as follows for the 2nd order dynamic system:

$$G(z)=\bullet\,[G_H(s)G(s)]=K\frac{\left(T+e^{-T}-1\right)z+\left(1-e^{-T}-Te^{-T}\right)}{(z-1)\left(z-e^{-T}\right)} \tag{3}$$

We investigate the stability of this system since the robots acquired the intelligence is considered stable. As the characteristic equation of the system is $1+G(z)=0$, the stability condition of the system in terms of Jury's criterion [7] is to satisfy $|a_2|<a_0$, $D(1)>0$, $D(-1)>0$ as defined in eq.(4). The stable area in terms of $K$ and $T$ is shown in Fig.10.

$$D(z)=1+G(z)=a_0z^2+a_1z+a_2 \tag{4}$$

$$a_0=1,\ \ a_1=K\left(T+e^{-T}-1\right)-\left(1+e^{-T}\right),\ \ a_2=K\left(1-e^{-T}-Te^{-T}\right)+e^{-T}$$

This result means that the sampling period $T$ must be small for the system to be stable when gain constant $K$ is large. It is considered that the gain constant $K$ corresponds to a coefficient to determine the control value in the system, and the sampling period $T$ corresponds to a magnitude of state-change. Therefore, the gain constant $K$ and the sampling period $T$ correspond to the locomotion step width *Lsw* and the mean radius of ellipsoid $\sqrt{S}$ for the case of the two-mobile robots system.

From these discussions and Figs.8 and 10, the intelligence, acquired by the STL learning algorithm, autonomously acquires the proper magnitude of state-change for the system.

## 6. Conclusion

In this paper, we offered the task to carry a long bar horizontally with keeping the bar-support length constant along the compound road to the two-mobile robots system whose modules are composed of crawler and 1dof prismatic arm. We characterized the robot by "size of body" and "locomotion step width" and employed 16 types of the robotic system with different values in these parameters. Then we compared the structural characteristics of the cooperative intelligence acquired through the STL learning trials that were done by computer simulation and investigated quantitatively the relationship between the cooperative intelligence and the robot embodiment. We found the followings:

- The robots with different size of body acquire the different cooperative intelligence that needs to execute the given task in the given environment. In the environment and task, the robots, which have the size of body less than 120, can execute the task by much simpler cooperative intelligence that needs fewer kinds of action
- Locomotion step width and the mean radius of ellipsoid in the state space correspond to the gain constant and the sampling period in the second order system. And the robots with STL learning algorithm acquires the proper magnitude of state-change for the system

Some of the future works are, we think, to investigate quantitatively structural characteristics of robotic intelligence and its embodiment specified by other characteristics of body such as material, mass, sensor and actuator morphology.

## Acknowledgements

This research work was partially supported by the JSPS project "Research for the Future" 96P00803. The authors express their thanks to those who are involved in the project.

## REFERENCES

1. R. Brooks. (1990) Elephants Don't Play Chess, Robotics and Autonomous Systems 6, 3-15.
2. R. Pheifer and C. Scheier. (1999) Understanding Intelligence, The MIT Press
3. K. Kikuchi and F. Hara. (2000) A Study on Evolutionary Design in Balancing Morphology and Intelligence of Robotic Systems, Journal of Robotics and Mechatronics Vol.12 No.2, 180-189.
4. M. Maris and R. Boekhorst. (1996) Exploiting Physical Constraints: Heap Formation through Behavioral Error in a Group of Robots, Proc. of the 1996 IEEE/RSJ Int. Conf. on Intelligent Robots and Systems, 1655-1660.
5. T. Sawada, S. Ichikawa and F. Hara. (1999) Autonomous Action-Mode Change in a Two-Mobile Robotic System –S-Temperature based On-line Learning–, Proc. of the 1999 IEEE/RSJ Int. Conf. on Intelligent Robots and Systems, 393-399
6. M. Asada, S. Noda and K. Hosoda. (1996) Action-Based Sensor Space Categorization for Robot Learning, Proc. of the 1996 IEEE/RSJ Int. Conf. on Intelligent Robots and Systems, 1502-1509
7. E. I. Jury. (1958) SAMPLED-DATACONTROL SYSTEMS, John Wiley and sons Inc

# Chapter 12

# Related Research Topics

# From distributed robot perception to human topology : a learning model

Nicolas Bredèche[1,2], Jean-Daniel Zucker[1]

[1] : Université Pierre et Marie Curie, LIP6-CNRS - 4, place Jussieu, F-75252, Paris Cedex 05, France.
[2] : LIMSI-CNRS, Bt 508, Université Paris-Sud, 91403 ORSAY cedex
nicolas.bredeche@[lip6|limsi].fr, jean-daniel.zucker@lip6.fr

**Abstract.** *This poster presents an approach to enable autonomous mobile robots to link perceived information from their environment to names of places (toponyms acquired through interaction with human beings).*

**Keywords.** *robot-human communication, distributed robotics, lexicon acquisition, numeric to symbolic link, toponyms.*

# 1 Introduction

Learning and knowledge acquisition in autonomous robotics proved to be a central field of investigation and building a representation of the world is a key aspect in autonomous systems, would it be for solving specific problems individually [1] and collectively [2] or exploring communication from a robot point of view [3]. In this paper, we are interested in the learning by a robot of human toponyms related to the topography of the environment and the building of a hierarchy of inclusion between these toponyms. The toponyms are then to be bound to what the robot perceives through its own representation of the world [4] according to Fig. 1.

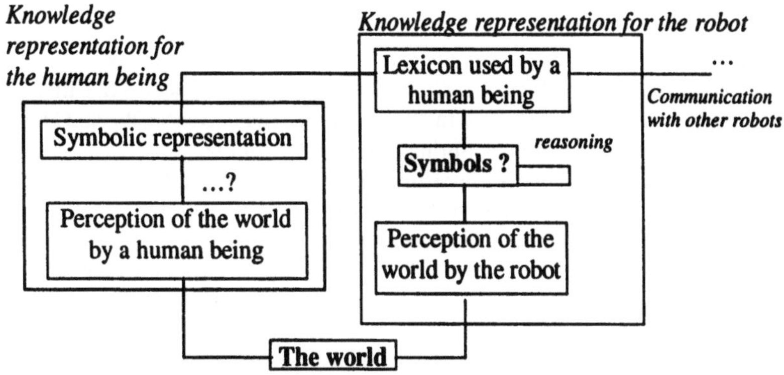


*Fig. 1. How the robot build its representation of the world.*

## 2 Context and model

Our research takes place in the Microbes project [5] and uses mobile robots with several sensors capabilities and cartography related information. Thanks to perception and description of the environment, the robot is supposed to build its own symbols to represent the world compatible with that of a human being. We want to study how a robot will be able to understand human concepts by just using its own perceptions and toponyms given by a human.
In order to store information from the sensors and from communication, we divide our model in three layers as seen in Fig. 1, each containing specific information :

- the numeric layer records all the information perceived in the environment and each leaf represents a specific context.
- the lexical layer records all the toponyms used to describe the environment. Relations between nodes are topologic inclusion and synonymy.
- the symbolic layer materializes the link between the two other layers since symbols emerge from the perceptions of the agent and the naming of humans.

The different layers are build during the " life " of the agent. Every time an agent comes to a new place (i.e. a new context), a new leaf is created in the numeric layer to which is linked all toponyms used to qualify this context.

## 3 Conclusion

What is addressed here is how a robot learns to build its own representation of the world from its sensor perceptions and human description. The toponyms and relations between them that describe the map help enriching what we call a lexical layer while perceptions are recorded on a numeric layer. The link between those two layers is made through an emergent internal symbolic layer.

## References


1. Maja J Mataric (1995) "Learning in Multi-Robot Systems", in Proceedings, IJCAI-95 Workshop on Adaptation and Learning in Multi-Agent Systems.
2. Peter Stone, Manuela Veloso (1999) Team-Partitioned, Opaque-Transition Reinforcement learning (TPOT-RL). In "RoboCup-98: Robot Soccer World Cup II", M. Asada and H. Kitano (eds.). Springer Verlag, Berlin.Luc.
3. Steels and Frederic Kaplan (1999) Bootstrapping Grounded Word Semantics. In: Briscoe, T. (ed.) Linguistic evolution through language acquisition: formal and computational models. Cambridge University Press.
4. Mustière, S., J.-D. Zucker , L. Saitta (1999). Cartographic Generalization as a Combination of Representing and Abstracting Knowledge. ACM/GIS'99, Texas, USA.
5. A. Drogoul, S. Picault (2000) Towards Situated Robots Collectivities in a Human Environment: the MICRobES Project, submitted to ECAI.

# Reactive Sensor Networks: Mobile Code Support for Autonomous Networks

R. R. Brooks, E. Grele, W. Klimkiewicz, J. Moore, C. Griffin, B. Kovak, J. Koch

The Pennsylvania State University Applied Research Laboratory
P.O. Box 30
State College, PA 16804-0030
E-mail: rrb@acm.org

**Abstract.** Sensor networks implement a distributed approach to automated perception. Autonomous sensor networks combine perception with actions in the form of internal network state reconfiguration. Our approach to autonomous sensor network design is an extension of the active networks paradigm [1], which provides an active network middleware for sensor networks. An innovative aspect of this middleware is the use of a mobile code infrastructure.

**Key Words**: Distributed sensing, Autonomous systems, Mobile code, Adaptation, Sensor networks

## 1 Distributed Sensing Paradigm

Distributed sensor networks use multiple autonomous sensor nodes to provide greater precision and dependability than the component sensor nodes [2]. When multiple sensor nodes survey the same region, redundancy reduces system sensitivity to single points of failure. At any point in time, a single sensor provides a single data point. Collaborative signal processing aggregates data points into a more reliable global estimate with dependability estimates. This is similar to the use of multiple experiments to statistically determine a parameter value and its variance. A number of military and commercial applications exist for this technology. Sensor nodes do local processing and relay information. They self-organize forming an ad hoc multi-hop network. Data is relayed from one node to another. Routing choices can be made dynamically. Master nodes determine a frequency-hopping schedule that slave nodes follow. Data can be forwarded from one cluster of nodes to the next, until a gateway to the Internet is reached, at which point, a number of user workstations can access the information simultaneously. The sensor network is essentially a large distributed computational device. Two conflicting forces determine the flow of data through the network:

- *Pull* – requests formulated at the user workstation indicate events of interest.
- *Push* – events occurring in the environment provide the raw data.

## 2 Active Network Middleware

The Reactive Sensor Network project at the Pennsylvania State University Applied

Research Laboratory implements a mobile code infrastructure that augments sensor network adaptability. This approach is inspired by the *active network* paradigm [1]. Active networks build communication networks that are re-programmed by packets passing through the network. The active network approach can be manifested by transmitting data encapsulated in packets that contain a program. As data moves through the network, routers can execute the program. In this manner the latency imposed by data transmission can be exploited. An alternative active network approach is to use the packets to reprogram network routing algorithms. This increases system adaptability in a number of ways. Sensor nodes are a type of embedded system. Embedded systems have more severe storage constraints than desktop systems. This usually requires system functionality to be limited to the absolutely necessary. Mobile code allows behaviors to be downloaded as necessary. The system has access to a virtually infinite amount of online storage. Functionality is no longer limited by memory. It is still constrained by time. A storage hierarchy is formed with high cost, fast, low volume storage on the node and low cost, slow, high volume storage elsewhere on the network. When networks are to be deployed for long periods of time, upgrading software is problematic. Without code mobility, nodes need to be physically removed from the field, their software changed, and then redeployed. The existence of a mobile code infrastructure eliminates these problems. As soon as new target detection and classification algorithms become available they can be fielded. This approach also helps in implementing a self-tasking network. Specific node work assignments need not be known in advance. The software can be reconfigured and modified as needs arise. Similarly if the battery fails on a node performing an essential task, another node can download the software needed to replace it.

## 3 Acknowledgements and Disclaimers

Efforts sponsored by the Defense Advance Research Projects Agency (DARPA) and Air Force Research Laboratory, Air Force Materiel Command, USAF, under agreement number F30602-99-2-0520 (Reactive Sensor Network). The U.S. Government is authorized to reproduce and distribute reprints for Governmental purposes notwithstanding any copyright annotation thereon. The views and conclusions contained herein are those of the authors and should not be interpreted to necessarily represent the official policies or endorsements, either expressed or implied, of the Defense Advanced Research Projects Agency (DARPA), the Air Force Research Laboratory, or the U.S. Government.

## 4 References

[1] D. L. Tennenhouse, J. M. Smith, W. D. Sincoskie, D. J. Wetherall, and G. J. Minden. (1997) A Survey of Active Network Research. IEEE Communications Magazine, vol. 35, no. 1, 80-86

[2] R. R. Brooks and S. S. Iyengar. (1998) Multi-Sensor Fusion: Fundamentals and Applications with Software. Prentice Hall PTR. Upper Saddle River, NJ

# Cooperation of Multiple Robots to Solve Maze Tasks

Claudio R. F. Costa[1] and Eliana P. L. Aude[2]

[1] IME - Military Institute of Engineering, Rio de Janeiro, Brazil
[2] NCE/Federal University of Rio de Janeiro, PO Box 2324, RJ, 20001-970, Brazil
elaude@nce.ufrj.br

**Abstract.** This paper discusses the use of cooperation and distributed intelligence within a robot team to solve a maze problem in which a set of ordered control point tasks has to be performed before the robots exit the maze. At first, the maze structure and the control point locations are unknown. While the maze is explored, each robot dynamically maps the maze structure with information provided by its sensors and sent by the other robots. Within each robot, a dynamic scheduler uses the gathered information to update the robot goal.
**Keywords:** distributed intelligence, maze problem, dynamic scheduling

Research on robot teams has led to great progress in the quality of the solution for many problems [1, 2, 3]. In this work, a maze exploration by a robot team is considered. The maze contains a set of ordered control points, each one with an associated task. At first, the maze structure and the control point locations are unknown. The robots cannot exit the maze before all control point tasks have been performed in the correct order. They cooperate to map the maze structure and to locate the control points by exchanging information. At each maze position, the information sent by a robot consists of the position coordinates and status in each direction (free or blocked). The message also contains information on the exit location and on the control point locations and status.

A simulator has been developed to study the problem. The maze structure is gradually displayed on the screen as the maze exploration by the robots evolves. Fig.1 shows a simulator screen snapshot. The robots are the small numbered squares, indicating the order they entered the maze. The control points are represented by larger numbered squares, which indicate their task execution order.

Each robot performs a depth-first maze search while the robot team as a whole performs a breadth-first search. In this process, each robot dynamically maps the maze structure and is able to find which is the closest robot to a target control point. When a robot reaches a junction and chooses a direction that leads it to a dead end, the status of all the maze cells along that way is set to blocked in the direction towards the blockage and to free in the opposite direction. So, if another robot reaches the same junction, it won't take the path leading to that dead end and any robot that had already taken that path moves back to the junction.

When a robot finds a control point which is not the one in turn to have its task performed, three behavior models can be adopted. In the first one, the robot continues the maze search and when a located control point is due to have its task performed, the closest robot goes to it along the shortest known path. The second behavior is similar except that a robot will stay still at a control point if it is the one associated with the next task to be performed. In the third behavior, a robot always stays at a located control point position waiting for the right time to perform its task. With the first two behaviors, when a located control point is due to have its task performed, the closest robot may have to interrupt the maze search and go to it. For that, each robot has its own dynamic task scheduler, which permanently updates the robot goal and avoids conflicts between different robot goals.

**Fig. 1.** Simulator screen snapshot

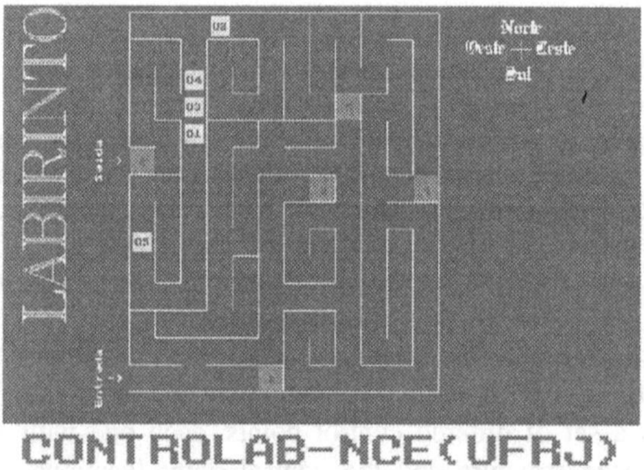


Simulation experiments have been performed with the use of different maze sizes, robot team sizes, maze search models and numbers of control points. Wireless robot communication bandwidth was always set to 2 Mbps and the robots were assumed to move at 0.5 m/s as they followed the Controlab AGV [4] model. Simulation results show that the third behavior model is superior when the number of robots is much larger than the number of control points. For any maze size, the time spent on solving the problem is reduced at a faster rate when the number of robots increases within a small sized team. When the maze problem requires all the robots to exit the maze instead of a single one, the solution becomes slower for large robot teams due to long exit queues and maze path saturation.

## Acknowledgements

The authors acknowledge the support given by CNPq/RHAE, Brazil, and the Brazilian Navy to the development of this research work.

## References

1. R. Alami, S. Fleury, M. Herrb, F. Ingrand, F. Robert. (1998) Multi Robot Cooperation in the Martha Project. IEEE Robotics and Automation Magazine, **5**, No. 1, 1-21.
2. B. Yamauchi. (1998) Frontier-Based Exploration Using Multiple Robots, Proceedings of the Autonomous Agents 98.
3. L. Parker. (1994) Heterogeneous Multi-Robot Cooperation, Ph.D. Thesis, MIT.
4. E. Aude, G. Carneiro, H. Serdeira, J. Silveira, M. Martins, E. Lopes. (1999) Controlab MUFA: A Multi-level Fusion Architecture for Intelligent Navigation of a Tele-robot, Proceedings of the IEEE Int'l Conference on Robotics and Automation, 465-472.

# Localizing a Wireless Node in a Multi-Agent System Using Kalman Filtering

Puneet Goel and Gaurav S. Sukhatme

University of Southern California, Los Angeles CA 90089, USA
*puneet|gaurav@robotics.usc.edu*

## 1 Introduction and Problem Definition

In order to autonomously navigate and perform useful tasks, a mobile robot often needs to know its exact position and orientation. The objective of this work is to estimate the position of a mobile node (we will call it the receiver, Rx) based on signals from other moving/stationary nodes (transmitters, Tx). The measurements are the straight line distances between the receiver and the transmitters. Our aim is to calculate the $(x, y)$ position of the receiver given the $(x, y)$ positions of the transmitters. A baseline approach is to use triangulation. If there are more than 3 transmitters then averaging the estimates obtained from various triangles provides a quick estimate. The key problem is that naive averaging treats each estimate equally. We propose instead to use Kalman filtering because it explicitly uses the the variances in the measurements. Indeed, the Kalman filter provides an optimum estimate in minimum mean-square error sense.

## 2 Kalman Filter

Kalman filtering is a recursive estimation procedure that uses sequential sets of measurements. In the experiments reported here the measurement vector used in the filtering is composed of the position estimates $x$ and $y$. We are using Kalman Filter in a scenario where we do not have any position estimates for the next step. Therefore the system matrix $\mathbf{\Phi}$ is the identity matrix. The system noise covariance matrix $\mathbf{Q}$ is determined empirically. The values of measurement noise matrix $\mathbf{R}$ are the most important in our experiments.

$$\mathbf{\Phi} = \begin{bmatrix} 1 & 0 \\ 0 & 1 \end{bmatrix} \quad \mathbf{Q} = \begin{bmatrix} 0.01 & 0 \\ 0 & 0.01 \end{bmatrix} \quad \mathbf{R} = \begin{bmatrix} var(x) & cov(x, y) \\ cov(x, y) & var(y) \end{bmatrix}$$

## 3 Experimental Results

Kalman filter (KF) is implemented in simulation. We assume that the position $(x, y)$ of each transmitter is known. Further we assume that the straight line distance between the transmitter and receiver is also known but it is corrupted by Gaussian noise. The variance in the noise depends linearly on the distance between the transmitter and receiver. To improve the performance of the KF we provide a good initial estimate of the receiver position. The initial estimate had a Gaussian noise of standard deviation 0.5 units added to the actual position of the receiver. After the initial estimate, we take two transmitters at a time, calculate the two possible positions of the receiver using the straight line distance measurements and use the one closer to our current estimate as the new measurement to the KF. So if there are '$n$' transmitters there will actually be ${}^{n}C_2 = \frac{n(n-1)}{2}$ such measurements.

Figure 1 shows the performance of the KF versus averaging when a good

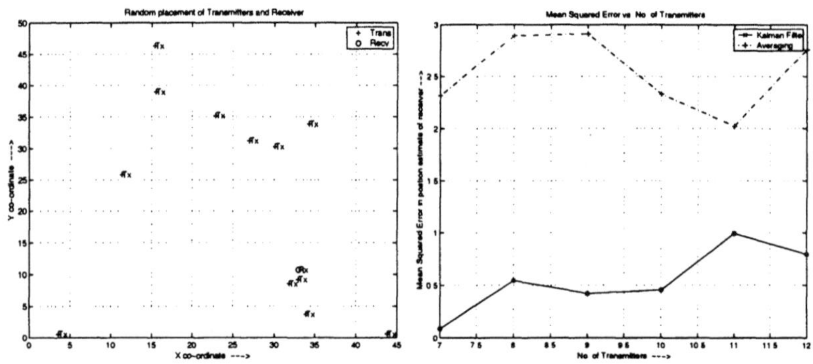


**Fig. 1.** A random distribution of nodes and performance of KF and averaging on that distribution. Standard deviation in the measurements is 10% ($s = 10$).

initial estimate was provided to the KF. Notice the oscillation in error as the number of transmitters increases: a new placement of transmitter can actually increase the error instead of decreasing it due to its bad positioning.

## 4 Discussion and Conclusion

We presented the application of Kalman filtering to a scenario where the measurements do not come as a time series. We described a technique for position estimation of a wireless node (e.g. mobile robot or any portable device) while using only the line-of-sight distance estimates from the node to other nodes in its environment. In reality we do not expect to have an even distribution of transmitters. Thus if we use averaging we might see that the error rises as number of transmitters increases, instead of going down. We have shown that the KF performs well with larger noise and a good initial estimate. However if we continue to get poor measurements then KF will keep disregarding them and stay close to its first estimate. Therefore, the need for a good first estimate is apparent.

The major motivation of this work is in distributed tasks. If the receiver is a mobile robot then we can use odometry to estimate its position in the propagation step of the KF. In that case we do not need to rely only on measurements, thus improving the performance of the KF based technique.

**Note: Full report is available at www-robotics.usc.edu/~puneet/research/research.html**

# Learning cooperation from human-robot interaction

Monica N. Nicolescu and Maja J. Matarić

Computer Science Department, University of Southern California
941 West 37th Place, Mailcode 0781, Los Angeles, CA 90089-0781

**Abstract.** We present an approach to human-robot interaction that facilitates the robot's interaction with humans, permitting the robots to benefit from the humans and learn about cooperation. The approach is based on an abstract behavior representation that allows humans to freely interact with the robots without the use of a predefined vocabulary of words, gestures or signs, and also enables the robots to get useful information from this interaction.

## 1 Motivation

Human-robot interaction has been addressed mostly from the perspective of using predefined, common *vocabularies* between a robot and a human. In this sense, the interaction has been limited to the robot *understanding* those *symbols* and executing some predetermined behavior for each case. Due to this, the human is forced to interact with the robot at a very low level of representation which is far from the robot's high-level goals. Furthermore, the interaction becomes difficult because the human and the robot need to share the same vocabulary in order to be able to help and understand the other. We are interested in allowing the robot to learn from the interaction at the human 's high level of abstraction up to the point at which the robot is able to request appropriate help or it can help other robots in turn.

## 2 Approach

Our approach uses behavior-based control (BBC)[1,2], which we extend in order to make possible the goals we have proposed. We introduce an abstract behavior representation that is derived from behaviors developed for any one or more specific tasks. We constitute the pairing of a behavior's activation conditions and its effects, without the specification of its inner workings, into an *abstract behavior*. The behaviors that do the work that achieves those effects under the given conditions are called *primitive behaviors*, and may involve one or an entire collection of sequential or concurrently executing behaviors. This representation allows behavior-based systems (BBS) to be used at a high-level, like operators in a plan, for complex problems that contain temporal sequences. Also, the introduction of abstract behaviors provides the ability of manipulating and changing the behavior's activation conditions such that solutions for new tasks can be dynamically generated without behavior re-compilation and re-design. We represent tasks in the form of abstract behavior networks in which the links represent postcondition-precondition dependencies. The generality of the network's components allows for behavior reuse both off-line (for system specification) and on-line (for system adaptation to a new task or directive).

Our approach to human-robot interaction is to allow the human to help the robot at his level of understanding and have the robot extract information from this interaction. By helping the robot, the human changes the environment, fulfilling thus some of the robot's (sub)goals, i.e. the behavior postconditions. The robot learns through continuous monitoring of the activation status of its behaviors and the state of the environment. If a goal of one of its currently not active behaviors is achieved, the robot infers that the human helped it with that. Thus, due to the abstract representation of behaviors, which defines the goals that each behavior achieves, the robot is able to relate the human 's helpful actions to its own behaviors that achieve the same (sub)goals. The interaction occurs through the environment, to which both the human and the robot have access.

After the demonstrated cooperation experience, the robot, faced with the same task, or even with a different task but involving some of the behaviors that have been categorized as suitable for cooperation, can ask another robot for help to achieve the goals of those behaviors. Help is requested for the most necessary behavior at that time, based on the behavior's activation level which is continuously monitored by the robot.

We have validated the proposed concepts on two Pioneer 2-DX mobile robots given an object delivery task, in which one robot has to move a box, through a closed door, from one side to the other of a two section, enclosed environment. First, one robot learns from a helpful human that it can get help with getting the box and opening the door. Second, it requests and gets assistance for those subtasks from a helper robot (having the same set of behaviors) that maps those requests into corresponding behavior networks that were dynamically activated and then executed.

This describes work in progress. We intend to generalize it to other tasks, to automate the process of on-line behavior-network construction and to incorporate it with the rest of the robots behaviors. Also, we plan to use the abstract behavior representation to enable multiple robots to share their acquired knowledge and experience of cooperation.

## References

1. Ronald C. Arkin (1998). Behavior-Based Robotics. MIT Press
2. Maja J. Matarić (1997). Behavior-Based Control: Examples from Navigation, Learning, and Group Behavior. Journal of Experimental and Theoretical Artificial Intelligence. H. Hexmoor and I. Horswill and D. Kortenkamp, **2-3**, 323–336

# Susceptibility of Swarm Control Algorithms to Agent Failure

Daniel W. Palmer[1], Daniel H. Steinberg[2], Chad M. Hantak[1], Michael A. Kovacina[1], and Sam N. Calabrese[1]

[1] John Carroll University, Cleveland OH, 44118, USA
[2] Oberlin College, Oberlin OH, 44074, USA

**Abstract.** Complex behaviors emerge from the interaction of swarms of independent agents. When an agent malfunctions, these interactions can magnify the failure, propagate incorrect information, and adversely alter the swarm's behavior. In this research we categorize swarm control algorithms based on the type and amount of *influence* agents exert on others. We investigate the resilience of algorithms by varying both the amount of influence and the probability of anomalous behavior. Algorithms that can best take advantage of the success of individual agents can also be the most susceptible to their failure. We strike a balance by dynamically shifting influence among agents based on local information.

## 1 Introduction

In 1990, a swarm of leaf-cutting ants on display at the National Zoo in Washington, D.C., accidentally decapitated their queen while moving her from one location to another. That single mistake led to the eventual demise of the colony, yet the ants continued to care for her headless remains until its royal scent diminished. Lacking global knowledge, a swarm cannot distinguish between the proper and errant behavior of its agents. Therefore it is critical that swarm algorithms inherently resist the propagated negative effects of erroneous actions.

## 2 Research Summary

We refer to a swarm consisting of agents with differing capabilities, behavior or influence as using *agent-critical* algorithms. These agents are designed to perform pre-defined sub-tasks and must communicate in order to leverage their differences. In contrast, all the agents in a *uniform* algorithm have the same capabilities and programming (more costly), but can self-select what sub-tasks to perform based on local information (more flexible). We developed both types of algorithms for several swarm scenarios including firefly synchronization, agent diffusion, object collection, and simulated attack on a tumor. To perform our experiments, we use SWEEP (SWarm Experimentation and Evaluation Platform)([3]) to encode agent behavior as a prioritized set of rules. The system animates the agents, and collects data as the swarm progresses towards its goal. We run each of several experiments under various introduced error rates and measure the time it takes to reach the goal. To simulate agent failure, we introduce probabilities to the rule set that temporarily dissociate a rule's conditions from its consequences.

## 3 Results and Conclusions

The results in Fig. 1 show that for a particular experiment based on [1], a uniform approach shows better performance under increasing amounts of introduced error. While these results are not uncommon, for some scenarios an agent-critical approach also exhibits robustness.

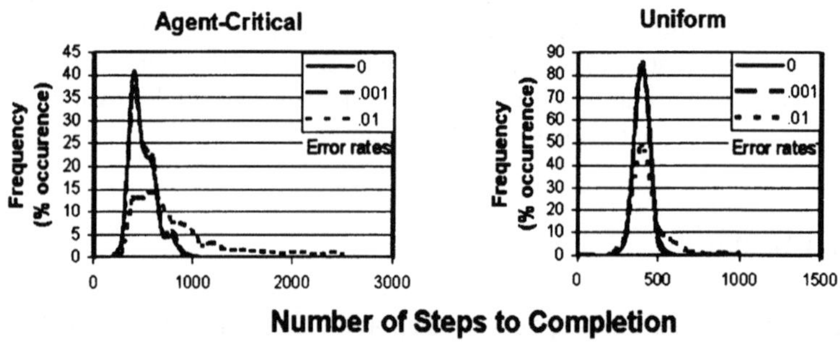


**Fig. 1.** 2000 runs of a simulated nanorobot attack on a tumor under various error rates.

Thus the development of swarm algorithms should include a feedback stage in which the effects of erroneous behavior can be observed and evaluated. Additional information and links to swarm animations are available on the website `www.jcu.edu/math/faculty/dpalmer/dpalmer.htm`.

## References


1. Lewis A., Bekey G., (1992) The Behavorial Self-Organization of Nanorobots Using Local Rules, In *Proceedings of IEEE/RSJ International Conference on Intelligent Robots and Systems*, IEEE **2** 1333-1338
2. Martinoli A., Mondada F., (1998) Probabilistic Modelling of a Bio-Inspired Collective Experiment with Real Robots, In *Distributed Autonomous Robotic Systems 3*, Springer, Berlin Heidelberg 289-298
3. Palmer D., Hantak C., Kovacina M., (1999) Impact of Behavior Influence on Decentralized Control Strategies for Swarms of Simple, Autonomous, Mobile Agents, In *Proceedings of Biomechanics Meets Robotics*, `www.iwr.uni-heidelberg.de/gk/Workshop`

# The MICRobES Project, an Experimental Approach towards "Open Collective Robotics"

Sébastien Picault and Alexis Drogoul

LIP6, Université Paris 6, case 169, 4 place Jussieu, F-75252 Paris Cedex, France.
{Sebastien.Picault, Alexis.Drogoul}@lip6.fr

**Abstract.** We focus here on "Open Collective Robotics", a research domain interested in studying concepts and techniques which could lead to true collectivities between humans and robots. We also present the MICRobES project[1], a long-term experiment intended to investigate how to design such collectivities. Therefore, the MICRobES robots have no prior specific functionality, they have to develop individual competences, a social organization, and sociability towards humans in order to "survive" in a dynamic environment.

## 1 Towards "Open Collective Robotics"

As an analogy to the definition of Hewitt's "Open Systems" – work groups mixing humans and software agents – we call "Open Collective Robotics" (OCR) the study of collective robotic systems that would be able to operate in an autonomous, adaptive way in a human environment.

Classical experiments, done either in a close, controlled environment, or in open ones completely adapted to the robots, neglect too much the huge dynamicity of the real world, in which those robots will have to operate sooner or later. Thus, the robots should rather be designed as ***physically and socially situated*** agents to have the ability to dynamically adapt themselves to a human, complex, environment. Social aptitudes already exist in many multi-robot systems but they usually apply to the organization of the robots only. Some experiments on human-robot interaction have been done, but concern immobile robots. Finally, the museum guides give interesting examples of single robots operating in dynamic environments. The MICRobES project is an attempt to integrate those three issues.

## 2 The MICRobES project

The MICRobES project is an OCR experiment that started in April'99 at the LIP6. It consists in studying the long-term adaptation (at least 2 years) of a micro-society composed by autonomous robots immersed in a "normal"

[1] This research is supported by a grant from the French Department of Higher Education and Research, by a grant from the LIP6, and by a research contract between the LIP6 and France Telecom R&D.

and inhabited environment: the laboratory. The robots colony does not have any specific, functional goal but "surviving" in a real ecosystem which is unknown and highly dynamic. Therefore, they have to make maps, localize themselves, organize themselves (to coordinate one with another, to exchange informations...), and interact with humans (source of help or information). Thus they need individual and collective competences that allow them to confront themselves to unexpected situations, environmental changes and new kinds of events, either in their physical environment or in their robotic and human social environments. We think that this "ontogenic" step is absolutely essential before trying to make the robots perform any functional task.

Robots are seen as individuals by humans: thus their social organization must be easily understood by people. However, human-specific features are not essential to build a complex social structure: for example, primate groups exhibit sophisticated social behaviors. Previous works on this subject [1] have shown that simple social cognition models could be successfully implemented to reproduce their behavioral characteristics. They can easily be applied to robotics to define adaptive hierarchies or affiliations between agents, in order to avoid or solve conflicts (e.g. access to energetic resources). Before being implemented on the robots, those behaviors are currently being investigated in simulation. It is also essential to provide the robots with social aptitudes towards the humans working in the lab, since the project is intended to last during a long period of time and therefore implies the elaboration of shared rules of life in society. Two general principles will be applied: to avoid conflicts (not to make the robots a nuisance for the humans, neither the contrary); to encourage interactions (to make the staff pleased or interested in interacting with the robots, and the robots "curious"). We will be helped in this task by sociologists from the 'École des Mines de Paris' and from the EHESS.

The originality of our project consists in going against the general trend regarding its working hypotheses and its purpose – to build a long-term social network between two populations (robotic and human) sharing the same environment. At present, a considerable work has been done to give the MICRobES robots physical aptitudes to navigate, make maps, recharge themselves, etc. [2]. A first model of human-robot interaction, applied to map learning, has also been designed [3]. We should go further in the next months by focussing on the social organization and basic interaction abilities with humans.

## References


1. Picault, S. and Collinot, A. (1998) Designing Social Cognition Models for Multi-Agent Systems through Simulating Primate Societies. Proc. of ICMAS'98.
2. Hugues, L. (2000) Collective Grounded Representations for Robots. Proc. of DARS'2000.
3. Bredèche, N. and Zucker, J.-D. (2000) From distributed robot perception to human topology: a learning model. Proc. of DARS'2000.

# Dynamic Task Selection: A Simple Structure for Multirobot Systems

Poj Tangamchit (poj@andrew.cmu.edu)[1], John M. Dolan (jmd@ri.cmu.edu)[2], and Pradeep K. Khosla (pkk@cs.cmu.edu)[1,2]

[1] Dept. of Electrical and Computer Engineering, Carnegie Mellon Univ. , Pittsburgh PA, USA
[2] The Robotics Institute, Carnegie Mellon Univ. , Pittsburgh PA, USA

**Abstract.** We present a simple architecture capable of use in a wide variety of applications by focusing on basic principles of efficient decentralized operation. The resulting robust architecture is called dynamic task selection. We present an experiment that shows the ability of the system to work properly when some robots fail. It also shows the ability to add new robots dynamically while other robots are already performing a mission.

## 1 Introduction

In the design of a multirobot system, two decisions of architecture must be made. One decision is at the robot level, between a reactive and deliberative architecture. The other is at the group level, between a centralized and decentralized architecture. Centralized systems, which have a central unit managing tasks for the group, have advantages over decentralized systems in terms of traffic control, resource management, and task optimization. Decentralized systems are superior in terms of robustness and scalability. We choose a hybrid architecture at the robot level because we want our system to be task-oriented while also having low-level reactive behavior. At the group level, we choose a decentralized architecture mainly because of its robustness. In this paper, we propose basic principles of operation for decentralized systems for general applications instead of providing a complex architecture for specific applications.

## 2 Dynamic Task Allocation and Robustness

In missions composed of several subtasks, a hybrid architecture enables us to deal with task and task allocation. We assume that the task decomposition is already done so that there is a list of subtasks known to all robots. We propose creating a virtual global list of subtasks among robots and let each robot actively pick a task to execute by itself (see Figure 1). This list is actually located on each robot with a communication mechanism to update and synchronize them. This mechanism is called "dynamic task selection" or "dynamic task allocation".

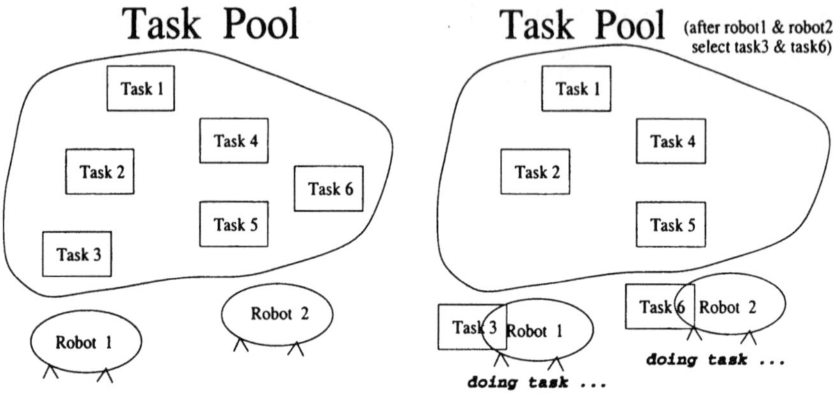


**Fig. 1.** Dynamic Task Selection

# 3 Experiment

We have demonstrated the robustness of dynamic task selection to a sudden failure of robots. We chose the multirobot patrolling problem as our testbed. The problem is composed of a field with several checkpoints to visit. Each checkpoint is considered as one subtask. The robots need to share these points and move to all of them cyclically. The pictures below are video clips of our experiment. There are two robots and four checkpoints. In the first clip, we start our system. The robots share the checkpoints equally, two for each robot. In the second clip, we power down one robot and move it away. The third clip shows that the other robot compensates for the absence of the dead robot by selecting and moving to all four points. In the fourth clip, we put the dead robot back into operation. The fifth clip shows the two robots operating as before, i.e., equally sharing the points.

**Fig. 2.** Dynamic Task Alloation Experiment

# References

1. Y.U. Cao, A.S. Fukunaga, A.B. Khang (1997) Cooperative Mobile Robotics: Antecedents and Directions, Autonomous Robots, 4, 1-23 Kluwer Academic Publishers
2. L. Parker (1994) Heterogeneous Multi-Robot Cooperation PhD. Thesis, Dept. of Electrical Eng. and Computer Science MIT

# Author Index

MIX
Papier aus verantwortungsvollen Quellen
Paper from responsible sources
FSC® C105338

If you have any concerns about our products,
you can contact us on
**ProductSafety@springernature.com**

In case Publisher is established outside the EU,
the EU authorized representative is:
**Springer Nature Customer Service Center GmbH**
**Europaplatz 3, 69115 Heidelberg, Germany**

Printed by Libri Plureos GmbH
in Hamburg, Germany